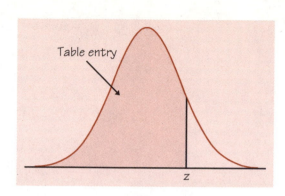

Table entry for z is the area under the standard normal curve to the left of z.

TABLE A	Standard normal probabilities (*continued*)								
z	.00	.01	.02	.03	.04	.05	.06	.07	.08	.09
0.0	.5000	.5040	.5080	.5120	.5160	.5199	.5239	.5279	.5319	.5359
0.1	.5398	.5438	.5478	.5517	.5557	.5596	.5636	.5675	.5714	.5753
0.2	.5793	.5832	.5871	.5910	.5948	.5987	.6026	.6064	.6103	.6141
0.3	.6179	.6217	.6255	.6293	.6331	.6368	.6406	.6443	.6480	.6517
0.4	.6554	.6591	.6628	.6664	.6700	.6736	.6772	.6808	.6844	.6879
0.5	.6915	.6950	.6985	.7019	.7054	.7088	.7123	.7157	.7190	.7224
0.6	.7257	.7291	.7324	.7357	.7389	.7422	.7454	.7486	.7517	.7549
0.7	.7580	.7611	.7642	.7673	.7704	.7734	.7764	.7794	.7823	.7852
0.8	.7881	.7910	.7939	.7967	.7995	.8023	.8051	.8078	.8106	.8133
0.9	.8159	.8186	.8212	.8238	.8264	.8289	.8315	.8340	.8365	.8389
1.0	.8413	.8438	.8461	.8485	.8508	.8531	.8554	.8577	.8599	.8621
1.1	.8643	.8665	.8686	.8708	.8729	.8749	.8770	.8790	.8810	.8830
1.2	.8849	.8869	.8888	.8907	.8925	.8944	.8962	.8980	.8997	.9015
1.3	.9032	.9049	.9066	.9082	.9099	.9115	.9131	.9147	.9162	.9177
1.4	.9192	.9207	.9222	.9236	.9251	.9265	.9279	.9292	.9306	.9319
1.5	.9332	.9345	.9357	.9370	.9382	.9394	.9406	.9418	.9429	.9441
1.6	.9452	.9463	.9474	.9484	.9495	.9505	.9515	.9525	.9535	.9545
1.7	.9554	.9564	.9573	.9582	.9591	.9599	.9608	.9616	.9625	.9633
1.8	.9641	.9649	.9656	.9664	.9671	.9678	.9686	.9693	.9699	.9706
1.9	.9713	.9719	.9726	.9732	.9738	.9744	.9750	.9756	.9761	.9767
2.0	.9772	.9778	.9783	.9788	.9793	.9798	.9803	.9808	.9812	.9817
2.1	.9821	.9826	.9830	.9834	.9838	.9842	.9846	.9850	.9854	.9857
2.2	.9861	.9864	.9868	.9871	.9875	.9878	.9881	.9884	.9887	.9890
2.3	.9893	.9896	.9898	.9901	.9904	.9906	.9909	.9911	.9913	.9916
2.4	.9918	.9920	.9922	.9925	.9927	.9929	.9931	.9932	.9934	.9936
2.5	.9938	.9940	.9941	.9943	.9945	.9946	.9948	.9949	.9951	.9952
2.6	.9953	.9955	.9956	.9957	.9959	.9960	.9961	.9962	.9963	.9964
2.7	.9965	.9966	.9967	.9968	.9969	.9970	.9971	.9972	.9973	.9974
2.8	.9974	.9975	.9976	.9977	.9977	.9978	.9979	.9979	.9980	.9981
2.9	.9981	.9982	.9982	.9983	.9984	.9984	.9985	.9985	.9986	.9986
3.0	.9987	.9987	.9987	.9988	.9988	.9989	.9989	.9989	.9990	.9990
3.1	.9990	.9991	.9991	.9991	.9992	.9992	.9992	.9992	.9993	.9993
3.2	.9993	.9993	.9994	.9994	.9994	.9994	.9994	.9995	.9995	.9995
3.3	.9995	.9995	.9995	.9996	.9996	.9996	.9996	.9996	.9996	.9997
3.4	.9997	.9997	.9997	.9997	.9997	.9997	.9997	.9997	.9997	.9998

THE BASIC PRACTICE
OF STATISTICS

THE BASIC PRACTICE
OF STATISTICS

David S. Moore
Purdue University

W. H. FREEMAN AND COMPANY
New York

Library of Congress Cataloging-in-Publication Data

Moore, David S.
  The basic practice of statistics / David S. Moore.
      p.   cm.
  Includes index.
  ISBN 0-7167-2628-9
  1. Statistics.   I. Title.
QA276.12.M648   1994
519.5—dc20   94-40756
                                                        CIP

Printed in the United States of America

Fourth printing 1996, RRD

CONTENTS

Starred sections are optional

PREFACE

The Basic Practice of Statistics is an introduction to statistics for students in two-year and four-year colleges and unversities that emphasizes working with data and statistical ideas. In this preface I describe the book in order to help teachers judge whether it is right for their students.

Guiding principles

The American Statistical Association and the Mathematical Association of America recently formed a joint committee to study the teaching of introductory statistics. Here are their main recommendations:[1]

- Emphasize statistical thinking.
- More data and concepts; less theory, fewer recipes.
- Foster active learning.

I was a member of the ASA/MAA committee, and I agree with their conclusions. Fostering active learning is the business of the teacher (though an emphasis on working with data helps). The first two recommendations are the guiding principles of this text. Although the book is elementary in the level of mathematics required and in the statistical procedures presented, it aims to give students both an understanding of the main ideas of statistics and useful skills for working with data. Examples and exercises, though intended for beginners, use real data and give enough background to allow students to consider the meaning of their calculations. I often ask for conclusions that are more than a number (or "reject H_0"). Some exercises require judgement in addition to right-or-wrong calculations and conclusions. Both students and teachers should recognize that not every part of every exercise has a single correct answer. I hope that teachers will encourage further discussion of results in class.

Chapters 1 and 2 present the methods and unifying ideas of data analysis. Students appreciate the usefulness of data analysis, and the fact that they can actually do it relieves a bit of their anxiety about statistics. I hope that they will grow accustomed to examining data and will continue to do so even when formal inference to answer a specific question is the ultimate goal. Chapter 3 discusses random sampling and randomized comparative experiments. These are among the most important ideas in statistics, and are often unjustly neglected in beginning instruction. Chapter 4 builds on the ideas of Chapter 3 and the data-analytic tools of Chapter 1 to present the central idea of a sampling distribution and (informally) the language of probability. Chapter 5 is the cornerstone of the rest of the book. It describes the reasoning of statistical inference. The remaining chapters present methods of inference for various settings, with a strong emphasis on practical aspects of using these methods. Chapters 6 and 7 discuss basic one-sample and two-sample procedures. Chapters 8, 9, and 10 (which can be read independently of each other in any order) offer a choice of somewhat more advanced topics.

Technology

Automating calculations increases students' ability to complete problems, reduces their frustration, and helps them concentrate on ideas and problem recognition rather than mechanics. This book requires that students have a **calculator** that will do statistical calculations through correlation and simple linear regression. As I write, several makers sell such calculators for less than $18. Because students have calculators, the text doesn't discuss "computing formulas" for the sample standard deviation or least-squares regression line. Exercises assume that students will use a "two-variable statistics" calculator rather than a more rudimentary model. Many scientific calculators are even more capable, offering simulation and some graphics as well as more statistical procedures. Course-wide use of a graphing calculator is somewhat similar to using software, and will probably require specific instruction.

    Statistical software has considerable advantages over calculators: easier data entry and editing, much better graphics, more statistical procedures. I encourage the use of software whenever time and facilities permit. This book does not, however, assume that students will use software. Output from four statistical software packages appears in the text, with Minitab most common. Students should be able to interpret output and use it for further work. Because of the great variety of software used for statistics, from spreadsheets to SAS, I have avoided instruction on how to use any one program. A supplementary Minitab handbook is available for teachers who want to introduce Minitab in a way closely tied to the text. Data sets for examples and exercises are available on a data disk, from which they can be read into any statistical software.

Although **video** is a poor medium for exposition, actually seeing statistics at work in a variety of settings is a powerful means of changing student attitudes about statistics. I show one of the short on-location documentary segments from the Annenberg/Corporation for Public Broadcasting telecourse *Against All Odds: Inside Statistics* about once a week. Because I was the content developer for this video series, it fits the style and sequence of this book closely. The Instructor's Guide contains suggestions about using video, a list of recommended excerpts, and information on how to obtain *Against All Odds*.

Probability

The difficult issue of how to present probability in a first course on statistical methods deserves separate comment. Experienced teachers recognize that students find probability difficult. Research on learning confirms our experience. I recommend the survey article "Difficulties in learning basic concepts in probability and statistics: implications for research," by Joan Garfield and Andrew Ahlgren.[2] Garfield and Ahlgren document the fragility of probability concepts even among students who can work formal probability problems. They conclude that "teaching a conceptual grasp of probability still appears to be a very difficult task, fraught with ambiguity and illusion." They recommend exploring "how useful ideas of statistical inference can be taught independently of technically correct probability."

Probability is a noble and useful subject. Attempting to present a substantial introduction to probability in a data-oriented statistics course for students who are not mathematically trained, however, is in my opinion unwise. Formal probability does not help these students master the ideas of inference (at least not as much as we teachers imagine), and it depletes reserves of mental energy that might better be applied to essentially statistical ideas.

I have therefore presented very little formal probability in the core of the text. The central idea is that of the sampling distribution of a statistic. Sampling distributions answer a question that leads to the heart of statistical inference, "What would happen if I repeated this random process many times?" Moreover, students can use tools from data analysis to understand distributions. Normal distributions appear already in Chapter 1 as models for the overall pattern of some distributions of data. In Section 1 of Chapter 4, sampling distributions motivate a brief and quite informal introduction to the language of probability. Sections 4.3 and 4.5 look at the sampling distributions of sample proportions and sample means in more detail. Such important probability facts as the central limit theorem and

the law of large numbers appear in this context. A few additional probability ideas, particularly independence, are treated informally when needed. This is the approach I recommend.

Instructors who want more depth in the study of probability distributions will find it in the optional Sections 4.2 and 4.4. Even these sections avoid the formal notation of set theory and general probability. *Introduction to the Practice of Statistics* offers more, and more traditionally presented, probability. I confess that when teaching from that text I omit most of its probability material in favor of additional statistics. Sections 4.1, 4.3, and 4.5 of this book (with the optional Section 4.6 on control charts) contain what I actually teach students without calculus.

What about that other book?

The Basic Practice of Statistics is kin to *Introduction to the Practice of Statistics* (IPS).[3] It shares both its guiding principles and some details (especially exercises) with the larger book. But this is not an abridgement of *IPS*. Much of the material, including all of Chapters 7 to 10, is completely new. I have omitted most of the optional material in *IPS,* as well as a number of topics that were not optional. The result is a shorter book.

I have also endeavored to make this book considerably easier to read. The language, the length of the exposition, the organization of exercises, and the choice of topics have been rethought throughout for a less advanced audience. Each main idea is followed by a short section of exercises for immediate reinforcement. Each chapter review includes a list of specific skills against which students may check their learning. An index of symbols (note that it isn't long—the density of symbols is as low as I could efficiently manage) and an index of procedures are placed ahead of the usual index. There is a removable gatefold insert that includes essential formulas and tables for easy reference and use in testing.

Instructors who appreciate the data-and-ideas approach of *IPS,* but found it hard going for their students, should like this book. Conversely, instructors who wish to present such topics as normal quantile plots, transformations of variables, formal probability, or brief introductions to multiple regression and two-way analysis of variance should consider *IPS.*

Supplements

A full range of supplements is available to help teachers and students use *The Basic Practice of Statistics.* In the *Instructor's Guide* I give an overview and teaching suggestions for each chapter, comments on the use of video, and sample examinations. I also present additional examples for classroom use, with data on the data disk. These helps are followed by complete solutions

(not just answers) for all exercises, prepared by Darryl Nester of Bluffton College. William Notz and Becky Busam of Ohio State University have prepared a *Student Study Guide* and a *Test Bank* for instructors. A set of *Transparency Masters* is available from the publisher. A *Data Disk,* available in both DOS/Windows and Macintosh formats, enables instructors to easily enter the data from examples and exercise sets into their software. Betsy Greenberg of the Unversity of Texas has written a *Minitab Handbook* that gives detailed instruction about the Minitab statistical software linked to the sequence and examples of the text.

Acknowledgments

I am grateful to many colleagues from two-year and four-year colleges and universities who commented on successive drafts of the manuscript:

Douglas M. Andrews
Wittenberg University

Rebecca Busam
The Ohio State University

Michael Butler
College of the Redwoods

Carolyn Pillers Dobler
Gustavus Adolphus College

Joel B. Greenhouse
Carnegie Mellon University

Larry Griffey
Florida Community College at Jacksonville

Brenda Gunderson
The University of Michigan

Catherine Cummins Hayes
University of Mobile

Tim Hesterberg
Franklin & Marshall College

Ronald La Porte
Macomb Community College

Ken McDonald
Northwest Missouri State University

William Notz
The Ohio State University

Mary Parker
Austin Community College

Calvin Schmall
Solano Community College

Frank Soler
De Anza College

Linda Sorensen
Algoma University

Tom Sutton
Mohawk College

I am also grateful to the excellent editorial and design professionals at W. H. Freeman and Company. Diane Cimino Maass and Carol Pritchard-Martinez in particular contributed a great deal to this book.

More generally, I am indebted to many statistics teachers with whom I have discussed the teaching of our subject over many years; to people from diverse fields with whom I have worked to understand data; and especially to students whose compliments and complaints have changed and improved my teaching. Working with teachers, colleagues in other disciplines, and students has reminded me of the importance of hands-on experience with data and of statistical ideas in an era when computer routines and professional advice quickly handle statistical details.

David S. Moore

Notes

1. George Cobb, "Teaching statistics," in L.A. Steen (ed.), *Heeding the Call for Change: Suggestions for Curricular Action,* MAA Notes No. 22, Mathematical Association of America, Washington D.C., 1992, pp. 3–43.

2. Joan Garfield and Andrew Ahlgren, "Difficulties in learning basic concepts in probability and statistics: implications for research," *Journal for Research in Mathematics Education,* 19 (1988), pp. 44–63.

3. David S. Moore and George P. McCabe, *Introduction to the Practice of Statistics,* 2d ed., W. H. Freeman, New York, 1993.

What Is Statistics?

Why study statistics?

Statistics is the science of gaining information from numerical data. We study statistics because the use of data has become ever more common in a growing number of professions, in public policy, and in everyday life. Here are some examples of statistical questions raised by a viewer of the nightly news.

- The Bureau of Labor Statistics reports that the unemployment rate last month was 6.5%. The government didn't ask me if I was unemployed. How did they obtain this information? How accurate is that 6.5%?

- Another news item describes restrictions on smoking in public places. I hear that much of the evidence that links smoking to lung cancer and other health problems is "statistical." What kind of evidence is statistical evidence?

- A medical reporter cites a study claiming to show that regular physical exercise leads to longer life. But a doctor she interviews casts doubt on the study's usefulness. How can I tell whether the data from a study really support the conclusions that are announced?

- Here's a special report on the international competitiveness of American industry. The experts interviewed talk about improving quality and productivity through better management, new technology, and effective use of statistics. What can statistics do except keep score?

We can no more escape data than we can avoid the use of words. Like words, data do not interpret themselves, but must be read with understanding. Just as a writer can arrange words into convincing arguments or incoherent nonsense, so can data be convincing, misleading, or just irrelevant. Numerical literacy, the ability to follow and understand arguments based on data, is important for everyone. The study of statistics is an essential part of a sound education.

What is statistics?

For most users of statistics, and even for most professional statisticians, statistics provides tools and ideas for using data to gain understanding of some other subject. Statistics in practice is applied to study the effectiveness of medical treatments, the reaction of consumers to television advertising, the attitudes of young people toward sex and marriage, and much else. Although statistics has an impressive mathematical theory, we are concerned with the *practice* of statistics. We can divide statistics in practice into three parts:

1. **Data analysis** concerns methods and ideas for organizing and describing data using graphs, numerical summaries, and more elaborate mathematical descriptions. The computer revolution has brought analysis of data back to the center of statistical practice. Statisticians have responded with new tools and (more important) new organizing ideas for exploring data. Chapters 1 and 2 discuss data analysis.

2. **Data production** provides methods for producing data that can give clear answers to specific questions. Basic concepts about how to select samples and design experiments are perhaps the most influential of all statistical ideas. These concepts are the subject of Chapter 3.

3. **Statistical inference** moves beyond the data in hand to draw conclusions about some wider universe. Statistical inference not only draws conclusions, but accompanies those conclusions with a statement about how trustworthy they are. Inference uses the language of probability, which we introduce in Chapter 4. Because we are concerned with practice rather than theory, we can function with a quite limited knowledge of probability. Chapter 5 discusses the reasoning of statistical inference, and Chapters 6 and 7 present inference as used in practice in several simple settings. Chapters 8, 9, and 10 offer brief introductions to inference in some more complex settings.

How will we approach the study of statistics?

The goal of statistics is to gain understanding from data. Data are numbers, but they are not "just numbers." *Data are numbers with a context.* The number 10.5, for example, carries no information by itself. But if we hear that a friend's new baby weighed 10.5 pounds at birth, we congratulate her on the healthy size of the child. The context engages our background knowledge and allows us to make judgments. We know that a baby weighing 10.5 pounds is quite large, and that it isn't possible for a human baby to weigh 10.5 ounces or 10.5 kilograms. The context makes the number informative.

Because data are numbers with a context, doing statistics means more than manipulating numbers. This book is full of data, and each set of data has some brief background to help you understand what the data say. Examples and exercises usually try to express briefly some understanding gained from the data. In practice, you would know much more about the background of the data you work with and about the questions you hope the data will answer. No textbook can be fully realistic. But it is very important to form the habit of asking "What do the data tell me?" rather than just concentrating on making graphs and doing calculations. This book tries to encourage good habits.

Nonetheless, statistics involves lots of calculating and graphing. The text presents the techniques you need, but you should use a calculator or computer software to automate calculations and graphs as much as possible. There are many kinds of statistical software, from spreadsheets to large programs for advanced users of statistics. The kind of computing available to learners varies a great deal from place to place—and the big ideas of statistics don't depend on any particular level of access to computing. We encourage use of statistical software, but this book does not require software and is not tied to any specific software.

The book does require that you have a calculator with some built-in statistical functions. Specifically, you need a calculator that will find means and standard deviations and calculate correlations and regression lines. Look for a calculator that claims to do "two-variable statistics" or mentions "regression." Advanced calculators will do much more, including some statistical graphs, but large screens and easy data editing make computers more suitable for elaborate statistical analyses.

Calculators and computers can follow recipes for graphs and calculations both more quickly and more accurately than humans. Because graphing and calculating are automated in statistical practice, the most important assets you can gain from the study of statistics are an understanding of the big ideas and the beginnings of good judgment in working with data. Ideas and judgment can't (at least yet) be automated. They guide you in telling the computer what to do and in interpreting its output. This book tries to explain the most important ideas of statistics, not just teach methods. Some examples of big ideas that you will meet (one from each of the three areas of statistics) are "always plot your data," "randomized comparative experiments," and "statistical significance."

These, then, are the principles that should guide your learning of statistics:

• Try to understand what data say in each specific context. All the methods you will learn are just tools to help understand data.

- Let a calculator or computer do as much of the calculating and graphing as possible, so that you can concentrate on what to do and why.
- Focus on the big ideas of statistics, not just on rules and recipes.

But perhaps the basic principle of all learning is persistence. The main ideas of statistics, like the main ideas of any important subject, took a long time to discover and take some time to master. The gain will be worth the pain.

The first step in understanding data is to organize and display the data, to "let the numbers speak for themselves." That's *data analysis*. The second step is careful attention to where the data come from. That's *data production*. Data analysis and data production are the starting points for *statistical inference*, in which we use data to draw conclusions that reach beyond the specific individuals that our data describe. The three chapters of Part I deal with data analysis and data production.

Chapters 1 and 2 reflect the strong emphasis on data analysis that characterizes modern statistical practice. Although careful exploration of data is an essential preparation for inference, data analysis isn't just preparation for inference. The point of view of inference carefully distinguishes between the data we actually have and the larger universe we want conclusions about. The Bureau of Labor Statistics, for example, has data about employment in the 60,000 households contacted by its Current Population Survey. The bureau wants to draw conclusions about employment in all 96 million U.S. households. That's a complex problem, as we will see in Part II of this book.

From the viewpoint of data analysis, things are simpler. We want simply to explore and understand the data in hand. The distinctions that inference requires don't concern us in Chapters 1 and 2. What does concern us is a systematic strategy for examining data and the tools that we use to carry out that strategy.

Of course, we often do want to use data to reach general conclusions. Whether we can do that depends most of all on how the data were produced. Good data rarely "just happen." They are products of human effort, like video games and nylon sweaters. Chapter 3 tells how to produce good data yourself and how to decide whether to trust data produced by others.

The study of data analysis and data production equips you with tools and ideas that are immediately useful whenever you deal with numbers. Inference is more specialized and more subtle. Inference requires more attention in a textbook, but that doesn't mean it is more important. Statistics is the science of data, and the three chapters of Part I deal directly with data.

PART **I**

UNDERSTANDING DATA

"Tonight we're going to let the statistics speak for themselves."

Drawing by Koren ©1974 *The New Yorker Magazine, Inc.*

FLORENCE NIGHTINGALE

Florence Nightingale (1820–1910) won fame as a founder of the nursing profession and as a reformer of health care. As chief nurse for the British army during the Crimean War, from 1854 to 1856, she found that lack of sanitation and disease killed large numbers of soldiers hospitalized by wounds. Her reforms reduced the death rate at her military hospital from 42.7% to 2.2%, and she returned from the war famous. She at once began a fight to reform the entire military health care system, with considerable success.

One of the chief weapons Florence Nightingale used in her efforts was data. She had the facts, because she reformed record keeping as well as medical care. She was a pioneer in using graphs to present data in a vivid form that even generals and members of Parliament could understand. Her inventive graphs are a landmark in the growth of the new science of statistics. She considered statistics essential to understanding any social issue and tried to introduce the study of statistics into higher education.

In beginning our study of statistics, we will follow Florence Nightingale's lead. This chapter and the next will stress the analysis of data as a path to understanding. Like her, we will start with graphs to see what data can teach us. Along with the graphs we will present numerical summaries, just as Florence Nightingale calculated detailed death rates and other summaries. Data for Florence Nightingale were not dry or abstract, because they showed her, and helped her show others, how to save lives. That remains true today.

CHAPTER **1**

Examining Distributions

INTRODUCTION

Statistics is the science of data. We therefore begin our study of statistics by mastering the art of examining data. Any set of data contains information about some group of *individuals*. The information is organized in *variables*.

INDIVIDUALS AND VARIABLES

Individuals are the objects described by a set of data. Individuals may be people, but they may also be animals or things.

A **variable** is any characteristic of an individual. A variable can take different values for different individuals.

Data for a study of a company's pay policies, for example, might contain data about every employee. These are the individuals described by the data set. For each individual, the data contain the values of variables such as the age in years, gender (female or male), job category, and annual salary in dollars. In practice, any set of data is accompanied by background information that helps us understand the data. When you meet a new set of data, ask yourself the following questions:

1. What **individuals** do the data describe? **How many** individuals appear in the data?

units
2. How many **variables** do the data contain? What are the **exact definitions** of these variables? In what *units* is each variable recorded? Weights, for example, might be recorded in pounds, in thousands of pounds, or in kilograms. Is there any reason to mistrust the values of any variable?

3. **What purpose** do the data have? Do we hope to answer some specific questions? Do we want to draw conclusions about individuals other than the ones we actually have data for?

The third set of questions is very important, so important that Chapter 3 addresses them in detail. For now, however, we are content to describe the individuals and the variables in a data set.

EXAMPLE 1.1

Here is a small part of a data set that describes public education in the States.

State	Region	Population (1000)	SAT Verbal	SAT Math	Percent taking	Dollars per pupil	Teacher's pay ($1000)
⋮							
CA	PAC	29,760	419	484	45	4,826	39.6
CO	MTN	3,294	456	513	28	4,809	31.8
CT	NE	3,287	430	471	74	7,914	43.8
⋮							

The *individuals* described are the states. There are 51 of them, the 50 states and the District of Columbia, but we give data for only three. Each row in the table describes one individual. Each column contains the values of one *variable* for all the individuals. This is the usual arrangement in data tables. The first column identifies the state by its two-letter post office code. We give data for California, Colorado, and Connecticut.

The second column says which region of the country each state is in. The Census Bureau divides the nation into nine regions. These three are Pacific, Mountain, and New England. The third column contains state populations, in thousands of people. Be sure to notice that the *units* are thousands of people. California's 29,760 stands for 29,760,000 people. The population data come from the 1990 census. They are therefore quite accurate as of April 1, 1990, but don't show later changes in population.

The remaining five variables are the average scores of the states' high school seniors on the Scholastic Assessment Test (SAT) verbal and mathematics exams, the percent of seniors who take the SAT, public school spending per pupil in dollars, and average teachers' salaries in thousands of dollars. Each of these variables needs more explanation before we can fully understand the data. For example, spending per pupil is total spending on public education divided by the number of pupils. Because the number of pupils in school varies from day to day, the data use average attendance. Attendance is reported by each school system. Attendance often determines how much state aid a school system gets, so some schools may report numbers that are too high. If this happens, the spending per pupil in the table will be too low. ◄

As Example 1.1 illustrates, a full understanding of the variables in a data set often requires some background study. Fortunately, when you work with your own data you usually understand them well. Statistical tools and ideas can help you examine your data in order to describe their *exploratory data analysis* main features. This examination is called ***exploratory data analysis***. Like an explorer crossing unknown lands, we want first to simply describe what we see. Each example we meet will have some background information to help us, but our emphasis is on examining the data. Here are two basic strategies that help us organize our exploration of a set of data:

- Begin by examining each variable by itself. Then move on to study the relationships among the variables.
- Begin with a graph or graphs. Then add numerical summaries of specific aspects of the data.

We will organize our learning the same way. This chapter concerns examination of a single variable, and Chapter 2 looks at relationships among variables. In each chapter we begin with graphs and then move on to numerical summaries.

1.1   DISPLAYING DISTRIBUTIONS WITH GRAPHS

Some variables, like gender and job title, simply place individuals into categories. Others, like height and annual income, take numerical values for which we can do arithmetic. It makes sense to give an average income for a company's employees, but it does not make sense to give an "average" gender. We can, however, count the numbers of female and male employees and do arithmetic with these counts.

CATEGORICAL AND QUANTITATIVE VARIABLES

A **categorical variable** records which of several groups or categories an individual belongs to.

A **quantitative variable** takes numerical values for which it makes sense to do arithmetic operations like adding and averaging.

The **distribution** of a variable tells us what values it takes and how often it takes these values.

Categorical variables

The values of a categorical variable are just labels for the categories, like "male" and "female." The distribution of a categorical variable lists the categories and gives either the count or the percent of individuals who fall in each category. For example, here is the distribution of marital status for all Americans age 18 and over.

Marital status	Count (millions)	Percent
Single	41.8	22.6
Married	113.3	61.1
Widowed	13.9	7.5
Divorced	16.3	8.8

bar chart

pie chart

To present such data to an audience, you may wish to use graphs like those in Figure 1.1. The **bar chart** in Figure 1.1(a) quickly compares the sizes of the four marital status groups. The heights of the four bars show the counts in the four categories. The **pie chart** in Figure 1.1(b) helps us see

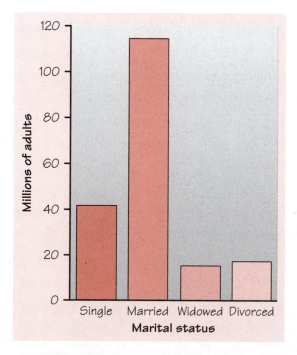

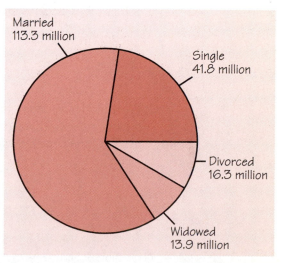

FIGURE 1.1(a)   Bar chart of the marital status of U.S. adults.

FIGURE 1.1(b)   Pie chart of the same data.

what part of the whole each group forms. For example, the "married" slice makes up 61% of the pie because 61% of adults are married. Bar charts and pie charts help an audience grasp the distribution quickly. But neither kind of graph is essential to understanding the data. Categorical data on a single variable such as marital status are easy to describe without a graph. You should be able to interpret bar charts and pie charts, but we will not draw them ourselves. We can move on at once to graphs for quantitative variables.

EXERCISES

1.1   Example 1.1 presents data on the States. The first column identifies the states. Each of the remaining seven columns contains values of a variable. Which of these variables are categorical and which are quantitative?

1.2   Data from a medical study contain values of many variables for each of the people who were the subjects of the study. Which of the following variables are categorical and which are quantitative?

(a) Gender (female or male)

(b) Age (years)

(c) Race (Asian, black, white, or other)

(d) Smoker (yes or no)

(e) Systolic blood pressure (millimeters of mercury)

(f) Level of calcium in the blood (micrograms per milliliter)

Drawing histograms

histogram   Quantitative variables often take so many values that a graph of the distribution is clearer if nearby values are grouped together. The most common graph of the distribution of one quantitative variable is a **histogram**.

EXAMPLE 1.2

Table 1.1 presents the percent of residents aged 65 years and over in each of the 50 states. To make a histogram of this distribution, proceed as follows.

1. Divide the range of the data into classes of equal width. The data in Table 1.1 range from 4.2 to 18.3, so we choose as our classes.

$$4.0 < \text{ percent over } 65 \leq 5.0$$
$$5.0 < \text{ percent over } 65 \leq 6.0$$
$$\vdots$$
$$18.0 < \text{ percent over } 65 \leq 19.0$$

Be sure to specify the classes precisely so that each observation falls into exactly one class. A state with 5.0% of its residents aged 65 or older would fall into the first class, but 5.1% falls into the second.

TABLE 1.1	Percent of population 65 years old and over, by state (1991)		
State	Percent	State	Percent
Alabama	12.9	Montana	13.4
Alaska	4.2	Nebraska	14.1
Arizona	13.2	Nevada	10.8
Arkansas	14.9	New Hampshire	11.6
California	10.5	New Jersey	13.4
Colorado	10.1	New Mexico	10.9
Connecticut	13.7	New York	13.1
Delaware	12.2	North Carolina	12.3
Florida	18.3	North Dakota	14.5
Georgia	10.1	Ohio	13.1
Hawaii	11.4	Oklahoma	13.5
Idaho	12.0	Oregon	13.7
Illinois	12.5	Pennsylvania	15.5
Indiana	12.6	Rhode Island	15.1
Iowa	15.4	South Carolina	11.4
Kansas	13.9	South Dakota	14.7
Kentucky	12.7	Tennessee	12.7
Louisiana	11.2	Texas	10.1
Maine	13.4	Utah	8.8
Maryland	10.9	Vermont	11.9
Massachusetts	13.7	Virginia	10.9
Michigan	12.1	Washington	11.8
Minnesota	12.5	West Virginia	15.1
Mississippi	12.4	Wisconsin	13.3
Missouri	14.1	Wyoming	10.6

SOURCE: *Statistical Abstract of the United States*, 1992.

2. Count the number of observations in each class. Here are the counts.

Class	Count	Class	Count	Class	Count
4.1 to 5.0	1	9.1 to 10.0	0	14.1 to 15.0	5
5.1 to 6.0	0	10.1 to 11.0	9	15.1 to 16.0	4
6.1 to 7.0	0	11.1 to 12.0	7	16.1 to 17.0	0
7.1 to 8.0	0	12.1 to 13.0	10	17.1 to 18.0	0
8.1 to 9.0	1	13.1 to 14.0	12	18.1 to 19.0	1

3. Draw the histogram. First mark the scale for the variable whose distribution you are displaying on the horizontal axis. That's "percent of residents over 65" in this example. The scale runs from 4 to 19 because that is the span of the classes we chose. Then mark the scale of counts on the vertical axis. Each bar represents a class. The base of the bar covers the class, and the bar height is the class count. Draw the graph with no horizontal space between the bars (unless a class is empty, so that its bar has height zero). Figure 1.2 is our histogram. ◀

The bars of a histogram should cover the entire range of values of a variable. When the possible values of a variable have gaps between them, extend the bases of the bars to meet halfway between two adjacent possible values. For example, in a histogram of the ages in years of university faculty, the bars representing 25 to 29 years and 30 to 34 years would meet at 29.5. There is no one right choice of the classes in a histogram. Too few classes will give a "skyscraper" graph, with all values in a few classes with tall bars. Too many will produce a "pancake" graph, with most classes having one or no observations. Neither choice will give a good picture of the shape of the distribution. You must use your judgment in choosing classes to display the shape. Our eyes respond to the *area* of the bars in a histogram, so be sure to choose classes that are all the same width. Then area is determined by height and all classes are fairly represented. If you use a computer, the software will choose the classes for you. The computer's choice is usually a good one, but you can change it if you want.

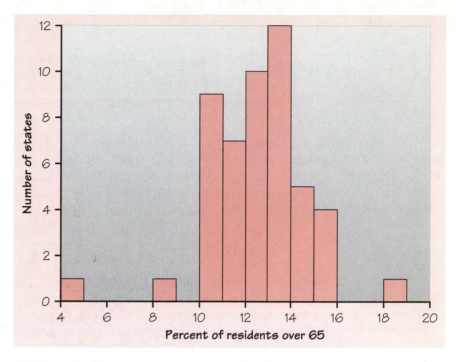

FIGURE 1.2   Histogram of the percent of residents over 65 years of age in the 50 states, from Table 1.1.

Interpreting histograms

Making a statistical graph is not an end in itself. After all, a computer can make graphs faster than we can. The purpose of the graph is to help us understand the data. After you (or your computer) make a graph, always ask, "What do I see?" Here is a general tactic for looking at graphs:

• Look for an overall pattern and also for striking deviations from that pattern.

In the case of a histogram, the overall pattern is the overall shape of the distribution. *Outliers* are an important kind of deviation from the overall pattern.

OUTLIERS

An **outlier** in any graph of data is an individual observation that falls outside the overall pattern of the graph.

Three states stand out in the histogram of Figure 1.2. You can find them in the table once the histogram has called attention to them. Florida has 18.3% of its residents over age 65, and Alaska has only 4.2%. These states are clear outliers. You might also call Utah, with 8.8% over 65, an outlier, though it is not as far from the overall pattern as Florida and Alaska. Whether an observation is an outlier is to some extent a matter of judgment. It is much easier to spot outliers in the histogram than in the data table.

Once you have spotted outliers, look for an explanation. Many outliers are due to mistakes, such as typing 4.0 as 40. Other outliers point to the special nature of some observations. Explaining outliers usually requires some background information. It is not surprising to an American that Florida, with its many retired people, has many residents over 65 and that Alaska, the northern frontier, has few.

What about the *overall pattern* of the distribution in Figure 1.2?

OVERALL PATTERN OF A DISTRIBUTION

To describe the overall pattern of a distribution:

- Give the **center** and the **spread**.

- See if the distribution has a simple **shape** that you can describe in a few words.

The next section tells in detail how to measure center and spread. For now, describe the center by finding a value that divides the observations so that about half take larger values and about half have smaller values. The center in Figure 1.2 is about 13%. That is, about 13% of the residents of a typical state are over 65 years old. You can describe the spread by the extent of the data from smallest to largest value. The spread in Figure 1.2 is about 10% to 16% if we ignore the outliers. The histogram in Figure 1.2 has an irregular shape that isn't easy to describe. Some distributions, however, do have simple shapes. Here are examples that illustrate some shapes to look for.

EXAMPLE 1.3

Look at the histograms in Figures 1.3 and 1.4. Figure 1.3 comes from a study of lightning storms in Colorado. It shows the distribution of the hour of the day during which the first lightning flash for that day occurred. The distribution has a single peak at noon and falls off on either side of this peak. The two sides of the histogram are roughly the same shape, so we call the distribution *symmetric*.

Figure 1.4 shows the distribution of lengths of words used in Shakespeare's plays.[1] This distribution is *skewed to the right*. That is, there are many short words (3 and 4 letters) and few very long words (10, 11, or 12 letters), so that the right tail of the histogram extends out much farther than the left tail.

Notice that the vertical scale in Figure 1.4 is not the *count* of words but the *percent* of all of Shakespeare's words that have each length. A histogram of percents rather than counts is convenient when the counts are very large or when we want to compare several distributions. Different kinds of writing have different distributions of word lengths, but all are right skewed because short words are common and very long words are rare. ◄

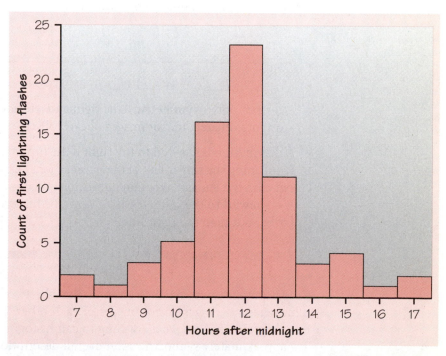

FIGURE 1.3 The distribution of the time of the first lightning flash each day at a site in Colorado.

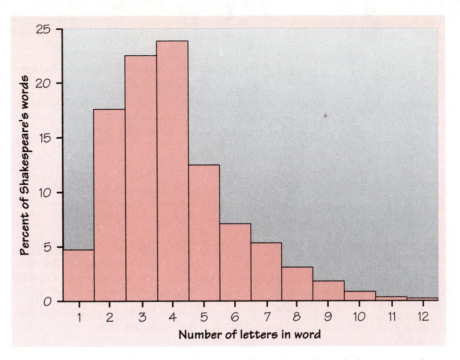

FIGURE 1.4   The distribution of lengths of words used in Shakespeare's plays.

SYMMETRIC AND SKEWED DISTRIBUTIONS

A distribution is **symmetric** if the right and left sides of the histogram are approximately mirror images of each other.

A distribution is **skewed to the right** if the right side of the histogram (containing the upper half of the observations) extends much farther out than the left side (containing the lower half of the observations). It is **skewed to the left** if the left side of the histogram extends much farther out than the right side.

In mathematics, symmetry means that the two sides of a figure like a histogram are exact mirror images of each other. Data are almost never exactly symmetric, so we are willing to call histograms like that in Figure 1.3 approximately symmetric as an overall description.

The overall shape of a distribution is important information about a variable. Some types of data regularly produce distributions that are symmetric or skewed. For example, the sizes of many living things of the same species (like lengths of cockroaches) tend to be symmetric. Data on incomes (whether of individuals, companies, or nations) are usually strongly skewed to the right. There are many moderate incomes, some large incomes, and a few very large incomes. Do remember that many histograms, like Figure 1.2, can't reasonably be called either symmetric or skewed. Some data show other patterns. Scores on an exam, for example, may have a cluster near the top of the scale if many students did well. Or they may show two distinct peaks if a tough problem divided the class into those who did and didn't solve it. Use your eyes and say what you see.

EXERCISES

1.3   The total return on a stock is the change in its market price plus any dividend payments made. Total return is usually expressed as a percent of the beginning price. Figure 1.5 is a histogram of the distribution of total returns for all 1,528 stocks listed on the New York Stock Exchange in one

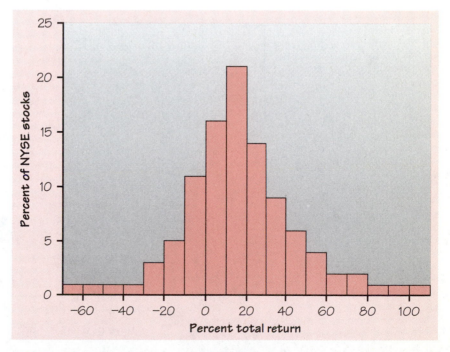

FIGURE 1.5 The distribution of percent total return for all New York Stock Exchange common stocks in one year, for Exercise 1.3.

year.[2] Like Figure 1.4, it is a histogram of the percents in each class rather than a histogram of counts.

(a) Describe the overall shape of the distribution of total returns.

(b) What is the approximate center of this distribution? (For now, take the center to be the value with roughly half the stocks having lower returns and half having higher returns.)

(c) Approximately what were the smallest and largest total returns? (This describes the spread of the distribution.)

(d) A return less than zero means that an owner of the stock lost money. About what percent of all stocks lost money?

1.4   Figure 1.6 is a histogram of the number of days in the month of April on which the temperature fell below freezing at Greenwich, England.[3] The data cover a period of 65 years.

(a) Describe the shape, center, and spread of this distribution. Are there any outliers?

(b) In what percent of these 65 years did the temperature never fall below freezing in April?

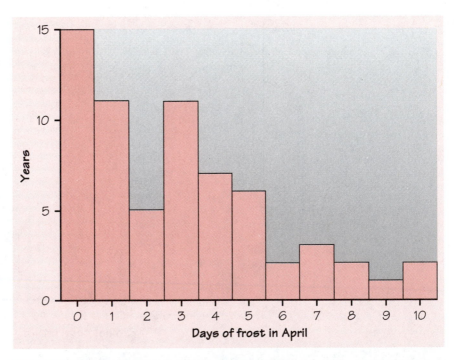

FIGURE 1.6   The distribution of the number of frost days during April at Greenwich, England, over a 65-year period, for Exercise 1.4.

1.5 How would you describe the center and spread of the distribution of first lightning flash times in Figure 1.3? Of the distribution of Shakespeare's word lengths in Figure 1.4?

1.6 The distribution of the ages of a nation's population has a strong influence on economic and social conditions. The following table shows the age distribution of U.S. residents in 1950 and 2075, in millions of persons. The 1950 data come from that year's census. The 2075 data are projections made by the Census Bureau.

(a) Because the total population in 2075 is much larger than the 1950 population, comparing percents in each age group is clearer than comparing counts. Make a table of the percent of the total population in each age group for both 1950 and 2075.

(b) Make a histogram of the 1950 age distribution (in percents). Then describe the main features of the distribution. In particular, look at the percent of children relative to the rest of the population.

(c) Make a histogram of the projected age distribution for the year 2075. Use the same scales as in (b) for easy comparison. What are the most important changes in the U.S. age distribution projected for the 125-year period between 1950 and 2075?

Age Group	1950	2075
Under 10 years	29.3	34.9
10 to 19 years	21.8	35.7
20 to 29 years	24.0	36.8
30 to 39 years	22.8	38.1
40 to 49 years	19.3	37.8
50 to 59 years	15.5	37.5
60 to 69 years	11.0	34.5
70 to 79 years	5.5	27.2
80 to 89 years	1.6	18.8
90 to 99 years	0.1	7.7
100 to 109 years	–	1.7
Total	151.1	310.6

Stemplots

Histograms are not the only graphical display of distributions. For small *stemplot* data sets, a **stemplot** is quicker to make and presents more detailed information.

EXAMPLE 1.4

stem

leaf

To make a stemplot, separate each observation into a **stem** consisting of all but the final (rightmost) digit and a **leaf**, the final digit. For the "over 65" percents in Table 1.1, the whole number part of the observation is the stem and the final digit (tenths) is the leaf. The Alabama entry, 12.9, has stem 12 and leaf 9. Stems can have as many digits as needed, but each leaf must consist of only a single digit.

- Write the stems vertically in increasing order from top to bottom, and draw a vertical line to the right of the stems.

- Go through the data, writing each leaf to the right of its stem.

- Write the stems again, and rearrange the leaves in increasing order out from each stem.

Here is the completed stemplot for the data in Table 1.1.

```
 4 | 2
 5 |
 6 |
 7 |
 8 | 8
 9 |
10 | 111568999
11 | 244689
12 | 01234556779
13 | 112344457779
14 | 11579
15 | 1145
16 |
17 |
18 | 3
```

A stemplot looks like a histogram turned on end. The stemplot in Example 1.4 is exactly like the histogram in Figure 1.2 because each stem is a class in the histogram. But the stemplot, unlike the histogram, preserves the actual value of each observation. We interpret stemplots like histograms, looking for the overall pattern and for any outliers.

You can choose the classes in a histogram. The classes (the stems) of a stemplot are given to you. There are two variations of a stemplot that give

rounding us more flexibility in graphing a distribution. First, you can **round** the data so that the final digit after rounding is suitable as a leaf. Do this when the data have too many digits. For example, data like

$$3.468 \quad 2.567 \quad 2.981 \quad 1.095 \ldots$$

would have too many stems if we took the first three digits as the stem and the final digit as the leaf. You would probably round these data to

$$3.5 \quad 2.6 \quad 3.0 \quad 1.1 \ldots$$

before making a stemplot.

splitting stems You can also **split stems** to double the number of stems when all the leaves would otherwise fall on just a few stems. Each stem then appears twice. Leaves 0 to 4 go on the upper stem and leaves 5 to 9 go on the lower stem. If you split the stems in the stemplot of Example 1.4, for example, the 11 and 12 stems become

```
11 | 244
11 | 689
12 | 01234
12 | 556779
```

Rounding and splitting stems are matters for judgment, like choosing the classes in a histogram. The stemplot in Example 1.4 does not need either change. Stemplots work well for small sets of data. When there are more than 100 observations, a histogram is almost always a better choice.

EXERCISES

1.7 The Survey of Study Habits and Attitudes (SSHA) is a psychological test that evaluates college students' motivation, study habits, and attitudes toward school. A private college gives the SSHA to a sample of 18 of its incoming first-year women students. Their scores are

$$154 \quad 109 \quad 137 \quad 115 \quad 152 \quad 140 \quad 154 \quad 178 \quad 101$$
$$103 \quad 126 \quad 126 \quad 137 \quad 165 \quad 165 \quad 129 \quad 200 \quad 148$$

Make a stemplot of these data. The overall shape of the distribution is irregular, as often happens when only a few observations are available. Are there any outliers? About where is the center of the distribution (the score with half the scores above it and half below)? What is the spread of the scores (ignoring any outliers)?

1.8   The Modern Language Association provides listening tests that measure understanding of spoken French. The range of scores is 0 to 36. Here are the scores of 20 high school French teachers at the beginning of an intensive summer course in French.[4]

32 31 29 10 30 33 22 25 32 20 30 20 24 24 31 30 15 32 23 23

(a) Make a stemplot of these scores. Use split stems.

(b) Describe in words the most important features of the distribution. Are there any outliers?

(c) About what score would place a teacher in the center of this distribution, with roughly half the scores lower and half higher?

Time plots

Many variables are measured at intervals over time. We might, for example, measure the height of a growing child or the price of a stock at the end of each month. In these examples, our main interest is change over time. To display change over time, make a *time plot*.

TIME PLOT

A **time plot** of a variable plots each observation against the time at which it was measured. Always mark the time scale on the horizontal axis and the variable of interest on the vertical axis. If there are not too many points, connecting the points by lines helps show the pattern of changes over time.

EXAMPLE 1.5

Here are data on the rate of deaths from cancer (deaths per 100,000 people) in the United States over the 50-year period 1940 to 1990.

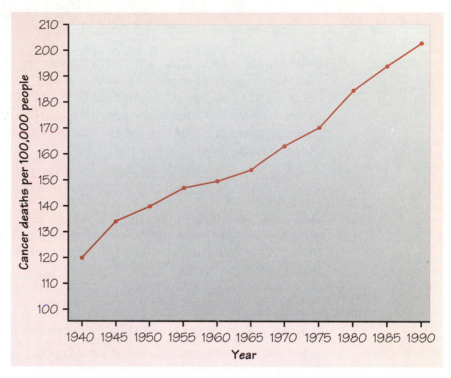

FIGURE 1.7   Time plot of the death rate from cancer (deaths per 100,000 people) from 1940 to 1990.

Year   1940 1945 1950 1955 1960 1965 1970 1975 1980 1985 1990
Deaths 120.3 134.0 139.8 146.5 149.2 153.5 162.8 169.7 183.9 193.3 201.7

Figure 1.7 is a time plot of these data. It shows the steady increase in the cancer death rate during the past half-century. This increase does not mean that we have made no progress in treating cancer. Because cancer is primarily a disease of old age, the death rate from cancer increases when people live longer even if treatment improves.   ◀

When you examine a time plot, look once again for an overall pattern and for strong deviations from the pattern. One common overall pattern is a ***trend***, a long-term upward or downward movement over time. Figure 1.7 shows an upward trend in the cancer death rate, with no striking deviations such as short-term drops.

trend

1.9   The buying power of a dollar changes over time. The Bureau of Labor Statistics measures the cost of a "market basket" of goods and services to compile its Consumer Price Index (CPI). If the CPI is 120, goods and services that cost $100 in the base period now cost $120. Here are the yearly average values of the CPI for the 1970s and 1980s. The base period is the years 1982 to 1984.

Year	CPI	Year	CPI	Year	CPI	Year	CPI
1970	38.8	1975	53.8	1980	82.4	1985	107.6
1971	40.5	1976	56.9	1981	90.9	1986	109.6
1972	41.8	1977	60.6	1982	96.5	1987	113.6
1973	44.4	1978	65.2	1983	99.6	1988	118.3
1974	49.3	1979	72.6	1984	103.9	1989	124.0

(a) Make a time plot of the CPI for these two decades.

(b) What was the overall trend in prices during this period? Were there any years in which this trend was reversed?

(c) In what period during these decades were prices rising fastest? In what period were they rising slowest?

SUMMARY

A data set contains information on a number of **individuals**. Individuals may be people, animals, or things. For each individual, the data give values for one or more **variables**. A variable describes some characteristic of an individual, such as a person's height, gender, or salary.

Exploratory data analysis uses graphs and numerical summaries to describe the variables in a data set and the relations among them.

Some variables are **categorical** and others are **quantitative**. A categorical variable places each individual into a category, like male or female. A quantitative variable has numerical values that measure some characteristic of each individual, like height in centimeters or annual salary in dollars.

The **distribution** of a variable describes what values the variable takes and how often it takes these values.

To describe a distribution, begin with a graph. **Histograms** and **stemplots** graph the distributions of quantitative variables.

When examining any graph, look for an **overall pattern** and for notable **deviations** from the pattern.

The **center**, **spread**, and **shape** describe the overall pattern of a distribution. Some distributions have simple shapes, such as **symmetric** and **skewed**. Not all distributions have a simple overall shape, especially when there are few observations.

Outliers are observations that lie outside the overall pattern of a distribution. Always look for outliers and try to explain them.

When observations on a variable are taken over time, make a **time plot** that graphs time horizontally and the values of the variable vertically. A time plot can reveal **trends** or other changes over time.

SECTION 1.1 EXERCISES

1.10   Here is a small part of the data set that a company keeps to record information about its employees.

Name	Age	Gender	Race	Salary	Job type
⋮					
Fleetwood, Delores	39	Female	White	52,100	Management
Fleming, LaVerne	27	Male	Black	37,500	Technical
Foo, Ruoh-Lin	22	Female	Asian	15,250	Clerical
⋮					

(a) What individuals does the complete data set describe?

(b) The data set records five variables in addition to the name. Which of these are categorical variables?

(c) Which of the variables are quantitative? Based on the data in the table, what do you think are the units of measurement for each of the quantitative variables?

1.11   The histogram in Figure 1.8 shows the number of hurricanes reaching the east coast of the United States each year over a 70-year period.[5] Give a brief description of the overall shape of this distribution. About where does the center of the distribution lie? (For now, take the center to be the value with roughly half the observations on either side of it.)

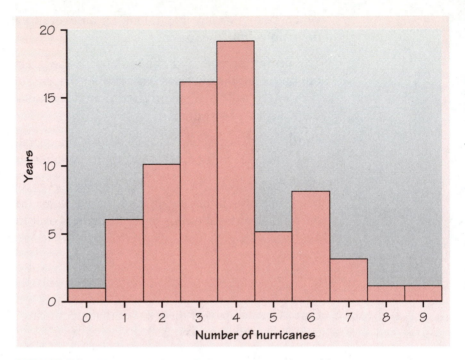

FIGURE 1.8 The distribution of the annual number of hurricanes on the U.S. east coast over a 70-year period, for Exercise 1.11.

1.12 Figure 1.9 displays the distribution of batting averages for all 167 American League baseball players who batted at least 200 times in the 1980 season. (The outlier is the .390 batting average of George Brett, the highest batting average in the major leagues since Ted Williams hit .406 in 1941.)

(a) Is the overall shape (ignoring the outlier) roughly symmetric or clearly skewed or neither?

(b) What was the approximate batting average of a typical American League player? About what were the highest and lowest batting averages, leaving out George Brett?

1.13 Here are the numbers of home runs that Babe Ruth hit in his 15 years with the New York Yankees, 1920 to 1934.[6]

54   59   35   41   46   25   47   60   54   46   49   46   41   34   22

Make a stemplot for these data. Is the distribution roughly symmetric, clearly skewed, or neither? About how many home runs did Ruth hit in a typical year? Is his famous 60 home runs in 1927 an outlier?

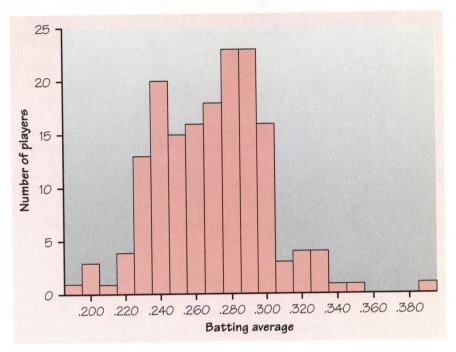

FIGURE 1.9   The distribution of batting averages of American League players in 1980, for Exercise 1.12.

1.14   **(Optional)** Babe Ruth's home run record for a single year was broken by another Yankee, Roger Maris, who hit 61 home runs in 1961. Here are Maris's home run totals for his 10 years in the American League.

13   23   26   16   33   61   28   39   14   8

A **back-to-back stemplot** helps us compare two distributions. Start with your stemplot for Ruth from Exercise 1.13. Draw a vertical line to the left of the stems to match the one you already drew to their right. Add Maris's data as leaves on the same stems, but going out to the left rather than to the right. (Be sure to arrange the leaves on each stem in increasing order out from the stem.)

   Is Maris's record 61 an outlier in his distribution of home runs hit? How does your plot show Ruth's superiority as a home run hitter?

1.15   Sometimes both a histogram or stemplot and a time plot give useful information about a variable. The following data are the *greatest* amounts of rain (in inches) that fell on one day in South Bend, Indiana, for each of

30 consecutive years. Successive years follow each other across the rows in the table.

1.88 2.23 2.58 2.07 2.94 2.29 3.14 2.15 1.95 2.51
2.86 1.48 1.12 2.76 3.10 2.05 2.23 1.70 1.57 2.81
1.24 3.29 1.87 1.50 2.99 3.48 2.12 4.69 2.29 2.12

(a) Make a stemplot for these data. Describe the general shape of the distribution and any prominent deviations from the overall pattern. (You are expected to round and split stems as needed in making stemplots.)

(b) Make a time plot of the data. Is there any suggestion of a long-term change in maximum rainfall at South Bend?

1.16   Sometimes both a histogram or stemplot and a time plot give useful information about a set of data. The data below are measurements of the tension on the wire grid behind the screen in a computer display. The display won't work properly if the tension is either too high or too low, so the manufacturer measures the tension of a small group of displays each hour. Here are the hourly averages for 20 consecutive hours of production, from left to right.

269.5 297.0 269.6 283.3 304.8 280.4 283.5 257.4 317.5 327.4
264.7 307.7 310.0 343.3 328.1 342.6 338.8 340.1 374.6 336.1

(a) Make a stemplot of these data.

(b) Describe the shape of the distribution. The distribution shows considerable variation in the tension, but nothing to alarm the manufacturer except for one outlier.

(c) Make a time plot of the data. (Mark the time scale in hours, 1 to 20.)

(d) Describe the overall pattern of the time plot and explain why the manufacturer should investigate right away.

1.17   (Optional) The impression that a time plot gives depends on the scales you use on the two axes. If you stretch the vertical axis and compress the time axis, change appears to be more rapid. Compressing the vertical axis and stretching the time axis make change appear slower. Make two more time plots of the data in Example 1.5, one that makes cancer death rates appear to increase very rapidly and one that shows only a gentle increase. The moral of this exercise is: pay close attention to the scales when you look at a time plot.

Table 1.2 presents data about the individual states that relate to education. The data for three states in Example 1.1 come from this table. Study of a data set with many variables begins by examining each variable by itself. Exercises 1.18 to 1.20 concern the data in Table 1.2.

TABLE 1.2   Education and related data for the states

State	Region*.	Population (1,000)	SAT verbal	SAT math	Percent taking	Dollars per pupil	Teachers' pay ($1,000)
AL	ESC	4,041	470	514	8	3,648	27.3
AK	PAC	550	438	476	42	7,887	43.4
AZ	MTN	3,665	445	497	25	4,231	30.8
AR	WSC	2,351	470	511	6	3,334	23.0
CA	PAC	29,760	419	484	45	4,826	39.6
CO	MTN	3,294	456	513	28	4,809	31.8
CT	NE	3,287	430	471	74	7,914	43.8
DE	SA	666	433	470	58	6,016	35.2
DC	SA	607	409	441	68	8,210	39.6
FL	SA	12,938	418	466	44	5,154	30.6
GA	SA	6,478	401	443	57	4,860	29.2
HI	PAC	1,108	404	481	52	5,008	32.5
ID	MTN	1,007	466	502	17	3,200	25.5
IL	ENC	11,431	466	528	16	5,062	34.6
IN	ENC	5,544	408	459	54	5,051	32.0
IA	WNC	2,777	511	577	5	4,839	28.0
KS	WNC	2,478	492	548	10	5,009	29.8
KY	ESC	3,685	473	521	10	4,390	29.1
LA	WSC	4,220	476	517	9	4,012	26.2
ME	NE	1,228	423	463	60	5,894	28.5
MD	SA	4,781	430	478	59	6,184	38.4
MA	NE	6,016	427	473	72	6,351	36.1
MI	ENC	9,295	454	514	12	5,257	38.3
MN	WNC	4,375	477	542	14	5,260	33.1
MS	ESC	2,573	477	519	4	3,322	24.4
MO	WNC	5,117	473	522	12	4,415	28.5
MT	MTN	799	464	523	20	5,184	26.7
NE	WNC	1,578	484	546	10	4,381	26.6
NV	MTN	1,202	434	487	24	4,564	32.2
NH	NE	1,109	442	486	67	5,504	31.3
NJ	MA	7,730	418	473	69	9,159	38.4
NM	MTN	1,515	480	527	12	4,446	26.2
NY	MA	17,990	412	470	70	8,500	42.1
NC	SA	6,629	401	440	55	4,802	29.2
ND	WNC	639	505	564	6	3,685	23.6
OH	ENC	10,847	450	499	22	5,639	32.6
OK	WSC	3,146	478	523	9	3,742	24.3
OR	PAC	2,842	439	484	49	5,291	32.3

TABLE 1.2	Education and related data for the states (continued)					
State	Region*	Population (1,000)	SAT verbal	SAT math	Percent taking	Dollars per pupil	Teachers' pay ($1,000)
PA	MA	11,882	420	463	64	6,534	36.1
RI	NE	1,003	422	461	62	6,989	37.7
SC	SA	3,487	397	437	54	4,327	28.3
SD	WNC	696	506	555	5	3,730	22.4
TN	ESC	4,877	483	525	12	3,707	28.2
TX	WSC	16,987	413	461	42	4,238	28.3
UT	MTN	1,723	492	539	5	2,993	25.0
VT	NE	563	431	466	62	5,740	31.0
VA	SA	6,187	425	470	58	5,360	32.4
WA	PAC	4,867	437	486	44	5,045	33.1
WV	SA	1,793	443	490	15	5,046	26.0
WI	ENC	4,892	476	543	11	5,946	33.1
WY	MTN	454	458	519	13	5,255	29.0

SOURCE: *Statistical Abstract of the United States*, 1992

*The census regions are: East North Central, East South Central, Mountain, New England, Pacific, South Atlantic, West North Central, and West South Central

1.18   Make a graph of the distribution of the percent of high school seniors who take the SAT in the various states. Briefly describe the overall pattern of the distribution and any outliers.

1.19   Make a graph to display the distribution of average teachers' salaries for the states. Is there a clear overall pattern? Are there any outliers or other notable deviations from the pattern?

1.20   Make a graph of the distribution of dollars per pupil spent on education for the states. Is there a clear overall pattern? Are there any outliers or other notable deviations from the pattern?

1.2   DESCRIBING DISTRIBUTIONS WITH NUMBERS

How old are presidents at their inauguration? Was Bill Clinton, at age 46, unusually young? Table 1.3 gives the data, the ages of all U.S. presidents when they took office. As the histogram in Figure 1.10 shows, there is a good deal of variation in the ages at which people become president. Teddy Roosevelt was the youngest, at age 42, and Ronald Reagan, at age 69, was the oldest. The distribution is roughly symmetric. It appears that the age

TABLE 1.3   Ages of the presidents at inauguration

President	Age	President	Age	President	Age
Washington	57	Buchanan	65	Harding	55
J. Adams	61	Lincoln	52	Coolidge	51
Jefferson	57	A. Johnson	56	Hoover	54
Madison	57	Grant	46	F. D. Roosevelt	51
Monroe	58	Hayes	54	Truman	60
J. Q. Adams	57	Garfield	49	Eisenhower	61
Jackson	61	Arthur	51	Kennedy	43
Van Buren	54	Cleveland	47	L. Johnson	55
W. H. Harrison	68	B. Harrison	55	Nixon	56
Tyler	51	Cleveland	55	Ford	61
Polk	49	McKinley	54	Carter	52
Taylor	64	T. Roosevelt	42	Reagan	69
Fillmore	50	Taft	51	Bush	64
Pierce	48	Wilson	56	Clinton	46

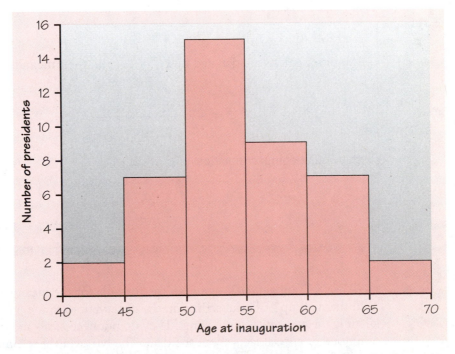

FIGURE 1.10   The distribution of the ages of presidents at their inauguration, from Table 1.3.

of a typical new president is about 55 years, because 55 is near the center of the histogram.

We just gave a brief description of a distribution that included its *shape* (roughly symmetric), a number describing its *center* (about 55), and numbers describing its *spread* (42 to 69). Shape, center, and spread provide a good description of the overall pattern of any distribution for a quantitative variable. Now we will learn specific ways to use numbers to measure the center and spread of a distribution. We can calculate these numerical measures for any quantitative variable. But to interpret measures of center and spread, and to choose among the several measures we will learn, you must think about the shape of the distribution and the meaning of the data. The numbers, like graphs, are aids to understanding, not "the answer" in themselves.

Measuring center: the mean

A description of a distribution almost always includes a measure of its center or average. The most common measure of center is the ordinary arithmetic average, or *mean*.

THE MEAN $\bar{x}$

If n observations are denoted by $x_1, x_2, \ldots, x_n$, their **mean** is

$$\bar{x} = \frac{x_1 + x_2 + \cdots + x_n}{n}$$

or in more compact notation

$$\bar{x} = \frac{1}{n} \sum x_i$$

The $\sum$ (capital Greek sigma) in the formula for the mean is short for "add them all up." The subscripts on the observations x_i are just a way of keeping the n observations distinct. They do not necessarily indicate order or any other special facts about the data. The bar over the x indicates the mean of all the x-values. Pronounce the mean $\bar{x}$ as "x-bar." This notation is very common. When writers who are discussing data use $\bar{x}$ or $\bar{y}$, they are talking about a mean.

EXAMPLE 1.6

Table 1.3 gives the ages of all 42 presidents when they were inaugurated. The mean age of a new president is

$$\bar{x} = \frac{x_1 + x_2 + \cdots + x_n}{n}$$

$$= \frac{57 + 61 + 57 + \cdots + 46}{42}$$

$$= \frac{2303}{42} = 54.8 \text{ years}$$

In practice, you can key the data into your calculator and hit the mean key. You don't have to actually add and divide. But you should know that this is what the calculator is doing.   ◄

EXAMPLE 1.7

A study in Switzerland examined the number of hysterectomies (removal of the uterus) performed in a year by doctors. Here are the data for a sample of 15 male doctors.

27 50 33 25 86 25 85 31 37 44 20 36 59 34 28

A stemplot shows that the distribution is skewed to the right and that there are two outliers on the high side.

```
2 | 05578
3 | 13467
4 | 4
5 | 09
6 |
7 |
8 | 56
```

Use your calculator to see that the mean number of operations that these 15 doctors performed is $\bar{x} = 41.3$. Then notice that only five of the 15 doctors performed more than the mean number of hysterectomies. That's because the two outliers (85 and 86) pull up the mean. Check that the mean of the other 13 observations is 34.5. The mean is the arithmetic average, but it is *not* the number of operations performed by a typical doctor.   ◄

Example 1.7 illustrates an important fact about the mean as a measure of center: it is sensitive to the influence of a few extreme observations. These may be outliers, but a skewed distribution that has no outliers will also pull the mean toward its long tail.

EXERCISE

1.21   Here are the scores of 18 first-year college women on the Survey of Study Habits and Attitudes (SSHA).

| 154 | 109 | 137 | 115 | 152 | 140 | 154 | 178 | 101 |
| 103 | 126 | 126 | 137 | 165 | 165 | 129 | 200 | 148 |

(a) Find the mean score from the formula for the mean. Then enter the data into your calculator and use the calculator's $\bar{x}$ button to obtain the mean. Verify that you get the same result.

(b) A stemplot (Exercise 1.7) suggests that the score 200 is an outlier. Use your calculator to find the mean for the 17 observations that remain when you drop the outlier. Briefly describe how the outlier changes the mean.

Measuring center: the median

The mean is not the only way to describe the center of a distribution. Another natural idea is to use the "middle value" in a histogram or stemplot.

THE MEDIAN M

To find the **median** of a distribution:

1. Arrange all observations in order of size, from smallest to largest.

2. If the number of observations n is odd, the median M is the center observation in the ordered list. Find the location of the median by counting $(n + 1)/2$ observations up from the bottom of the list.

3. If the number of observations n is even, the median M is the mean of the two center observations in the ordered list. The location of the median is again $(n + 1)/2$ from the bottom of the list.

That is, find the number such that half the observations are smaller and the other half are larger. This is the *median* of the distribution. We will call the median M for short. Although the idea of the median as the midpoint of a distribution is simple, we need a precise rule for putting the idea into practice. The rule appears in the box on the previous page.

Basic calculators do not include the median, so you will have to do the work step-by-step unless you are using a computer or an advanced calculator. All statistical computer packages calculate medians. Because arranging many observations in order is a lot of work, use a computer if you have many observations. When working by hand, be sure to write down each individual observation in the data set, even if several observations repeat the same value. And be sure to arrange the observations in order of size before locating the median. The middle observation in the haphazard order in which the observations first come has no importance. Here is an example that shows how the rule for the median works for both odd and even numbers of observations.

EXAMPLE 1.8

To find the median number of hysterectomies performed by the 15 doctors in Example 1.7, first arrange the observations in increasing order:

20   25   25   27   28   31   33   **34**   36   37   44   50   59   85   86

There is an odd number of observations, so there is one center observation. This is the median. The bold 34 in the list is the center observation, because there are seven observations to its left and seven to its right. So the median is $M = 34$.

It is faster to use the rule for locating the median in the list. Because $n = 15$, the location is

$$\frac{n+1}{2} = \frac{16}{2} = 8$$

That is, the median is the 8th observation in the ordered list.

The Swiss study also looked at a sample of 10 female doctors. The numbers of hysterectomies performed by these doctors (arranged in order) were

5   7   10   14   **18**   **19**   25   29   31   33

Here $n = 10$ is even. There is no center observation, but there is a center pair. These are the bold 18 and 19 in the list, which have four observations to their left in the list and four to their right. The median is midway between these two observations:

$$M = \frac{18 + 19}{2} = 18.5$$

The rule for locating the median in the list gives

$$\frac{n+1}{2} = \frac{11}{2} = 5.5$$

The location 5.5 means "halfway between the fifth and sixth observations in the ordered list." That agrees with what we found by eye.

The typical female doctor performed many fewer hysterectomies than the typical male doctor. This was one of the important conclusions of the study. ◀

Comparing the mean and the median

Examples 1.7 and 1.8 illustrate an important difference between the mean and the median. Example 1.7 shows how the two outliers pull the mean up. But they have no effect at all on the median. The outliers are just two observations in the upper half of the data. The median would stay the same even if one doctor had performed 1000 operations. Don't conclude, however, that because the median is not influenced by a few extreme observations, it is always preferred to the mean. The mean and median measure center in different ways, and both are useful. Use the median if you want the number of hysterectomies performed by a typical doctor. Use the mean if you are also interested in the total number of operations performed by all the doctors. The total is the mean times the number of doctors, but it has no connection with the median.

The mean and median of a symmetric distribution are close together. If the distribution is *exactly* symmetric, the mean and median are exactly the same. In a skewed distribution, the mean is farther out in the long tail than is the median. For example, the distribution of house prices is strongly skewed to the right. There are many moderately priced houses and a few very expensive mansions. The few expensive houses pull the mean up but do not affect the median. For example, the mean price of all houses sold in 1993 was $139,400, but the median price for these same houses was only $117,000. Reports about house prices, incomes, and other strongly skewed distributions usually give the median ("typical value") rather than the mean ("arithmetic average value").

EXERCISES

1.22   Here are the number of home runs Babe Ruth hit in each of his 15 years with the New York Yankees.

54 59 35 41 46 25 47 | 60 | 54 46 49 46 41 34 22

Roger Maris, who broke Ruth's single-year record, had these home run totals in his 10 years in the American League.

13   23   26   16   33 | 61   28   39   14   8

Find the median number of home runs hit in a season by each player.

1.23 In Exercise 1.21 you found the mean of the SSHA scores of 18 first-year college students. Now find the median of these scores. Is the median smaller or larger than the mean? Explain why this is so.

1.24 Last year a small accounting firm paid each of its five clerks $22,000, two junior accountants $50,000 each, and the firm's owner $270,000. What is the mean salary paid at this firm? How many of the employees earn less than the mean? What is the median salary?

1.25 The mean and median salaries paid to major league baseball players in 1993 were $490,000 and $1,160,000. Which of these numbers is the mean, and which is the median? Explain your answer.

Measuring spread: the quartiles

The mean and median provide two different measures of the center of a distribution. A measure of center alone can be misleading. Two nations with the same median family income are very different if one has extremes of wealth and poverty and the other has little variation among families. A drug with the correct mean concentration of active ingredient is dangerous if some batches are much too high and others much too low. We are interested in the *spread* or *variability* of incomes and drug potencies as well as their centers. The simplest useful numerical description of a distribution consists of both a measure of center and a measure of spread.

range   One way to measure spread is to calculate the **range**, which is the difference between the largest and smallest observations.

EXAMPLE 1.9

Example 1.7 reports data on the number of hysterectomies performed in a year by 15 male doctors. The smallest number was 20 and the largest was 86, so the range is

$$\text{range} = 86 - 20 = 66$$ ◄

The range shows the full spread of the data. But it depends only on the smallest observation and the largest observation, which may be outliers.

quartiles   We can improve our description of spread by also looking at the spread of the middle half of the data. The **quartiles** mark out the middle half. Count up the ordered list of observations, starting from the smallest. The *first quartile* lies one-quarter of the way up the list. The *third quartile* lies three-quarters of the way up the list. In other words, the first quartile is larger than 25% of the observations, and the third quartile is larger than 75% of the observations. The second quartile is the median, which is larger than 50% of the observations. That is the idea of quartiles. We need a rule to make the idea exact. The rule for calculating the quartiles uses the rule for the median.

THE QUARTILES Q_1 AND Q_3

To calculate the **quartiles**:

1. Arrange the observations in increasing order and locate the median M in the ordered list of observations.
2. The first quartile Q_1 is the median of the observations whose position in the ordered list is to the left of the location of the overall median.
3. The third quartile Q_3 is the median of the observations whose position in the ordered list is to the right of the location of the overall median.

Here is an example that shows how the rules for the quartiles work for both odd and even numbers of observations.

EXAMPLE 1.10

The numbers of hysterectomies performed by our sample of 15 male doctors are (arranged in order)

20   25   25   27   28   31   33   **34**   36   37   44   50   59   85   86

There is an odd number of observations, so the median is the middle one, the bold 34 in the list. The first quartile is the median of the 7 observations to the left of the median. This is the 4th of these 7 observations, so $Q_1 = 27$. If you want, you can use the recipe for the location of the median with $n = 7$:

$$\frac{n+1}{2} = \frac{7+1}{2} = 4$$

The third quartile is the median of the 7 observations to the right of the median, $Q_3 = 50$. The overall median is left out of the calculation of the quartiles when there is an odd number of observations.

For the 10 female doctors, the data are (again arranged in increasing order)

$$5 \quad 7 \quad 10 \quad 14 \quad 18 \quad | \quad 19 \quad 25 \quad 29 \quad 31 \quad 33$$

There is an even number of observations, so the median lies midway between the middle pair. Its location is between the 5th and 6th values, marked by | in the list. The first quartile is the median of the first 5 observations, because these are the observations to the left of the location of the median. Check that $Q_1 = 10$ and $Q_3 = 29$. When the number of observations is even, all the observations enter into the calculation of the quartiles.   ◀

Be careful when several observations take the same numerical value. Write down all of the observations and apply the rules just as if they all had distinct values. For example, the median of

$$4 \quad 7 \quad 7 \quad \mathbf{7} \quad 8 \quad 9 \quad 9$$

is $M = 7$, because the bold 7 is the center observation in the list. The first quartile is the median of the three observations to the left of the bold 7, which are 4, 7, 7. So the first quartile is also 7. The third quartile is 9.

Some computer packages use a slightly different rule to find the quartiles, so computer results may be a bit different from your own work. Don't worry about this. The differences will always be too small to be important.

The five-number summary and boxplots

A convenient way to describe both the center and spread of a data set is to give the median to measure center and the quartiles and the smallest and largest individual observations to show the spread.

> ### THE FIVE-NUMBER SUMMARY
>
> The **five-number summary** of a data set consists of the smallest observation, the first quartile, the median, the third quartile, and the largest observation, written in order from smallest to largest. In symbols, the five-number summary is
>
> $$\text{Minimum} \quad Q_1 \quad M \quad Q_3 \quad \text{Maximum}$$

These five numbers offer a reasonably complete description of center and spread. The five-number summaries from Example 1.10 are

$$20 \quad 27 \quad 34 \quad 50 \quad 86$$

for the male doctors and

$$5 \quad 10 \quad 18.5 \quad 29 \quad 33$$

for the female doctors.

boxplot

The five-number summary of a distribution leads to a new graph, the **boxplot**. Figure 1.11 shows boxplots for the Swiss doctors. The central box in a boxplot has its ends at the quartiles and therefore spans the middle half of the data. The line within the box marks the median. The "whiskers" at either end extend to the smallest and largest observations. We can see at once that female doctors perform far fewer hysterectomies than male doctors, and also that there is less variation among female doctors.

You can draw boxplots either horizontally or vertically. Do be sure to include a numerical scale in the graph. When you look at a boxplot, first locate the median, which marks the center of the distribution. Then look at the spread. The quartiles show the spread of the middle half of the data, and the extremes (the smallest and largest observations) show the spread of the entire data set. The spacing of the quartiles and the extremes about the median gives an indication of the symmetry or skewness of the distribution. In a symmetric distribution, the first and third quartiles are equally distant from the median. In most distributions that are skewed to the right, on the other hand, the third quartile will be farther above the median than the first quartile is below it. The extremes behave the same way, but remember that they are just single observations and may say little about the distribution as a whole. Because boxplots show less detail than histograms or stemplots, they are best used for side-by-side comparison of more than one distribution, as in Figure 1.11.

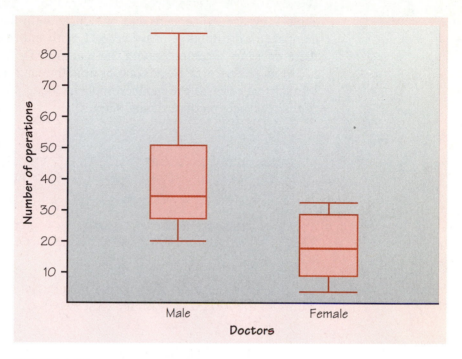

FIGURE 1.11 Side-by-side boxplots comparing the number of hysterectomies performed by female and male doctors.

EXERCISES

1.26 Exercise 1.22 (page 40) gives the number of home runs hit each year by Babe Ruth and Roger Maris.

(a) Find the five-number summary for each player.

(b) Make side-by-side boxplots to compare Ruth and Maris as home run hitters. What do you conclude?

1.27 Return to the data on presidential ages in Table 1.3 (page 35). In Example 1.6 we saw that the mean age is 54.8 years.

(a) From the shape of the histogram (Figure 1.10, on page 35), do you expect the median to be much less than the mean, about the same as the mean, or much greater than the mean?

(b) Find the five-number summary and verify your expectation about the median.

(c) What is the range of the middle half of the ages of new presidents?

1.28   Table 1.2 (page 33) contains data on education in the states. We want to compare the distributions of average SAT math and verbal scores. We enter these data into a computer with the names SATV for verbal scores and SATM for math scores. Here is output from the statistical software package Minitab that gives the five-number summary along with other information. (Other packages produce similar output.)

```
SATV
  N    MEAN   MEDIAN   STDEV     MIN      MAX       Q1       Q3
 51  448.16   443.00   30.82  397.00   511.00   422.00   476.00


SATM
  N    MEAN   MEDIAN   STDEV     MIN      MAX       Q1       Q3
 51  497.39   490.00   34.57  437.00   577.00   470.00   523.00
```

Use the output to make side-by-side boxplots of SAT math and verbal scores for the states. Briefly compare the two distributions in words.

Measuring spread: the standard deviation

The five-number summary is the most widely useful numerical description of a distribution, but it is not the most common. That distinction belongs to the combination of the mean to measure center with the *standard deviation* as a measure of spread. The standard deviation measures spread by looking at how far the observations are from their mean.

THE STANDARD DEVIATION s

The **variance s^2** of a set of observations is the average of the squares of the deviations of the observations from their mean. In symbols, the variance of n observations $x_1, x_2, \ldots, x_n$ is

$$s^2 = \frac{(x_1 - \overline{x})^2 + (x_2 - \overline{x})^2 + \cdots + (x_n - \overline{x})^2}{n - 1}$$

or, more compactly,

$$s^2 = \frac{1}{n - 1} \sum (x_i - \overline{x})^2$$

(continued on next page)

(continued from previous page)

The **standard deviation** s is the square root of the variance s^2:

$$s = \sqrt{\frac{1}{n-1} \sum (x_i - \bar{x})^2}$$

In practice, use software or your calculator to obtain the standard deviation from keyed-in data. Doing a few examples step-by-step will help you understand how the variance and standard deviation work, however. Here is such an example.

EXAMPLE 1.11

A person's metabolic rate is the rate at which the body consumes energy. Metabolic rate is important in studies of weight gain, dieting, and exercise. Here are the metabolic rates of 7 men who took part in a study of dieting. (The units are calories per 24 hours. These are the same calories used to describe the energy content of foods.)

$$1792 \quad 1666 \quad 1362 \quad 1614 \quad 1460 \quad 1867 \quad 1439$$

The researchers reported $\bar{x}$ and s for these men.

First find the mean:

$$\bar{x} = \frac{1792 + 1666 + 1362 + 1614 + 1460 + 1867 + 1439}{7} = \frac{11,200}{7}$$
$$= 1600$$

To see clearly the nature of the variance, start with a table of the deviations of the observations from this mean.

Observations x_i	Deviations $x_i - \bar{x}$			Squared deviations $(x_i - \bar{x})^2$		
1792	$1792 - 1600$	=	192	192^2	=	36,864
1666	$1666 - 1600$	=	66	66^2	=	4,356
1362	$1362 - 1600$	=	-238	$(-238)^2$	=	56,644
1614	$1614 - 1600$	=	14	14^2	=	196
1460	$1460 - 1600$	=	-140	$(-140)^2$	=	19,600
1867	$1867 - 1600$	=	267	267^2	=	71,289
1439	$1439 - 1600$	=	-161	$(-161)^2$	=	25,921
	sum	=	0	sum	=	214,870

The variance is the sum of the squared deviations divided by one less than the number of observations:

$$s^2 = \frac{1}{n-1}\sum (x_i - \bar{x})^2$$
$$= \frac{1}{6}(214{,}870) = 35{,}811.67$$

The standard deviation is the square root of the variance:

$$s = \sqrt{35{,}811.67} = 189.24 \qquad \blacktriangleleft$$

Figure 1.12 displays the data of Example 1.11 as points above the number line, with their mean marked by an asterisk (*). The arrows show two of the deviations from the mean. These deviations show how spread out the data are about their mean. Some of the deviations will be positive and some negative because some of the observations fall on each side of the mean. In fact, *the sum of the deviations of the observations from their mean will always be zero.* Check that this is true in Example 1.11. So we cannot simply add the deviations to get an overall measure of spread. Squaring the deviations makes them all positive, so that observations far from the mean in either direction will have large positive squared deviations. The variance s^2 is the average squared deviation. The variance is large if the observations are widely spread about their mean; it is small if the observations are all close to the mean.

Because the variance involves squaring the deviations, it does not have the same unit of measurement as the original observations. Lengths measured in centimeters, for example, have a variance measured in squared centimeters. Taking the square root remedies this. The standard deviation s measures spread about the mean in the original scale.

The idea of the variance is the average of the squares of the deviations of the observations from their mean. Why do we average by dividing by

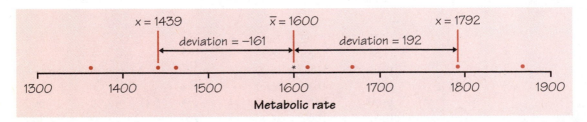

FIGURE 1.12 Metabolic rates for seven men, with their mean (*) and the deviations of two observations from the mean.

$n - 1$ rather than n? Because the sum of the deviations is always zero, the last deviation can be found once we know the other $n - 1$. So we are not averaging n unrelated numbers. Only $n - 1$ of the squared deviations can vary freely, and we average by dividing the total by $n - 1$. The number $n - 1$ is called the ***degrees of freedom*** of the variance or standard deviation. Many calculators offer a choice between dividing by n and dividing by $n - 1$, so be sure to use $n - 1$.

*degrees
of freedom*

Leaving the arithmetic to a calculator allows us to concentrate on what we are doing and why. What we are doing is measuring spread. Here are the basic properties of the standard deviation s as a measure of spread.

PROPERTIES OF THE STANDARD DEVIATION

- s measures spread about the mean and should be used only when the mean is chosen as the measure of center.

- $s = 0$ only when there is *no spread*. This happens only when all observations have the same value. Otherwise $s > 0$. As the observations become more spread out about their mean, s gets larger.

- s, like the mean $\bar{x}$, is strongly influenced by extreme observations. A few outliers can make s very large.

A skewed distribution with a few observations in the single long tail will have a large standard deviation. The number s does not give much helpful information in such a case. Because the two sides of a strongly skewed distribution have different spreads, no single number describes the spread well. The five-number summary, with its two quartiles and two extremes, does a better job. *The five-number summary is usually better than the mean and standard deviation for describing a skewed distribution. Use $\bar{x}$ and s only for reasonably symmetric distributions.*

You may rightly feel that the importance of the standard deviation is not yet clear. It is a bit complicated and is not a good description for skewed distributions. We will see in the next section that the standard deviation is the natural measure of spread for an important class of symmetric distributions, the normal distributions. The usefulness of many statistical procedures is tied to distributions of particular shapes. This is certainly true of the standard deviation.

Do remember that a graph gives the best overall picture of a distribution. Numerical measures of center and spread report specific facts about a distribution, but they do not describe its entire shape. Numerical summaries do not disclose the presence of multiple peaks or gaps, for example. Exercise 1.31 gives an example of a distribution for which numerical summaries alone are misleading. ALWAYS PLOT YOUR DATA.

EXERCISES

1.29   The level of various substances in the blood influences our health. Here are measurements of the level of phosphate in the blood of a patient, in milligrams of phosphate per deciliter of blood, made on 6 consecutive visits to a clinic.

$$5.6 \quad 5.2 \quad 4.6 \quad 4.9 \quad 5.7 \quad 6.4$$

A graph of only 6 observations gives little information, so we proceed to compute the mean and standard deviation.

(a) Find the mean from its definition. That is, find the sum of the 6 observations and divide by 6.

(b) Find the standard deviation from its definition. That is, find the deviations of each observation from the mean, square the deviations, then obtain the variance and the standard deviation. Example 1.11 shows the method.

(c) Now enter the data into your calculator and use the mean and standard deviation buttons to obtain $\bar{x}$ and s. Do the results agree with your hand calculations?

1.30   The number of hysterectomies performed in a year by the male doctors in the study described in Example 1.7 (page 31) were

$$27 \quad 50 \quad 33 \quad 25 \quad 86 \quad 25 \quad 85 \quad 31 \quad 37 \quad 44 \quad 20 \quad 36 \quad 59 \quad 34 \quad 28$$

The stemplot shows that the highest two observations are outliers.

(a) The mean number of observations is $\bar{x} = 41.3$. Find the standard deviation s from its definition. That is, find the deviation of each observation from 41.3, square these deviations, then obtain the variance s^2 and the standard deviation s. Follow the model of Example 1.11.

(b) Enter the data into your calculator and verify your results. Then use your calculator to find $\bar{x}$ and s for the 13 observations that remain when you leave out the two outliers. How do the outliers affect the values of $\bar{x}$ and s?

1.31   Exercise 1.28 (page 46) gives numerical summaries for the average SAT scores for the states. These numerical summaries (and the boxplots based on them) do *not* show one of the most important features of the distributions. Make a stemplot of the SAT math scores from Table 1.2 (page 33). What is the overall shape of the distribution? Remember to always start with a graph of your data—numerical summaries are not a complete description.

SUMMARY

A numerical summary of a distribution should report its **center** and its **spread** or **variability**.

The **mean $\bar{x}$** and the **median M** describe the center of a distribution in different ways. The mean is the arithmetic average of the observations, and the median is the midpoint of the values.

When you use the median to indicate the center of the distribution, describe its spread by giving the **quartiles**. The **first quartile Q_1** has one-fourth of the observations below it, and the **third quartile Q_3** has three-fourths of the observations below it.

The **five-number summary** consisting of the median, the quartiles, and the high and low extremes provides a quick overall description of a distribution. The median describes the center, and the quartiles and extremes show the spread.

Boxplots based on the five-number summary are useful for comparing several distributions. The box spans the quartiles and shows the spread of the central half of the distribution. The median is marked within the box. The whiskers extend to the extremes and show the full spread of the data.

The **variance s^2** and especially its square root, the **standard deviation s**, are common measures of spread about the mean as center. The standard deviation s is zero when there is no spread and gets larger as the spread increases.

The mean and standard deviation are strongly influenced by outliers or skewness in a distribution. They are good descriptions for symmetric distributions and are most useful for the normal distributions introduced in the next section.

The median and quartiles are not affected by outliers, and the two quartiles and two extremes describe the two sides of a distribution separately.

The five-number summary is the preferred numerical summary for skewed distributions.

SECTION 1.2 EXERCISES

1.32   Here are the percents of the popular vote won by the successful candidate in each of the presidential elections from 1948 to 1992.

Year	1948	1952	1956	1960	1964	1968	1972	1976	1980	1984	1988	1992
Percent	49.6	55.1	57.4	49.7	61.1	43.4	60.7	50.1	50.7	58.8	53.9	43.2

(a) Make a stemplot of the winners' percents. (Round to whole numbers and use split stems.)

(b) What is the median percent of the vote won by the successful candidate in presidential elections? (Work with the unrounded data.)

(c) Call an election a landslide if the winner's percent falls at or above the third quartile. Find the third quartile. Which elections were landslides?

1.33   Some people worry about how many calories they consume. *Consumer Reports* magazine, in a story on hot dogs, measured the calories in 20 brands of beef hot dogs, 17 brands of meat hot dogs, and 17 brands of poultry hot dogs.[7] Here is computer output describing the beef hot dogs,

```
Mean = 156.8    Standard deviation = 22.64    Min = 111    Max = 190
N = 20    Median = 152.5    Quartiles = 140, 178.5
```

the meat hot dogs,

```
Mean = 158.7    Standard deviation = 25.24    Min = 107    Max = 195
N = 17    Median = 153    Quartiles = 139, 179
```

and the poultry hot dogs,

```
Mean = 122.5    Standard deviation = 25.48    Min = 87    Max = 170
N = 17    Median = 129    Quartiles = 102, 143
```

Use this information to make side-by-side boxplots of the calorie counts for the three types of hot dogs. Write a brief comparison of the distributions. Will eating poultry hot dogs usually lower your calorie consumption compared with eating beef or meat hot dogs?

1.34   We want to compare spending on public education in the northeastern and the southern states. Because states vary in population, we should compare dollars spent per pupil rather than total dollars spent. Table 1.2 (page 33) gives the dollars spent per pupil in each state. Take the northeastern states to be those in the MA (Middle Atlantic) and NE (New England) regions.

The southern states are those in the SA (South Atlantic) and ESC (East South Central) regions. Leave out the District of Columbia, which is a city rather than a state.

(a) List the dollars-per-pupil data for the northeastern and for the southern states from Table 1.2. These are the two data sets we want to compare.

(b) Make numerical summaries and graphs to compare the two distributions. Write a brief statement of what you find.

1.35   In 1798 the English scientist Henry Cavendish measured the density of the earth with great care. It is common practice to repeat careful measurements several times and use the mean as the final result. Cavendish repeated his work 29 times. Here are his results (the data give the density of the earth as a multiple of the density of water).[8]

5.50	5.61	4.88	5.07	5.26	5.55	5.36	5.29	5.58	5.65
5.57	5.53	5.62	5.29	5.44	5.34	5.79	5.10	5.27	5.39
5.42	5.47	5.63	5.34	5.46	5.30	5.75	5.68	5.85	

Present these measurements with a graph of your choice. Does the shape of the distribution allow the use of $\bar{x}$ and s to describe it? Find $\bar{x}$ and s. What is your estimate of the density of the earth based on these measurements?

1.36   Table 1.1 (page 15) records the percent of people over age 65 living in each of the states. Figure 1.2 (page 17) is a histogram of these data. Do you prefer the five-number summary or $\bar{x}$ and s as a brief numerical description? Why? Calculate your preferred description.

1.37   The 1993 New York Mets had the worst won-lost record in major league baseball. They were paid well for playing poorly. Here are the Mets' salaries, in thousands of dollars. (For example, 6200 stands for Bobby Bonilla's salary of $6,200,000.)[9]

6200	5917	4000	3375	3000	2312	2300	2150	2100
1500	1012	850	650	635	500	475	220	205
195	195	158	145	109	109	109	109	109

Describe this salary distribution both with a graph and with an appropriate numerical summary. Then write a brief description of the important features of the distribution.

1.38   A study of the size of jury awards in civil cases (such as injury, product liability, and medical malpractice) in Chicago showed that the median award was about $8000. But the mean award was about $69,000. Explain how this great difference between two measures of center can occur.

1.39   Which measure of center, the mean or the median, should you use in each of the following situations?

(a) Middletown is considering imposing an income tax on citizens. The city government wants to know the average income of citizens so that it can estimate the total tax base.

(b) In a study of the standard of living of typical families in Middletown, a sociologist estimates the average family income in that city.

1.40   You want to measure the average speed of vehicles on the interstate highway on which you are driving. You adjust your speed until the number of vehicles passing you equals the number you are passing. Have you found the mean speed or the median speed of vehicles on the highway?

1.41   This is a standard deviation contest. You must choose four numbers from the whole numbers 0 to 10, with repeats allowed.

(a) Choose four numbers that have the smallest possible standard deviation.

(b) Choose four numbers that have the largest possible standard deviation.

(c) Is more than one choice possible in either (a) or (b)? Explain.

1.3   THE NORMAL DISTRIBUTIONS

We now have a kit of graphical and numerical tools for describing distributions. What is more, we have a clear strategy for exploring data on a single quantitative variable:

- Start with a graph, usually a stemplot or histogram.
- Look for the overall pattern and for striking deviations such as outliers.
- Choose a numerical summary to briefly describe center and spread.

Here is one more step to add to this strategy:

- Sometimes the overall pattern of a large number of observations is so regular that we can describe it by a smooth curve.

Density curves

Look at Figure 1.13. The histogram in that figure displays the distribution of the scores of all 947 seventh-grade students in Gary, Indiana, on the vocabulary part of the Iowa Test of Basic Skills.[10] Scores of many students on this national test have a quite regular distribution. The histogram is symmetric, and both tails fall off quite smoothly from a single center peak. There are no large gaps or obvious outliers, The smooth curve drawn

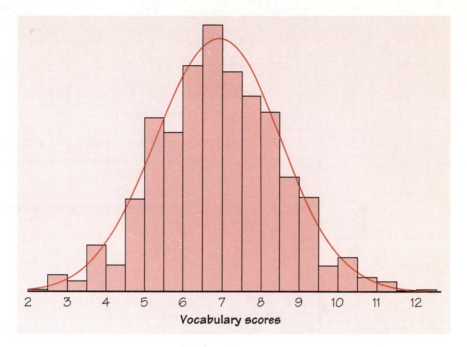

FIGURE 1.13 Histogram of the vocabulary scores of all seventh-grade students in Gary, Indiana. The smooth curve shows the overall shape of the distribution.

through the tops of the histogram bars is a good description of the overall pattern of the data.

We will see that it is easier to work with the smooth curve in Figure 1.13 than with the histogram. The reason is that the histogram depends on our choice of classes, while with a little care we can use a curve that does not depend on any choices we make. Here's how we do it.

- We will use a smooth curve to describe what *proportions* of the observations fall in each range of values, not the counts of observations.

- Our eyes respond to the areas of the bars in a histogram. The bar areas represent the proportions of observations falling into each class. The same is true of the smooth curve: *areas under the curve represent proportions of the observations.*

- Adjust the scale of the graph so that *the total area under the curve is exactly 1.* This area represents the proportion 1, that is, all the observations. The area under the curve and above any range of values on the horizontal axis is equal to the proportion of observations falling in this range, thought of as a fraction of the whole. The curve is a *density curve* for the distribution.

> ## DENSITY CURVE
>
> A **density curve** is a curve that
>
> - is always on or above the horizontal axis, and
> - has area exactly 1 underneath it.
>
> A density curve describes the overall pattern of a distribution. The area under the curve and above any range of values is the proportion of all observations that fall in that range.

EXAMPLE 1.12

Figure 1.14 shows the density curve for a distribution that is slightly skewed to the left. The smooth curve makes the overall shape of the distribution clearly visible. The shaded area under the curve covers the range of values from 7 to 8. This area is 0.12. This means that the proportion 0.12 of all observations from this distribution have values between 7 and 8. ◀

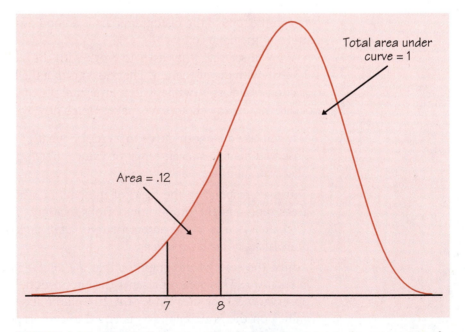

Total area under curve = 1

Area = .12

7   8

FIGURE 1.14   The shaded area under this density curve is the proportion of observations taking values between 7 and 8.

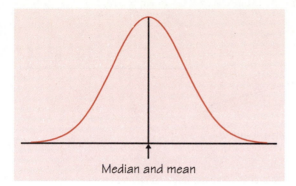
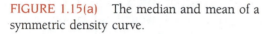

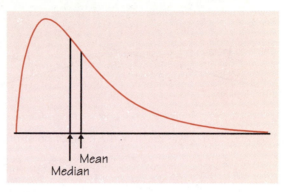

FIGURE 1.15(a)   The median and mean of a symmetric density curve.

FIGURE 1.15(b)   The median and mean of a right-skewed density curve.

Density curves, like distributions, come in many shapes. Figure 1.15 shows two density curves: a symmetric density curve and a right-skewed curve. A density curve is often an adequate description of the overall pattern of a distribution. The curve doesn't describe outliers, which are deviations from the overall pattern. Of course, no set of real data is exactly described by a density curve. The curve is an approximation that is easy to use and accurate enough for practical use.

The median and mean of a density curve

Our measures of center and spread apply to density curves as well as to actual sets of observations. The median and quartiles are easy. Areas under a density curve represent proportions of the total number of observations. The median is the point with half the observations on either side. So *the median of a density curve is the equal-areas point*, the point with half the area under the curve to its left and the remaining half of the area to its right. The quartiles divide the area under the curve into quarters. One-fourth of the area under the curve is to the left of the first quartile, and three-fourths of the area is to the left of the third quartile. You can roughly locate the median and quartiles of any density curve by eye by dividing the area under the curve into four equal parts.

A symmetric density curve, unlike most distributions of real data, is exactly symmetric. The median of a symmetric density curve is therefore at its center. Figure 1.15(a) shows the median of a symmetric curve. It isn't so easy to spot the equal-areas point on a skewed curve. There are mathematical ways of finding the median for any density curve. We did that to mark the median on the skewed curve in Figure 1.15(b).

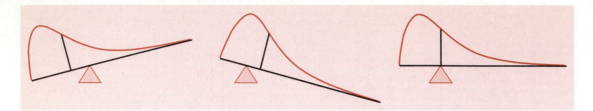

FIGURE 1.16   The mean is the balance point of a density curve.

What about the mean? The mean of a set of observations is their arithmetic average. If we think of the observations as weights strung out along a thin rod, the mean is the point at which the rod would balance. This fact is also true of density curves. *The mean is the point at which the curve would balance if made of solid material.* Figure 1.16 illustrates this fact about the mean. A symmetric curve balances at its center because the two sides are identical. So *the mean and median of a symmetric density curve are equal*, as in Figure 1.15(a). We know that the mean of a skewed distribution is pulled

---

MEDIAN AND MEAN OF A DENSITY CURVE

The **median** of a density curve is the equal-areas point, the point that divides the area under the curve in half.

The **mean** of a density curve is the balance point, at which the curve would balance if made of solid material.

The median and mean are the same for a symmetric density curve. They both lie at the center of the curve.

---

toward the long tail. Figure 1.15(b) shows how the mean of a skewed density curve is pulled toward the long tail more than is the median. It's hard to locate the balance point by eye on a skewed curve. There are mathematical ways of calculating the mean for any density curve, so we are able to mark the mean as well as the median in Figure 1.15(b).

We can roughly locate the mean, median, and quartiles of any density curve by eye. This is not true of the standard deviation. Remember the earlier remark that the standard deviation is not a natural measure for most distributions. When necessary, we can once again call on more advanced mathematics to learn the value of the standard deviation. The study of mathematical methods for doing calculations with density curves is part of theoretical statistics. Though we are concentrating on statistical practice, we often make use of the results of mathematical study.

Because a density curve is an idealized description of the distribution of data, we need to distinguish between the mean and standard deviation of the density curve and the mean $\bar{x}$ and standard deviation s computed from the actual observations. The usual notation for the mean of an idealized distribution is μ (the Greek letter mu). We write the standard deviation of a density curve as σ (the Greek letter sigma).

mean μ

standard deviation σ

EXERCISES

1.42   **(a)** Sketch a density curve that is symmetric but has a shape different from that of the curve in Figure 1.15(a).

(b) Sketch a density curve that is strongly skewed to the left.

1.43   Figure 1.17 displays the density curve of a *uniform distribution*. The curve takes the constant value 1 over the interval from 0 to 1 and is zero outside that range of values. This means that data described by this distribution take values that are uniformly spread between 0 and 1. Use areas under this density curve to answer the following questions.

(a) What percent of the observations lie above 0.8?

(b) What percent of the observations lie below 0.6?

(c) What percent of the observations lie between 0.25 and 0.75?

1.44   Figure 1.18 displays three density curves, each with three points marked on them. At which of these points on each curve do the mean and the median fall?

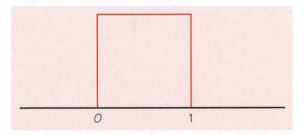

FIGURE 1.17   The density curve of a uniform distribution, for Exercise 1.43.

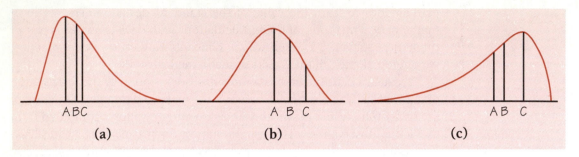

FIGURE 1.18   Three density curves, for Exercise 1.44.

Normal distributions

One particularly important class of density curves has already appeared in Figures 1.13 and 1.15(a). These density curves are symmetric, single-peaked, and bell-shaped. They are called *normal curves*, and they describe ***normal distributions***. All normal distributions have the same overall shape. The exact density curve for a particular normal distribution is described by giving its mean μ and its standard deviation σ. The mean is located at the center of the symmetric curve, and is the same as the median. Changing μ without changing σ moves the normal curve along the horizontal axis without changing its spread. The standard deviation σ controls the spread of a normal curve. Figure 1.19 shows two normal curves with different values of σ. The curve with the larger standard deviation is more spread out.

normal distributions

The standard deviation σ is the natural measure of spread for normal distributions. Not only do μ and σ completely determine the shape of a normal curve, but we can locate σ by eye on the curve. Here's how. As we move out in either direction from the center μ, the curve changes from falling ever more steeply

to falling ever less steeply

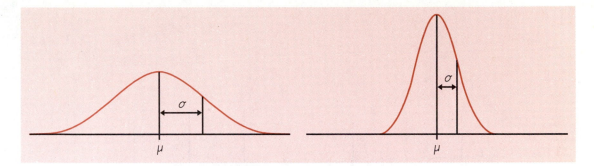

FIGURE 1.19   Two normal curves, showing the mean μ and standard deviation σ.

The points at which this change of curvature takes place are located at distance σ on either side of the mean μ. You can feel the change as you run a pencil along a normal curve, and so find the standard deviation. Figure 1.19 shows σ for two different normal curves.

Although there are many normal curves, they all have common properties. In particular, all normal distributions have the properties described by the following rule.

THE 68–95–99.7 RULE

In the normal distribution with mean μ and standard deviation σ:

- **68%** of the observations fall within σ of the mean μ.

- **95%** of the observations fall within 2σ of μ.

- **99.7%** of the observations fall within 3σ of μ.

Figure 1.20 illustrates the 68–95–99.7 rule. By remembering these three numbers, you can think about normal distributions without constantly making detailed calculations.

EXAMPLE 1.13

The distribution of heights of young women aged 18 to 24 is approximately normal with mean $\mu = 64.5$ inches and standard deviation $\sigma = 2.5$ inches. Figure 1.21 shows the application of the 68–95–99.7 rule in this example.

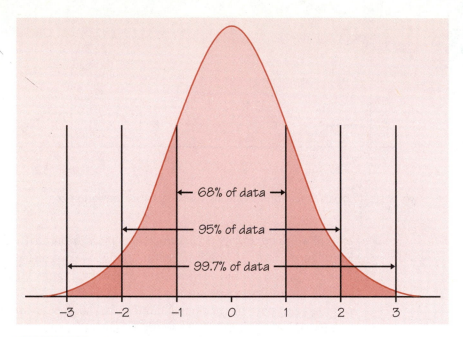

FIGURE 1.20   The 68–95–99.7 rule for normal distributions.

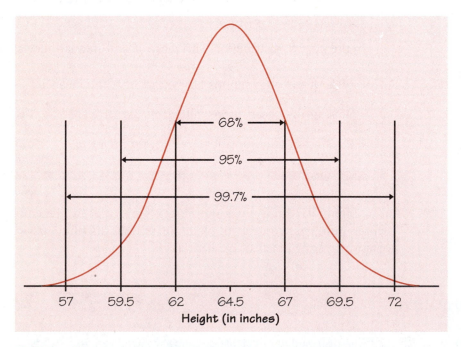

FIGURE 1.21   The 68–95–99.7 rule applied to the distribution of the heights of young women.

Two standard deviations is 5 inches for this distribution. The 95 part of the 68–95–99.7 rule says that the middle 95% of young women are between $64.5 - 5$ and $64.5 + 5$ inches tall, that is, between 59.5 inches and 69.5 inches. This fact is exactly true for an exactly normal distribution. It is approximately true for the heights of young women because the distribution of heights is approximately normal.

The other 5% of young women have heights outside the range from 59.5 to 69.5 inches. Because the normal distributions are symmetric, half of these women are on the tall side. So the tallest 2.5% of young women are taller than 69.5 inches.

The 99.7 part of the 68–95–99.7 rule says that almost all young women (99.7% of them) have heights between $\mu - 3\sigma$ and $\mu + 3\sigma$. This range of heights is 57 to 72 inches. ◀

Because we will mention normal distributions often, a short notation is helpful. We abbreviate the normal distribution with mean μ and standard deviation σ as $N(\mu, \sigma)$. For example, the distribution of young women's heights is $N(64.5, 2.5)$.

Why are the normal distributions important in statistics? Here are three reasons. First, normal distributions are good descriptions for some distributions of *real data*. Distributions that are often close to normal include scores on tests taken by many people (such as SAT exams and many psychological tests), repeated careful measurements of the same quantity, and characteristics of biological populations (such as lengths of cockroaches and yields of corn). Second, normal distributions are good approximations to the results of many kinds of *chance outcomes*, such as tossing a coin many times. Third, and most important, we will see that many *statistical inference* procedures based on normal distributions work well for other roughly symmetric distributions. HOWEVER . . . even though many sets of data follow a normal distribution, many do not. Most income distributions, for example, are skewed to the right and so are not normal. Nonnormal data, like nonnormal people, are not only common, but are sometimes more interesting than their normal counterparts.

EXERCISES

1.45   The distribution of heights of adult men is approximately normal with mean 69 inches and standard deviation 2.5 inches. Draw a normal curve on which this mean and standard deviation are correctly located. (Hint:

Draw the curve first, locate the points where the curvature changes, then mark the horizontal axis.)

1.46   The distribution of heights of adult men is approximately normal with mean 69 inches and standard deviation 2.5 inches. Use the 68–95–99.7 rule to answer the following questions.

(**a**) What percent of men are taller than 74 inches?

(**b**) Between what heights do the middle 95% of men fall?

(**c**) What percent of men are shorter than 66.5 inches?

1.47   Scores on the Wechsler Adult Intelligence Scale (a standard "IQ test") for the 20 to 34 age group are approximately normally distributed with $\mu = 110$ and $\sigma = 25$. Use the 68–95–99.7 rule to answer these questions.

(**a**) About what percent of people in this age group have scores above 110?

(**b**) About what percent have scores above 160?

(**c**) In what range do the middle 95% of all IQ scores lie?

The standard normal distribution

As the 68–95–99.7 rule suggests, all normal distributions share many common properties. In fact, all normal distributions are the same if we measure in units of size σ about the mean μ as center. Changing to these units is called *standardizing*. To standardize a value, subtract the mean of the distribution and then divide by the standard deviation.

STANDARDIZED OBSERVATIONS

If x is an observation from a distribution that has mean μ and standard deviation σ, the **standardized value** of x is

$$z = \frac{x - \mu}{\sigma}$$

---

z-scores   The letter z is commonly used for standardized observations. In fact, standardized observations are sometimes called **z-scores**. We often standardize observations from symmetric distributions to express them in a common scale. A standardized observation tells us how many standard deviations the original observation falls away from the mean, and in which direction. Observations larger than the mean are positive when standardized, and observations smaller than the mean are negative.

EXAMPLE 1.14

The heights of young women are approximately normal with $\mu = 64.5$ inches and $\sigma = 2.5$ inches. The standardized height is

$$z = \frac{\text{height} - 64.5}{2.5}$$

A woman's standardized height is the number of standard deviations by which her height differs from the mean height of all young women. A woman 68 inches tall, for example, has standardized height

$$z = \frac{68 - 64.5}{2.5} = 1.4$$

or 1.4 standard deviations above the mean. Similarly, a woman 5 feet (60 inches) tall has standardized height

$$z = \frac{60 - 64.5}{2.5} = -1.8$$

or 1.8 standard deviations less than the mean height.   ◀

If the variable we standardize has a normal distribution, standardizing does more than give a common scale. It makes all normal distributions into a single distribution, and this distribution is still normal. Standardizing a variable that has any normal distribution produces a new variable that has the *standard normal distribution*.

STANDARD NORMAL DISTRIBUTION

The **standard normal distribution** is the normal distribution $N(0, 1)$ with mean 0 and standard deviation 1.

If a variable x has any normal distribution $N(\mu, \sigma)$ with mean μ and standard deviation σ, then the standardized variable

$$z = \frac{x - \mu}{\sigma}$$

has the standard normal distribution.

1.48   Eleanor scores 680 on the mathematics part of the SAT. The distribution
of SAT scores in a reference population is normal, with mean 500 and
standard deviation 100. Gerald takes the American College Testing (ACT)
mathematics test and scores 27. ACT scores are normally distributed with
mean 18 and standard deviation 6. Find the standardized scores for both
students. Assuming that both tests measure the same kind of ability, who
has the higher score?

Normal distribution calculations

An area under a density curve is a proportion of the observations in a distribution. Any question about what proportion of observations lie in some
range of values can be answered by finding an area under the curve. Because all normal distributions are the same when we standardize, we can
find areas under any normal curve from a single table, a table that gives
areas under the curve for the standard normal distribution.

EXAMPLE 1.15

What proportion of all young women are less than 68 inches tall? This proportion is the area under the $N(64.5, 2.5)$ curve to the left of the point 68. Because
the standardized height corresponding to 68 inches is

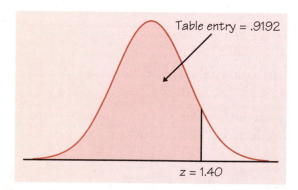

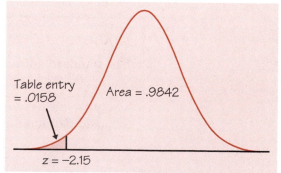

FIGURE 1.22(a)   The area under a standard
normal curve to the left of the point $z = 1.4$ is
0.9192. Table A gives areas under the standard
normal curve.

FIGURE 1.22(b)   Areas under the standard
normal curve to the right and left of $z =$
-2.15. Table A gives only areas to the left.

$$z = \frac{x - \mu}{\sigma} = \frac{68 - 64.5}{2.5} = 1.4$$

this area is the same as the area under the *standard normal* curve to the left of the point $z = 1.4$. Figure 1.22(a) shows this area. ◀

Many calculators will give you areas under the standard normal curve. In case your calculator does not, Table A in the back of the book gives some of these areas. Table A also appears on the inside front cover.

THE STANDARD NORMAL TABLE

Table A is a table of areas under the standard normal curve. The table entry for each value z is the area under the curve to the left of z.

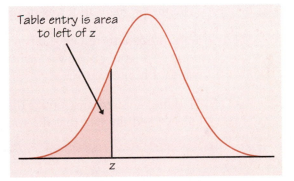

Table entry is area to left of z

z

EXAMPLE 1.16

Problem: Find the proportion of observations from the standard normal distribution that are less than 1.4.

Solution: To find the area to the left of 1.40, locate 1.4 in the left-hand column of Table A, then locate the remaining digit 0 as .00 in the top row. The entry opposite 1.4 and under .00 is 0.9192. This is the area we seek. Figure 1.22(a) illustrates the relationship between the value $z = 1.40$ and the area 0.9192. Because $z = 1.40$ is the standardized value of height 68 inches, the proportion of young women who are less than 68 inches tall is 0.9192 (about 92%).

Problem: Find the proportion of observations from the standard normal distribution that are greater than -2.15.

Solution: Enter Table A under $z = -2.15$. That is, find -2.1 in the left-hand column and .05 in the top row. The table entry is 0.0158. This is the area to the *left* of -2.15. Because the total area under the curve is 1, the area lying to the *right* of -2.15 is $1 - .0158 = .9842$. Figure 1.22(b) illustrates these areas.

◀

We can answer any question about proportions of observations in a normal distribution by standardizing and then using the standard normal table. Here is an outline of the method for finding the proportion of the distribution in any region.

FINDING NORMAL PROPORTIONS

- State the problem in terms of the observed variable x.

- Standardize x to restate the problem in terms of a standard normal variable z. Draw a picture to show the area under the standard normal curve.

- Find the required area under the standard normal curve, using Table A and the fact that the total area under the curve is 1.

EXAMPLE 1.17

The level of cholesterol in the blood is important because high cholesterol levels may increase the risk of heart disease. The distribution of blood cholesterol levels in a large population of people of the same age and sex is roughly normal. For 14-year-old boys, the mean is $\mu = 170$ milligrams of cholesterol per deciliter of blood (mg/dl) and the standard deviation is $\sigma = 30$ mg/dl.[11] Levels above 240 mg/dl may require medical attention. What percent of 14-year-old boys have more than 240 mg/dl of cholesterol?

- *State the problem*. Call the level of cholesterol in the blood x. The variable x has the $N(170, 30)$ distribution. We want the proportion of boys with $x > 240$.

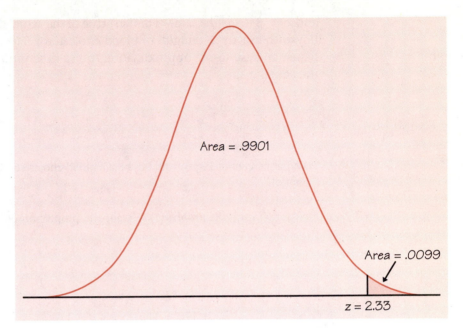

Area = .9901

Area = .0099

z = 2.33

FIGURE 1.23   Areas under the standard normal curve for Example 1.17.

- *Standardize*. Subtract the mean, then divide by the standard deviation, to turn x into a standard normal z:

$$x \quad > \quad 240$$

$$\frac{x - 170}{30} \quad > \quad \frac{240 - 170}{30}$$

$$z \quad > \quad 2.33$$

Figure 1.23 shows the standard normal curve with the area of interest shaded.

- *Use the table*. From Table A, we see that the proportion of observations less than 2.33 is 0.9901. About 99% of boys have cholesterol levels less than 240. The area to the right of 2.33 is therefore $1 - .9901 = .0099$. This is about 0.01, or 1%. Only about 1% of boys have high cholesterol. ◀

In a normal distribution, the proportion of observations with $x > 240$ is the same as the proportion with $x \geq 240$. There is no area under the curve and exactly over 240, so the areas under the curve with $x > 240$

and $x \geq 240$ are the same. This isn't true of the actual data. There may be a boy with exactly 240 mg/dl of blood cholesterol. The normal distribution is just an easy-to-use approximation, not a description of every detail in the actual data.

EXAMPLE 1.18

What percent of 14-year-old boys have blood cholesterol between 170 and 240 mg/dl?

- **State the problem.** We want the proportion of boys with $170 \leq x \leq 240$.

- **Standardize:**

$$
\begin{array}{ccccc}
170 & \leq & x & \leq & 240 \\[4pt]
\dfrac{170 - 170}{30} & \leq & \dfrac{x - 170}{30} & \leq & \dfrac{240 - 170}{30} \\[4pt]
0 & \leq & z & \leq & 2.33
\end{array}
$$

Figure 1.24 shows the area under the standard normal curve.

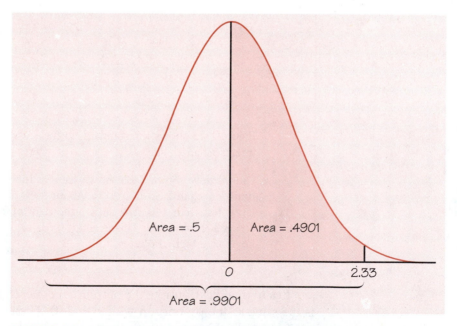

FIGURE 1.24   Areas under the standard normal curve for Example 1.18.

- *Use the table*. The area between 2.33 and 0 is the area below 2.33 *minus* the area below 0. Look at Figure 1.24 to check this. From Table A,

$$\begin{aligned} \text{area between 0 and 2.33} \\ = \text{ area below } 2.33 \ - \ \text{area below } 0.00 \\ = .9901 - .5000 = .4901 \end{aligned}$$

About 49% of boys have cholesterol levels between 170 and 240 mg/dl.   ◀

What if we meet a z that falls outside the range covered by Table A? For example, the area to the left of $z = -4$ does not appear in the table. But since -4 is less than -3.4, this area is smaller than the entry for $z = -3.40$, which is 0.0003. There is very little area under the standard normal curve outside the range covered by Table A. You can take this area to be zero with little loss of accuracy.

Finding a value given a proportion

Examples 1.17 and 1.18 illustrate the use of Table A to find what proportion of the observations satisfies some condition, such as "blood cholesterol between 170 mg/dl and 240 mg/dl." We may instead want to find the observed value with a given proportion of the observations above or below it. To do this, use Table A backward. Find the given proportion in the body of the table, read the corresponding z from the left column and top row, then "unstandardize" to get the observed value. Here is an example.

EXAMPLE 1.19

Scores on the SAT for verbal ability of high school seniors follow the $N(430, 100)$ distribution. (This is approximately true for current scores. The range of possible scores is 200 to 800.) How high must a student score in order to place in the top 10% of all students taking the SAT?

- *State the problem*. We want to find the SAT score x with area 0.1 to its right under the normal curve with mean $\mu = 430$ and standard deviation $\sigma = 100$. That's the same as finding the SAT score x with area 0.9 to its left. Figure 1.25 poses the question in graphical form. Because Table A gives the areas to the *left* of z-values, always state the problem in terms of the area to the left of x.

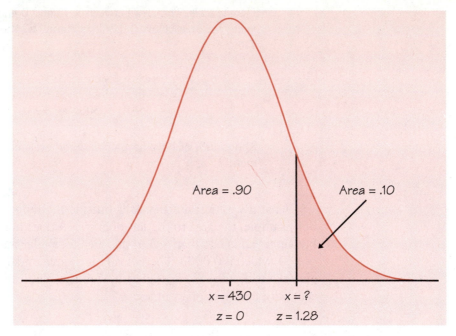

FIGURE 1.25   Locating the point on a normal curve with area 0.10 to its right.

- **Use the table**. Look in the body of Table A for the entry closest to 0.9. It is 0.8997. This is the entry corresponding to $z = 1.28$. So $z = 1.28$ is the standardized value with area 0.9 to its left.

- **Unstandardize**. We now know that the standardized value of the unknown x is $z = 1.28$. So x itself satisfies

$$\frac{x - 430}{100} = 1.28$$

Solving this equation for x gives

$$x = 430 + (1.28)(100) = 558$$

A student must score at least 558 to place in the highest 10%.   ◀

Here is the general formula for unstandardizing. To find the value x from the normal distribution with mean μ and standard deviation σ corresponding to a given standard normal value z, use

$$x = \mu + z\sigma$$

EXERCISES

1.49 Use Table A to find the proportion of observations from a standard normal distribution that satisfies each of the following statements. In each case, sketch a standard normal curve and shade the area under the curve that is the answer to the question.

(a) $z < 2.85$

(b) $z > 2.85$

(c) $z > -1.66$

(d) $-1.66 < z < 2.85$

1.50 Use Table A to find the value z of a standard normal variable that satisfies each of the following conditions. (Use the value of z from Table A that comes closest to satisfying the condition.) In each case, sketch a standard normal curve with your value of z marked on the axis.

(a) The point z with 25% of the observations falling below it.

(b) The point z with 40% of the observations falling above it.

1.51 The distribution of heights of adult American men is approximately normal with mean 69 inches and standard deviation 2.5 inches.

(a) What percent of men are at least 6 feet (72 inches) tall?

(b) What percent of men are between 5 feet (60 inches) and 6 feet tall?

(c) How tall must a man be to be in the tallest 10% of all adult men?

1.52 Scores on the Wechsler Adult Intelligence Scale (a standard "IQ test") for the 20 to 34 age group are approximately normally distributed with $\mu = 110$ and $\sigma = 25$.

(a) What percent of people age 20 to 34 have IQ scores above 100?

(b) What percent have scores above 150?

(c) How high an IQ score is needed to be in the highest 25%?

Assessing normality*

The decision to describe a distribution by a normal curve may determine the later steps in our analysis of the data. How can we judge whether data are approximately normal?

*This section is not needed to read the rest of the book.

A histogram or stemplot can reveal distinctly nonnormal features of a distribution, such as outliers, pronounced skewness, or gaps and clusters. You can improve the effectiveness of these plots for assessing whether a distribution is normal by marking the points $\bar{x}$, $\bar{x} \pm s$, and $\bar{x} \pm 2s$ on the x axis. This gives the scale natural to normal distributions. Then compare the count of observations in each interval with the 68–95–99.7 rule.

EXAMPLE 1.20

The histogram in Figure 1.13 suggests that the distribution of the 947 Gary vocabulary scores is close to normal. It is hard to assess by eye how close to normal a histogram is. Let's use the 68–95–99.7 rule to check more closely. We enter the scores into a statistical computing system and ask for the mean and standard deviation. The computer replies,

$$\text{MEAN} = 6.8585$$

$$\text{STDEV} = 1.5952$$

Now that we know that $\bar{x} = 6.8585$ and $s = 1.5952$, we check the 68–95–99.7 rule by finding the actual counts of Gary vocabulary scores in intervals of length s about the mean $\bar{x}$. The computer will also do this for us. Here are the counts.

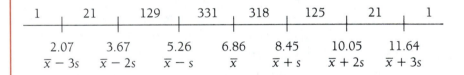

The distribution is very close to symmetric. It also follows the 68–95–99.7 rule closely: there are 68.5% of the scores (649 out of 947) within one standard deviation of the mean, 95.4% (903 of 947) within two standard deviations, and 99.8% (945 of the 947 scores) within three. These counts confirm that the normal distribution with $\mu = 6.86$ and $\sigma = 1.595$ fits these data well. ◀

Smaller data sets rarely fit the 68–95–99.7 rule as well as the Gary vocabulary scores. This is true even of observations taken from a larger population that really has a normal distribution. There is more chance variation in small data sets. Statistical software packages offer more elaborate ways to check whether a distribution is close to normal. How closely you should check normality depends on how you plan to analyze the data. "This distribution looks roughly normal" is good enough when you are describing the overall pattern of a distribution. It is also good enough for the situations in which we will study statistical inference methods. We will avoid methods that require very normal data, so we do not need more precise ways to check normality.

1.53 Repeated careful measurements of the same physical quantity often have a distribution that is close to normal. Here are Henry Cavendish's 29 measurements of the density of the earth, made in 1798. (The data give the density of the earth as a multiple of the density of water.)

5.50	5.61	4.88	5.07	5.26	5.55	5.36	5.29	5.58	5.65
5.57	5.53	5.62	5.29	5.44	5.34	5.79	5.10	5.27	5.39
5.42	5.47	5.63	5.34	5.46	5.30	5.75	5.68	5.85	

A stemplot (Exercise 1.35, page 53) shows that the data are reasonably symmetric. Now check how closely they follow the 68–95–99.7 rule. Find $\bar{x}$ and s, then count the number of observations that fall between $\bar{x} - s$ and $\bar{x} + s$, between $\bar{x} - 2s$ and $\bar{x} + 2s$, and between $\bar{x} - 3s$ and $\bar{x} + 3s$. Compare the percents of the 29 observations in each of these intervals with the 68–95–99.7 rule.

We expect that when we have only a few observations from a normal distribution, the percents will show some deviation from 68, 95, and 99.7. Cavendish's measurements are in fact close to normal.

We can sometimes describe the overall pattern of a distribution by a **density curve**. A density curve always remains on or above the horizontal axis and has total area 1 underneath it. An area under a density curve gives the proportion of observations that fall in a range of values.

A density curve is an idealized description of the overall pattern of a distribution that smooths out the irregularities in the actual data. Write the mean of a density curve as μ and the standard deviation of a density curve as σ to distinguish them from the mean $\bar{x}$ and standard deviation s of the actual data.

The **mean**, the **median**, and the **quartiles** of a density curve can be located by eye. The mean μ is the balance point of the curve. The median divides the area under the curve in half. The quartiles with the median divide the area under the curve into quarters. The **standard deviation** σ cannot be located by eye on most density curves.

The mean and median are equal for symmetric density curves. The mean of a skewed curve is located farther toward the long tail than is the median.

The **normal distributions** are described by a special family of bell-shaped symmetric density curves, called **normal curves**. The mean μ and standard deviation σ completely specify a normal distribution $N(\mu, \sigma)$. The mean is the center of the curve and σ is the distance from μ to the change-of-curvature points on either side.

All normal distributions are the same when measurements are made in units of σ about the mean. These are called **standardized observations**. The standardized value z of an observation x is

$$z = \frac{x - \mu}{\sigma}$$

In particular, all normal distributions satisfy the **68–95–99.7 rule**, which describes what percent of observations lie within one, two, and three standard deviations of the mean.

If x has the $N(\mu, \sigma)$ distribution, then the **standardized variable** $z = (x - \mu)/\sigma$ has the **standard normal distribution** $N(0, 1)$ with mean 0 and standard deviation 1. Table A gives the proportions of standard normal observations that are less than z for many values of z. By standardizing, we can use Table A for any normal distribution.

SECTION 1.3 EXERCISES

1.54   Figure 1.26 shows two normal curves, both with mean 0. Approximately what is the standard deviation of each of these curves?

1.55   The army reports that the distribution of head circumference among male soldiers is approximately normal with mean 22.8 inches and standard deviation 1.1 inches. Use the 68–95–99.7 rule to answer the following questions.

(a) What percent of soldiers have head circumference greater than 23.9 inches?

(b) What percent of soldiers have head circumference between 21.7 inches and 23.9 inches?

1.56   The length of human pregnancies from conception to birth varies according to a distribution that is approximately normal with mean 266 days

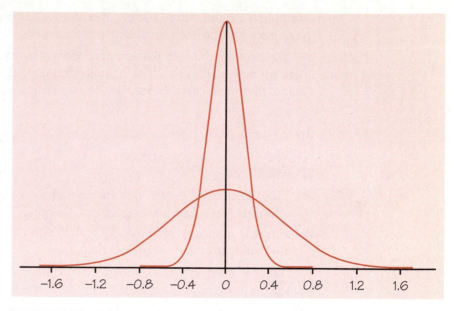

FIGURE 1.26 Two normal curves with the same mean but different standard deviations, for Exercise 1.54.

and standard deviation 16 days. Use the 68–95–99.7 rule to answer the following questions.

(a) Between what values do the lengths of the middle 95% of all pregnancies fall?

(b) How short are the shortest 2.5% of all pregnancies?

1.57 Three landmarks of baseball achievement are Ty Cobb's batting average of .420 in 1911, Ted Williams's .406 in 1941, and George Brett's .390 in 1980. These batting averages cannot be compared directly because the distribution of major league batting averages has changed over the years. The distributions are quite symmetric and (except for outliers such as Cobb, Williams, and Brett) reasonably normal. While the mean batting average has been held roughly constant by rule changes and the balance between hitting and pitching, the standard deviation has dropped over time. Here are the facts.

Decade	Mean	Std. dev.
1910s	.266	.0371
1940s	.267	.0326
1970s	.261	.0317

Compute the standardized batting averages for Cobb, Williams, and Brett to compare how far each stood above his peers.[12]

1.58   Use Table A to find the proportion of observations from a standard normal distribution that falls in each of the following regions. In each case, sketch a standard normal curve and shade the area representing the region.

(a) $z \leq -2.25$

(b) $z \geq -2.25$

(c) $z > 1.77$

(d) $-2.25 < z < 1.77$

1.59   (a) Find the number z such that the proportion of observations that are less than z in a standard normal distribution is 0.8.

(b) Find the number z such that 35% of all observations from a standard normal distribution are greater than z.

1.60   The rate of return on stock indexes (which combine many individual stocks) is approximately normal. Since 1945, the Standard & Poor's 500 index has had a mean yearly return of 11.8%, with a standard deviation of 16.6%. Take this normal distribution to be the distribution of yearly returns over a long period.

(a) In what range do the middle 95% of all yearly returns lie?

(b) The market is down for the year if the return on the index is less than zero. In what percent of years is the market down?

(c) In what percent of years does the index gain 25% or more?

1.61   The length of human pregnancies from conception to birth varies according to a distribution that is approximately normal with mean 266 days and standard deviation 16 days.

(a) What percent of pregnancies last less than 240 days (that's about 8 months)?

(b) What percent of pregnancies last between 240 and 270 days (roughly between 8 months and 9 months)?

(c) How long do the longest 20% of pregnancies last?

1.62   The quartiles of any density curve are the points with area 0.25 and 0.75 to their left under the curve.

(a) What are the quartiles of a standard normal distribution?

(b) How many standard deviations away from the mean do the quartiles lie in any normal distribution? What are the quartiles for the lengths of human pregnancies? (Use the distribution in the previous exercise.)

1.63   The *deciles* of any distribution are the points that mark off the lowest 10% and the highest 10%. The deciles of a density curve are therefore the points with area 0.1 and 0.9 to their left under the curve.

(a) What are the deciles of the standard normal distribution?

(b) The heights of young women are approximately normal with mean 64.5 inches and standard deviation 2.5 inches. What are the deciles of this distribution?

CHAPTER REVIEW

Data analysis is the art of describing data using graphs and numerical summaries. The purpose of data analysis is to describe the most important features of a set of data. This chapter introduces data analysis by presenting statistical ideas and tools for describing the distribution of a single variable. To describe a distribution, we begin with a graph, add numerical descriptions of specific aspects of the distribution, and sometimes use a density curve to describe the overall pattern. Here is a review list of the most important skills you should have acquired from your study of this chapter.

A. DATA

1. Identify the individuals and variables in a set of data.

2. Identify each variable as categorical or quantitative. Identify the units in which each quantitative variable is measured.

B. HISTOGRAMS

1. Make a histogram of the distribution of a quantitative variable when you are given counts for classes of equal width.

2. Make a histogram from a set of observations by choosing classes of equal widths, finding the class counts, and drawing the histogram.

C. STEMPLOTS

1. Make a stemplot of the distribution of a small set of observations.

2. Round leaves or split stems as needed to make an effective stemplot.

D. INSPECTING DISTRIBUTIONS

1. Look for the overall pattern and for major deviations from the pattern.

2. Describe the overall pattern by giving numerical measures of center and spread and a verbal description of shape.

3. Assess from a stemplot or histogram whether the shape of a distribution is roughly symmetric, distinctly skewed, or neither.

4. Decide which measures of center and spread are more appropriate: the mean and standard deviation (especially for symmetric distributions) or the five-number summary (especially for skewed distributions).

5. Recognize outliers.

E. TIME PLOTS

1. Make a time plot of data, with the time of each observation on the horizontal axis and the value of the observed variable on the vertical axis.

2. Recognize strong trends or other patterns in a time plot.

F. MEASURING CENTER

1. Calculate the mean $\bar{x}$ of a set of observations using a calculator.

2. Calculate the median M of a set of observations.

3. Understand that the median is less affected by extreme observations than the mean. Recognize that skewness in a distribution moves the mean away from the median toward the long tail.

G. MEASURING SPREAD

1. Calculate the quartiles Q_1 and Q_3 for a set of observations.

2. Give the five number summary and draw a boxplot; assess symmetry and skewness from a boxplot.

3. Calculate the standard deviation s for a set of observations using a calculator.

4. Know the basic properties of s: $s \geq 0$ always; $s = 0$ only when all observations are identical and increases as the spread increases; s has the same units as the original measurements; s is pulled strongly up by outliers or skewness.

H. DENSITY CURVES

1. Know that areas under a density curve represent proportions of all observations and that the total area under a density curve is 1.

2. Approximately locate the median (equal-areas point) and the mean (balance point) on a density curve.

3. Know that the mean and median both lie at the center of a symmetric density curve and that the mean moves farther toward the long tail of a skewed curve.

I. NORMAL DISTRIBUTIONS

1. Recognize the shape of normal curves and be able to estimate both the mean and standard deviation from such a curve.

2. Use the 68–95–99.7 rule and symmetry to state what percent of the observations from a normal distribution fall between two points when both points lie at the mean or one, two, or three standard deviations on either side of the mean.

3. Given that a variable has the normal distribution with a stated mean μ and standard deviation σ, calculate the proportion of values above a stated number, below a stated number, or between two stated numbers.

4. Given that a variable has the normal distribution with a stated mean μ and standard deviation σ, calculate the point having a stated proportion of all values above it. Also calculate the point having a stated proportion of all values below it.

CHAPTER 1 REVIEW EXERCISES

1.64 The Canadian Province of Ontario carries out statistical studies of the working of Canada's national health care system in the province. The bar charts in Figure 1.27 come from a study of admissions and discharges from community hospitals in Ontario.[13] They show the number of heart attack patients admitted and discharged on each day of the week during 1992 and 1993.

(a) Explain why you expect the number of patients admitted with heart attacks to be roughly the same for all days of the week. Do the data show that this is true?

(b) Describe how the distribution of the day on which patients are discharged from the hospital differs from that of the day on which they are admitted. What do you think explains the difference?

1.65 Professor Moore, who lives a few miles outside a college town, records the time he takes to drive to the college each morning. Here are the times (in minutes) for 42 consecutive weekdays, with the dates in order along the rows.

```
8.25   7.83  8.30  8.42  8.50  8.67  8.17  9.00  9.00  8.17  7.92
9.00   8.50  9.00  7.75  7.92  8.00  8.08  8.42  8.75  8.08  9.75
8.33   7.83  7.92  8.58  7.83  8.42  7.75  7.42  6.75  7.42  8.50
8.67  10.17  8.75  8.58  8.67  9.17  9.08  8.83  8.67
```

(a) Make a stemplot of these drive times. (Round to the nearest tenth of a minute and use split stems.) Is the distribution roughly symmetric, clearly skewed, or neither? Are there any clear outliers?

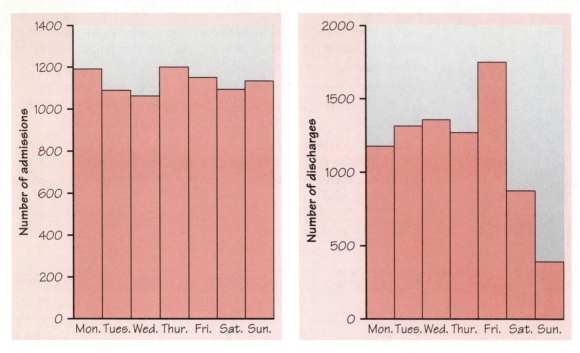

FIGURE 1.27   Bar charts of the number of heart attack victims admitted and discharged on each day of the week by hospitals in Ontario, Canada.

(b) Make a time plot of the drive times. (Label the horizontal axis in days, 1 to 42.) The plot shows no clear trend, but it does show one unusually low drive time and two unusually high drive times. Circle these observations on your plot.

(c) All three unusual observations can be explained. The low time is the day after Thanksgiving (no traffic on campus). The two high times reflect delays due to an accident and icy roads. Remove these three observations, and find the mean $\bar{x}$ and standard deviation s of the remaining 39 drive times.

(d) Count the number of observations between $\bar{x} - s$ and $\bar{x} + s$, between $\bar{x} - 2s$ and $\bar{x} + 2s$, and between $\bar{x} - 3s$ and $\bar{x} + 3s$. Find the percent of observations in each of these intervals and compare with the 68–95–99.7 rule. As real data go, these data are reasonably close to having a normal distribution.

1.66   Corn is an important animal food. Normal corn lacks certain amino acids, which are building blocks for protein. Plant scientists have developed new corn varieties that have more of these amino acids. To test a new corn as an animal food, a group of 20 one-day-old male chicks was fed a ration containing the new corn. A control group of another 20 chicks was fed a ration that was identical except that it contained normal corn. Here are the weight gains (in grams) after 21 days.[14]

Normal corn				New corn			
380	321	366	356	361	447	401	375
283	349	402	462	434	403	393	426
356	410	329	399	406	318	467	407
350	384	316	272	427	420	477	392
345	455	360	431	430	339	410	326

(a) Compute five-number summaries for the weight gains of the two groups of chicks. Then make boxplots to compare the two distributions. What do the data show about the effect of the new corn?

(b) The researchers actually reported means and standard deviations for the two groups of chicks. What are they? How much larger is the mean weight gain of chicks fed the new corn?

1.67   Joe DiMaggio played center field for the Yankees for 13 years. He was succeeded by Mickey Mantle, who played for 18 years. Here is the number of home runs hit each year by DiMaggio.

29   46   32   30   31   30   21   25   20   39   14   32   12

Here are Mantle's home run counts.

13 23 21 27 37 52 34 42 31 40 54 30 15 35 19 23 22 18

Compute the five-number summary for each player, and make side-by-side boxplots of the home run distributions. What does your comparison show about DiMaggio and Mantle as home run hitters?

1.68   Environmental Protection Agency regulations require auto makers to give the city and highway gas mileages for each model of car. Here are the highway mileages (miles per gallon) for 30 midsize and large 1994 car models.[15]

(a) Make a graph that displays the distribution. Describe its main features (overall pattern and any outliers) in words.

(b) The government imposes a "gas guzzler" tax on cars with low gas mileage. Which of these cars do you think are certainly subject to the gas guzzler tax? (In fact, four of these cars are subject to the tax, but highway mileage alone doesn't point to all four.)

(c) Find the five-number summary for these data. What is the highway gas mileage of a "typical" 1994 midsize or large car? How many miles per gallon must a car achieve to be in the top quarter of these models?

Model	MPG	Model	MPG	Model	MPG
BMW 740i	23	Chrysler New Yorker	26	Mazda 626	31
Buick Century	31	Dodge Spirit	27	Mazda 929	24
Buick LeSabre	28	Ford LTD	25	Mercedes-Benz S320	24
Buick Park Avenue	27	Ford Taurus	29	Mercedes-Benz S420	20
Buick Regal	29	Ford Thunderbird	26	Nissan Maxima	26
Buick Roadmaster	25	Hyundai Sonata	27	Rolls-Royce Silver Spur	15
Cadillac DeVille	25	Infiniti Q45	22	Saab 900	26
Chevrolet Caprice	26	Lexus LS400	23	Saab 9000	27
Chevrolet Lumina	29	Lincoln Continental	26	Toyota Camry	28
Chrysler Concorde	28	Lincoln Mark VIII	25	Volvo 850	26

(d) Find the mean highway gas mileage for these cars. What feature of the distribution explains the difference between the mean and the median?

1.69   The data below are the survival times in days of 72 guinea pigs after they were injected with tubercle bacilli in a medical experiment.[16] Survival times, whether of machines under stress or cancer patients after treatment, usually have distributions that are skewed to the right.

43	45	53	56	56	57	58	66	67	73
74	79	80	80	81	81	81	82	83	83
84	88	89	91	91	92	92	97	99	99
100	100	101	102	102	102	103	104	107	108
109	113	114	118	121	123	126	128	137	138
139	144	145	147	156	162	174	178	179	184
191	198	211	214	243	249	329	380	403	511
522	598								

(a) Graph the distribution and describe its main features. Does it show the expected right skew?

(b) Here is output from the computer statistical package Data Desk for these data.

```
Summary statistics for days

Mean 141.84722
Median 102.50000
Cases 72
StdDev 109.20863
Min 43
Max 598
25th%ile 82.250000
75th%ile 153.75000
```

(Data Desk uses "Cases" for the number of observations, and "25th%ile" for the first quartile, which is also called the 25th percentile because 25% of the data lie below it. Similarly, "75th%ile" is the third quartile.) Explain how the relationship between the mean and the median reflects the skewness of the data.

(c) Give the five-number summary and explain briefly how it reflects the skewness of the data.

1.70   The rate of return on a stock is its change in price plus any dividends paid, usually measured in percent of the starting value. We have data on the monthly rate of return for the stock of Wal-Mart stores for the years 1973 to 1991, the first 19 years Wal-Mart was listed on the New York Stock Exchange. There are 228 observations. Here is output from the computer statistical package SPLUS that describes the distribution of these data.

```
Mean     =    3.064
Standard deviation   =    11.49

N = 228    Median = 3.4691
Quartiles = -2.950258, 8.4511

Decimal point is 1 place to the right of the colon

Low:   -34.04255 -31.25000 -27.06271 -26.61290
```

```
-1 : 985
-1 : 444443322222110000
-0 : 999988777666666665555
-0 : 44444444333333322222222222221111111100
 0 : 000001111111111112222223333333344444444
 0 : 55555555555555555555566666666666777777788888888888899999
 1 : 0000000000111111111122233334444
 1 : 55566667889
 2 : 011334
```

```
High: 32.01923 41.80531 42.05607 57.89474 58.67769
```

SPLUS gives high and low outliers separately from the stemplot rather than spreading out the stemplot to include them. Notice that the stems in the plot are the tens digits of the percent returns. The leaves are the ones digits.

(a) Give the five-number summary for monthly returns on Wal-Mart stock.

(b) If you had $1000 worth of Wal-Mart stock at the beginning of the best month during these 19 years, how much would your stock be worth at the end of the month? If you had $1000 worth of stock at the beginning of the worst month, how much would your stock be worth at the end of the month?

(c) Describe in words the main features of the distribution.

1.71   (Optional) A common criterion for detecting suspected outliers in a set of data is as follows:

1. Find the quartiles Q_1 and Q_3 and the **interquartile range** $IQR = Q_3 - Q_1$. The interquartile range is the spread of the central half of the data.

2. Call an observation an outlier if it falls more than $1.5 \times IQR$ above the third quartile or below the first quartile.

Find the interquartile range IQR for the Wal-Mart data in the previous exercise. Are there any outliers according to the $1.5 \times IQR$ criterion? Does it appear to you that SPLUS uses this criterion in choosing which observations to report separately as outliers?

1.72   Has the behavior of Wal-Mart stock changed over the 19 years 1973 to 1991? In Exercise 1.70 we saw the distribution of all 228 monthly returns. That display can't answer questions about change over time. Figure 1.28 is a variation of a time plot. Rather than plotting all 228 observations, we give

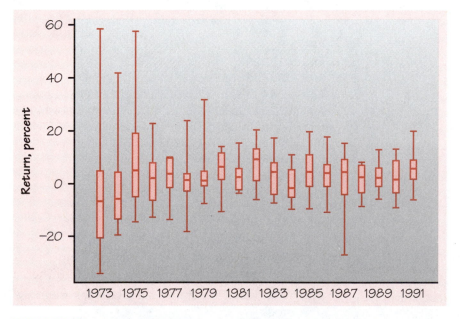

FIGURE 1.28   Side-by-side boxplots comparing the distributions of monthly rates of return on Wal-Mart stock for 19 years.

side-by-side boxplots that compare the 19 years with each other. There are 12 monthly returns in each year.

(a) Is there a long term trend in the typical monthly return over time?

(b) Is there a trend in the spread of the monthly returns?

(c) The stemplot in Exercise 1.70 reports several outliers. Which of these can you spot in the boxplots? In what years did they occur? Does this reinforce your conclusions from part (b)? Are there any outliers that are especially striking after taking your result from (b) into account?

1.73   Julie says, "People are living longer now, so presidents are likely to be older than in the past." John replies, "No—modern voters like youth and don't respect age, so presidents are likely to be younger now than earlier in our history."

Make a time plot of the presidential ages in Table 1.3 (page 35). Take the horizontal axis to be the order of the presidents, from 1 for Washington to 42 for Clinton. Are there any clear trends over time? Do the data support either Julie or John?

1.74   The years around 1970 brought unrest to many U.S. cities. Here are government data on the number of civil disturbances in each 3-month period during the years 1968 to 1972.

Period	Count	Period	Count
1968, Jan.–Mar.	6	1970, July–Sept.	20
Apr.–June	46	Oct.–Dec.	6
July–Sept.	25	1971, Jan.–Mar.	12
Oct.–Dec.	3	Apr.–June	21
1969, Jan.–Mar.	5	July–Sept.	5
Apr.–June	27	Oct.–Dec.	1
July–Sept.	19	1972, Jan.–Mar.	3
Oct.–Dec.	6	Apr.–June	8
1970, Jan.–Mar.	26	July–Sept.	5
Apr.–June	24	Oct.–Dec.	5

(a) Make a time plot of these counts. Connect the points in your plot with straight line segments to make the pattern clearer.

(b) Describe the trend and any other striking pattern in your time plot. Can you suggest an explanation for the pattern in civil disorders?

1.75   Table 1.2 reports data on the states. Much more information is available. Do your own exploration of differences among the states. Find in the library a recent edition of the annual *Statistical Abstract of the United States*. Look up data on either

(a) marriage rates per thousand inhabitants or

(b) death rates per thousand inhabitants

for the 50 states. Make a graph and a numerical summary to display the distribution and write a brief description of the most important characteristics. Suggest an explanation for the outliers you see.

1.76   The NASDAQ Composite Index describes the average price of common stock traded over the counter, that is, not on one of the stock exchanges. In 1991, the mean capitalization of the companies in the NASDAQ index was $80 million and the median capitalization was $20 million. (A company's capitalization is the total market value of its stock.) Explain why the mean capitalization is much higher than the median.

1.77   The Acculturation Rating Scale for Mexican Americans (ARSMA) is a psychological test that measures the degree to which Mexican Americans are adapted to Mexican/Spanish versus Anglo/English culture. The range of possible scores is 1.0 to 5.0, with higher scores showing more Anglo/English acculturation. The distribution of ARSMA scores in a population used to develop the test is approximately normal with mean 3.0 and standard deviation 0.8. A researcher believes that Mexicans will have an average score near 1.7 and that first-generation Mexican Americans will average about 2.1 on the ARSMA scale. What proportion of the population used to develop the test has scores below 1.7? Between 1.7 and 2.1?

1.78   The army reports that the distribution of head circumference among soldiers is approximately normal with mean 22.8 inches and standard deviation 1.1 inches. Helmets are mass-produced for all except the smallest 5% and the largest 5% of head sizes. Soldiers in the smallest or largest 5% get custom-made helmets. What head sizes get custom-made helmets?

1.79   The ARSMA test is described in Exercise 1.77. How high a score on this test must a Mexican American obtain to be among the 30% of the population used to develop the test who are most Anglo/English in cultural orientation? What scores make up the 30% who are most Mexican/Spanish in their acculturation?

NOTES AND DATA SOURCES

1. The Shakespeare data appear in C. B. Williams, *Style and Vocabulary: Numerical Studies*, Griffin, London, 1970.

2. Data from John K. Ford, "Diversification: how many stocks will suffice?" *American Association of Individual Investors Journal*, January 1990, pp. 14–16.

3. Data on frosts from C. E. Brooks and N. Carruthers, *Handbook of Statistical Methods in Meteorology*, H. M. Stationery Office, 1953.

4. Data provided by Joseph A. Wipf, Department of Foreign Languages and Literatures, Purdue University.

5. Hurricane data from H. C. S. Thom, *Some Methods of Climatological Analysis*, World Meteorological Organization, Geneva, Switzerland, 1966.

6. Data from *The Baseball Encyclopedia*, 3rd ed., Macmillan, New York, 1976. Maris's home run data are from the same source.

7. *Consumer Reports*, June 1986, pp. 366–367. A more recent study of hot dogs appears in *Consumer Reports*, July 1993, pp. 415–419. The newer data cover few brands of poultry hot dogs and take calorie counts mainly from the package labels, resulting in suspiciously round numbers.

8. Cavendish's data and the background information about his work appear in S. M. Stigler, "Do robust estimators work with real data?" *Annals of Statistics*, 5 (1977), pp. 1055–1078.

9. The Mets' salaries were reported in the *New York Times*, April 11, 1993.

10. Data from Gary Community School Corporation, courtesy of Celeste Foster, Department of Education, Purdue University.

11. Detailed data appear in P. S. Levy *et al.*, "Total serum cholesterol values for youths 12–17 years," *Vital and Health Statistics Series 11*, No. 150 (1975), U.S. National Center for Health Statistics.

12. Data from Stephen Jay Gould, "Entropic homogeneity isn't why no one hits .400 any more," *Discover*, August 1986, pp. 60–66. Gould does not standardize but gives a speculative discussion instead.

13. Based on Antoni Basinski, "Almost never on Sunday: implications of the patterns of admission and discharge for common conditions," Institute for Clinical Evaluative Sciences in Ontario, October 18, 1993.

14. Based on data summaries in G. L. Cromwell *et al.*, "A comparison of the nutritive value of *opaque-2, floury-2* and normal corn for the chick," *Poultry Science*, 57 (1968), pp. 840–847.

15. The gas mileage data are from the U.S. Department of Energy's *1994 Gas Mileage Guide*, October 1993. In the table, the data are for the basic engine/transmission combination for each model, and models that are essentially identical (such as the Ford Taurus and Mercury Sable) appear only once.

16. Data from T. Bjerkedal, "Acquisition of resistance in guinea pigs infected with different doses of virulent tubercle bacilli," *American Journal of Hygiene*, 72 (1960), pp. 130–148.

JOHN W. TUKEY

He started as a chemist, became a mathematician, and was converted to statistics by what he called "the real problems experience and the real data experience" of war work during the Second World War. John W. Tukey (1915-) came to Princeton University in 1937 to study chemistry but took a doctorate in mathematics in 1939. During the war, he worked on the accuracy of range finders and of gunfire from bombers, among other problems. After the war he divided his time between Princeton and nearby Bell Labs, perhaps the world's most eminent industrial research group.

Tukey devoted much of his attention to the statistical study of messy problems with complex data: the safety of anesthetics used by many doctors in many hospitals on many patients, the Kinsey studies of human sexual behavior, monitoring compliance with a nuclear test ban, and air quality and environmental pollution.

From this "real problems experience and real data experience," John Tukey developed exploratory data analysis. He invented some of the simple tools we have met, such as boxplots and stemplots. More important, he changed the way we approach data by emphasizing the need for a flexible, exploratory approach that does not just seek to answer specific questions but asks, "What do the data say?" This chapter, like Chapter 1, follows Tukey's path by presenting ideas and tools for examining data.

CHAPTER **2**

Examining Relationships

INTRODUCTION

Most statistical studies involve more than one variable. Sometimes we want to compare the distributions of the same variable for several groups. For example, we might compare the distributions of Scholastic Assessment Test (SAT) scores among students at several colleges. Side-by-side box-plots, stemplots, or histograms make the comparison visible. In this chapter, however, we concentrate on relationships among several variables for the same group of individuals. For example, Table 2.1 records seven variables that describe education in the states. We have already examined some of these variables one at a time. Now we might ask how SAT mathematics scores are related to SAT verbal scores or to the percent of a state's high school seniors who take the SAT or to what region a state is in.

When you examine the relationship between two or more variables, first ask the preliminary questions that are familiar from Chapter 1:

- What *individuals* do the data describe?
- What exactly are the *variables*? How are they measured?
- Are all the variables *quantitative* or is at least one a *categorical* variable?

TABLE 2.1	Education and related data for the states						
State	Region*	Population (1,000)	SAT verbal	SAT math	Percent taking	Dollars per pupil	Teachers' pay ($1,000)
AL	ESC	4,041	470	514	8	3,648	27.3
AK	PAC	550	438	476	42	7,887	43.4
AZ	MTN	3,665	445	497	25	4,231	30.8
AR	WSC	2,351	470	511	6	3,334	23.0
CA	PAC	29,760	419	484	45	4,826	39.8
CO	MTN	3,294	456	513	28	4,809	31.8
CT	NE	3,287	430	471	74	7,914	43.8
DE	SA	666	433	470	58	6,016	35.2
DC	SA	607	409	441	68	8,210	39.6
FL	SA	12,938	418	466	44	5,154	30.6
GA	SA	6,478	401	443	57	4,860	29.2
HI	PAC	1,108	404	481	52	5,008	32.5
ID	MTN	1,007	466	502	17	3,200	25.5
IL	ENC	11,431	466	528	16	5,062	34.6
IN	ENC	5,544	408	459	54	5,051	32.0
IA	WNC	2,777	511	577	5	4,839	28.0
KS	WNC	2,478	492	548	10	5,009	29.8

TABLE 2.1   Education and related data for the states (*continued*)

State	Region*	Population (1,000)	SAT verbal	SAT math	Percent taking	Dollars per pupil	Teachers' pay ($1,000)
KY	ESC	3,685	473	521	10	4,390	29.1
LA	WSC	4,220	476	517	9	4,012	26.2
ME	NE	1,228	423	463	60	5,894	28.5
MD	SA	4,781	430	478	59	6,164	38.4
MA	NE	6,016	427	473	72	6,351	36.1
MI	ENC	9,295	454	514	12	5,257	38.3
MN	WNC	4,375	477	542	14	2,260	33.1
MS	ESC	2,573	477	519	4	3,322	24.4
MO	WNC	5,117	473	522	12	4,415	28.5
MT	MTN	799	464	523	20	5,184	26.7
NE	WNC	1,578	484	546	10	4,381	26.6
NV	MTN	1,202	434	487	24	4,564	32.2
NH	NE	1,109	442	486	67	5,504	31.3
NJ	MA	7,730	418	473	69	9,159	38.4
NM	MTN	1,515	480	527	12	4,446	26.2
NY	MA	17,990	412	470	70	8,500	42.1
NC	SA	6,629	401	440	55	4,802	29.2
ND	WNC	639	505	564	6	3,685	23.6
OH	ENC	10,847	450	499	22	5,639	32.6
OK	WSC	3,146	478	523	9	3,742	24.3
OR	PAC	2,842	439	484	49	5,291	32.3
PA	MA	11,882	420	463	64	6,534	36.1
RI	NE	1,003	422	461	62	6,989	37.7
SC	SA	3,487	397	437	54	4,327	28.3
SD	WNC	696	506	555	5	3,730	22.4
TN	ESC	4,877	483	525	12	3,707	28.2
TX	WSC	16,987	413	461	42	4,238	28.3
UT	MTN	1,723	492	539	5	2,993	25.0
VT	NE	563	431	466	62	5,740	31.0
VA	SA	6,187	425	470	58	5,360	32.4
WA	PAC	4,867	437	486	44	5,045	33.1
WV	SA	1,793	443	490	15	5,046	26.0
WI	ENC	4,892	476	543	11	5,946	33.1
WY	MTN	454	458	519	13	5,255	29.0

SOURCE: *Statistical Abstract of the United States,* 1992

*The census regions are: East North Central, East South Central, Mountain, New England, Pacific, South Atlantic, West North Central, and West South Central

We have concentrated on quantitative variables until now. When we have data on several variables, however, categorical variables are often present and help organize the data. Categorical variables will play a larger role in this chapter. There is one more question you should ask when you are interested in relations among several variables:

- Do you want simply to explore the nature of the relationship, or do you think that some of the variables explain or even cause changes in others? That is, are some of the variables *response variables* and others *explanatory variables*?

RESPONSE VARIABLE, EXPLANATORY VARIABLE

A **response variable** measures an outcome of a study. An **explanatory variable** attempts to explain the observed outcomes.

It is easiest to identify explanatory and response variables when we actually set values of one variable in order to see how it affects another variable.

EXAMPLE 2.1

Alcohol has many effects on the body. One effect is a drop in body temperature. To study this effect, researchers give several different amounts of alcohol to mice, then measure the change in each mouse's body temperature in the 15 minutes after taking the alcohol. Amount of alcohol is the explanatory variable, and change in body temperature is the response variable.   ◀

When you don't set the values of either variable but just observe both variables, there may or may not be explanatory and response variables. Whether there are depends on how you plan to use the data.

EXAMPLE 2.2

Jim wants to know how the median SAT math and verbal scores in the 51 states (including the District of Columbia) are related to each other. He doesn't think that either score explains or causes the other. Jim has two related variables, and neither is an explanatory variable.

Julie looks at the same data. She asks, "Can I predict a state's SAT math score if I know its SAT verbal score?" Julie is treating the verbal score as the explanatory variable and the math score as the response variable. ◀

In Example 2.1, alcohol actually *causes* a change in body temperature. There is no cause-and-effect relationship between SAT math and verbal scores in Example 2.2. Because the scores are closely related, we can nonetheless use a state's SAT verbal score to predict its math score. We will learn how to do the prediction in Section 2.3. Prediction requires that we identify an explanatory variable and a response variable. Other statistical techniques ignore this distinction. Do remember that calling one variable explanatory and the other response doesn't necessarily mean that changes in one *cause* changes in the other.

You will often find explanatory variables called *independent variables*, and response variables called *dependent variables*. The idea behind this language is that the response variable depends on the explanatory variable. Because the words "independent" and "dependent" have other, unrelated meanings in statistics, we won't use them here.

The statistical techniques used to study relations among variables are more complex than the one-variable methods in Chapter 1. Fortunately, analysis of several-variable data builds on the tools used for examining individual variables. The principles that guide examination of data are also the same:

1. Start with a graph.
2. Look for an overall pattern and deviations from the pattern.
3. Add numerical descriptions of specific aspects of the data.
4. Sometimes there is a way to describe the overall pattern very briefly.

EXERCISES

2.1 How well does a child's height at age 6 predict height at age 16? To find out, measure the heights of a large group of children at age 6, wait until they reach age 16, then measure their heights again. What are the explanatory and response variables here? Are these variables categorical or quantitative?

2.2   There may be a "gender gap" in political party preference in the United States, with women more likely than men to prefer Democratic candidates. A political scientist selects a large sample of registered voters, both men and women. She asks each voter whether they voted for the Democratic or for the Republican candidate in the last congressional election. What are the explanatory and response variables in this study? Are they categorical or quantitative variables?

2.3   The most common treatment for breast cancer was once removal of the breast. It is now usual to remove only the tumor and nearby lymph nodes, followed by radiation. The change in policy was due to a large medical experiment that compared the two treatments. Some breast cancer patients, chosen at random, were given each treatment. The patients were closely followed to see how long they lived following surgery. What are the explanatory and response variables? Are they categorical or quantitative variables?

2.1   SCATTERPLOTS

The most effective way to display the relation between two quantitative variables is a *scatterplot*. Here is an example of a scatterplot.

EXAMPLE 2.3

Some people use median SAT scores to rank state or local school systems. This is not proper, because the percent of high school seniors who take the SAT varies from place to place. Let us examine the relationship between the percent in a state who take the exam and the state median SAT mathematics score, using data from Table 2.1.

We think that "percent taking" will help explain "median score." Therefore, "percent taking" is the explanatory variable and "median score" is the response variable. We want to see how median score changes when percent taking changes, so we put percent taking (the explanatory variable) on the horizontal axis. Figure 2.1 is the scatterplot. Each point represents a single state. In Alabama, for example, 8% take the SAT, and the median SAT math score is 514. Find 8 on the x (horizontal) axis and 514 on the y (vertical) axis. Alabama appears as the point (8, 514) above 8 and to the right of 514. Alaska appears as the point (42, 476), and so on. ◀

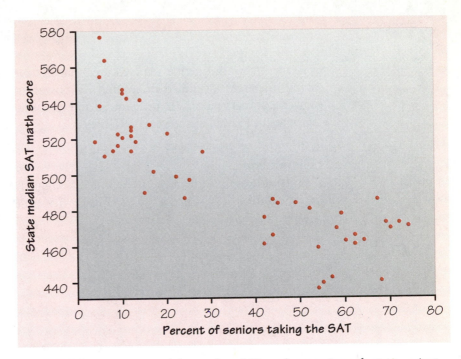

FIGURE 2.1   Scatterplot of the median SAT math score in each state against the percent of that state's high school seniors who take the SAT, from Table 2.1.

SCATTERPLOT

A **scatterplot** shows the relationship between two quantitative variables measured on the same individuals. The values of one variable appear on the horizontal axis, and the values of the other variable appear on the vertical axis. Each individual in the data appears as the point in the plot fixed by the values of both variables for that individual.

Always plot the explanatory variable, if there is one, on the horizontal axis (the x axis) of a scatterplot. As a reminder, we usually call the explanatory variable x and the response variable y. If there is no explanatory-response distinction, either variable can go on the horizontal axis.

2.4   Manatees are large, gentle sea creatures that live along the Florida coast. Many manatees are killed or injured by powerboats. Here are data on powerboat registrations (in thousands) and the number of manatees killed by boats in Florida in the years 1977–1990.

Year	Powerboat registrations (1000)	Manatees killed	Year	Powerboat registrations (1000)	Manatees killed
1977	447	13	1984	559	34
1978	460	21	1985	585	33
1979	481	24	1986	614	33
1980	498	16	1987	645	39
1981	513	24	1988	675	43
1982	512	20	1989	711	50
1983	526	15	1990	719	47

(a) We want to examine the relationship between number of powerboats and number of manatees killed by boats. Which is the explanatory variable?

(b) Make a scatterplot of these data. (Be sure to label the axes with the variable names, not just x and y.) What does the scatterplot show about the relationship between these variables?

Interpreting scatterplots

To interpret a scatterplot, look first for an overall pattern. This pattern should reveal the *direction*, *form*, and *strength* of the relationship between the two variables.

EXAMPLE 2.4

clusters

The *form* of the relationship in Figure 2.1 strikes us at once: there are two distinct **clusters** of states. In one cluster, more than 40% of high school seniors take the SAT, and the state median score is low. Less than 30% of seniors in states in the other cluster take the SAT, and these states have higher median scores. What explains the clusters? There are two widely used college entrance exams: the SAT and the American College Testing (ACT) exam. Each state favors one or the other. The left cluster in Figure 2.1 contains the ACT states, and the

SAT states make up the right cluster. In ACT states, most students who take the SAT are applying to a selective college that requires SAT scores. This selected group of students has a higher median score than the much larger group of students who take the SAT in SAT states.

The *direction* of the overall pattern in Figure 2.1 is also clear. Even within the clusters, states where a higher percent of seniors take the SAT tend to have lower median scores. We say that percent taking and median score are *negatively associated*. ◀

POSITIVE ASSOCIATION, NEGATIVE ASSOCIATION

Two variables are **positively associated** when above-average values of one tend to accompany above-average values of the other and below-average values also tend to occur together. Two variables are **negatively associated** when above-average values of one accompany below-average values of the other, and vice versa.

The relationship in Figure 2.1 has a clear *direction*, the negative association. The *form* is dominated by the clusters and may have a curved shape overall. We say "may have" because, aside from the clusters, the *strength* of the relationship is weak. There is a wide spread in SAT scores among states with about the same percent of students taking the exam. Here is an example of a stronger relationship with a clearer form.

EXAMPLE 2.5

The Sanchez household is about to install solar panels to reduce the cost of heating their house. In order to know how much the solar panels help, they record their consumption of natural gas before the panels are installed. Gas consumption is higher in cold weather, so the relationship between outside temperature and gas consumption is important.

Table 2.2 gives data for 16 months.[1] The response variable y is the average amount of natural gas consumed each day during the month, in hundreds of cubic feet. The explanatory variable x is the average number of heating degree-days each day during the month. (Heating degree-days are the usual measure of demand for heating. One degree-day is accumulated for each degree a day's average temperature falls below 65° F. An average temperature of 20° F, for example, corresponds to 45 degree-days.)

The scatterplot in Figure 2.2 shows a strong positive association. More degree-days means colder weather and so more gas consumed. The form of the

TABLE 2.2	Average degree-days and natural gas consumption				
Month	Degree-days	Gas (100 cu ft)	Month	Degree-days	Gas (100 cu ft)
Nov.	24	6.3	July	0	1.2
Dec.	51	10.9	Aug.	1	1.2
Jan.	43	8.9	Sept.	6	2.1
Feb.	33	7.5	Oct.	12	3.1
Mar.	26	5.3	Nov.	30	6.4
Apr.	13	4.0	Dec.	32	7.2
May	4	1.7	Jan.	52	11.0
June	0	1.2	Feb.	30	6.9

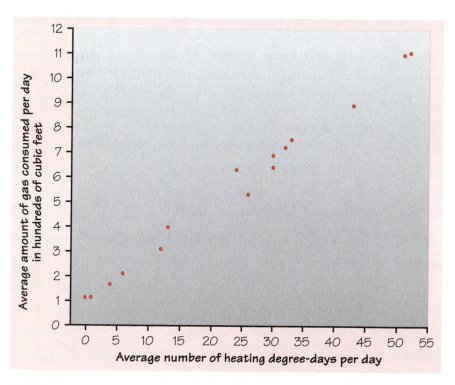

FIGURE 2.2   Scatterplot of the average amount of natural gas used per day by the Sanchez household in 16 months against the average number of heating degree-days per day in those months, from Table 2.2.

linear
relationship

relationship is **linear**. That is, the points lie in a straight-line pattern. It is a strong relationship because the points lie close to a line, with little scatter. If we know how cold a month is, we can predict gas consumption quite accurately from the scatterplot. That strong relationships make accurate predictions possible is an important point that we will soon discuss in more detail. ◀

Of course, not all relationships are linear in form. What is more, not all relationships have a clear direction that we can describe as positive association or negative association. Exercise 2.6 gives an example that is not linear and has no clear direction.

EXERCISES

2.5   In Exercise 2.4 you made a scatterplot of powerboats registered in Florida and manatees killed by boats.

(a) Describe the direction of the relationship. Are the variables positively or negatively associated?

(b) Describe the form of the relationship. Is it linear?

(c) Describe the strength of the relationship. Can the number of manatees killed be predicted accurately from powerboat registrations? If Florida decided to freeze powerboat registrations at 700,000, about how many manatees would be killed by boats each year?

2.6   How does the fuel consumption of a car change as its speed increases? Here are data for a British Ford Escort. Speed is measured in kilometers per hour, and fuel consumption is measured in liters of gasoline used per 100 kilometers traveled.[2]

Speed (km/h)	Mileage (liters/100 km)	Speed (km/h)	Mileage (liters/100 km)
10	21.00	90	7.57
20	13.00	100	8.27
30	10.00	110	9.03
40	8.00	120	9.87
50	7.00	130	10.79
60	5.90	140	11.77
70	6.30	150	12.83
80	6.95		

(a) Make a scatterplot. (Which is the explanatory variable?)

(b) Describe the form of the relationship. Why is it not linear? Explain why the form of the relationship makes sense.

(c) It does not make sense to describe the variables as either positively associated or negatively associated. Why?

(d) Is the relationship reasonably strong or quite weak? Explain your answer.

Adding categorical variables to scatterplots

After examining the overall pattern of a scatterplot, look for deviations from the pattern such as *outliers*.

OUTLIERS

An **outlier** in any graph of data is an individual observation that falls outside the overall pattern of the graph.

Neither of the scatterplots in Figures 2.1 and 2.2 has strong outliers. Figure 2.1 does show a group of four points at the bottom right that have unusually low SAT math scores. Let's investigate.

EXAMPLE 2.6

One of the four low points is the District of Columbia, which is a city rather than a state. The other three are Georgia, North Carolina, and South Carolina. This finding suggests that perhaps the South as a whole has low SAT scores even after we take into account the effect of the percent taking the test.

Figure 2.3 explores that guess by plotting the southern states with a different plot symbol. (We took the South to be the states in the East South Central and South Atlantic regions.) The states we already noticed, and perhaps West Virginia, do have low SAT scores. The other southern states blend in with the rest of the country. The data defeat our guess that the South as a whole has low scores. ◀

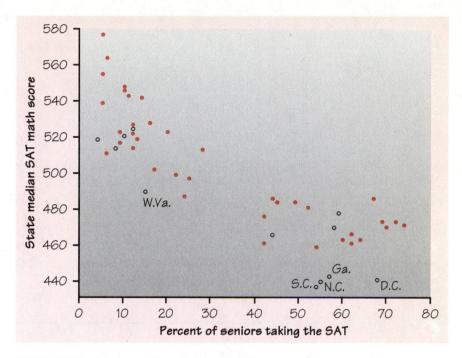

FIGURE 2.3 Median SAT math score and percent of high school seniors who took the test, by state, with the southern states highlighted.

In dividing the states into "southern" and "nonsouthern," we introduced a third variable into the scatterplot. This is a categorical variable that has only two values. The two values are displayed by the two different plotting symbols. *Use different colors or symbols to plot points when you want to add a categorical variable to a scatterplot.*[3]

EXAMPLE 2.7

After the Sanchez household gathered the information recorded in Table 2.2 and Figure 2.2, they added solar panels to their house. They then measured their natural gas consumption for 23 more months. To see how the solar panels affected gas consumption, add the new data (including degree-days for these months) to the scatterplot. Figure 2.4 is the result. We use different symbols to distinguish before from after. The "after" data form a linear pattern that is close to the "before" pattern in warm months (few degree-days). In colder months, with more degree-days, gas consumption after installing the solar panels is less than in similar months before the panels were added. The scatterplot shows the energy savings from the panels.   ◄

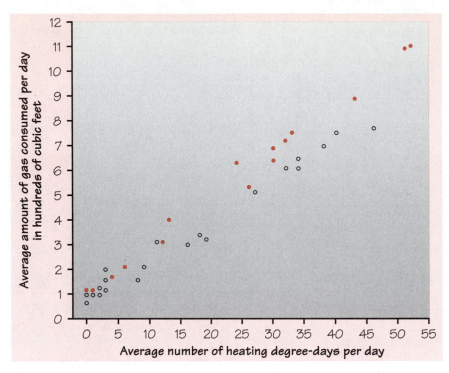

FIGURE 2.4 Natural gas consumption against degree-days for the Sanchez household. The solid red observations are for 16 months before installing solar panels. The open black observations are for 23 months with the panels in use.

Both of our examples suffer from a common problem in drawing scatterplots that you may not notice when a computer does the work. When several individuals have exactly the same data, they occupy the same point on the scatterplot. Look at Delaware and Virginia in Table 2.1, or at June and July in Table 2.2. Table 2.2 contains data for 16 months, but there are only 15 points in Figure 2.2. June and July both occupy the same point. You can use a different plotting symbol to call attention to points that stand for more than one individual. Some computer software does this automatically, but some—including the software used for our scatterplots—does not. We recommend that you do use a different symbol for repeated observations when you plot a small number of observations by hand.

EXERCISE

2.7   Metabolic rate, the rate at which the body consumes energy, is important in studies of weight gain, dieting, and exercise. The table below gives data on the lean body mass and resting metabolic rate for 12 women and 7 men who are subjects in a study of dieting. Lean body mass, given in kilograms, is a person's weight leaving out all fat. Metabolic rate is measured in calories burned per 24 hours, the same calories used to describe the energy content of foods. The researchers believe that lean body mass is an important influence on metabolic rate.

Subject	Sex	Mass	Rate	Subject	Sex	Mass	Rate
1	M	62.0	1792	11	F	40.3	1189
2	M	62.9	1666	12	F	33.1	913
3	F	36.1	995	13	M	51.9	1460
4	F	54.6	1425	14	F	42.4	1124
5	F	48.5	1396	15	F	34.5	1052
6	F	42.0	1418	16	F	51.1	1347
7	M	47.4	1362	17	F	41.2	1204
8	F	50.6	1502	18	M	51.9	1867
9	F	42.0	1256	19	M	46.9	1439
10	M	48.7	1614				

(a) Make a scatterplot of the data for the female subjects. Which is the explanatory variable?

(b) Is the association between these variables positive or negative? What is the form of the relationship? How strong is the relationship?

(c) Now add the data for the male subjects to your graph, using a different color or a different plotting symbol. Does the pattern of relationship that you observed in (b) hold for men also? How do the male subjects as a group differ from the female subjects as a group?

SUMMARY

When we think that changes in a variable x explain or even cause changes in a second variable y, we call x an **explanatory variable** and y a **response variable**.

A **scatterplot** displays the relationship between two quantitative variables measured on the same individuals. Mark values of one variable on the horizontal axis (x axis) and values of the other variable on the vertical axis (y axis). Plot each individual's data as a point on the graph.

Values of the explanatory variable, if there is one, always go on the horizontal axis of a scatterplot.

In examining a scatterplot, look for an overall pattern showing the **direction**, **form**, and **strength** of the relationship, and then for **outliers** or other deviations from this pattern.

If the relationship has a clear direction, we speak of either **positive association** (high values of the two variables tend to occur together) or **negative association** (high values of one variable tend to occur with low values of the other variable).

Linear relationships, where the points show a straight-line pattern, are an important form of relationship between two variables. Curved relationships and clusters are other forms to watch for.

The **strength** of a relationship is determined by how close the points in the scatterplot lie to a simple form such as a line.

You can show the effect of a **categorical variable** by plotting points on a scatterplot with different colors or symbols.

SECTION 2.1 EXERCISES

2.8   Figure 2.5 is a scatterplot that displays the heights of 53 pairs of parents. The mother's height is plotted on the vertical axis and the father's height on the horizontal axis.[4]

   (a) What is the smallest height of any mother in the group? How many mothers have that height? What are the heights of the fathers in these pairs?

   (b) What is the greatest height of any father in the group? How many fathers have that height? How tall are the mothers in these pairs?

   (c) Are there clear explanatory and response variables, or could we freely choose which variable to plot horizontally?

   (d) Say in words what a positive association between these variables means. The scatterplot shows a weak positive association. Why do we say the association is weak?

2.9   Are hot dogs that are high in calories also high in salt? Figure 2.6 is a scatterplot of the calories and salt content (measured as milligrams of sodium) in 17 brands of meat hot dogs.[5]

   (a) Roughly what are the lowest and highest calorie counts among these brands? Roughly what is the sodium level in the brands with the fewest and with the most calories?

   (b) Does the scatterplot show a clear positive or negative association? Say in words what this association means about calories and salt in hot dogs.

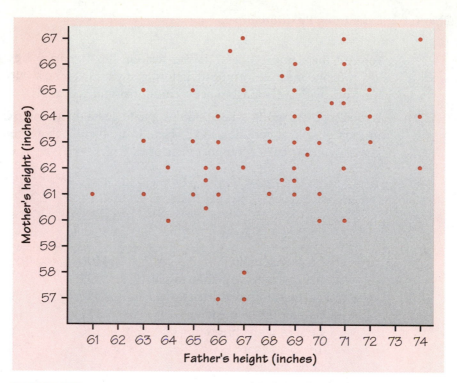

FIGURE 2.5   Scatterplot of the heights of the mother and father in 53 pairs of parents.

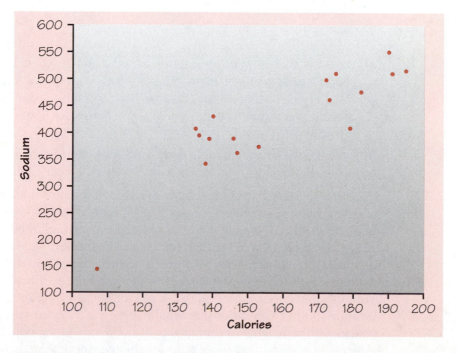

FIGURE 2.6   Scatterplot of milligrams of sodium and calories in each of 17 brands of meat hot dogs.

(c) Are there any outliers? Is the relationship (ignoring any outliers) roughly linear in form? Still ignoring any outliers, how strong would you say the relationship between calories and sodium is?

2.10   A food industry group asked 3368 people to guess the number of calories in each of several common foods. Here is a table of the average of their guesses and the correct number of calories.[6]

Food	Guessed calories	Correct calories
8 oz. whole milk	196	159
5 oz. spaghetti with tomato sauce	394	163
5 oz. macaroni with cheese	350	269
One slice wheat bread	117	61
One slice white bread	136	76
2-oz. candy bar	364	260
Saltine cracker	74	12
Medium-size apple	107	80
Medium-size potato	160	88
Cream-filled snack cake	419	160

(a) We think that how many calories a food actually has helps explain people's guesses of how many calories it has. With this in mind, make a scatterplot of these data. (Because both variables are measured in calories, you should use the same scale on both axes. Your plot will be square.)

(b) Describe the relationship. Is there a positive or negative association? Is the relationship approximately linear? Are there any outliers?

2.11   *Archaeopteryx* is an extinct beast having feathers like a bird but teeth and a long bony tail like a reptile. Only six fossil specimens are known. Because these specimens differ greatly in size, some scientists think they are different species rather than individuals from the same species. We will examine some data. If the specimens belong to the same species and differ in size because some are younger than others, there should be a positive linear relationship between the lengths of a pair of bones from all individuals. An outlier from this relationship would suggest a different species. Here are data on the lengths in centimeters of the femur (a leg bone) and the humerus (a bone in the upper arm) for the five specimens that preserve both bones.[7]

Femur	38	56	59	64	74
Humerus	41	63	70	72	84

Make a scatterplot. Do you think that all five specimens come from the same species?

2.12   How much corn per acre should a farmer plant to obtain the highest yield? Too few plants will give a low yield. On the other hand, if there are too many plants, they will compete with each other for moisture and nutrients, and yields will fall. To find the best planting rate, plant at different rates on several plots of ground and measure the harvest. (Be sure to treat all the plots the same except for the planting rate.) Here are data from such an experiment.[8]

Plants per acre	Yield (bushels per acre)			
12,000	150.1	113.0	118.4	142.6
16,000	166.9	120.7	135.2	149.8
20,000	165.3	130.1	139.6	149.9
24,000	134.7	138.4	156.1	
28,000	119.0	150.5		

(a) Is yield or planting rate the explanatory variable?

(b) Make a scatterplot of yield and planting rate.

(c) Describe the overall pattern of the relationship. Is it linear? Is there a positive or negative association, or neither?

(d) Find the mean yield for each of the five planting rates. Plot each mean yield against its planting rate on your scatterplot and connect these five points with lines. This combination of numerical description and graphing makes the relationship clearer. What planting rate would you recommend to a farmer whose conditions were similar to those in the experiment?

2.13   Table 2.1 gives educational data for the states. We are interested in the relationship between how much states spend on education (dollars per pupil) and how much they pay their teachers (median teacher salaries, in thousands of dollars).

(a) Explain why you expect a positive association between these variables.

(b) Make a scatterplot, with education spending (dollars per pupil) as the explanatory variable.

(c) Describe the relationship. Is there a positive association? Is the relationship approximately linear?

(d) On the plot, identify a state where teacher salaries are unusually high relative to the state's education spending. (This state is an outlier, though not an extreme outlier.) What state is this?

(e) How do the Mountain states compare with the rest of the country in education spending and teacher salaries? Mark the points for states in the MTN region with a different color on your scatterplot. Based on the plot, briefly answer the question.

2.14   (Optional) Data analysts often look for a simple *transformation* of data that simplifies the overall pattern. Here is an example of how transforming the response variable can simplify the pattern of a scatterplot.

The population of Europe grew as follows between 1750 and 1950.

Year	1750	1800	1850	1900	1950
Population (millions)	125	187	274	423	594

(a) Make a scatterplot of population against year. Briefly describe the pattern of Europe's growth.

(b) Now take the logarithm of the population in each year (use the log button on your calculator). Plot the logarithms against year. What is the overall pattern on this plot?

2.15   (Optional) A scatterplot shows the relationship between two quantitative variables. Here is a similar plot to study the relationship between a categorical explanatory variable and a quantitative response variable.

The presence of harmful insects in farm fields is detected by putting up boards covered with a sticky material and then examining the insects trapped on the board. Which colors attract insects best? Experimenters placed six boards of each of four colors in a field of oats and measured the number of cereal leaf beetles trapped.[9]

Board color	Insects trapped					
Lemon yellow	45	59	48	46	38	47
White	21	12	14	17	13	17
Green	37	32	15	25	39	41
Blue	16	11	20	21	14	7

(a) Make a plot of the counts of insects trapped against board color (space the four colors equally on the horizontal axis). Compute the mean count for each color, add the means to your plot, and connect the means with line segments.

(b) Based on the data, what do you conclude about the attractiveness of these colors to the beetles?

(c) Does it make sense to speak of a positive or negative association between board color and insect count?

2.2   CORRELATION

A scatterplot displays the direction, form, and strength of the relationship between two quantitative variables. Linear relations are particularly important because a straight line is a simple pattern that is quite common. We say a linear relation is strong if the points lie close to a straight line, and weak if they are widely scattered about a line. Our eyes are not good judges of how strong a linear relationship is. The two scatterplots in Figure 2.7 depict exactly the same data, but the lower plot is drawn smaller in a large field. The lower plot seems to show a stronger linear relationship. Our eyes can be fooled by changing the plotting scales or the amount of white space around the cloud of points in a scatterplot.[10] We need to follow our strategy for data analysis by using a numerical measure to supplement the graph. *Correlation* is the measure we use.

The correlation *r*

CORRELATION

The **correlation** measures the strength and direction of the linear relationship between two quantitative variables. Correlation is usually written as r.

Suppose that we have data on variables x and y for n individuals. The values for the first individual are x_1 and y_1, the values for the second individual are x_2 and y_2, and so on. The means and standard deviations of the two variables are $\bar{x}$ and s_x for the x-values, and $\bar{y}$ and s_y for the y-values. The correlation r between x and y is

$$r = \frac{1}{n-1}\sum\left(\frac{x_i - \bar{x}}{s_x}\right)\left(\frac{y_i - \bar{y}}{s_y}\right)$$

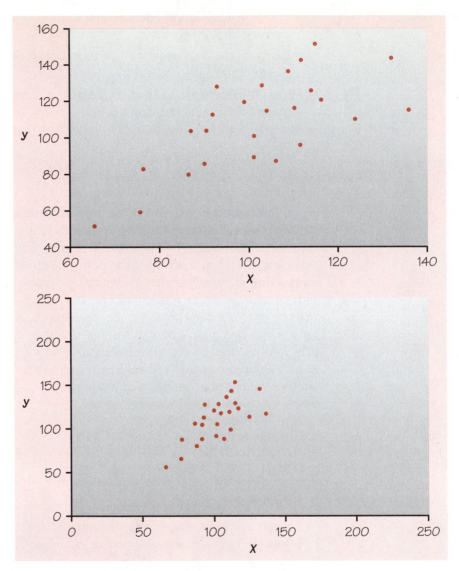

FIGURE 2.7 Two scatterplots of the same data; the straight-line pattern in the lower plot appears stronger because of the surrounding white space.

As always, the summation sign $\sum$ means "add these terms for all the individuals." The formula for the correlation r is a bit complex. It helps us see what correlation is, but in practice you should use software or a calculator that finds r from keyed-in values of two variables x and y.

The formula for r begins by standardizing the observations. Suppose, for example, that x is height in centimeters and y is weight in kilograms and that we have height and weight measurements for n people. Then $\bar{x}$ and s_x

are the mean and standard deviation of the n heights, both in centimeters. The value

$$\frac{x_i - \overline{x}}{s_x}$$

is the standardized height of the ith person, familiar from Chapter 1. The standardized height says how many standard deviations above or below the mean a person's height lies. Standardized values have no units—in this example, they are no longer measured in centimeters. Standardize the weights also. The correlation r is an average of the products of the standardized height and the standardized weight for the n people.

Correlation makes no use of the distinction between explanatory and response variables. It makes no difference which variable you call x and which you call y in calculating the correlation. Correlation does require that both variables be quantitative, so that it makes sense to do the arithmetic required by the formula for r. You can't calculate a correlation between the incomes of a group of people and what city they live in, because city is a categorical variable.

EXERCISES

2.16   We will calculate the correlation r step-by-step in some very simple examples to see how the formula works. For each set of data:

- Draw a scatterplot. Briefly describe in words the association between x and y.
- Find the mean and standard deviation for both x and y. (Use your calculator.)
- Find the standardized values of x and y, then use the formula to find the correlation r.

(a)	x	4	4	-4	-4	
	y	-4	4	4	-4	
(b)	x	4	3	0	-3	-4
	y	-4	-2	0	2	4
(c)	x	4	2	-2	-4	
	y	4	-2	2	-4	

2.17   Exercise 2.11 (page 108) gives the lengths of two bones in five fossil specimens of the extinct beast *Archaeopteryx*.

(a) Find the correlation r step-by-step. That is, find the mean and standard deviation of the femur lengths and of the humerus lengths. Then find the five standardized values for each variable and use the formula for r.

(b) Now enter these data into your calculator and use the calculator's correlation function to find r. Check that you get the same result as in (a).

Facts about correlation

The formula for correlation helps us see that r is positive when there is a positive association between the variables. Height and weight, for example, have a positive association. People who are above average in height tend to also be above average in weight. Both the standardized height and the standardized weight for such a person are positive. People who are below average in height tend to also have below-average weight. Then both standardized height and standardized weight are negative. In both cases, the products in the formula for r are mostly positive and so r is positive. In the same way, we can see that r is negative when the association between x and y is negative. More detailed study of the formula gives more detailed properties of r. Here is what you need to know in order to interpret correlation.

1. Positive r indicates positive association between the variables, and negative r indicates negative association.

2. The correlation r always falls between -1 and 1. Values of r near 0 indicate a very weak linear relationship. The strength of the linear relationship increases as r moves away from 0 toward either -1 or 1. Values of r close to -1 or 1 indicate that the points lie close to a straight line. The extreme values $r = -1$ and $r = 1$ occur only in the case of a perfect linear relationship, when the points in a scatterplot lie exactly along a straight line.

3. Because r uses the standardized values of the observations, r does not change when we change the units of measurement of x, y, or both. Changing from centimeters to inches and from kilograms to pounds does not change the correlation between height and weight. The correlation r itself has no unit of measurement; it is just a number between -1 and 1.

4. Correlation measures the strength of only a linear relationship between two variables. Correlation does not describe curved relationships between variables, no matter how strong they are.

5. Like the mean and standard deviation, the correlation is strongly affected by a few outlying observations. Use r with caution when outliers appear in the scatterplot.

The scatterplots in Figure 2.8 illustrate how values of r closer to 1 or -1 correspond to stronger linear relationships. To make the essential meaning of r clear, the standard deviations of both variables in these plots are equal and the horizontal and vertical scales are the same. In general, it is not so easy to guess the value of r from the appearance of a scatterplot. Remember that changing the plotting scales in a scatterplot may mislead our eyes, but it does not change the correlation.

The real data we have examined also illustrate how correlation measures the strength and direction of linear relationships. Figure 2.2 shows a very strong positive linear relationship between degree-days and natural gas consumption. The correlation is $r = .9953$. Check this on your calculator using the data in Table 2.2. Figure 2.1 shows a clear but weaker negative association between percent of students taking the SAT and the median SAT math score in a state. The correlation is $r = -.8581$.

Do remember that *correlation is not a complete description of two-variable data*, even when the relationship between the variables is linear. You should give the means and standard deviations of both x and y along with the correlation. (Because the formula for correlation uses the means and standard deviations, these measures are the proper choice to accompany a correlation.) Conclusions based on correlations alone may require rethinking in the light of a more complete description of the data.

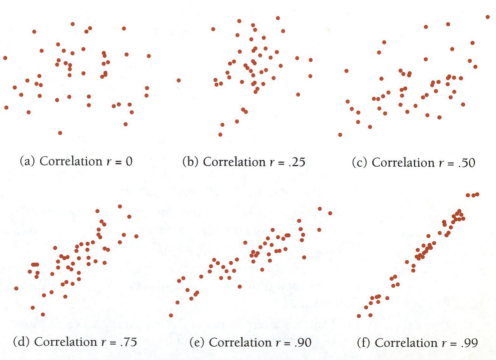

(a) Correlation $r = 0$ (b) Correlation $r = .25$ (c) Correlation $r = .50$

(d) Correlation $r = .75$ (e) Correlation $r = .90$ (f) Correlation $r = .99$

FIGURE 2.8   How the correlation coefficient measures the strength of linear association. Patterns closer to a straight line have correlations closer to 1 or −1.

EXAMPLE 2.8

Competitive divers are scored on their form by a panel of judges who use a scale from 1 to 10. The subjective nature of the scoring often results in controversy. We have the scores awarded by two judges, Ivan and George, on a large number of dives. How well do they agree? We do some calculation and find that the correlation between their scores is $r = .9$. But the mean of Ivan's scores is 3 points lower than George's mean.

These facts do not contradict each other. They are simply different kinds of information. The mean scores show that Ivan awards much lower scores than George. But because Ivan gives *every* dive a score about 3 points lower than George, the correlation remains high. Adding or subtracting the same number to all values of either x or y does not change the correlation. If Ivan and George both rate several divers, the contest is fairly scored because Ivan and George agree on which dives are better than others. The high r shows their agreement. But if Ivan scores one diver and George another, we must add 3 points to Ivan's scores to arrive at a fair comparison. ◀

EXERCISES

2.18   Figure 2.5 (page 107) is a scatterplot that displays the heights of 53 pairs of parents. Do you think the correlation r for these data is near -1, clearly negative but not near -1, near 0, clearly positive but not near 1, or near 1? Explain your answer.

2.19   Figure 2.6 (page 107) is a scatterplot of the calories and sodium content in 17 brands of meat hot dogs. Do you think the correlation r for these data is near -1, clearly negative but not near -1, near 0, clearly positive but not near 1, or near 1? Explain your answer.

2.20   If women always married men who were 2 years older than themselves, what would be the correlation between the ages of husband and wife? (Hint: Draw a scatterplot for several ages.)

2.21   Exercise 2.11 (page 108) gives the lengths of two bones in five fossil specimens of the extinct beast *Archaeopteryx*. You found the correlation r in Exercise 2.17.

(a)   Make a scatterplot if you did not do so earlier. Explain why the value of r matches the scatterplot.

(b)   The lengths were measured in centimeters. If we changed to inches, how would r change? (There are 2.54 centimeters in an inch.)

2.22   The gas mileage of an automobile first increases and then decreases as the speed increases. Suppose that this relationship is very regular, as shown by the following data on speed (miles per hour) and mileage (miles per gallon):

Speed	20	30	40	50	60
MPG	24	28	30	28	24

Make a scatterplot of mileage versus speed. Show that the correlation between speed and mileage is $r = 0$. Explain why the correlation is 0 even though there is a strong relationship between speed and mileage.

SUMMARY

The **correlation r** measures the strength and direction of the linear association between two quantitative variables x and y. Although you can calculate a correlation for any scatterplot, r measures only straight-line relationships.

Correlation indicates the direction of a linear relationship by its sign: $r > 0$ for a positive association and $r < 0$ for a negative association.

Correlation always satisfies $-1 \le r \le 1$ and indicates the strength of a relationship by how close it is to -1 or 1. Perfect correlation, $r = \pm 1$, occurs only when the points lie exactly on a straight line.

Correlation ignores the distinction between explanatory and response variables. The value of r is not affected by changes in the unit of measurement of either variable. But r can be strongly affected by outlying observations.

SECTION 2.2 EXERCISES

2.23   Exercise 2.7 (page 104) gives data on the lean body mass and metabolic rate for 12 women and 7 men.

   (a) Make a scatterplot if you did not do so in Exercise 2.7. Use different symbols or colors for women and men. Do you think the correlation will be about the same for men and women or quite different for the two groups? Why?

   (b) Calculate r for women alone and also for men alone. (Use your calculator.)

(c) Calculate the mean body mass for the women and for the men. Does the fact that the men are heavier than the women on the average influence the correlations? If so, in what way?

(d) Lean body mass was measured in kilograms. How would the correlations change if we measured body mass in pounds? (There are about 2.2 pounds in a kilogram.)

2.24   Exercise 2.10 (page 108) gives data on the true calorie counts in ten foods and the average guesses made by a large group of people.

(a) Make a scatterplot if you did not do so in Exercise 2.10. Then calculate the correlation r (use your calculator). Explain why your r is reasonable based on the scatterplot.

(b) The guesses are all higher than the true calorie counts. Does this fact influence the correlation in any way? How would r change if every guess were 100 calories higher?

(c) The guesses are much too high for spaghetti and snack cake. Circle these points on your scatterplot. Calculate r for the other eight foods, leaving out these two points. Explain why r changed in the direction that it did.

2.25   Changing the units of measurement can dramatically alter the appearance of a scatterplot. Consider the following data:

x	-4	-4	-3	3	4	4
y	.5	$-.6$	$-.5$	.5	.5	$-.6$

(a) Draw x and y axes each extending from -6 to 6. Plot the data on these axes.

(b) Calculate the values of new variables $x^* = x/10$ and $y^* = 10y$, starting from the values of x and y. Plot y^* against x^* on the same axes using a different plotting symbol. The two plots are very different in appearance.

(c) Use your calculator to find the correlation between x and y. Then find the correlation between x^* and y^*. How are the two correlations related? Explain why this isn't surprising.

2.26   A college newspaper interviews a psychologist about student ratings of the teaching of faculty members. The psychologist says, "The evidence indicates that the correlation between the research productivity and teaching rating of faculty members is close to zero." The paper reports this as "Professor McDaniel said that good researchers tend to be poor teachers, and vice versa." Explain why the paper's report is wrong. Write a statement in plain language (don't use the word "correlation") to explain the psychologist's meaning.

2.27   The data in Exercise 2.22 were made up to create an example of a strong curved relationship for which, nonetheless, $r = 0$. Exercise 2.6 (page 101) gives actual data on gas used versus speed for a British Ford Escort. Make a scatterplot if you did not do so in Exercise 2.6. Calculate the correlation, and explain why r is close to 0 despite a strong relationship between speed and gas used.

2.28   Each of the following statements contains a blunder. Explain in each case what is wrong.

(a) "There is a high correlation between the sex of American workers and their income."

(b) "We found a high correlation ($r = 1.09$) between students' ratings of faculty teaching and ratings made by other faculty members."

(c) "The correlation between planting rate and yield of corn was found to be $r = .23$ bushel."

2.3   LEAST-SQUARES REGRESSION

Correlation measures the strength and direction of the linear relationship between any two quantitative variables. If a scatterplot shows a linear relationship, we would like to summarize this overall pattern by drawing a line through the scatterplot. *Least-squares regression* is a method for finding a line that summarizes the relationship between two variables, but only in a specific setting.

REGRESSION LINE

A **regression line** is a straight line that describes how a response variable y changes as an explanatory variable x changes. We often use a regression line to predict the value of y for a given value of x. Regression, unlike correlation, requires that we have an explanatory variable and a response variable.

EXAMPLE 2.9

A scatterplot shows that there is a strong linear relationship between the average outside temperature (measured by heating degree-days) in a month and the average amount of natural gas that the Sanchez household uses per day during the month. The Sanchez household wants to use this relationship to predict their natural gas consumption. "If a month averages 20 degree-days per day (that's 45° F), how much gas will we use?"

prediction

In Figure 2.9 we have drawn a regression line on the scatterplot. To use this line to **predict** gas consumption at 20 degree-days, first locate 20 on the x axis. Then go "up and over" as in the figure to find the gas consumption y that corresponds to $x = 20$. We predict that the Sanchez household will use about 4.9 hundreds of cubic feet of gas each day in such a month.   ◀

The least-squares regression line

Different people might draw different lines by eye on a scatterplot. This is especially true when the points are more widely scattered than those in Figure 2.9. We need a way to draw a regression line that doesn't depend on our guess as to where the line should go. No line will pass exactly through all the points, so we want one that is as close as possible. We will use the

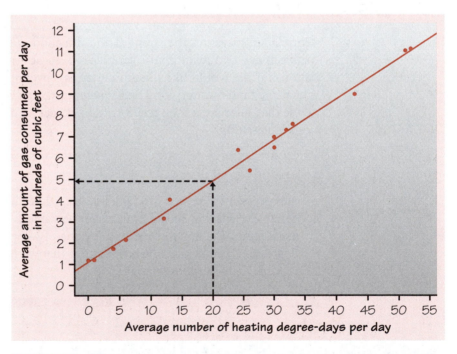

FIGURE 2.9   The Sanchez household gas consumption data, with a regression line for predicting gas consumption from degree-days. The dashed lines illustrate how to use the regression line to predict gas consumption for a month averaging 20 degree-days per day.

line to predict y from x, so we want a line that is as close as possible to the points in the *vertical* direction. That's because the prediction errors we make are errors in y, which is the vertical direction in the scatterplot. If we predict 4.9 hundreds of cubic feet for a month with 20 degree-days and the actual usage turns out to be 5.1 hundreds of cubic feet, our error is

$$\text{error} = \text{observed} - \text{predicted}$$
$$= 5.1 - 4.9 = .2$$

We want a regression line that makes the vertical distances of the points in a scatterplot from the line as small as possible. Figure 2.10 illustrates the idea. For clarity, the plot shows only three of the points from Figure 2.9, along with the line, on an expanded scale. The line passes above two of the points and below one of them. The vertical distances of the data points from the line appear as vertical line segments. A "good" regression line makes these distances as small as possible. There are many ways to make "as small as possible" precise. The most common is the *least-squares* idea.

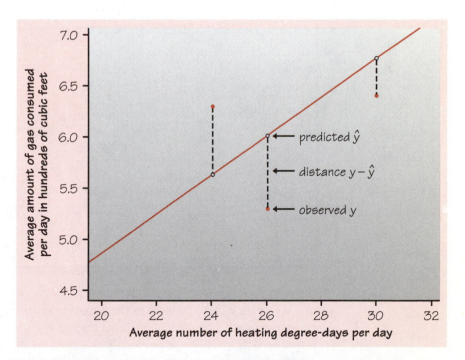

FIGURE 2.10   The least-squares idea. For each observation, find the vertical distance of each point on the scatterplot from a regression line. The least-squares regression line makes the sum of the squares of these distances as small as possible.

LEAST-SQUARES REGRESSION LINE

The **least-squares regression line** of y on x is the line that makes the sum of the squares of the vertical distances of the data points from the line as small as possible.

One reason for the popularity of the least-squares regression line is that the problem of finding the line has a simple answer. We can give the recipe for the least-squares line in terms of the means and standard deviations of the two variables and their correlation.

EQUATION OF THE LEAST-SQUARES REGRESSION LINE

We have data on an explanatory variable x and a response variable y for n individuals. From the data, calculate the means $\overline{x}$ and $\overline{y}$ and the standard deviations s_x and s_y of the two variables, and their correlation r. The least-squares regression line is the line

$$\hat{y} = a + bx$$

with **slope**

$$b = r\frac{s_y}{s_x}$$

and **intercept**

$$a = \overline{y} - b\overline{x}$$

We write $\hat{y}$ (read "y hat") in the equation of the regression line to emphasize that the line gives a *predicted* response $\hat{y}$ for any x. The predicted response will usually not be exactly the same as the actually *observed* response y. We will see that the equation gives insight into the behavior of least-squares regression. But in practice, you don't need to calculate the means, standard deviations, and correlation first. Statistical software or your calculator will give the slope b and intercept a of the least-squares line from keyed-in values of the variables x and y. You can then concentrate on understanding and using the regression line.

EXAMPLE 2.10

The line in Figure 2.9 is in fact the least-squares regression line of gas consumption on degree-days. Enter the data from Table 2.2 into your calculator and check that the equation of this line is

$$\hat{y} = 1.0892 + .1890x$$

slope

The **slope** of a regression line is usually important for the interpretation of the data. The slope is the rate of change, the amount of change in $\hat{y}$ when x increases by 1. The slope $b = .1890$ in this example says that, on the average, each additional degree-day predicts consumption of 0.1890 more hundreds of cubic feet of natural gas per day.

intercept

The **intercept** of the regression line is the value of $\hat{y}$ when $x = 0$. Although we need the value of the intercept to draw the line, it is statistically meaningful only when x can actually take values close to zero. In our example, $x = 0$ occurs when the average outdoor temperature is at least 65° F. We predict that the Sanchez household will use an average of $a = 1.0892$ hundreds of cubic feet of gas per day when there are no degree-days. They use this gas for cooking and heating water, which continue in warm weather.

The equation of the regression line makes prediction easy. Just substitute an x-value into the equation. To predict gas consumption at 20 degree-days, substitute $x = 20$.

$$\hat{y} = 1.0892 + (.1890)(20)$$
$$= 1.0892 + 3.78 = 4.869$$

plotting a line

To **plot the line** on the scatterplot, use the equation to find $\hat{y}$ for two values of x, one near each end of the range of x in the data. Plot each $\hat{y}$ above its x and draw the line through the two points. ◀

EXERCISES

2.29   Example 2.10 gives the equation of the regression line of gas consumption y on degree-days x for the data in Table 2.2 as

$$\hat{y} = 1.0892 + .1890x$$

Enter the data from Table 2.2 into your calculator.

(a) Use your calculator's regression function to find the equation of the least-squares regression line.

(b) Use your calculator to find the mean and standard deviation of both x and y and their correlation r. Find the slope b and intercept a of the regression line from these, using the facts in the box *Equation of the least-squares regression line*. Verify that in both part (a) and part (b) you get the equation in Example 2.10. (Results may differ slightly because of rounding off.)

2.30   Researchers studying acid rain measured the acidity of precipitation in a Colorado wilderness area for 150 consecutive weeks. Acidity is measured by pH. Lower pH values show higher acidity. The acid rain researchers observed a linear pattern over time. They reported that the least-squares regression line

$$\text{pH} = 5.43 - (.0053 \times \text{weeks})$$

fit the data well.[11]

(a) Draw a graph of this line. Is the association positive or negative? Explain in plain language what this association means.

(b) According to the regression line, what was the pH at the beginning of the study (weeks $= 1$)? At the end (weeks $= 150$)?

(c) What is the slope of the regression line? Explain clearly what this slope says about the change in the pH of the precipitation in this wilderness area.

2.31   Concrete road pavement gains strength over time as it cures. Highway builders use regression lines to predict the strength after 28 days (when curing is complete) from measurements made after 7 days. Let x be strength after 7 days (in pounds per square inch) and y the strength after 28 days. One set of data gives this least-squares regression line:

$$\hat{y} = 1389 + .96x$$

(a) Draw a graph of this line, with x running from 3000 to 4000 pounds per square inch.

(b) Explain what the slope $b = .96$ in this equation says about how concrete gains strength as it cures.

(c) A test of some new pavement after 7 days shows that its strength is 3300 pounds per square inch. Use the equation of the regression line to predict the strength of this pavement after 28 days. Also draw the "up and over" lines from $x = 3300$ on your graph (as in Figure 2.9).

Facts about least-squares regression

Least-squares regression looks at the distances of the data points from the line only in the y direction. So the two variables x and y play different roles in regression.

EXAMPLE 2.11

Figure 2.11 is a scatterplot of data that played a central role in the discovery that the universe is expanding. They are the distances from earth of 24 spiral

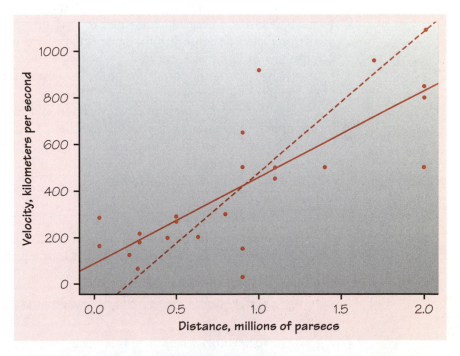

FIGURE 2.11   Scatterplot of Hubble's data on the distance from earth of 24 galaxies and the velocity at which they are moving away from us. The two lines are the two least-squares regression lines: of velocity on distance (solid) and of distance on velocity (dashed).

galaxies and the speed at which these galaxies are moving away from us, re-ported by the astronomer Edwin Hubble in 1929.[12] There is a positive linear relationship, $r = .7842$, so that more distant galaxies are moving away more rapidly. Astronomers believe that there is in fact a perfect linear relationship, and that the scatter is caused by imperfect measurements.

The two lines on the plot are the two least-squares regression lines. The re-gression line of velocity on distance is solid. The regression line of distance on velocity is dashed. *Regression of velocity on distance and regression of distance on velocity give different lines*. In the regression setting you must know clearly which variable is explanatory. ◀

Although the correlation r ignores the distinction between explanatory and response variables, there is a close connection between correlation and regression. The slope of the least-squares regression line is

$$b = r\frac{s_y}{s_x}$$

This equation says that along the regression line, *a change of one standard deviation in x corresponds to a change of r standard deviations in y*. When the variables are perfectly correlated ($r = 1$ or $r = -1$), the change in the predicted response $\hat{y}$ is the same (in standard deviation units) as the change in x. Otherwise, because $-1 \le r \le 1$, the change in $\hat{y}$ is less than the change in x. As the correlation grows less strong, the prediction $\hat{y}$ moves less in response to changes in x.

Another connection between correlation and regression is even more important. In fact, the numerical value of r as a measure of the strength of a linear relationship is best interpreted by thinking about regression. Here is the fact we need.

---

r^2 IN REGRESSION

The **square of the correlation**, r^2, is the fraction of the variation in the values of y that is explained by the least-squares regression of y on x.

The idea is that when there is a linear relationship, some of the vari-ation in y is accounted for by the fact that as x changes it pulls y along with it. Look again at Figure 2.9. There is a lot of variation in the observed y's, the gas consumption data. They range from a low of about 1 to a high of 11. The scatterplot shows that most of this variation in y is accounted for by the fact that outdoor temperature (measured by degree-days x) was changing and pulled gas consumption along with it. There is only a little remaining variation in y, which appears in the scatter of points about the line. The points in Figure 2.11, on the other hand, are more scattered. Linear dependence on distance does explain some of the observed variation in velocity. You would guess a higher value for the velocity y knowing that $x = 2$ than you would if you were told that $x = 0$. But there is still considerable variation in y even when x is held fixed—look at the four points in Figure 2.11 with $x = 2$.

This idea can be expressed in algebra, though we won't do it. It is possible to break the total variation in the observed values of y, measured by their variance, into two parts. One part is the variation we expect as x moves and $\hat{y}$ moves with it along the regression line. The other measures the variation of the data points about the line. The squared correlation r^2 is the first of these as a fraction of the whole:

$$r^2 = \frac{\text{variation in } \hat{y} \text{ as } x \text{ pulls it along the line}}{\text{total variation in observed values of } y}$$

EXAMPLE 2.12

In Figure 2.9, $r = .9953$ and $r^2 = .9906$. Over 99% of the variation in gas consumption is accounted for by the linear relationship with degree-days. In Figure 2.11, $r = .7842$ and $r^2 = .6150$. The linear relationship between distance and velocity explains 61.5% of the variation *in either variable*. There are two regression lines, but just one correlation, and r^2 helps interpret both regressions. ◀

When you report a regression, give r^2 as a measure of how successful the regression was in explaining the response. When you see a correlation, square it to get a better feel for the strength of the association. Perfect correlation ($r = -1$ or $r = 1$) means the points lie exactly on a line. Then $r^2 = 1$ and all of the variation in one variable is accounted for by the linear relationship with the other variable. If $r = -.7$ or $r = .7$, $r^2 = .49$ and about half the variation is accounted for by the linear relationship. In the r^2 scale, correlation $\pm.7$ is about halfway between 0 and ± 1.

These connections with correlation are special properties of least-squares regression. They are not true for other methods of fitting a line to data. Another reason that least-squares is the most common method for fitting a regression line to data is that it has many convenient special properties.

EXERCISES

2.32   In Professor Smith's economics course the correlation between the students' total scores prior to the final examination and their final examination scores is $r = .6$. The pre-exam totals for all students in the course have mean 280 and standard deviation 30. The final exam scores have mean 75 and standard deviation 8. Professor Smith has lost Julie's final exam but knows that her total before the exam was 300. He decides to predict her final exam score from her pre-exam total.

(a) What is the slope of the least-squares regression line of final exam scores on pre-exam total scores in this course? What is the intercept?

(b) Use the regression line to predict Julie's final exam score.

(c) Julie doesn't think this method accurately predicts how well she did on the final exam. Calculate r^2 and use the value you get to argue that her actual score could have been much higher (or much lower) than the predicted value.

2.33   Good runners take more steps per second as they speed up. Here are the average numbers of steps per second for a group of top female runners at different speeds. The speeds are in feet per second.[13]

Speed (ft/s)	15.86	16.88	17.50	18.62	19.97	21.06	22.11
Steps per second	3.05	3.12	3.17	3.25	3.36	3.46	3.55

(a) You want to predict steps per second from running speed. Make a scatterplot of the data with this goal in mind.

(b) Describe the pattern of the data and find the correlation.

(c) Find the least-squares regression line of steps per second on running speed. Draw this line on your scatterplot.

(d) Does running speed explain most of the variation in the number of steps that a runner takes per second? Calculate r^2 and use it to answer this question.

(e) If you wanted to predict running speed from a runner's steps per second, would you use the same line? Explain your answer. Would r^2 stay the same?

2.34   A study of class attendance and grades among first-year students at a state university showed that in general students who attended a higher percent of their classes earned higher grades. Class attendance explained 16% of the variation in grade index among the students. What is the numerical value of the correlation between percent of classes attended and grade index?

Residuals

A regression line is a simple way to describe the overall pattern of a linear relationship between an explanatory variable and a response variable. Deviations from the overall pattern are also important. In the regression setting, we see deviations by looking at the scatter of the data points about the regression line. The vertical distances from the points to the least-squares regression line are as small as possible, in the sense that they have the smallest possible sum of squares. So we give these distances a special name, *residuals*.

RESIDUALS

A **residual** is the difference between an observed value of the response variable and the value predicted by the regression line. That is,

$$\text{residual} = \text{observed } y - \text{predicted } y$$
$$= y - \hat{y}$$

EXAMPLE 2.13

Does the age at which a child begins to talk predict later score on a test of mental ability? A study of the development of young children recorded the age in months at which each of 21 children spoke their first word and their Gesell Adaptive Score, the result of an aptitude test taken much later. The data appear in Table 2.3.[14]

Figure 2.12 is a scatterplot, with age at first word as the explanatory variable x and Gesell score as the response variable y. Children 3 and 13, and also Children 16 and 21, have identical values of both variables. We use a different plotting symbol to show that one point stands for two individuals. The plot shows a negative association. That is, children who begin to speak later tend to have lower test scores than early talkers. The overall pattern is moderately

Child	Age	Score	Child	Age	Score
1	15	95	12	9	96
2	26	71	13	10	83
3	10	83	14	11	84
4	9	91	15	11	102
5	15	102	16	10	100
6	20	87	17	12	105
7	18	93	18	42	57
8	11	100	19	17	121
9	8	104	20	11	86
10	20	94	21	10	100
11	7	113			

TABLE 2.3   Age at first word and Gesell score

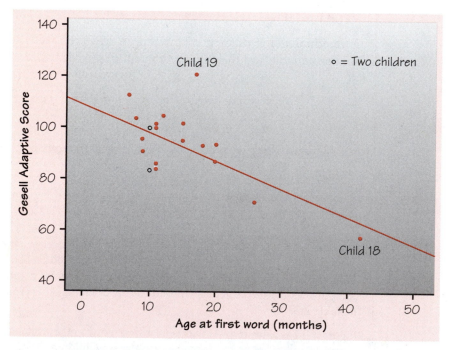

FIGURE 2.12 Scatterplot of Gesell Adaptive Scores versus the age at first word for 21 children, from Table 2.3. The line is the least-squares regression line for predicting Gesell score from age at first word.

linear. The correlation describes both the direction and strength of the linear relationship. It is $r = -.640$.

The line on the plot is the least-squares regression line of Gesell score on age at first word. Its equation is

$$\hat{y} = 109.8738 - 1.1270x$$

For Child 1, who first spoke at 15 months, we predict the score

$$\hat{y} = 109.8738 - (1.1270)(15) = 92.97$$

This child's actual score was 95. The residual is

$$\text{residual} = \text{observed } y - \text{predicted } y$$
$$= 95 - 92.97 = 2.03$$

The residual is positive because the data point lies above the line. ◄

There is a residual for each data point. Finding the residuals with a calculator is a bit unpleasant, because you must first find the predicted response for every x. Statistical software gives you the residuals all at once. Here are the 21 residuals for the Gesell data, as output by a statistical software package:

```
residuals:
   2.0310   -9.5721  -15.6040   -8.7309    9.0310   -0.3341    3.4120
   2.5230    3.1421    6.6659   11.0151   -3.7309  -15.6040  -13.4770
   4.5230    1.3960    8.6500   -5.5403   30.2850  -11.4770    1.3960
```

residual plot

Examine the residuals carefully to check how well the regression line fits the data. You can do this by looking at the vertical deviations of the points from the line in a scatterplot of the original data like Figure 2.12. A **residual plot** plots the residuals on the vertical axis against the explanatory variable on the horizontal axis. Such a plot magnifies the residuals and makes patterns easier to see. You should make a residual plot if computer software makes it easy to do so. Figure 2.13 is a residual plot for the data in Figure 2.12. Residuals from least-squares regression have a special property: *the mean of the residuals is always zero.* You can check that the sum of the residuals above is -0.0002. The sum is not exactly 0 because the

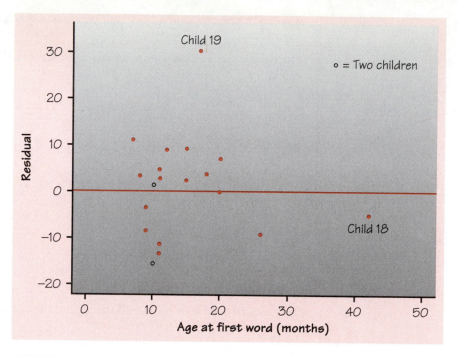

FIGURE 2.13 Residual plot for the regression of Gesell score on age at first word. Child 19 is an outlier. Child 18 is an influential observation that does not have a large residual.

*roundoff
error* software rounded the residuals to four decimal places. This is **roundoff error**. The horizontal line at zero in Figure 2.13 helps orient us. It corresponds to the regression line in Figure 2.12.

We would like a residual plot to look something like the simplified pattern in Figure 2.14(a). That plot shows a uniform scatter of the points about the fitted line, with no unusual individual observations. Here are some things to look for when you examine the residuals, using either a scatterplot of the data or a residual plot.

• A curved pattern, which shows that the overall pattern is not linear. Figure 2.14(b) is a simplified example. A straight line is not a good summary for such data.

• Increasing or decreasing spread about the line as *x* increases. Figure 2.14(c) is a simplified example. Prediction of *y* will be less accurate for larger *x* in that example.

• Individual points with large residuals, like Child 19 in Figures 2.12 and 2.13. These points are *outliers* because they lie outside the straight-line pattern.

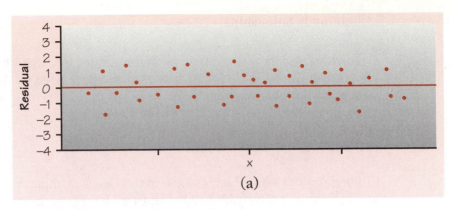

(a)

FIGURE 2.14(a) Idealized patterns in plots of least-squares residuals. Plot (a) indicates that the regression line fits the data well.

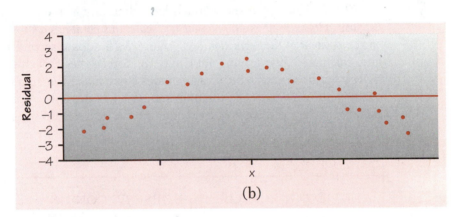

(b)

FIGURE 2.14(b) The data in plot (b) have a curved pattern, so a straight line fits poorly.

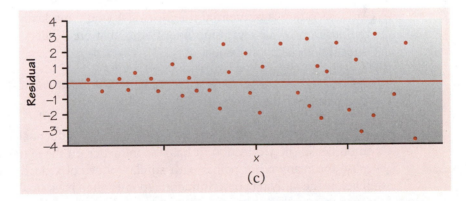

(c)

FIGURE 2.14(c) The response variable y in plot (c) has more spread for larger values of the explanatory variable x, so prediction will be less accurate when x is large.

- Individual points that are extreme in the x direction, like Child 18 in Figures 2.12 and 2.13. Such points may not have large residuals, but they can be very important.

Influential observations

Children 18 and 19 are both unusual in the Gesell example. They are unusual in different ways. Child 19 lies far from the regression line. Child 18 is close to the line but far out in the x direction. Child 19 is an *outlier*, with a Gesell score so high that we should check for a mistake in recording it. In fact, the score is correct.

Child 18 began to speak much later than any of the other children. *Because of its extreme position on the age scale, this point has a strong influence on the position of the regression line.* Figure 2.15 adds a second regression line, calculated after leaving out Child 18. You can see that this one point moves the line quite a bit. Least-squares lines make the sum of squares of the vertical distances to the points as small as possible. A point that is extreme in the x direction with no other points near it pulls the line toward itself. We call such points *influential*.

> **OUTLIERS AND INFLUENTIAL OBSERVATIONS IN REGRESSION**
>
> An **outlier** is an observation that lies outside the overall pattern of the other observations in a scatterplot. An observation can be an outlier in the x direction, in the y direction, or in both directions.
>
> An observation is **influential** if removing it would markedly change the position of the regression line. Points that are outliers in the x direction are often influential.

Child 18 is an outlier in the x direction and is influential. Child 19 is an outlier in the y direction. It has less influence on the regression line because the many other points with similar values of x anchor the line well below the outlying point. Influential points often have small residuals, because they pull the regression line toward themselves. If you just look at residuals, you will miss influential points. Influential observations can greatly change the interpretation of data.

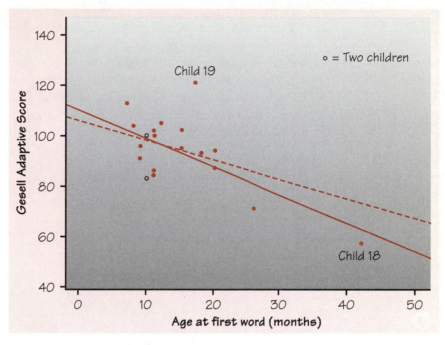

FIGURE 2.15 Two least-squares regression lines of Gesell score on age at first word. The solid line is calculated from all the data. The dashed line was calculated leaving out Child 18. Child 18 is an influential observation because leaving out this point moves the regression line quite a bit.

EXAMPLE 2.14

Correlation and least-squares regression are strongly influenced by extreme observations. In the Gesell example, the original data have $r^2 = 0.41$. That is, the age at which a child begins to talk explains 41% of the variation on a later test of mental ability. This relationship is strong enough to be interesting to parents. But if we leave out Child 18, r^2 drops to only 11%.

What should the child development researcher do? She must decide whether Child 18 is so slow to speak that this individual should not be allowed to influence the analysis. If she excludes Child 18, much of the evidence for a connection between the age at which a child begins to talk and later ability score vanishes. If she keeps Child 18, she needs data on other children who were also slow to begin talking, so that the analysis no longer depends so heavily on just one child.   ◀

2.35   Exercise 2.6 (page 101) gives data on the fuel consumption y of a car at various speeds x. Fuel consumption is measured in liters of gasoline per 100 kilometers driven and speed is measured in kilometers per hour. A statistical software package gives the least-squares regression line and also the residuals. The regression line is

$$\hat{y} = 11.058 - .01466x$$

The residuals, in the same order as the observations, are

10.09	2.24	−0.62	−2.47	−3.33	−4.28	−3.73	−2.94
−2.17	−1.32	−0.42	0.57	1.64	2.76	3.97	

(a) Make a scatterplot of the observations and draw the regression line on your plot.

(b) Would you use the regression line to predict y from x? Explain your answer.

(c) Check that the residuals have sum zero (up to roundoff error).

(d) Make a plot of the residuals against the values of x. Draw a horizontal line at height zero on your plot. Notice that the residuals show the same pattern about this line as the data points show about the regression line in the scatterplot in (a).

2.36   Exercise 2.10 (page 108) gives data on the true calories in ten foods and the average guesses made by a large group of people. Exercise 2.24 (page 118) explored the influence of two outlying observations on the correlation.

(a) Make a scatterplot suitable for predicting guessed calories from true calories. Circle the points for spaghetti and snack cake on your plot. These points lie outside the linear pattern of the other eight points.

(b) Use your calculator to find the least-squares regression line of guessed calories on true calories. Do this twice, first for all ten data points and then leaving out spaghetti and snack cake.

(c) Plot both lines on your graph. (Make one dashed so you can tell them apart.) Are spaghetti and snack cake, taken together, influential observations? Explain your answer.

2.37   The discussion of Example 2.13 shows that Child 18 in the Gesell data in Table 2.3 is an influential observation. Now we will examine the effect of Child 19, who is also an outlier in Figure 2.12.

(a) Find the least-squares regression line of Gesell score on age at first word, leaving out Child 19. Example 2.13 gives the regression line from all the children. Plot both lines on the same graph. (You do not have to make a scatterplot of all the points—just plot the two lines.) Would you call Child 19 very influential? Why?

(b) How does removing Child 19 change the r^2 for this regression? Explain why r^2 changes in this direction when you drop Child 19.

SUMMARY

A **regression line** is straight line that describes how a response variable y changes as an explanatory variable x changes.

The **least-squares regression line** is the straight line $\hat{y} = a + bx$ that minimizes the sum of the squares of the vertical distances of the observed y-values from the line.

You can use a regression line to **predict** the value of y for any value of x by substituting this x into the equation of the line.

The **slope** b of a regression line $\hat{y} = a + bx$ is the rate at which the predicted response $\hat{y}$ changes along the line as the explanatory variable x changes. Specifically, b is the change in $\hat{y}$ when x increases by 1.

The **intercept** a of a regression line $\hat{y} = a + bx$ is the predicted response $\hat{y}$ when the explanatory variable $x = 0$. This prediction is of no statistical use unless x can actually take values near 0.

Correlation and regression are closely connected. The correlation r is the slope of the least-squares regression line when we measure both x and y in standardized units. The square of the correlation r^2 is the fraction of the variance of one variable that is explained by least-squares regression on the other variable.

You can examine the fit of a regression line by studying the **residuals**, which are the differences between the observed and predicted values of y. Be on the lookout for outlying points with unusually large residuals and also for nonlinear patterns and uneven variation about the line.

Also look for **influential observations**, individual points that substantially change the regression line. Influential observations are often outliers in the x direction, but they need not have large residuals.

SECTION 2.3 EXERCISES

2.38   (Review of straight lines) Fred keeps his savings in his mattress. He began with $500 from his mother and adds $100 each year. His total savings y after x years are given by the equation

$$y = 500 + 100x$$

(a) Draw a graph of this equation. (Choose two values of x, such as 0 and 10. Compute the corresponding values of y from the equation. Plot these two points on graph paper and draw the straight line joining them.)

(b) After 20 years, how much will Fred have in his mattress?

(c) If Fred had added $200 instead of $100 each year to his initial $500, what is the equation that describes his savings after x years?

2.39   (Review of straight lines) During the period after birth, a male white rat gains exactly 40 grams (g) per week. (This rat is unusually regular in his growth, but 40 g per week is a realistic rate.)

(a) If the rat weighed 100 g at birth, give an equation for his weight after x weeks. What is the slope of this line?

(b) Draw a graph of this line between birth and 10 weeks of age.

(c) Would you be willing to use this line to predict the rat's weight at age 2 years? Do the prediction and think about the reasonableness of the result. (There are 454 grams in a pound. To help you assess the result, note that a large cat weighs about 10 pounds.)

2.40   The solid line in Figure 2.11 (page 125) is the least-squares regression line of a galaxy's speed on its distance. Use this graph to predict the speed of a galaxy that is 1.5 million parsecs away. Do you expect your prediction to be very accurate? Why?

2.41   Sarah's parents are concerned that she seems short for her age. Their doctor has the following record of Sarah's height:

Age (months)	36	48	51	54	57	60
Height (cm)	86	90	91	93	94	95

(a) Make a scatterplot of these data. Note the strong linear pattern.

(b) Using your calculator, find the equation of the least-squares regression line of height on age.

(c) Predict Sarah's height at 40 months and at 60 months. Use your results to draw the regression line on your scatterplot.

(d) What is Sarah's rate of growth, in centimeters per month? Normally growing girls gain about 6 cm in height between ages 4 (48 months) and 5 (60 months). What rate of growth is this in centimeters per month? Is Sarah growing more slowly than normal?

2.42   Investors ask about the relationship between returns on investments in the United States and on investments overseas. Here are data on the total returns on U.S. and overseas common stocks over a 22-year period. (The total return is change in price plus any dividends paid, converted into U.S. dollars. Both returns are averages over many individual stocks.)[15]

Year	Overseas % return	U.S. % return	Year	Overseas % return	U.S. % return
1971	29.6	14.6	1982	−1.9	21.5
1972	36.3	18.9	1983	23.7	22.4
1973	−14.9	−14.8	1984	7.4	6.1
1974	−23.2	−26.4	1985	56.2	31.6
1975	35.4	37.2	1986	69.4	18.6
1976	2.5	23.6	1987	24.6	5.1
1977	18.1	−7.4	1988	28.5	16.8
1978	32.6	6.4	1989	10.6	31.5
1979	4.8	18.2	1990	−23.0	−3.1
1980	22.6	32.3	1991	12.8	30.4
1981	−2.3	−5.0	1992	−12.1	7.6

(a) Make a scatterplot suitable for predicting overseas returns from U.S. returns.

(b) Find the correlation and r^2. Describe the relationship between U.S. and overseas returns in words, using r and r^2 to make your description more precise.

(c) Find the least-squares regression line of overseas returns on U.S. returns. Draw the line on the scatterplot.

(d) In 1993, the return on U.S. stocks was 10.1%. Use the regression line to predict the return on overseas stocks. The actual overseas return was 33.1%. Are you confident that predictions using the regression line will be quite accurate? Why?

(e) Circle the point that has the largest residual (either positive or negative). What year is this? Are there any points that seem likely to be very influential?

2.43   Use the least-squares regression line for the data in Exercise 2.41 to predict Sarah's height at age 40 years (480 months). Your prediction is in centimeters. Convert it to inches using the fact that a centimeter is 0.3937 inch.

The prediction is impossibly large. It is not reasonable to use data for 36 to 60 months to predict height at 480 months. Don't use regression lines for prediction for values of x far outside the range used to find the line.

2.44   Exercise 2.42 examined the relationship between returns on U.S. and overseas stocks. Investors also want to know what typical returns are and how much year-to-year variability (called *volatility* in finance) there is. Regression and correlation don't answer questions about center and spread.

(a) Find the five-number summaries for both U.S. and overseas returns, and make side-by-side boxplots to compare the two distributions.

(b) Were returns generally higher in the United States or overseas during this period? Explain your answer.

(c) Were returns more volatile (more variable) in the United States or overseas during this period? Explain your answer.

2.45   The mathematics department of a large state university would like to use the number of freshmen entering the university x to predict the number of students y who will sign up for freshman-level math courses in the fall semester. Here are data for recent years.[16]

Year	1986	1987	1988	1989	1990	1991	1992	1993
x	4595	4827	4427	4258	3995	4330	4265	4351
y	7364	7547	7099	6894	6572	7156	7232	7450

Computer software gives the correlation $r = .8333$ and the least-squares regression line

$$\hat{y} = 2492.69 + 1.0663x$$

The software also gives a table of the residuals:

Year	1986	1987	1988	1989	1990	1991	1992	1993
Residual	−28.44	−92.83	−114.30	−139.09	−180.65	46.13	191.44	3.17.74

(a) Make a scatterplot and draw the regression line on it. The regression line does not predict very accurately. What percent of variation in class enrollment is explained by the linear relationship with the count of freshmen?

(b) Check that the residuals have sum zero (at least up to roundoff error).

(c) Plots of the residuals against other variables are often revealing. Plot the residuals against year. One of the schools in the university recently changed its program to require that entering students take another mathematics course. How does the residual plot show the change? In what year was the change effective?

2.46   Table 2.4 presents three sets of data prepared by the statistician Frank Anscombe to illustrate the dangers of calculating without first plotting the data. *All three sets have the same correlation and the same least-squares regression line* to several decimal places.

TABLE 2.4 **Three data sets with the same correlation and least-squares regression line**										
Data Set A										
x	10	8	13	9	11	14	6	4	12	7	5
y	8.04	6.95	7.58	8.81	8.33	9.96	7.24	4.26	10.84	4.82	5.68
Data Set B										
x	10	8	13	9	11	14	6	4	12	7	5
y	9.14	8.14	8.74	8.77	9.26	8.10	6.13	3.10	9.13	7.26	4.74
Data Set C										
x	8	8	8	8	8	8	8	8	8	8	19
y	6.58	5.76	7.71	8.84	8.47	7.04	5.25	5.56	7.91	6.89	12.50

(a) Calculate the correlation and the least-squares regression line for all three data sets and verify that they agree.

(b) Make a scatterplot for each of the three data sets and draw the regression line on each of the plots.

(c) In which of the three cases would you be willing to use the regression line to predict y given that $x = 14$? Explain your answer in each case. The moral: ALWAYS PLOT YOUR DATA.

2.4   INTERPRETING CORRELATION AND REGRESSION

Correlation and regression are powerful tools for describing the relationship between two variables. When you use these tools, you must be aware of their limitations, beginning with the fact that they describe *only linear* relationships. Also remember that *both r and the least-squares regression line can be strongly influenced by a few extreme observations*. One influential observation or incorrectly entered data point can greatly change these measures. Always plot your data before interpreting regression or correlation. Here are some other cautions to keep in mind when you apply correlation and regression or read accounts of their use.

Extrapolation

Suppose that you have data on a child's growth between 3 and 8 years of age. You find a strong linear relationship between age x and height y. If you fit a regression line to these data and use it to predict height at age 25 years, you will predict that the child will be 8 feet tall. Growth slows down and stops at maturity, so extending the straight line to adult ages is foolish. Few relationships are linear for all values of x. So don't stray far from the range of x that actually appears in your data.

EXTRAPOLATION

Extrapolation is the use of a regression line for prediction outside the range of values of the explanatory variable x that you used to obtain the line. Such predictions cannot be trusted.

2.47   The number of people living on American farms has declined steadily during this century. Here are data on the farm population (millions of persons) from 1935 to 1980.

Year	1935	1940	1945	1950	1955	1960	1965	1970	1975	1980
Population	32.1	30.5	24.4	23.0	19.1	15.6	12.4	9.7	8.9	7.2

(a) Make a scatterplot of these data and find the least-squares regression line of farm population on year.

(b) According to the regression line, how much did the farm population decline each year on the average during this period? What percent of the observed variation in farm population is accounted for by linear change over time?

(c) Use the regression equation to predict the number of people living on farms in 1990. Is this result reasonable? Why?

Lurking variables

In our study of correlation and regression we looked at just two variables at a time. Often the relationship between two variables is strongly influenced by other variables. More advanced statistical methods allow the study of many variables together, so that we can take other variables into account. But sometimes the relationship between two variables is influenced by other variables that we did not measure or even think about. Because these variables are lurking in the background, we call them *lurking variables*.

> **LURKING VARIABLE**
>
> A **lurking variable** is a variable that has an important effect on the relationship among the variables in a study but is not included among the variables studied.

A lurking variable can falsely suggest a strong relationship between x and y, or it can hide a relationship that is really there. Here are examples of each of these effects.

EXAMPLE 2.15

The National Halothane Study was a major investigation of the safety of the anesthetics used in surgery. Records of over 850,000 operations performed in 34 major hospitals showed the following death rates for four common anesthetics.[17]

Anesthetic	A	B	C	D
Death rate	1.7%	1.7%	3.4%	1.9%

There is a clear association between the anesthetic used and the death rate of patients. Anesthetic C appears dangerous. But there are obvious lurking variables: the age and condition of the patient and the seriousness of the surgery. In fact, anesthetic C was more often used in serious operations on older patients in poor condition. The death rate would be higher among these patients no matter what anesthetic they received. After measuring the lurking variables and adjusting for their effect, the apparent relationship between anesthetic and death rate is very much weaker.   ◄

EXAMPLE 2.16

A study of housing conditions in the city of Hull, England, measured a large number of variables for each of the wards in the city. Two of the variables were a measure x of overcrowding and a measure y of the lack of indoor toilets. Because x and y are both measures of inadequate housing, we expect a high correlation. In fact the correlation was only $r = .08$. How can this be?

Investigation found that some poor wards had a lot of public housing. These wards had high values of x but low values of y because public housing always includes indoor toilets. Other poor wards lacked public housing, and these wards had high values of both x and y. Within wards of each type, there *was* a strong positive association between x and y. Analyzing all wards together ignored the lurking variable—amount of public housing—and hid the nature of the relationship between x and y.[18]

Figure 2.16 shows in simplified form how groups formed by a lurking variable can make correlation and regression misleading. The groups appear as clusters of points in the scatterplot. There is a strong relationship between x and y within each of the clusters. In fact, $r = .85$ and $r = .91$ in the two clusters. However, because similar values of x correspond to quite different values of y in the two clusters, x alone is of little value for predicting y. The correlation for all the points together is only $r = .14$.   ◄

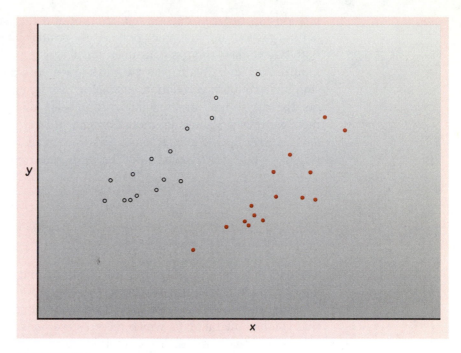

FIGURE 2.16 The variables in this scatterplot have a small correlation even though there is a strong correlation within each of the two clusters.

EXERCISE

2.48 A group of college students believes that herbal tea has remarkable powers. To test this belief, they make weekly visits to a local nursing home, where they visit with the residents and serve them herbal tea. The nursing home staff reports that after several months many of the residents are more cheerful and healthy. A skeptical sociologist commends the students for their good deeds but scoffs at the idea that herbal tea helped the residents. Identify the explanatory and response variables in this informal study. Then explain what lurking variables account for the observed association.

Using averaged data

Many regression or correlation studies work with averages or other measures that combine information from many individuals. You should note this carefully and resist the temptation to apply the results of such studies to individuals. We have seen, starting with Figure 2.2 (page 100), a strong relationship between outside temperature and the Sanchez household's natural gas consumption. Each point on the scatterplot represents a month. Both degree-days and gas consumed are averages over all the days in the month. Data for individual days would show more scatter about the regression line and lower correlation. Averaging over an entire month smoothes out the day-to-day variation due to doors left open, houseguests using more gas to heat water, and so on. *Correlations based on averages are usually too high when applied to individuals.* This is another reminder that it is important to note exactly what variables were measured in a statistical study.

EXERCISE

2.49   The data in Exercise 2.33 (page 128) give the average steps per second for a group of top female runners at each of several running speeds. There is a high positive correlation between steps per second and speed. Suppose that you had the full data, which record steps per second for each runner separately at each speed. If you plotted each individual observation and computed the correlation, would you expect the correlation to be lower than, about the same as, or higher than the correlation for the published data? Why?

Association is not causation

When we study the relationship between two variables, we often want to show that changes in the explanatory variable *cause* changes in the response variable. But a strong association between two variables is not enough to draw conclusions about cause and effect. Sometimes an observed association really does reflect cause and effect. The Sanchez household uses more natural gas in colder months because cold weather requires burning more gas to stay warm. In other cases, an association is explained by lurking variables, and the conclusion that x causes y is either wrong or not proved.

EXAMPLE 2.17

An article in a women's magazine reported that mothers who nurse their babies feel more receptive toward their infants than mothers who bottle-feed. The author concluded that breast-feeding (x) leads to a more positive attitude (y)

toward the child. But women choose whether to nurse or bottle-feed, and this choice may reflect already existing attitudes toward their infants. Perhaps mothers who already feel more positive about the child choose to nurse, while those to whom the baby is a nuisance more often choose the bottle. The mothers' already established attitude is a lurking variable that prevents conclusions about whether breast-feeding itself changes mothers' attitudes. ◄

experiment   The best way to get good evidence that x causes y is to do an ***experiment*** in which x is changed and lurking variables are kept under control. We might, for example, flip a coin to decide which new mothers would nurse their infants and which would use a bottle. That eliminates the influence of the mothers' attitudes and helps us see the effects of the feeding method. Of course, this experiment is morally and legally impossible.

When experiments cannot be done, finding the explanation for an observed association is often difficult and controversial. Data show that a company pays women less than men. The issue goes to court, where the plaintiffs claim that the pay differences are the result of sex discrimination. The company replies that pay depends on other variables, including education, experience, and seniority in the company. These are lurking variables if we only look at sex and pay. Both sides hire statisticians, who try to show that the difference in pay between men and women is or is not explained by the other variables. Which is true depends on the facts in the particular case.

> ### ASSOCIATION DOESN'T IMPLY CAUSATION
>
> An association between an explanatory variable x and a response variable y, even if it is very strong, is not by itself good evidence that changes in x actually cause changes in y.

EXAMPLE 2.18

Sometimes claims that x causes y are clearly wrong. For example, there is a strong negative correlation between teachers' salaries and students' scores on the SAT over time. The columnist William F. Buckley pointed this out and seemed to claim that paying teachers more causes lower SAT scores. "Teachers' salaries in 1945 averaged $14,770, and by 1992 had risen to $35,334. The results? In 1950, SAT math scores were 493; in 1992, down to 476."[19] But the

1992 SAT was taken by a much broader group of students because many more students plan to go to college than in 1950. And think of the social changes between 1950 and 1992 that probably reduce SAT scores: television, more divorce, more births to unmarried mothers, and so on. Society changes over time, so that a correlation between variables that are both changing over time is often meaningless. (Buckley did restate 1950 salaries in terms of 1992 buying power. Otherwise the effect of inflation in driving up salaries would make the negative correlation even stronger.) ◀

EXERCISES

2.50 Someone says, "There is a strong positive correlation between the number of firefighters at a fire and the amount of damage the fire does. So sending lots of firefighters just causes more damage." Explain why this reasoning is wrong.

2.51 A study of elementary school children, ages 6 to 11, finds a high positive correlation between shoe size x and score y on a test of reading comprehension. What explains this correlation?

2.52 A study shows that there is a positive correlation between the size of a hospital (measured by its number of beds x) and the median number of days y that patients remain in the hospital. Does this mean that you can shorten a hospital stay by choosing a small hospital?

SUMMARY

Correlation and regression should be **interpreted with caution**. Remember to **plot the data**. Watch out for the effects of **extreme observations** and remember that correlation and regression describe **only linear** relations.

Avoid **extrapolation**, which is use of a regression line for prediction for values of the explanatory variable outside the range of the data from which the line was calculated.

Lurking variables that you did not measure can explain the relations between the variables you did measure. Correlation and regression can be misleading if you ignore important lurking variables.

Remember that **correlations based on averages** are usually too high when applied to individuals.

Most of all, be careful not to conclude that there is a cause-and-effect relationship between two variables just because they are strongly associated.

High correlation does not imply causation. The best evidence that an association is due to causation comes from an **experiment** in which the explanatory variable is directly changed and other influences on the response are controlled.

SECTION 2.4 EXERCISES

2.53   Table 2.1 (page 92) gives education data for the states. The correlation between the median SAT math scores and the median SAT verbal scores for the states is $r = .9620$.

(a) Find r^2 and explain in simple language what this number tells us.

(b) If you calculated the correlation between the SAT math and verbal scores of a large number of individual students, would you expect the correlation to be about 0.96 or quite different? Explain your answer.

2.54   There is a strong positive correlation between years of education and income for economists employed by business firms. (In particular, economists with doctorates earn more than economists with only a bachelor's degree.) There is also a strong positive correlation between years of education and income for economists employed by colleges and universities. But when all economists are considered, there is a *negative* correlation between education and income. The explanation for this is that business pays high salaries and employs mostly economists with bachelor's degrees, while colleges pay lower salaries and employ mostly economists with doctorates. Sketch a scatterplot with two groups of cases (business and academic) illustrating how a strong positive correlation within each group and a negative overall correlation can occur together. (Hint: Begin by studying Figure 2.16.)

2.55   Members of a high school language club believe that study of a foreign language improves a student's command of English. From school records, they obtain the scores on an English achievement test given to all seniors. The mean score of seniors who studied a foreign language for at least two years is much higher than the mean score of seniors who studied no foreign language. The club's advisor says that these data are not good evidence that language study strengthens English skills. Identify the explanatory and response variables in this study. Then explain what lurking variable prevents the conclusion that language study improves students' English scores.

2.56   There is a strong positive correlation between years of schooling completed x and lifetime earnings y for American men. One possible reason for this association is causation: more education leads to higher-paying jobs. But lurking variables may explain some of the correlation. Suggest some lurking variables that would explain why men with more education earn more.

2.57   A study of London double-decker bus drivers and conductors found that drivers had twice the death rate from heart disease as conductors. Because drivers sit while conductors climb up and down stairs all day, it was at first thought that this association reflected the effect of physical activity on heart disease. Then a look at bus company records showed that drivers were issued consistently larger-size uniforms when hired than were conductors. This fact suggested an alternative explanation of the observed association between job type and deaths. What is it?

2.5   RELATIONS IN CATEGORICAL DATA*

To this point we have concentrated on relationships in which at least the response variable was quantitative. Now we will shift to describing relationships between two or more categorical variables. Some variables—such as sex, race, and occupation—are inherently categorical. Other categorical variables are created by grouping values of a quantitative variable into classes. Published data are often reported in grouped form to save space. To analyze categorical data, we use the *counts* or *percents* of individuals that fall into various categories.

EXAMPLE 2.19

two-way table

row and column variables

Table 2.5 presents Census Bureau data on the years of school completed by Americans of different ages. Many people under 25 years of age have not completed their education, so they are left out of the table. Both variables, age and education, are grouped into categories. This is a *two-way table* because it describes two categorical variables. Education is the **row variable** because each row in the table describes people with one level of education. Age is the **column variable** because each column describes one age group. The entries in the table are the counts of persons in each age-by-education class. Although both age and education in this table are categorical variables, both have a natural order from least to most. The order of the rows and the columns in Table 2.5 reflects the order of the categories. ◄

---

*This material is important in statistics, but it is needed later in this book only for Chapter 8. You can omit it if you do not plan to read Chapter 8, or delay it until Chapter 8 is reached.

| Education | Age Group | | | | | |
	25–34	35–44	45–54	55–64	≥ 65	Total
TABLE 2.5 Years of school completed by age, 1991 (thousands of persons)						
Did not complete high school	5,965	4,755	4,829	5,999	12,702	34,251
Completed high school	17,505	14,498	10,300	8,645	10,310	61,259
1–3 years of college	9,267	8,777	4,598	3,094	3,428	29,165
4 or more years of college	10,168	10,633	5,959	3,607	3,652	34,019
Total	42,905	38,665	25,686	21,346	30,092	158,694

Marginal distributions

How can we best grasp the information contained in Table 2.5? First, *look at the distribution of each variable separately*. The distribution of a categorical variable just says how often each outcome occurred. The "Total" column at the right of the table contains the totals for each of the rows. These row totals give the distribution of education level (the row variable) among all people over 25 years of age: 34,251,000 did not complete high school, 61,259,000 finished high school but did not attend college, and so on. In the same way, the "Total" row at the bottom gives the age distribution. If the row and column totals are missing, the first thing to do in studying a two-way table is to calculate them. The distributions of education alone and age alone are often called **marginal distributions** because they appear at the right and bottom margins of the two-way table.

marginal distributions

If you check the row and column totals in Table 2.5, you will notice a few discrepancies. For example, the sum of the entries in the "Did not complete high school" row is 34,250. The entry in the "Total" column for that row is 34,251. The explanation is **roundoff error**. The table entries are in thousands of persons, and each is rounded to the nearest thousand. The Census Bureau obtained the "Total" entry by rounding the exact number of people who did not finish high school to the nearest thousand. The result was 34,251,000. Adding the row entries, each of which is already rounded, gives a slightly different result.

roundoff error

Percents are often easier to grasp than counts. We can display the marginal distribution of education level in terms of percents by dividing each row total by the table total and converting to a percent.

EXAMPLE 2.20

The percent of people over 25 years of age who have at least 4 years of college is

$$\frac{\text{total with 4 years of college}}{\text{table total}} = \frac{34{,}019}{158{,}694} = .214 = 21.4\%$$

Do three more such calculations to obtain the marginal distribution of education level in percents. Here it is.

Education	Did not finish high school	Completed high school	1–3 years of college	≥ 4 years of college
Percent	21.6	38.6	18.4	21.4

The total is 100% because everyone is in one of the four education categories. ◄

In working with two-way tables, you must calculate percents—lots of them. Here's a tip to help decide what fraction gives the percent you want. Ask, "What group represents the total that I want a percent of?" The count for that group is the denominator of the fraction that leads to the percent. In Example 2.20, we wanted a percent "of people over 25 years of age," so the count of people over 25 (the table total) is the denominator.

EXERCISES

2.58   Sum the counts in the "35 to 44" age column in Table 2.5. Then explain why the sum is not the same as the entry for this column in the "Total" row.

2.59   Give the marginal distribution of age among people over 25 years of age in percents, starting from the counts in Table 2.5.

2.60   Here are data from eight high schools on smoking among students and among their parents.[20]

	Student smokes	Student does not smoke
Both parents smoke	400	1380
One parent smokes	416	1823
Neither parent smokes	188	1168

(a) How many students do these data describe?

(b) What percent of these students smoke?

(c) Give the marginal distribution of parents' smoking behavior, both in counts and in percents.

Describing relationships

The marginal distributions of age and of education separately do not tell us how the two variables are related. That information is in the body of the table. How can we describe the relationship between age and years of school completed? No single graph (such as a scatterplot) portrays the form of the relationship between categorical variables, and no single numerical measure (such as the correlation) summarizes the strength of an association. *To describe relationships among categorical variables, calculate appropriate percents from the counts given.* We use percents because counts are often hard to compare. For example, 10,168,000 people aged 25 to 34 have completed college, and only 3,607,000 people in the 55 to 64 age group have done so. But the younger age group is about twice as large, so we can't directly compare these counts.

EXAMPLE 2.21

What percent of people aged 25 to 34 have completed 4 years of college? This is the count who are 25 to 34 and have 4 years of college as a percent of the age group total:

$$\frac{10,168}{42,905} = .237 = 23.7\%$$

"People aged 25 to 34" is the group we want a percent of, so the count for that group is the denominator. In the same way, the percent of people in the 55 to 64 age group who completed college is

$$\frac{3607}{21,346} = .169 = 16.9\%$$

Here are the results for all five age groups:

Age group	25–34	35–44	45–54	55–64	≥ 65
Percent with 4 years of college	23.7	27.5	23.2	16.9	12.1

These percents help us see how the education of Americans varies with age. Older people are less likely to have completed college, but the data don't show a steady drop in college graduates as age increases. The percent of college graduates does not drop off sharply until we reach ages over 55.   ◀

bar chart

Although graphs are not as useful for describing categorical variables as they are for quantitative variables, a graph still helps an audience to grasp the data quickly. You can use a **bar chart** like Figure 2.17 to present the information in Example 2.21. Each bar represents one age group. The height of the bar is the percent of that age group with at least 4 years of college. Although bar charts look a bit like histograms, their details and uses are different. A histogram shows the distribution of the values of a quantitative variable. A bar chart compares the sizes of different items. The horizontal axis of a bar chart need not have any measurement scale but may simply identify the items being compared. The items compared in Figure 2.17 are the five age groups. Because each bar in a bar chart describes a different item, we draw the bars with space between them.

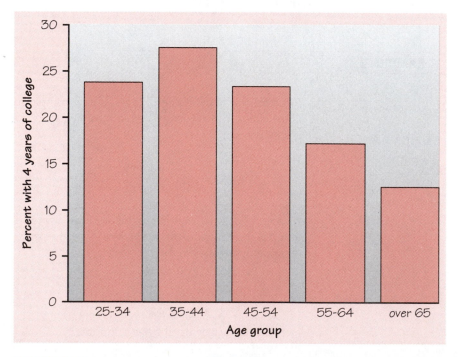

FIGURE 2.17   Bar chart comparing the percents of several age groups who have completed 4 or more years of college. The height of each bar is the percent of people in one age group who have 4 years of college.

Example 2.21 does not compare the complete distributions of years of schooling in the five age groups. It compares only the percents who finished college. Let's look at the complete picture.

EXAMPLE 2.22

Information about the 25 to 34 age group occupies the first column in Table 2.5. To find the complete distribution of education in this age group, look only at that column. Compute each count as a percent of the column total 42,905. Here is the distribution.

Education	Did not finish high school	Completed high school	1–3 years of college	≥ 4 years of college
Percent	13.9	40.8	21.6	23.7

conditional distribution

These percents add to 100%, because all 25 to 34 year-olds fall in one of the educational categories. The four percents together are the **conditional distribution** of education, given that a person is 25 to 34 years of age. We use the term "conditional" because the distribution refers only to people who satisfy the condition that they are 25 to 34 years old.

For comparison, here is the conditional distribution of years of school completed among people over 65 years of age. To find these percents, look only at the "≥ 65" column in Table 2.5. The column total is the denominator for each percent calculation.

Education	Did not finish high school	Completed high school	1–3 years of college	≥ 4 years of college
Percent	42.2	34.3	11.4	12.1

The percent who did not finish high school is much higher in the older age group, and the percents with some college and who finished college are much lower. Comparing the conditional distributions of education in different age groups describes the association between age and education. There are five different conditional distributions of education given age, one for each of the five age groups. All of these conditional distributions differ from the marginal distribution of education found in Example 2.20. ◄

In Example 2.22 we compared the education of different age groups. That is, we thought of age as the explanatory variable and education as the response variable. We might also be interested in the distribution of age among persons having a certain level of education. To do this, look only at one *row* in Table 2.5. Calculate each entry in that row as a percent of the row total, the total of that education group. The result is another conditional distribution, the conditional distribution of age given a certain level of education.

A two-way table contains a great deal of information in compact form. Making that information clear almost always requires finding percents. You must decide which percents you need. If you are studying trends in the training of the American workforce, comparing the distributions of education for different age groups reveals the more extensive education of younger people. If, on the other hand, you are planning a program to improve the skills of people who did not finish high school, the age distribution within this educational group is important information.

EXERCISES

2.61   Using the counts in Table 2.5, find the percent of people in each age group who did not complete high school. Draw a bar chart that compares these percents. State briefly what the data show.

2.62   Set up the four fractions that lead to the conditional distribution of education level among people over 65 years of age given in Example 2.22. Carry out the divisions and check that the results in the example are correct.

2.63   Example 2.22 gives the conditional distributions of education level among 25- to 34-year-olds and among people over 65. Find the conditional distribution of education level among 35- to 44-year-olds in percents, starting with the counts in Table 2.5. Is this distribution more like the distribution for 25- to 34-year-olds or the distribution for people over 65?

2.64   Find the conditional distribution of age among people with at least 4 years of college, starting from the counts in Table 2.5.

2.65   Here are the row and column totals for a two-way table with two rows and two columns.

	Col. 1	Col. 2	Total
Row 1			50
Row 2			50
Total	60	40	100

Find *two different* sets of four entries for the body of the table that give these same totals. This shows that the relationship between the two variables can't be described by their marginal distributions.

2.66   Is high blood pressure dangerous? Medical researchers classified each of a group of men as "high" or "low" blood pressure, then watched them for 5 years. (Men with systolic blood pressure 140 mm Hg or higher were "high"; the others, "low.") The following two-way table gives the results of the study.[21]

	Died	Survived
Low blood pressure	21	2655
High blood pressure	55	3283

(a) How many men took part in the study? What percent of these men died during the 5 years of the study?

(b) The two categorical variables in the table are blood pressure (high or low) and outcome (died or survived). Which is the explanatory variable?

(c) Is high blood pressure associated with a higher death rate? Calculate and compare percents to answer this question.

2.67   Do the smoking habits of parents help explain whether or not their children smoke? Exercise 2.60 gives these data from eight Arizona high schools.

	Student smokes	Student does not smoke
Both parents smoke	400	1380
One parent smokes	416	1823
Neither parent smokes	188	1168

(a) What percent of students smoke among those with two smoking parents, among those with one smoking parent, and among those with neither parent smoking?

(b) Draw a bar chart that compares the three percents of students who smoke that you found in (a).

(c) Briefly describe the relationship between parents' smoking and students' smoking.

Simpson's paradox

As is the case with quantitative variables, the effects of lurking variables can change or even reverse relationships between two categorical variables. Here is a hypothetical example that demonstrates the surprises that can await the unsuspecting user of data.

EXAMPLE 2.23

To help consumers make informed decisions about health care, the government releases data about patient outcomes in hospitals. You want to compare Hospital A and Hospital B, which serve your community. Here is a two-way table of data on the survival of patients after surgery in these two hospitals. All patients undergoing surgery in a recent time period are included. "Survived" means that the patient lived at least 6 weeks following surgery.

	Hospital A	Hospital B
Died	63	16
Survived	2037	784
Total	2100	800

The evidence seems clear: Hospital A loses 3% (63/2100) of its surgery patients, and Hospital B loses only 2% (16/800). It seems that you should choose Hospital B if you need surgery.

Not all surgery cases are equally serious, however. Later in the government report you find data on the outcome of surgery broken down by the condition of the patient before the operation. Patients are classified as being in either "poor" or "good" condition. Here are the more detailed data. Check that the entries in the original two-way table are just the sums of the "poor" and "good" entries in this pair of tables.

	Good condition		Poor condition	
	Hospital A	Hospital B	Hospital A	Hospital B
Died	6	8	57	8
Survived	594	592	1443	192
Total	600	600	1500	200

Aha! Hospital A beats Hospital B for patients in good condition: only 1% (6/600) died in Hospital A, compared with 1.3% (8/600) in Hospital B. And Hospital A wins again for patients in poor condition, losing 3.8% (57/1500) to Hospital B's 4% (8/200). So Hospital A is safer for both patients in good condition and patients in poor condition. If you are facing surgery, you should choose Hospital A. ◄

The patient's condition is a lurking variable when we compare the death rates at the two hospitals. When we ignore the lurking variable, Hospital B seems safer, even though Hospital A does better for both classes of patients. How can A do better in each group, yet do worse overall? Look at the data. Hospital A is a medical center that attracts seriously ill patients from a wide region. It had 1500 patients in poor condition. Hospital B had only 200 such cases. Because patients in poor condition are more likely to die, Hospital A has a higher death rate despite its superior performance for each class of patients. The original two-way table, which did not take account of the condition of the patients, was misleading. Example 2.23 illustrates *Simpson's paradox*.

SIMPSON'S PARADOX

Simpson's paradox refers to the reversal of the direction of a comparison or an association when data from several groups are combined to form a single group.

The lurking variables in Simpson's paradox are categorical. That is, they break the individuals into groups, as when surgery patients are classified as "good condition" or "poor condition." Simpson's paradox is just an extreme form of the fact that observed associations can be misleading when there are lurking variables. Although the hospital data in Example 2.23 were made up, the setting is real. The hospital death rates compiled and released by the Federal Health Care Financing Agency do allow for the patient's condition and diagnosis. Hospitals continue to complain that there are differences among patients that can make the government's mortality data misleading.

EXERCISES

2.68 Upper Wabash Tech has two professional schools, business and law. Here are two-way tables of applicants to both schools, categorized by sex and admission decision. (Although these data are made up, similar situations occur in reality.[22])

	Business			Law	
	Admit	Deny		Admit	Deny
Male	480	120	Male	10	90
Female	180	20	Female	100	200

(a) Make a two-way table of sex by admission decision for the two professional schools together by summing entries in this table.

(b) From the two-way table, calculate the percent of male applicants who are admitted and the percent of female applicants who are admitted. Wabash admits a higher percent of male applicants.

(c) Now compute separately the percents of male and female applicants admitted by the business school and by the law school. Each school admits a higher percent of female applicants.

(d) This is Simpson's paradox: both schools admit a higher percent of the women who apply, but overall Wabash admits a lower percent of female applicants than of male applicants. Explain carefully, as if speaking to a skeptical reporter, how it can happen that Wabash appears to favor males when each school individually favors females.

2.69 Whether a convicted murderer gets the death penalty seems to be influenced by the race of the victim. Here are data on 326 cases in which the defendant was convicted of murder.[23]

White Defendant

	Death penalty	
	Yes	No
White victim	19	132
Black victim	0	9

Black Defendant

	Death penalty	
	Yes	No
White victim	11	52
Black victim	6	97

(a) Use these data to make a two-way table of defendant's race (white or black) versus death penalty (yes or no).

(b) Show that Simpson's paradox holds: a higher percent of white defendants are sentenced to death overall, but for both black and white victims a higher percent of black defendants are sentenced to death.

(c) Use the data to explain why the paradox holds in language that a judge could understand.

2.70 Most baseball hitters perform differently against right-handed and left-handed pitching. Consider two players, Joe and Moe, both of whom bat right-handed. The table below records their performance against right-handed and left-handed pitchers.

Player	Pitcher	Hits	At bats
Joe	Right	40	100
	Left	80	400
Moe	Right	120	400
	Left	10	100

(a) Make a two-way table of player (Joe or Moe) versus outcome (hit or no hit) by summing over both kinds of pitcher.

(b) Find the overall batting average (hits divided by total times at bat) for each player. Who has the higher batting average?

(c) Make a separate two-way table of player versus outcome for each kind of pitcher. From these tables, find the batting averages of Joe and Moe against right-handed pitching. Who does better? Do the same for left-handed pitching. Who does better?

(d) The manager doesn't believe that one player can hit better against both left-handers and right-handers yet have a lower overall batting average. Explain in simple language why this happens to Joe and Moe.

2.71 A study by the National Science Foundation[24] found that the median salary of newly graduated female engineers and scientists was only 73% of the median salary for males. When the new graduates were broken down by field, however, the picture changed. Women earned at least 84% as much as men in *every* field of engineering and science. The median salary for women was higher than that of men in many engineering disciplines. How can women do nearly as well as men in every field yet fall far behind men when we look at all young engineers and scientists?

2.72 Recent studies have shown that earlier reports underestimated the health risks associated with being overweight. The error was due to overlooking some important variables. In particular, smoking tends both to reduce weight and to lead to earlier death. Illustrate Simpson's paradox by a simplified version of this situation. That is, make up a table of overweight (yes or no) by early death (yes or no) by smoker (yes or no) such that

- Overweight smokers and overweight nonsmokers both tend to die earlier than those not overweight.
- But when smokers and nonsmokers are combined into a two-way table of overweight by early death, persons who are not overweight tend to die earlier.

SUMMARY

A **two-way table** of counts describes the relationship between two categorical variables. Values of the **row variable** label the rows of the table, and values of the **column variable** label the columns. Two-way tables are often used to summarize large amounts of data by grouping outcomes into categories.

You can present various aspects of the information in a two-way table by calculating and comparing **percents** from the counts in the table. Start each calculation by deciding what count represents the total that you want a percent of. That count is the denominator of the fraction you must calculate.

The **row totals** and **column totals** in a two-way table give the **marginal distributions** of the two variables separately. They do not give any information about the relationship between the variables.

To find the **conditional distribution** of the row variable for one specific value of the column variable, look only at that one column in the table. Find each entry in the column as a percent of the column total. There is a conditional distribution of the row variable for each column in the table. Comparing these conditional distributions is one way of showing the association between the row and the column variables.

Compare the conditional distributions of the row variable for the different categories of the column variable when you think of the column variable as the explanatory variable. If the row variable is your explanatory variable, look at each row separately and show the association by comparing the conditional distributions of the column variable in the rows.

A comparison between two variables that holds for each individual value of a third variable can be changed or even reversed when the data for all values of the third variable are combined. This is **Simpson's paradox**.

SECTION 2.5 EXERCISES

Exercises 2.73 to 2.76 are based on Table 2.6. This two-way table reports Census Bureau data on undergraduate students enrolled in U.S. colleges and universities in the fall of 1991.

TABLE 2.6	Undergraduate college enrollment by age of students—autumn, 1991 (thousands of students)			
Age	2-year full-time	2-year part-time	4-year full-time	4-year part-time
15–17	44	4	79	0
18–21	1345	456	3869	159
22–29	489	690	1358	494
30–44	287	704	289	627
≥ 45	49	209	62	160
Total	2212	2065	5657	1440

SOURCE: Current Population Survey, October 1991

2.73   (a) How many students aged 30 to 44 were enrolled part-time at 2-year colleges?

(b) How many students of all ages were enrolled part-time at 2-year colleges?

(c) The sum of the entries in the "2-year part-time" column is not equal to the "Total" entry for that column. How do you explain this difference?

2.74   (a) How many undergraduate students were enrolled in colleges and universities?

(b) What percent of all undergraduate students were 18 to 21 years old in the fall of the academic year?

(c) Find the percent of the undergraduates enrolled in each of the four types of program who were 18 to 21 years old. Make a bar chart to compare these percents.

(d) The 18 to 21 group is the "traditional" age group for college students. Briefly summarize what you have learned from the data about the extent to which this group predominates in different kinds of college programs.

2.75   (a) What percent of students enrolled part-time at 2-year colleges are 30 to 44 years old?

(b) What percent of all 30- to 44-year-old students are enrolled part-time at 2-year colleges?

2.76   (a) Find the marginal distribution of age among all undergraduate students, first in counts and then in percents.

(b) Find the conditional distribution of age (in percents) among students enrolled part-time in 2-year colleges.

(c) Briefly describe the most important differences between the two age distributions.

2.77   Do child restraints and seat belts prevent injuries to young passengers in automobile accidents? Here are data on the 26,971 passengers under the age of 15 in accidents reported in North Carolina during two years before the law required restraints.[25]

	Restrained	Unrestrained
Injured	197	3,844
Uninjured	1,749	21,181

(a) What percent of these young passengers were restrained?

(b) Do the data provide evidence that young passengers are less likely to be injured in an accident if they wear restraints? Calculate and compare percents to answer this question.

2.78   A business school conducted a survey of companies in its state. They mailed a questionnaire to 200 small companies, 200 medium-sized companies, and 200 large companies. The rate of nonresponse is important in deciding how reliable survey results are. Here are the data on response to this survey.

	Response	No response	Total
Small	125	75	200
Medium	81	119	200
Large	40	160	200

(a) What was the overall percent of nonresponse?

(b) Describe how nonresponse is related to the size of the business. (Use percents to make your statements precise.)

(c) Draw a bar chart to compare the nonresponse percents for the three size categories.

2.79   Cocaine addiction is hard to break. Addicts need cocaine to feel any pleasure, so perhaps giving them an antidepressant drug will help. A 3-year study with 72 chronic cocaine users compared an antidepressant drug called desipramine with lithium and a placebo. (Lithium is a standard drug to treat cocaine addiction. A placebo is a dummy drug, used so that the effect of being in the study but not taking any drug can be seen.) One-third of the subjects, chosen at random, received each drug. Here are the results.[26]

	Cocaine relapse?	
	Yes	No
Desipramine	10	14
Lithium	18	6
Placebo	20	4

(a) Compare the effectiveness of the three treatments in preventing relapse. Use percents and draw a bar chart.

(b) Do you think that this study gives good evidence that desipramine actually *causes* a reduction in relapses?

2.80   The following two-way table describes the age and marital status of American women in 1991. The table entries are in thousands of women.

| | Marital status | | | | |
	Single	Married	Widowed	Divorced	Total
18–24	9,008	3,352	8	257	12,627
25–39	6,658	21,769	248	3,224	31,899
40–64	1,975	24,462	2,570	4,755	33,762
≥ 65	900	7,255	8,464	925	17,545
Total					95,833

Age (years) labels rows 18–24 through ≥ 65.

(a) Find the sum of the entries in the 18–24 row. Why does this sum differ from the "Total" entry for that row?

(b) Give the marginal distribution of marital status for all adult women (use percents). Draw a bar chart to display this distribution.

(c) Compare the conditional distributions of marital status for women aged 18 to 24 and women aged 40 to 64. Briefly describe the most important differences between the two groups of women, and back up your description with percents.

(d) You are planning a magazine aimed at single women who have never been married. (That's what "single" means in government data.) Find the conditional distribution of ages among single women.

CHAPTER REVIEW

Chapter 1 dealt with data analysis for a single variable. In this chapter, we have studied analysis of data for two or more variables. The proper analysis depends on whether the variables are categorical or quantitative and on whether one is an explanatory variable and the other a response variable.

Data analysis begins with graphs and then adds numerical summaries of specific aspects of the data. When you have a categorical explanatory variable and a quantitative response variable, use the tools of Chapter 1 to compare the distributions of the response variable for the different categories of the explanatory variable. Make side-by-side boxplots, stemplots, or histograms and compare medians or means. If both variables are categorical, there is no satisfactory graph (though bar charts can help). We describe the relationship numerically by comparing percents. The optional Section 2.5 explains how to do this.

Most of this chapter concentrates on relations between two quantitative variables. Scatterplots show the relationship, whether or not there is a explanatory-response distinction. Correlation describes the strength of a linear relationship, and least-squares regression fits a line to data that have an explanatory-response relation.

Relationships among several variables can be complex. We therefore paid attention to cautions about interpreting data. In particular, be slow to conclude that a strong observed association between two variables is due to a cause-and-effect link between them.

Here is a review list of the most important skills you should have gained from studying this chapter.

A. DATA

1. Recognize whether each variable is quantitative or categorical.
2. Identify the explanatory and response variables in situations where one variable explains or influences another.

B. SCATTERPLOTS

1. Make a scatterplot for two quantitative variables, placing the explanatory variable (if any) on the horizontal scale.
2. Add a categorical variable to the scatterplot by using a different plotting symbol.
3. Recognize positive or negative association, a linear pattern, and outliers in a scatterplot.

C. CORRELATION

1. Compute the correlation coefficient r for small sets of observations, using a calculator.
2. Know the basic properties of correlation: r measures the strength and direction of linear relations only; $-1 \leq r \leq 1$ always; $r = \pm 1$ only for perfect straight-line relations; r moves away from 0 toward ± 1 as the linear relation gets stronger.

D. STRAIGHT LINES

1. Explain what the slope b and the intercept a mean in the equation $y = a + bx$ of a straight line.
2. Draw a graph of the straight line when you are given its equation.

E. REGRESSION

1. Calculate the least-squares regression line of a response variable y on an explanatory variable x from data, using a calculator.

2. Find the slope and intercept of the least-squares regression line from the means and standard deviations of x and y and their correlation.

3. Use the regression line to predict y for a given x. Recognize extrapolation and be aware of its dangers.

4. Use r^2 to describe how much of the variation in one variable can be accounted for by a straight-line relationship with another variable.

5. Recognize outliers and potentially influential observations from a scatterplot with the regression line drawn on it.

6. Calculate the residuals and plot them against the explanatory variable x or against other variables. Recognize unusual patterns.

F. LIMITATIONS OF CORRELATION AND REGRESSION

1. Understand that both r and the least-squares regression line can be strongly influenced by a few extreme observations.

2. Recognize possible lurking variables that may explain the observed association between two variables x and y.

3. Understand that even a strong correlation does not mean that there is a cause-and-effect relationship between x and y.

G. CATEGORICAL DATA (OPTIONAL)

1. From a two-way table of counts, find the marginal distributions of both variables by obtaining the row sums and column sums.

2. Express any distribution in percents by dividing the category counts by their total.

3. Describe the relationship between two categorical variables by computing and comparing percents. Often this involves comparing the conditional distributions of one variable for the different categories of the other variable.

CHAPTER 2 REVIEW EXERCISES

2.81   Here are data on 18 people who fell ill from an incident of food poisoning.[27] The data give each person's age in years, the incubation period (the time in hours between eating the infected food and the first signs of illness), and whether the victim survived (S) or died (D).

				Person				
	1	2	3	4	5	6	7	8	9
Age	29	39	44	37	42	17	38	43	51
Incubation	13	46	43	34	20	20	18	72	19
Outcome	D	S	S	D	D	S	D	S	D

				Person				
	10	11	12	13	14	15	16	17	18
Age	30	32	59	33	31	32	32	36	50
Incubation	36	48	44	21	32	86	48	28	16
Outcome	D	D	S	D	D	S	D	S	D

(a) Make a scatterplot of incubation period against age, using different symbols for people who died and those who survived.

(b) Is there an overall relationship between age and incubation period? If so, describe it.

(c) More important, is there a relationship between either age or incubation period and whether the victim survived? Describe any relations that seem important here.

(d) Are there any unusual cases that may require individual investigation?

2.82 Nematodes are microscopic worms. Here are data from an experiment to study the effect of nematodes in the soil on plant growth. The experimenter prepared 16 planting pots and introduced different numbers of nematodes. Then he placed a tomato seedling in each pot and measured its growth (in centimeters) after 16 days.[28]

Nematodes	Seedling growth (cm)			
0	10.8	9.1	13.5	9.2
1,000	11.1	11.1	8.2	11.3
5,000	5.4	4.6	7.4	5.0
10,000	5.8	5.3	3.2	7.5

Analyze these data and give your conclusions about the effects of nematodes on plant growth.

2.83 (Optional) Here are data on infant mortality (deaths per thousand live births) in the New England states.[29]

Year	1974	1975	1976	1977	1978	1979	1980	1981
Mortality	14.5	13.8	12.7	11.9	11.4	10.8	10.3	9.7

Make a plot of infant mortality against time. Then make a plot of the logarithm of infant mortality against time. Does the logarithm transformation make the pattern more linear?

2.84   The mean height of American women in their early twenties is about 64.5 inches and the standard deviation is about 2.5 inches. The mean height of men the same age is about 68.5 inches, with standard deviation about 2.7 inches. If the correlation between the heights of husbands and wives is about $r = .5$, what is the slope of the regression line of the husband's height on the wife's height in young couples? Draw a graph of this regression line. Predict the height of the husband of a woman who is 67 inches tall.

2.85   Exercise 2.4 (page 98) gives data on the number of powerboats registered in Florida and the number of manatees killed by boats. Here is part of the output from the regression command in the Minitab statistical software.

```
The regression equation is

Killed = - 41.4 + 0.125 Boats


Unusual Observations

Obs.    Boats    Killed        Fit Stdev.Fit   Residual    St.Resid

  7      526     15.00      24.25     1.26       -9.25       -2.26R


R denotes an obs. with a large st. resid.
```

(a) Make a scatterplot of boats registered and manatees killed. Is there a strong straight-line pattern? What is r^2 for these data?

(b) Draw the regression line given by Minitab on your scatterplot. Predict how many manatees would be killed each year if Florida decided to freeze the number of boats at 700,000. (Use Minitab's work.)

(c) Minitab checks for large residuals and influential observations. It calls attention to one observation that has a somewhat large residual. Circle this observation on your plot. We have no reason to remove it.

(d) Residuals from least-squares regression often have a distribution that is roughly normal. So Minitab reports the *standardized* residuals— that's what St.Resid means. Use the 68–95–99.7 rule for normal distributions to say how surprising a residual with standardized value −2.26 is.

2.86   The Franklin National Bank failed in 1974. Franklin was one of the 20 largest banks in the nation, and the largest ever to fail. Could Franklin's weakened condition have been detected in advance by simple data analysis? The table below gives the total assets (in billions of dollars) and net income (in millions of dollars) for the 20 largest banks in 1973, the year before Franklin failed.[30] Franklin is bank number 19.

	1	2	3	4	Bank 5	6	7	8	9	10
Assets	49.0	42.3	36.3	16.4	14.9	14.2	13.5	13.4	13.2	11.8
Income	218.8	265.6	170.9	85.9	88.1	63.6	96.9	60.9	144.2	53.6

	11	12	13	14	Bank 15	16	17	18	19	20
Assets	11.6	9.5	9.4	7.5	7.2	6.7	6.0	4.6	3.8	3.4
Income	42.9	32.4	68.3	48.6	32.2	42.7	28.9	40.7	13.8	22.2

(a) We expect banks with more assets to earn higher income. Make a scatterplot of these data that displays the relation between assets and income. Mark Franklin (Bank 19) with a separate symbol.

(b) Describe the overall pattern of your plot. Are there any banks with unusually high or low income relative to their assets? Does Franklin stand out from other banks in your plot?

(c) Find the least-squares regression line for predicting a bank's income from its assets. Draw the regression line on your scatterplot.

(d) Use the regression line to predict Franklin's income. Was the actual income higher or lower than predicted? What is the residual?

2.87   **(Optional)** Does taking aspirin regularly help prevent heart attacks? The Physicians' Health Study tried to find out. The subjects were 22,071 healthy male doctors at least 40 years old. Half the subjects, chosen at random, took aspirin every other day. The other half took a placebo, a dummy pill that looked and tasted like aspirin. Here are the results.[31] (The row for "None of these" is left out of the two-way table.)

	Aspirin group	Placebo group
Fatal heart attacks	10	26
Other heart attacks	129	213
Strokes	119	98
Total	11,037	11,034

What do the data show about the association between taking aspirin and heart attacks and stroke? Use percents to make your statements precise. Do you think the study provides evidence that aspirin actually reduces heart attacks (cause and effect)?

2.88 **(Optional)** Here is a two-way table of suicides committed in 1990, categorized by the sex of the victim and the method used. Based on these data, write a brief account of differences in suicide between men and women. Be sure to cite appropriate counts or percents to justify your statements.

	Male	Female
Firearms	16,285	2,600
Poison	3,221	2,203
Hanging	3,688	756
Other	1,530	623
Total	24,724	6,182

2.89 Environmental Protection Agency regulations require auto makers to give the city and highway gas mileages for each model of car. Table 2.7 gives the mileages (miles per gallon) for 30 midsize and large 1994 car models.[32] Use computer software to analyze these data.

(a) Plot highway mileage y against city mileage x. Use different plotting symbols for midsize and large cars. Describe the overall pattern. Do midsize and large cars differ? How?

(b) There is one clear outlier. Which car model is this? Drop this unusual car from the data for all further work.

(c) How strong is the linear relationship between city and highway mileage for midsize cars? For large cars? (Give a numerical measure that makes your statements precise.)

| TABLE 2.7   City and highway gas mileage for 1994 car models | | | |
Model	Size	City MPG	Highway MPG
BMW 740i	Midsize	16	23
Buick Century	Midsize	25	31
Buick LeSabre	Large	19	28
Buick Park Avenue	Large	19	27
Buick Regal	Midsize	19	29
Buick Roadmaster	Large	17	25
Cadillac DeVille	Large	16	25
Chevrolet Caprice	Large	18	26
Chevrolet Lumina	Midsize	19	29
Chrysler Concorde	Large	20	28
Chrysler New Yorker	Large	18	26
Dodge Spirit	Midsize	22	27
Ford LTD	Large	18	25
Ford Taurus	Midsize	20	29
Ford Thunderbird	Midsize	19	26
Hyundai Sonata	Midsize	21	27
Infiniti Q45	Midsize	17	22
Lexus LS400	Midsize	18	23
Lincoln Continental	Large	18	26
Lincoln Mark VIII	Midsize	18	25
Mazda 626	Midsize	23	31
Mazda 929	Midsize	19	24
Mercedes-Benz S320	Large	17	24
Mercedes-Benz S420	Large	15	20
Nissan Maxima	Midsize	19	26
Rolls-Royce Silver Spur	Midsize	10	15
Saab 900	Midsize	19	26
Saab 9000	Large	17	27
Toyota Camry	Midsize	21	28
Volvo 850	Midsize	19	26

(d) Find the two least-squares regression lines of highway mileage on city mileage for midsize cars and for large cars. Draw both lines on your scatterplot. How do the regression lines reflect the difference you noted in (a) between midsize and large cars?

(e) If a midsize car's city mileage is 20 MPG, predict its highway mileage. Do the same for a large car.

2.90 Table 2.1 gives data about education in the states. Use computer software to examine the relationship between the median SAT verbal and mathematics scores as follows.

(a) You want to predict a state's SAT math score from its verbal score. Find the least-squares regression line for this purpose. You learn that a state's median verbal score the following year was 455. Use your regression line to predict its median math score.

(b) Plot the residuals from your regression against the SAT verbal score. (The software should do this for you.) One state is an outlier. What state is this? Does this state have a median math score higher or lower than would be predicted on the basis of its median verbal score?

NOTES AND DATA SOURCES

1. Data provided by Robert Dale, Purdue University.

2. Based on T. N. Lam, "Estimating fuel consumption from engine size," *Journal of Transportation Engineering*, 111 (1985), pp. 339–357. The data for 10 to 50 km/h are measured; those for 60 and higher are calculated from a model given in the paper and are therefore smoothed.

3. A sophisticated treatment of improvements and additions to scatterplots is W. S. Cleveland and R. McGill, "The many faces of a scatterplot," *Journal of the American Statistical Association*, 79 (1984), pp. 807–822.

4. The data are a random sample of 53 from the 1079 pairs recorded by K. Pearson and A. Lee, "On the laws of inheritance in man," *Biometrika*, November 1903, p. 408.

5. Data from *Consumer Reports*, June 1986, pp. 366–367.

6. From a survey by the Wheat Industry Council reported in *USA Today*, October 20, 1983.

7. The data are from M. A. Houck, et al. "Allometric scaling in the earliest fossil bird, *Archaeopteryx lithographica*," *Science*, 247 (1990), pp. 195–198. The authors conclude from a variety of evidence that all specimens represent the same species.

8. The data are from W. L. Colville and D. P. McGill, "Effect of rate and method of planting on several plant characters and yield of irrigated corn," *Agronomy Journal*, 54 (1962), pp. 235–238.

9. Modified from M. C. Wilson and R. E. Shade, "Relative attractiveness of various luminescent colors to the cereal leaf beetle and the meadow spittlebug," *Journal of Economic Entomology*, 60 (1967), pp. 578–580.

10. A careful study of this phenomenon is W. S. Cleveland, P. Diaconis, and R. McGill, "Variables on scatterplots look more highly correlated when the scales are increased," *Science*, 216 (1982), pp. 1138–1141.

11. From W. M. Lewis and M. C. Grant, "Acid precipitation in the western United States," *Science*, 207 (1980), pp. 176–177.

12. Data from E. P. Hubble, "A relation between distance and radial velocity among extra-galactic nebulae," *Proceedings of the National Academy of Sciences*, 15 (1929), pp. 168–173.

13. Data from R. C. Nelson, C. M. Brooks, and N. L. Pike, "Biomechanical comparison of male and female distance runners," in P. Milvy (ed.), *The Marathon: Physiological, Medical, Epidemiological, and Psychological Studies*, New York Academy of Sciences, 1977, pp. 793–807.

14. These data were originally collected by L. M. Linde of UCLA but were first published by M. R. Mickey, O. J. Dunn, and V. Clark, "Note on the use of stepwise regression in detecting outliers," *Computers and Biomedical Research*, 1 (1967), pp. 105–111. The data have been used by several authors. I found them in N. R. Draper and J. A. John, "Influential observations and outliers in regression," *Technometrics*, 23 (1981), pp. 21–26.

15. The U.S. returns are for the Standard & Poor's 500 stock index. The overseas returns are for the Morgan Stanley Europe, Australia, Far East (EAFE) index.

16. Data provided by Peter Cook, Purdue University.

17. L. E. Moses and F. Mosteller, "Safety of anesthetics," in J. Tanur *et al.* (eds.), *Statistics: A Guide to the Unknown*, 3rd ed., Wadsworth, 1989, pp. 15–24.

18. This example is drawn from M. Goldstein, "Preliminary inspection of multivariate data," *The American Statistician*, 36 (1982), pp. 358–362.

19. From his column of June 28, 1993.

20. From S. V. Zagona (ed.), *Studies and Issues in Smoking Behavior*, University of Arizona Press, 1967, pp. 157–180.

21. From J. Stamler, "The mass treatment of hypertensive disease: defining the problem," *Mild Hypertension: To Treat or Not to Treat*, New York Academy of Sciences, 1978, pp. 333–358.

22. See P. J. Bickel and J. W. O'Connell, "Is there a sex bias in graduate admissions?" *Science*, 187 (1975), pp. 398–404.

23. From M. Radelet, "Racial characteristics and imposition of the death penalty," *American Sociological Review*, 46 (1981), pp. 918–927.

24. National Science Board, *Science and Engineering Indicators, 1991*, U.S. Government Printing Office, 1991. The detailed data appear in Appendix Table 3-5, p. 274.

25. Adapted from data of Williams and Zador in *Accident Analysis and Prevention*, 9 (1977), pp. 69–76.

26. From D. M. Barnes, "Breaking the cycle of addiction," *Science*, 241 (1988), pp. 1029–1030.

27. Modified from data provided by Dana Quade, University of North Carolina.

28. Data provided by Matthew Moore.

29. From National Center for Health Statistics, *Monthly Vital Statistics Report*, December 1983.

30. Data from D. E. Booth, *Regression Methods and Problem Banks*, COMAP, Inc., 1986.

31. Reported in the *New York Times*, July 20, 1989, from an article appearing that day in the *New England Journal of Medicine*.

32. The gas mileage data are from the U.S. Department of Energy's *1994 Gas Mileage Guide*, October 1993. In the table, the data are for the basic engine/transmission combination for each model, and models that are essentially identical (such as the Ford Taurus and Mercury Sable) appear only once.

SIR RONALD A. FISHER

The ideas and methods that we study as "statistics" were mostly invented in the nineteenth and twentieth centuries by people working on problems that required analysis of data. Astronomy, biology, social science, and even surveying can claim a role in the birth of statistics. But if anyone can claim to be "the father of statistics," that honor belongs to Sir Ronald A. Fisher (1890–1962).

Fisher's writings helped organize statistics as a distinct field of study whose methods apply to practical problems across many disciplines. He systematized the mathematical theory of statistics and invented many new techniques. But the randomized comparative experiment is perhaps Fisher's greatest contribution.

Like many statistical pioneers, Fisher was driven by the demands of practical problems. Beginning in 1919, he worked on agricultural field experiments at Rothamsted in England. How should we arrange the planting of different crop varieties or the application of different fertilizers to get a fair comparison among them? Because fertility and other variables change as we move across a field, experimenters used elaborate checkerboard planting arrangements to obtain fair comparisons. Fisher had a better idea: "arrange the plots deliberately at random."

This chapter explores statistical designs for producing data to answer specific questions like "Which crop variety has the highest mean yield?" Fisher's innovation, the deliberate use of chance in producing data, is the central theme of the chapter and one of the most important ideas in statistics.

CHAPTER **3**

Producing Data

INTRODUCTION

Exploratory data analysis seeks to discover and summarize what data say by using graphs and numerical summaries. The conclusions we draw from data analysis apply to the specific data that we examine. Often we want to extend those conclusions to some larger group of individuals. If our data don't fairly represent the larger group, our conclusions from the data don't apply to the larger group.

EXAMPLE 3.1

The advice columnist Ann Landers once asked her readers, "If you had it to do over again, would you have children?" A few weeks later, her column was headlined "70% OF PARENTS SAY KIDS NOT WORTH IT." Indeed, 70% of the nearly 10,000 parents who wrote in said they would not have children if they could make the choice again.

These data are worthless as indicators of opinion among all American parents. The people who responded felt strongly enough to take the trouble to write Ann Landers. Their letters showed that many of them were angry at their children. These people don't fairly represent all parents. It is not surprising that a statistically designed opinion poll on the same issue a few months later found that 91% of parents *would* have children again. Ann Landers announced a 70% "No" result when the truth about parents was close to 90% "Yes." ◄

sampling   Both Ann Landers' write-in poll and the more careful opinion poll that contradicted her are examples of **sampling**. The idea is to study a part—a sample—in order to gain information about an entire group. Ann Landers used a *voluntary response sample*.

> **VOLUNTARY RESPONSE SAMPLE**
>
> A **voluntary response sample** consists of people who choose themselves by responding to a general appeal.

Voluntary response samples overrepresent people with strong opinions, most often negative opinions. If sample data are to give reliable information about a larger group of people or things, we must think carefully about how to choose the sample.

experiment   In other settings, we gather data from an **experiment**. In doing an experiment, we don't just observe individuals or ask them questions. We actively impose some treatment in order to observe the response. If the data we gather are to be trustworthy, we must design the experiment carefully.

EXAMPLE 3.2

In 1940, a psychologist conducted an experiment to study the effect of propaganda on attitude toward a foreign government. He made up a test of attitude toward the German government and administered it to a group of American students. After reading German propaganda for several months, he tested the students again to see if their attitudes had changed.

Unfortunately, Germany attacked and conquered France while the experiment was in progress. The students did change their attitudes toward the German government between the test and the retest—but we shall never know how much of the change was due to the explanatory variable (reading propaganda) and how much to the historical events of that time. The data give no information about the effect of reading propaganda. ◀

In Example 3.2, the effects of the explanatory variable are hopelessly mixed up with the effects of the events of history. We say that the variables are *confounded*. Unless experiments are carefully designed, the effects of the explanatory variables can't be seen because of confounding with lurking variables.

CONFOUNDING

Two variables (explanatory variables or lurking variables) are **confounded** when their effects on a response variable cannot be distinguished from each other.

In both Examples 3.1 and 3.2, the purpose of producing data was to answer specific questions: What percent of parents would have children again? Does reading propaganda change attitudes toward a foreign gov-

statistical   ernment? **Statistical inference** provides ways to answer specific questions
inference   from data with some guarantee that the answers are good ones. Inference is the topic of much of the rest of this book. Voluntary response and

confounding can both make it impossible to answer the questions we had in mind. These examples make it clear that when we do statistical inference we must think about how to *produce* data as well as about how to *analyze* data. If we design our production of data well, inference will be straightforward and the results will be convincing. Section 3.1 presents statistical ideas for choosing samples that will yield trustworthy data. Section 3.2 discusses how to design experiments. The ideas in these sections are simple, but they are among the most important in statistics.

3.1   DESIGNING SAMPLES

An opinion poll wants to know what fraction of the public approves the president's performance in office. A quality engineer must estimate what fraction of the bearings rolling off an assembly line are defective. Government economists inquire about household income. In all these situations, we want to gather information about a large group of people or things. Time, cost, and inconvenience usually forbid inspecting every bearing or contacting every household. In such cases, we gather information about only part of the group in order to draw conclusions about the whole.

POPULATION, SAMPLE

The entire group of individuals that we want information about is called the **population**.

A **sample** is a part of the population that we actually examine in order to gather information.

---

sample design

convenience sampling

Notice that the population is defined in terms of our desire for knowledge. If we wish to draw conclusions about all U.S. college students, that group is our population even if only local students are available for questioning. The sample is the part from which we draw conclusions about the whole. The ***design*** of a sample refers to the method used to choose the sample from the population. Poor sample designs can produce misleading conclusions. Voluntary response (Example 3.1) is one common type of bad sample design. Another is ***convenience sampling***, which chooses the individuals easiest to reach. Here is an example of convenience sampling.

EXAMPLE 3.3

> Manufacturers and advertising agencies often use interviews at shopping malls to gather information about the habits of consumers and the effectiveness of ads. A sample of mall shoppers is fast and cheap. "Mall interviewing is being propelled primarily as a budget issue," one expert told the *New York Times*. But people contacted at shopping malls are not representative of the entire U.S. population. They are richer, for example, and more likely to be teenagers or retired. Moreover, mall interviewers tend to select neat, safe-looking individuals from the stream of customers. Decisions based on mall interviews may not reflect the preferences of all consumers.[1]   ◄

Both voluntary response samples and convenience samples choose a sample that is almost guaranteed not to represent the entire population. These sampling methods display *bias*, or systematic error, in favoring some parts of the population over others.

BIAS

The design of a study is **biased** if it systematically favors certain outcomes.

EXERCISES

3.1   A sociologist wants to know the opinions of employed adult women about government funding for day care. She obtains a list of the 520 members of a local business and professional women's club and mails a questionnaire to 100 of these women selected at random. Only 48 questionnaires are returned. What is the population in this study? What is the sample?

3.2   For each of the following sampling situations, identify the population as exactly as possible. That is, say what kind of individuals the population consists of and say exactly which individuals fall in the population. If the information given is not complete, complete the description of the population in a reasonable way.

(a) Each week, the Gallup Poll questions a sample of about 1500 adult U.S. residents to determine national opinion on a wide variety of issues.

(b) The 1990 census tried to gather basic information from every household in the United States. But a "long form" requesting much additional information was sent to a sample of about 17% of households.

(c) A machinery manufacturer purchases voltage regulators from a supplier. There are reports that variation in the output voltage of the regulators is affecting the performance of the finished products. To assess the quality of the supplier's production, the manufacturer sends a sample of 5 regulators from the last shipment to a laboratory for study.

3.3   A newspaper advertisement for *USA Today: The Television Show* once said:

Should handgun control be tougher? You call the shots in a special call-in poll tonight. If yes, call 1-900-720-6181. If no, call 1-900-720-6182. Charge is 50 cents for the first minute.

Explain why this opinion poll is almost certainly biased.

3.4   You are on the staff of a member of Congress who is considering a bill that would provide government-sponsored insurance for nursing home care. You report that 1128 letters have been received on the issue, of which 871 oppose the legislation. "I'm surprised that most of my constituents oppose the bill. I thought it would be quite popular," says the congresswoman. Are you convinced that a majority of the voters oppose the bill? How would you explain the statistical issue to the congresswoman?

Simple random samples

In a voluntary response sample, people choose whether to respond. In a convenience sample, the interviewer makes the choice. In both cases, personal choice produces bias. The statistician's remedy is to allow impersonal chance to choose the sample. A sample chosen by chance allows neither favoritism by the sampler nor self-selection by respondents. Choosing a sample by chance attacks bias by giving all individuals an equal chance to be chosen. Rich and poor, young and old, black and white, all have the same chance to be in the sample.

The simplest way to use chance to select a sample is to place names in a hat (the population) and draw out a handful (the sample). This is the idea of *simple random sampling*.

SIMPLE RANDOM SAMPLE

A **simple random sample (SRS)** of size n consists of n individuals from the population chosen in such a way that every set of n individuals has an equal chance to be the sample actually selected.

An SRS not only gives each individual an equal chance to be chosen (thus avoiding bias in the choice) but also gives every possible sample an equal chance to be chosen. There are other random sampling designs that give each individual, but not each sample, an equal chance. Exercise 3.22 describes one such design, called systematic random sampling.

The idea of an SRS is to choose our sample by drawing names from a hat. In practice, computer software can choose an SRS almost instantly from a list of the individuals in the population. If you don't use software, you can randomize by using a *table of random digits*.

RANDOM DIGITS

A **table of random digits** is a long string of the digits 0, 1, 2, 3, 4, 5, 6, 7, 8, 9 with these two properties:

1. Each entry in the table is equally likely to be any of the 10 digits 0 through 9.
2. The entries are independent of each other. That is, knowledge of one part of the table gives no information about any other part.

Table B at the back of the book and inside the rear cover is a table of random digits. You can think of Table B as the result of asking an assistant (or a computer) to mix the digits 0 to 9 in a hat, draw one, then replace the digit drawn, mix again, draw a second digit, and so on. The assistant's mixing and drawing save us the work of mixing and drawing when we need to choose at random. The entries in Table B appear in groups of five, but this is only to make the table easier to read. The rows are numbered, but this just makes it easier to say where you started in the table. These conveniences do not affect the nature of the table as a long string of randomly chosen digits. Because the digits in Table B are random:

- Each entry is equally likely to be any of the 10 possibilities 0, 1,..., 9.
- Each pair of entries is equally likely to be any of the 100 possible pairs 00, 01, ..., 99.
- Each triple of entries is equally likely to be any of the 1000 possibilities 000, 001, ..., 999, and so on.

These "equally likely" facts make it easy to use Table B to choose an SRS. Here is an example that shows how.

EXAMPLE 3.4

Joan's small accounting firm serves 30 business clients. Joan wants to interview a sample of 5 clients in detail to find ways to improve client satisfaction. To avoid bias, she chooses an SRS of size 5.

Step 1: Label. Give each client a numerical label, using as few digits as possible. Two digits are needed to label 30 clients, so we use labels

$$01, 02, 03, \ldots, 29, 30$$

It is also correct to use labels 00 to 29 or even another choice of 30 two-digit labels. Here is the list of clients, with labels attached.

01	A-1 Plumbing	02	Accent Printing
03	Action Sport Shop	04	Anderson Construction
05	Bailey Trucking	06	Balloons Inc.
07	Bennett Hardware	08	Best's Camera Shop
09	Blue Print Specialties	10	Central Tree Service
11	Classic Flowers	12	Computer Answers
13	Darlene's Dolls	14	Fleisch Realty
15	Hernandez Electronics	16	Johnson Commodities
17	JL Records	18	Keiser Construction
19	Liu's Chinese Restaurant	20	MagicTan
21	Peerless Machine	22	Photo Arts
23	River City Books	24	Riverside Tavern
25	Rustic Boutique	26	Satellite Services
27	Scotch Wash	28	Sewer's Center
29	Tire Specialties	30	Von's Video Store

Step 2: Table. Enter Table B anywhere and read two-digit groups. Suppose we enter at line 130, which is

69051   64817   87174   09517   84534   06489   87201   97245

The first 10 two-digit groups in this line are

69   05   16   48   17   87   17   40   95   17

Each successive two-digit group is a label. The labels 00 and 31 to 99 are not used in this example, so we ignore them. The first 5 labels between 01 and 30 that we encounter in the table choose our sample. Of the first 10 labels in line 130, we ignore 5 because they are too high (over 30). The others are 05, 16, 17, 17, and 17. The clients labeled 05, 16, and 17 go into the sample. Ignore the second and third 17s because that client is already in the sample. Now run your finger across line 130 (and continue to line 131 if needed) until 5 clients are chosen.

The sample is the clients labeled 05, 16, 17, 20, 19. These are Bailey Trucking, Johnson Commodities, JL Records, MagicTan, and Liu's Chinese Restaurant. ◀

CHOOSING AN SRS

Choose an SRS in two steps:

Step 1: Label. Assign a numerical label to every individual in the population.

Step 2: Table. Use Table B to select labels at random.

Don't try to scramble the labels as you assign them. Table B will do the required randomizing. You can assign labels in any convenient manner, such as alphabetical order for names of people. Be certain that all labels have the same number of digits. Only then will all individuals have the same chance to be chosen. Use the shortest possible labels: one digit for a population of up to 10 members, 2 digits for 11 to 100 members, three digits for 101 to 1000 members, and so on. As standard practice, we recommend that you begin with label 1 (or 01, or 001, as needed). You can read digits from Table B in any order—across a row, down a column, and so on—because the table has no order. As standard practice, we recommend reading across rows.

3.5   A firm wants to understand the attitudes of its minority managers toward its system for assessing management performance. Below is a list of all the firm's managers who are members of minority groups. Use Table B at line 139 to choose 6 to be interviewed in detail about the performance appraisal system.

Agarwal	Gates	Peters
Anderson	Goel	Pliego
Baxter	Gomez	Puri
Bonds	Hernandez	Richards
Bowman	Huang	Rodriguez
Castillo	Kim	Santiago
Cross	Liao	Shen
Dewald	Mourning	Vega
Fernandez	Naber	Wang
Fleming		

3.6   Your class in ancient Ugaritic religion is poorly taught and wants to complain to the dean. The class decides to choose 4 of its members at random to carry the complaint. The class list appears below. Choose an SRS of 4 using the table of random digits, beginning at line 145.

Anderson	Gupta	Patnaik
Aspin	Gutierrez	Pirelli
Bennett	Harter	Rao
Bock	Henderson	Rider
Breiman	Hughes	Robertson
Castillo	Johnson	Rodriguez
Dixon	Kempthorne	Sosa
Edwards	Laskowsky	Tran
Gonzalez	Liang	Trevino
Green	Olds	Wang

3.7   You must choose an SRS of 10 of the 440 retail outlets in New York that sell your company's products. How would you label this population? Use Table B, starting at line 105, to choose your sample.

Other sampling designs

The general framework for sampling is a *probability sample*.

> **PROBABILITY SAMPLE**
>
> A **probability sample** gives each member of the population a known chance (greater than zero) to be selected.

Some probability sampling designs (such as an SRS) give each member of the population an equal chance to be selected. This may not be true in more elaborate sampling designs. In every case, however, the use of chance to select the sample is the essential principle of statistical sampling.

Designs for sampling from large populations spread out over a wide area are usually more complex than an SRS. For example, it is common to sample important groups within the population separately, then combine these samples. This is the idea of a *stratified sample*.

> **STRATIFIED RANDOM SAMPLE**
>
> To select a **stratified random sample**, first divide the population into groups of similar individuals, called **strata**. Then choose a separate SRS in each stratum and combine these SRSs to form the full sample.

Choose the strata based on facts known before the sample is taken. For example, a population of election districts might be divided into urban, suburban, and rural strata. A stratified design can produce more exact information than an SRS of the same size by taking advantage of the fact that individuals in the same stratum are similar to one another. If all individuals in each stratum are identical, for example, just one individual from each stratum is enough to completely describe the population.

EXAMPLE 3.5

A radio or television station that broadcasts a piece of music owes a royalty to the composer. ASCAP (the American Society of Composers, Authors and

Publishers) sells licenses that permit broadcast of works by any of its members. ASCAP must then pay the proper royalties to the composers whose music was played. Television networks keep program logs of all music played, but local radio and television stations do not. Because there are over a billion ASCAP-licensed performances each year, a detailed accounting is too expensive and cumbersome. Here is a case for sampling.

ASCAP divides its royalties among its members by taping a stratified sample of broadcasts. The sample of local commercial radio stations, for example, consists of 60,000 hours of broadcast time each year. Radio stations are stratified by type of community (metropolitan, rural), geographic location (New England, Pacific, etc.), and the size of the license fee paid to ASCAP, which reflects the size of the audience. In all, there are 432 strata. Tapes are made at random hours for randomly selected members of each stratum. The tapes are reviewed by experts who can recognize almost every piece of music ever written, and the composers are then paid according to their popularity.[2]   ◀

Another common means of restricting random selection is to choose the sample in stages. This is usual practice for national samples of households or people. For example, government data on employment and unemployment are gathered by the Current Population Survey, which conducts interviews in about 60,000 households each month. It is not practical to maintain a list of all U.S. households from which to select an SRS. Moreover, the cost of sending interviewers to the widely scattered households in an SRS would be too high. The Current Population Survey therefore uses a ***multistage sample design***. The final sample consists of clusters of nearby households. Most opinion polls and other national samples are also multistage, though interviewing in most national samples today is done by telephone rather than in person, eliminating the economic need for clustering.

multistage
sample

A national multistage sample proceeds somewhat as follows:

Stage 1. Take a sample from the 3000 counties in the United States.

Stage 2. Select a sample of townships within each of the counties chosen.

Stage 3. Select a sample of blocks within each chosen township.

Stage 4. Take a sample of households within each block.

Lists of counties and townships are easily available. From that point, the sampling can be based on a local map if necessary. We don't need a national list of households. Moreover, the sample households occur in clusters in the same block and so are easy for a single interviewer to contact. The sample at any stage of a multistage design may be an SRS. Stratified samples are also common—for example, we might group counties into rural, suburban, and urban strata before sampling.

Analysis of data from sampling designs more complex than an SRS takes us beyond basic statistics. But the SRS is the building block of more elaborate designs, and analysis of other designs differs more in complexity of detail than in fundamental concepts.

EXERCISES

3.8 A club has 30 student members and 10 faculty members. The students are

Abel	Fisher	Huber	Miranda	Reinmann
Carson	Ghosh	Jimenez	Moskowitz	Santos
Chen	Griswold	Jones	Neyman	Shaw
David	Hein	Kim	O'Brien	Thompson
Deming	Hernandez	Klotz	Pearl	Utts
Elashoff	Holland	Liu	Potter	Varga

The faculty members are

Andrews	Fernandez	Kim	Moore	West
Besicovitch	Gupta	Lightman	Phillips	Yang

The club can send 4 students and 2 faculty members to a convention. It decides to choose those who will go by random selection. Use Table B to choose a stratified random sample of 4 students and 2 faculty members.

3.9 Accountants often use stratified samples during audits to verify a company's records of such things as accounts receivable. The stratification is based on the dollar amount of the item and often includes 100% sampling of the largest items. One company reports 5000 accounts receivable. Of these, 100 are in amounts over $50,000; 500 are in amounts between $1000 and $50,000; and the remaining 4400 are in amounts under $1000. Using these groups as strata, you decide to verify all of the largest accounts and to sample 5% of the midsize accounts and 1% of the small accounts. How would you label the two strata from which you will sample? Use Table B, starting at line 115, to select *only the first 5* accounts from each of these strata.

Cautions about sample surveys

Random selection eliminates bias in the choice of a sample from a list of the population. When the population consists of human beings, however, accurate information from a sample requires much more than a good sampling design.[3] To begin, we need an accurate and complete list of the population. Because such a list is rarely available, most samples suffer from some degree of *undercoverage*. A sample survey of households, for example, will miss not only homeless people but prison inmates and students in dormitories. An opinion poll conducted by telephone will miss the 7% to 8% of American households without residential phones. The results of national sample surveys therefore have some bias if the people not covered—who most often are poor people—differ from the rest of the population.

A more serious source of bias in most sample surveys is *nonresponse*, which occurs when a selected individual cannot be contacted or refuses to cooperate. Nonresponse to sample surveys often reaches 30% or more, even with careful planning and several callbacks. Because nonresponse is higher in urban areas, most sample surveys substitute other people in the same area to avoid favoring rural areas in the final sample. If the people contacted differ from those who are rarely at home or who refuse to answer questions, some bias remains.

UNDERCOVERAGE AND NONRESPONSE

Undercoverage occurs when some groups in the population are left out of the process of choosing the sample.

Nonresponse occurs when an individual chosen for the sample can't be contacted or refuses to cooperate.

EXAMPLE 3.6

Even the 1990 census, backed by the resources and authority of the federal government, suffered from undercoverage and nonresponse. The census begins by mailing forms to every household in the country. The Census Bureau buys lists of addresses from private firms, then tries to fill in missing addresses. The final list is still incomplete, resulting in undercoverage. Despite special efforts to count homeless people (who can't be reached at any address), homelessness causes more undercoverage.

About 35% of the households who were mailed census forms did not mail them back. In New York City, 47% did not return the form. That's nonresponse. The Census Bureau sends interviewers to these households. In central cities, the interviewers could not contact about one in five of the nonresponders, even after six tries.

The Census Bureau estimates that the 1990 census missed about 1.6% of the total population due to undercoverage and nonresponse. Because the undercount was greater in the poorer sections of large cities, the Census Bureau estimates that it failed to count 4.6% of blacks and 5.0% of Hispanics.[4]   ◀

response bias

In addition, the behavior of the respondent or of the interviewer can cause **response bias** in sample results. Respondents may lie, especially if asked about illegal or unpopular behavior. The sample then underestimates the presence of such behavior in the population. An interviewer whose attitude suggests that some answers are more desirable than others will get these answers more often. The race or sex of the interviewer can influence responses to questions about race relations or attitudes toward feminism. Answers to questions that ask respondents to recall past events are often inaccurate because of faulty memory. For example, many people "telescope" events in the past, bringing them forward in memory to more recent time periods. "Have you visited a dentist in the last 6 months?" will often draw a "Yes" from someone who last visited a dentist 8 months ago.[5] Careful training of interviewers and careful supervision to avoid variation among the interviewers can greatly reduce response bias. Good interviewing technique is another aspect of a well-done sample survey.

wording effects

The **wording of questions** is the most important influence on the answers given to a sample survey. Confusing or leading questions can introduce strong bias, and even minor changes in wording can change a survey's outcome. Leading questions are common in surveys sponsored by companies and intended to persuade rather than inform. Here are two examples.[6]

EXAMPLE 3.7

When Levi Strauss & Co. asked college students to choose the most popular clothing item from a list, 90% chose Levi's 501 jeans—but they were the only jeans listed.

A survey paid for by makers of disposable diapers found that 84% of the sample opposed banning disposable diapers. Here is the actual question:

It is estimated that disposable diapers account for less than 2% of the trash in today's landfills. In contrast, beverage containers, third-class mail and yard wastes are estimated to account for about 21% of the trash in landfills. Given this, in your opinion, would it be fair to ban disposable diapers?

> This question gives information on only one side of an issue, then asks an opin-
> ion. That's a sure way to bias the responses. A different question that described
> how long disposable diapers take to decay and how many tons they contribute
> to landfills each year would draw a quite different response.   ◀

Never trust the results of a sample survey until you have read the exact questions posed. The sampling design, the amount of nonresponse, and the date of the survey are also important. Good statistical design is a part, but only a part, of a trustworthy survey.

EXERCISES

3.10 The list of individuals from which a sample is actually selected is called the *sampling frame*. Ideally, the frame should list every individual in the population, but in practice this is often difficult. A frame that leaves out part of the population is a common source of undercoverage.

(a) Suppose that a sample of households in a community is selected at random from the telephone directory. What households are omitted from this frame? What types of people do you think are likely to live in these households? These people will probably be underrepresented in the sample.

(b) It is more common in telephone surveys to use random digit dialing equipment that selects the last four digits of a telephone number at random after being given the exchange (the first three digits). Which of the households you mentioned in your answer to (a) will be included in the sampling frame by random digit dialing?

3.11 A common form of nonresponse in telephone surveys is "ring-no-answer." That is, a call is made to an active number but no one answers. The Italian National Statistical Institute looked at nonresponse to a government survey of households in Italy during the periods January 1 to Easter and July 1 to August 31. All calls were made between 7 and 10 p.m., but 21.4% gave "ring-no-answer" in one period versus 41.5% "ring-no-answer" in the other period.[7] Which period do you think had the higher rate of no answers? Why? Explain why a high rate of nonresponse makes sample results less reliable.

3.12 Here are two wordings for the same question:

A. Should laws be passed to eliminate all possibilities of special interests giving huge sums of money to candidates?

B. Should laws be passed to prohibit interest groups from contributing to campaigns, or do groups have a right to contribute to the candidates they support?

One of these questions drew 40% favoring banning contributions; the other drew 80% with this opinion.[8] Which question produced the 40% and which got 80%? Explain why the results were so different.

Inference about the population

Despite the many practical difficulties in carrying out a sample survey, using chance to choose a sample does eliminate bias in the actual selection of the sample from the list of available individuals. But it is unlikely that results from a sample are exactly the same as for the entire population. Sample results, like the official unemployment rate obtained from the monthly Current Population Survey, are only estimates of the truth about the population. If we select two samples at random from the same population, we will draw different individuals. So the sample results will almost certainly differ somewhat. Two runs of the Current Population Survey would produce somewhat different unemployment rates. Properly designed samples avoid systematic bias, but their results are rarely exactly correct and they vary from sample to sample.

probability   How accurate is a sample result like the monthly unemployment rate? We can't say for sure, because the result would be different if we took another sample. But the results of random sampling don't change haphazardly from sample to sample. Because we deliberately use chance, the results obey the laws of **probability** that govern chance behavior. We can say how large an error we are likely to make in drawing conclusions about the population from a sample. Results from a sample survey usually come with a margin of error that sets bounds on the size of the likely error. How to do this is part of the business of statistical inference. We will describe the reasoning in Chapter 5.

One point is worth making now: *larger samples give more accurate results than smaller samples*. By taking a very large sample, you can be confident that the sample result is very close to the truth about the population. The Current Population Survey's sample of 60,000 households estimates the national unemployment rate very accurately. Of course, only probability samples carry this guarantee. Ann Landers' voluntary response sample is worthless even though 10,000 people wrote in. Using a probability sampling design and taking care to deal with practical difficulties reduce bias in a sample. The size of the sample then determines how close to the population truth the sample result is likely to fall.

EXERCISE

3.13   Just before a presidential election, a national opinion polling firm increases the size of its weekly sample from the usual 1500 people to 4000 people. Why do you think the firm does this?

SUMMARY

A sample survey selects a **sample** from the **population** of all individuals about which we desire information. We base conclusions about the population on data about the sample.

The **design** of a sample refers to the method used to select the sample from the population. **Probability sampling designs** use random selection to give each member of the population a known chance (greater than zero) to be selected for the sample.

The basic probability sample is a **simple random sample (SRS)**. An SRS gives every possible sample of a given size the same chance to be chosen.

Choose an SRS by labeling the members of the population and using a **table of random digits** to select the sample.

To choose a **stratified random sample**, divide the population into **strata**, groups of individuals that are similar in some way that is important to the response. Then choose a separate SRS from each stratum.

Multistage samples select successively smaller groups within the population in stages. Each stage may employ an SRS, a stratified sample, or another type of sample.

Failure to use probability sampling often results in **bias**, or systematic errors in the way the sample represents the population. **Voluntary response samples**, in which the respondents choose themselves, are particularly prone to large bias.

In human populations, even probability samples can suffer from bias due to **undercoverage** or **nonresponse**, from **response bias**, or from misleading results due to **poorly worded questions**. Sample surveys must deal expertly with these potential problems in addition to using a probability sampling design.

SECTION 3.1 EXERCISES

3.14   Different types of writing can sometimes be distinguished by the lengths of the words used. A student interested in this fact wants to study the lengths of words used by Tom Wolfe in his novels. She opens a Wolfe novel at random and records the lengths of each of the first 250 words on the page.

What is the population in this study? What is the sample? What is the variable measured?

3.15   For each of the following sampling situations, identify the population as exactly as possible. That is, say what kind of individuals the population consists of and say exactly which individuals fall in the population. If the information given is not complete, complete the description of the population in a reasonable way.

(a) A business school researcher wants to know what factors affect the survival and success of small businesses. She selects a sample of 150 eating-and-drinking establishments from those listed in the telephone directory Yellow Pages for a large city.

(b) A member of Congress wants to know whether his constituents support proposed legislation on health care. His staff reports that 228 letters have been received on the subject, of which 193 oppose the legislation.

(c) An insurance company wants to monitor the quality of its procedures for handling loss claims from its auto insurance policyholders. Each month the company selects an SRS of all auto insurance claims filed that month to examine them for accuracy and promptness.

3.16   The author Shere Hite undertook a study of women's attitudes toward sex and love by distributing 100,000 questionnaires through women's groups. Only 4.5% of the questionnaires were returned. Based on this sample of women, Hite wrote *Women and Love*, a best-selling book claiming that women are fed up with men. For example, 91% of the divorced women who responded said that they had initiated the divorce, and 70% of the married women said that they had committed adultery.

Explain briefly why Hite's sampling method is nearly certain to produce a strong bias. Are the sample results cited (91% and 70%) much higher or much lower than the truth about the population of all adult American women?

3.17   Some television stations take quick polls of public opinion by announcing a question on the air and asking viewers to call one of two telephone numbers to register their opinion as "Yes" or "No." Telephone companies make available "900" numbers for this purpose. Dialing a 900 number results in a small charge to your telephone bill. The first major use of call-in polling was by the ABC television network in October 1980. At the end of the first Reagan-Carter presidential election debate, ABC asked its viewers which candidate won. The call-in poll proclaimed that Reagan had won the debate by a 2 to 1 margin. But a random survey by CBS News showed only a 44% to 36% margin for Reagan, with the rest undecided. Why are call-in polls likely to be biased? Can you suggest why this bias might have favored the Republican Reagan over the Democrat Carter?

3.18   A flour company wants to know what fraction of Toronto households bake some or all of their own bread. The company selects an SRS of 500 residential addresses in Toronto and sends interviewers to these addresses. The interviewers visit households only during regular working hours on weekdays. Explain why this sampling method is biased. Is the percent of the sample who bake bread probably higher or lower than the percent of the population who bake bread?

3.19   A manufacturer of chemicals chooses 3 from each lot of 25 containers of a reagent to test for purity and potency. Below are the control numbers stamped on the bottles in the current lot. Use Table B at line 111 to choose an SRS of 3 of these bottles.

A1096	A1097	A1098	A1101	A1108
A1112	A1113	A1117	A2109	A2211
A2220	B0986	B1011	B1096	B1101
B1102	B1103	B1110	B1119	B1137
B1189	B1223	B1277	B1286	B1299

3.20   Figure 3.1 is map of a census tract in Cedar Rapids, Iowa. Census tracts are small, homogeneous areas averaging 4000 in population. On the map, each block is marked with a Census Bureau identification number. An SRS of blocks from a census tract is often the next-to-last stage in a multistage sample. Use Table B beginning at line 125 to choose an SRS of 5 blocks from this census tract.

3.21   Which of the following statements are true of a table of random digits, and which are false? Briefly explain your answers.

(a) There are exactly four 0s in each row of 40 digits.

(b) Each pair of digits has chance 1/100 of being 00.

(c) The digits 0000 can never appear as a group, because this pattern is not random.

3.22   Sample surveys often use a *systematic random sample* to choose a sample of apartments in a large building or dwelling units in a block at the last stage of a multistage sample. An example will illustrate the idea of a systematic sample.

    Suppose that we must choose 4 addresses out of 100. Because 100/4 = 25, we can think of the list as four lists of 25 addresses. Choose 1 of the first 25 addresses at random using Table B. The sample contains this address and the addresses 25, 50, and 75 places down the list from it. If the table

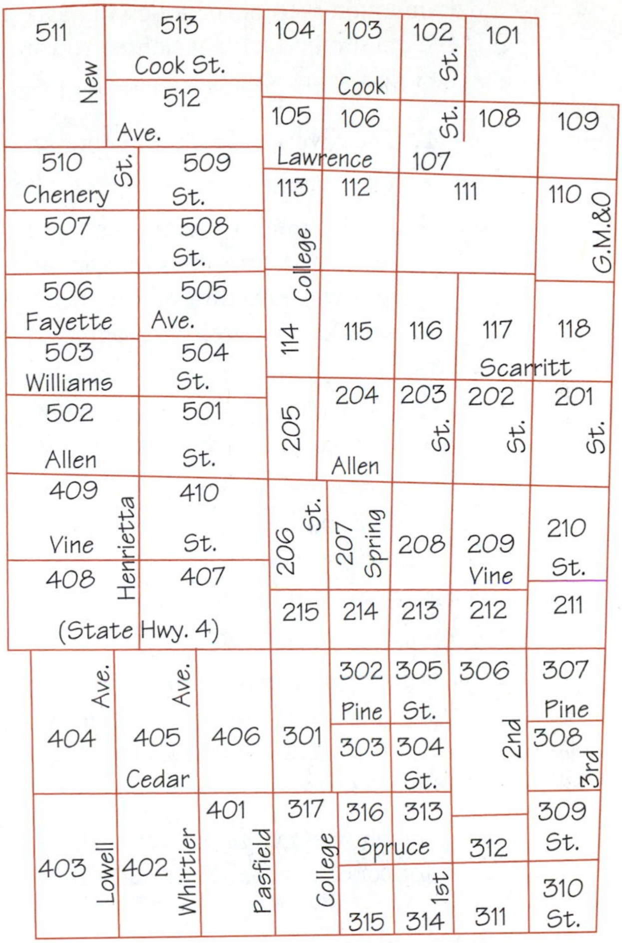

FIGURE 3.1   Map of a census tract in Cedar Rapids, Iowa, for Exercise 3.20.

gives 13, for example, then the systematic random sample consists of the addresses numbered 13, 38, 63, and 88.

(a) Use Table B to choose a systematic random sample of 5 addresses from a list of 200. Enter the table at line 120.

(b) Like an SRS, a systematic random sample gives all individuals the same chance to be chosen. Explain why this is true. Then explain carefully why a systematic sample is nonetheless *not* an SRS.

3.23   A corporation employs 2000 male and 500 female engineers. A stratified random sample of 200 male and 50 female engineers gives each engineer 1 chance in 10 to be chosen. This sample design gives every individual in the population the same chance to be chosen for the sample. Is it an SRS? Explain your answer.

3.24 Comment on each of the following as a potential sample survey question. Is the question clear? Is it slanted toward a desired response?

(a) Which of the following best represents your opinion on gun control?

1. The government should confiscate our guns.

2. We have the right to keep and bear arms.

(b) A freeze in nuclear weapons should be favored because it would begin a much-needed process to stop everyone in the world from building nuclear weapons now and reduce the possibility of nuclear war in the future. Do you agree or disagree?

(c) In view of escalating environmental degradation and incipient resource depletion, would you favor economic incentives for recycling of resource-intensive consumer goods?

3.25 A *New York Times* opinion poll on women's issues contacted a sample of 1025 women and 472 men by randomly selecting telephone numbers. The *Times* publishes complete descriptions of its polling methods. Here is part of the description for this poll.[9]

> In theory, in 19 cases out of 20 the results based on the entire sample will differ by no more than three percentage points in either direction from what would have been obtained by seeking out all adult Americans.
>
> The potential sampling error for smaller subgroups is larger. For example, for men it is plus or minus five percentage points.

Explain why the margin of error is larger for conclusions about men alone than for conclusions about all adults.

3.2 DESIGNING EXPERIMENTS

A sample survey collects information about a population by selecting and measuring a sample from the population. The goal is a picture of the population, disturbed as little as possible by the act of gathering information. Sample surveys are one kind of *observational study*.

OBSERVATION AND EXPERIMENT

An **observational study** observes individuals and measures variables of interest but does not attempt to influence the responses.

An **experiment**, on the other hand, deliberately imposes some treatment on individuals in order to observe their responses.

An observational study, even one based on a sound statistical sample, is a poor way to gauge the effect of an intervention. To see how nature responds to a change, we must actually impose the change. When our goal is to understand cause and effect, experiments are the only source of fully convincing data. Here is the basic vocabulary of experiments.

> ### EXPERIMENTAL UNITS, SUBJECTS, TREATMENT
>
> The individuals on which the experiment is done are the **experimental units**. When the units are human beings, they are called **subjects**. A specific experimental condition applied to the units is called a **treatment**.

factor

level

Because the purpose of an experiment is to reveal the response of one variable to changes in other variables, the distinction between explanatory and response variables is essential. The explanatory variables in an experiment are often called *factors*. Many experiments study the joint effects of several factors. In such an experiment, each treatment is formed by combining a specific value (often called a *level*) of each of the factors.

EXAMPLE 3.8

Researchers studying the absorption of a drug into the bloodstream inject the drug (the treatment) into 25 people (the subjects). The response variable is the concentration of the drug in a subject's blood, measured 30 minutes after the injection. This experiment has a single factor with only one level. If three different doses of the drug are injected, there is still a single factor (the dosage of the drug), now with three levels. The three levels of the single factor are the treatments that the experiment compares.   ◀

EXAMPLE 3.9

What are the effects of repeated exposure to an advertising message? The answer may depend both on the length of the ad and on how often it is repeated. An experiment investigated this question using undergraduate students as subjects. All subjects viewed a 40-minute television program that included ads for a 35 mm camera. Some subjects saw a 30-second commercial; others, a 90-second version. The same commercial was repeated either 1, 3, or 5 times during the program. After viewing, all of the subjects answered questions about their recall of the ad, their attitude toward the camera, and their intention to purchase it. These are the response variables.[10]

This experiment has two factors: length of the commercial, with 2 levels, and repetitions, with 3 levels. The 6 combinations of one level of each factor form 6 treatments. Figure 3.2 shows the layout of the treatments.   ◀

Examples 3.8 and 3.9 illustrate the advantages of experiments over observational studies. Experimentation allows us to study the effects of the specific treatments we are interested in. Moreover, we can control the environment of the experimental units to hold constant factors that are of no interest to us, such as the specific product advertised in Example 3.9. The ideal case is a laboratory experiment in which we control all outside factors. Like most ideals, such control is not always realized in practice. Nonetheless, a well-designed experiment makes it possible to draw conclusions about the effect of one variable on another.

Another advantage of experiments is that we can study the combined effects of several factors simultaneously. The interaction of several factors can produce effects that could not be predicted from looking at the effect of each factor alone. Perhaps longer commercials increase interest in a product, and more commercials also increase interest, but if we both make a commercial longer and show it more often, viewers get annoyed and their interest in the product drops. The two-factor experiment in Example 3.9 will help us find out.

| | | Factor B Repetitions | | |
		1 time	3 times	5 times
Factor A Length	30 seconds	1	2	3
	90 seconds	4	5	6

FIGURE 3.2   The treatments in the experimental design of Example 3.9. Combinations of levels of the two factors form six treatments.

3.26 There may be a "gender gap" in political party preference in the United States, with women more likely than men to prefer Democratic candidates. A political scientist selects a large sample of registered voters, both men and women. She asks each voter whether they voted for the Democratic or the Republican candidate in the last congressional election. Is this study an experiment? Why or why not? What are the explanatory and response variables?

3.27 A manufacturer of food products uses package liners that are sealed at the top by applying heated jaws after the package is filled. The customer peels the sealed pieces apart to open the package. What effect does the temperature of the jaws have on the force needed to peel the liner? To answer this question, engineers obtain 20 pairs of pieces of package liner. They seal five pairs of each at 250° F, 275° F, 300° F, and 325° F. Then they measure the force needed to peel each seal.

(a) What are the experimental units?

(b) There is one factor (explanatory variable). What is it, and what are its levels?

(c) What is the response variable?

3.28 An educator wants to compare the effectiveness of computer software that teaches reading with that of a standard reading curriculum. She tests the reading ability of each student in a class of fourth graders, then divides them into two groups. One group uses the computer regularly, while the other studies a standard curriculum. At the end of the year, she retests all the students and compares the increase in reading ability in the two groups.

(a) Is this an experiment? Why or why not?

(b) What are the explanatory and response variables?

3.29 A chemical engineer is designing the production process for a new product. The chemical reaction that produces the product may have higher or lower yield, depending on the temperature and the stirring rate in the vessel in which the reaction takes place. The engineer decides to investigate the effects of combinations of two temperatures (50° C and 60° C) and three stirring rates (60 rpm, 90 rpm, and 120 rpm) on the yield of the process. She will process two batches of the product at each combination of temperature and stirring rate.

(a) What are the experimental units and the response variable in this experiment?

(b) How many factors are there? How many treatments? Use a diagram like that in Figure 3.2 to lay out the treatments.

(c) How many experimental units are required for the experiment?

Comparative experiments

The design of an experiment first describes the response variable or variables, the factors (explanatory variables), and the specific treatments. Figure 3.2 illustrates this aspect of the design of a marketing experiment. Laboratory experiments in the sciences and engineering often have a simple design with only one treatment, which is applied to all of the units. The design of such an experiment looks like this

$$\textbf{Treatment} \longrightarrow \textbf{Observation}$$

or, if before-and-after measurements are made, like this

$$\textbf{Observation 1} \longrightarrow \textbf{Treatment} \longrightarrow \textbf{Observation 2}$$

For example, we may subject a beam to a load (treatment) and measure how far it bends (observation). Alas, when experiments are conducted in the field or with living subjects, these simple designs often yield worthless data because of confounding. Example 3.2 (page 179) shows what can go wrong. Here is another example.

EXAMPLE 3.10

Ulcers in the upper intestine are unfortunately common in modern society. "Gastric freezing" is a clever treatment for ulcers. The patient swallows a deflated balloon with tubes attached, then a refrigerated solution is pumped through the balloon for an hour. The idea is that cooling the stomach will reduce its production of acid and so relieve ulcers. An experiment reported in the *Journal of the American Medical Association* showed that gastric freezing did reduce acid production and relieve ulcer pain. The treatment was safe and easy and was widely used for several years.

placebo effect

The gastric freezing experiment was poorly designed. The patients' response may have been due to the ***placebo effect***. A placebo is a dummy treatment that can have no physical effect. Many patients respond favorably to *any* treatment, even a placebo, presumably because of trust in the doctor and expectations of a cure. This response to a dummy treatment is the placebo effect.

A second experiment, done several years later, divided ulcer patients into two groups. One group was treated by gastric freezing as before. The other group received a placebo treatment in which the solution in the balloon was at body temperature rather than freezing. The results: 34% of the 82 patients in the treatment group improved, but so did 38% of the 78 patients in the placebo group. This and other properly designed experiments showed that gastric freezing was no better than a placebo, and its use was abandoned.[11]   ◀

The original gastric freezing experiment had the simple design

$$\text{Observe pain} \; \longrightarrow \; \text{Gastric freezing} \; \longrightarrow \; \text{Observe pain}$$

The experimental data were misleading because of the placebo effect. The data also reflect any special features of that particular study, such as a physician with a soothing manner. The effect of gastric freezing is confounded with these lurking variables.

Fortunately, the remedy is simple. Experiments should *compare* treatments rather than attempt to assess a single treatment in isolation. When we compare the two groups of patients in the second gastric freezing experiment, the placebo effect and other lurking variables operate on both groups. The only difference between the groups is the actual effect of gastric freezing. The group of patients who received a sham treatment is called a **control group**, because it enables us to control the effects of lurking variables on the outcome. *Control of the effects of lurking variables is the first principle of statistical design of experiments.* Comparison of several treatments is the simplest form of control.

control group

Without comparison of treatments, experimental results in such areas as consumer behavior and medicine can be dominated by the details of the experimental arrangement, the selection of subjects, and the placebo effect. The result is often **bias**, systematic favoritism toward one outcome. An uncontrolled study of a new ulcer treatment, for example, is biased in favor of finding the treatment effective because of the placebo effect. It should not surprise you to learn that uncontrolled studies in medicine give new therapies a much higher success rate than proper comparative experiments.

bias

Completely randomized experiments

The first step in an experimental design is the choice of treatments, with comparison as the leading principle. Next, we must say how we will assign the experimental units to the treatments. Comparison of the effects of several treatments is valid only if we apply all treatments to similar groups of experimental units. If one corn variety is planted on more fertile ground, or if one TV commercial is shown to more receptive subjects, comparisons among several corn varieties or among several commercials are meaningless. Systematic differences among the groups of experimental units in a comparative experiment are a possible source of bias. How should we allocate the available units or subjects among the treatments?

matching   Experimenters often try to **match** the treatment groups in a systematic way. A comparison of two TV ads, for example, may try to match the subjects in the experimental group that will see the new ad with the subjects in the control group that will see the current ad by age, sex, race, and so on. If there is a 27-year-old black female accountant in one group, there should be a matching young black female professional in the other group. Attempts at matching are helpful but not adequate—there are too many lurking variables that might affect the outcome. The experimenter can't measure some of these variables and will not think of others until after the experiment.

The remedy is to use impersonal chance to make the assignment. Then the groups don't depend on any characteristic of the experimental units or on the judgment of the experimenter. The use of chance can be combined with matching, but the simplest design creates groups by chance alone. Here is an example.

EXAMPLE 3.11

A food company assesses the nutritional quality of a new "instant breakfast" product by feeding it to newly weaned male white rats. The response variable is a rat's weight gain over a 28-day period. A control group of rats eats a standard diet but otherwise receives exactly the same treatment as the experimental group.

This experiment has one factor (the diet) with two levels. The researchers use 30 rats for the experiment and so must divide them into two groups of 15. To do this in an unbiased fashion, put the cage numbers of the 30 rats in a hat, mix them up, and draw 15. These rats form the experimental group and the remaining 15 make up the control group. That is, *each group is an SRS of the available rats.*

In practice, we use the table of random digits to randomize. Label the rats 01 to 30. Enter Table B at (say) line 130. Run your finger along this line (and continue to lines 131 and 132 as needed) until 15 rats are chosen. They are the rats labeled

05, 16, 17, 20, 19, 04, 25, 29, 18, 07, 13, 02, 23, 27, 21

These rats form the experimental group; the remaining 15 are the control group.

◄

The use of chance to divide experimental units into groups is called *randomization*. Randomization is the second major principle of statistical de-sign of experiments. Combining comparison and randomization, we arrive at the simplest randomized comparative design:

randomization

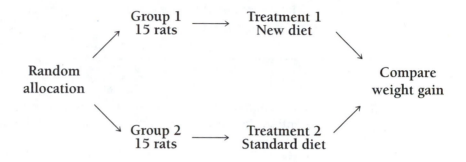

EXAMPLE 3.12

Many utility companies have introduced programs to encourage energy con-servation among their customers. An electric company considers placing elec-tronic indicators in households to show what the cost would be if the electricity use at that moment continued for a month. Will indicators reduce electricity use? Would cheaper methods work almost as well? The company decides to design an experiment.

One cheaper approach is to give customers a chart and information about mon-itoring their electricity use. The experiment compares these two approaches (indicator, chart) and also a control. The control group of customers receives information about energy conservation but no help in monitoring electricity use. The response variable is total electricity used in a year. The company finds 60 single-family residences in the same city willing to participate, so it assigns 20 residences at random to each of the 3 treatments. The outline of the design is

Group 1
20 houses ⟶ Indicator

Random
allocation ⟶ Group 2
20 houses ⟶ Chart ⟶ Compare
electricity used

Group 3
20 houses ⟶ Control

To carry out the random assignment, label the 60 households 01 to 60. Enter Table B and select an SRS of 20 to receive the indicators. Continue in Table B, selecting 20 more to receive charts. The remaining 20 form the control group.

◀

The designs used for the experiments in Examples 3.11 and 3.12 are *completely randomized*.

COMPLETELY RANDOMIZED DESIGN

In a **completely randomized** experimental design, all the experimental units are allocated at random among all the treatments.

Completely randomized designs can compare any number of treatments. To do this, assign the experimental units at random to as many groups as there are treatments, as in Example 3.12. The diagrams of the experiments in Examples 3.11 and 3.12 display the randomization and comparison used in the design. They also display other essential information: the treatments, the response variable, and the number of units in each group. We will see later that for statistical inference there are advantages to having equal numbers of units in all groups. You should choose groups of equal size unless there is some compelling reason not to do so.

Examples 3.11 and 3.12 involve completely randomized experimental designs to compare levels of a single factor. In Example 3.11, the factor is the diet fed to the rats. In Example 3.12, it is the method used to encourage energy conservation. Completely randomized designs can have more than one factor. The advertising experiment of Example 3.9 has two factors: the length and the number of repetitions of a television commercial. Their combinations form the six treatments outlined in Figure 3.2 (page 200). A completely randomized design assigns subjects at random to these six treatments. Once the layout of treatments is set, the randomization needed for a completely randomized design is tedious but straightforward.

EXERCISES

3.30   A large study used records from Canada's national health care system to compare the effectiveness of two ways to treat prostate disease. The two treatments are traditional surgery and a new method that does not require surgery. The records described many patients whose doctors had chosen each method. The study found that patients treated by the new method were significantly more likely to die within 8 years.[12]

(a) Further study of the data showed that this conclusion was wrong. The extra deaths among patients who got the new method could be explained by lurking variables. What lurking variables might be confounded with a doctor's choice of surgical or nonsurgical treatment?

(b) You have 300 prostate patients who are willing to serve as subjects in an experiment to compare the two methods. Use a diagram to outline the design of a randomized comparative experiment. (When using a diagram to outline the design of an experiment, be sure to indicate the size of the treatment groups and the response variable. The diagrams in Examples 3.11 and 3.12 are models.)

3.31   Use a diagram to describe a completely randomized experimental design for the package liner experiment of Exercise 3.27. (When using a diagram to outline the design of an experiment, be sure to indicate the size of the treatment groups and the response variable. The diagrams in Examples 3.11 and 3.12 are models.)

3.32   Will providing child care for employees make a company more attractive to women, even those who are unmarried? You are designing an experiment to answer this question. You prepare recruiting material for two fictitious companies, both in similar businesses in the same location. Company A's brochure does not mention child care. There are two versions of Company B's material, identical except that one describes the company's on-site child-care facility. Your subjects are 40 unmarried women who are college seniors seeking employment. Each subject will read recruiting material for both companies and choose the one she would prefer to work for. You will give each version of Company B's brochure to half the women. You expect that a higher percentage of those who read the description that includes child care will choose Company B.

(a) Outline an appropriate design for the experiment.

(b) The names of the subjects appear below. Use Table B, beginning at line 131, to do the randomization required by your design. List the subjects who will read the version that mentions child care.

Abrams	Danielson	Gutierrez	Lippman	Rosen
Adamson	Durr	Howard	Martinez	Sugiwara
Afifi	Edwards	Hwang	McNeill	Thompson
Brown	Fluharty	Iselin	Morse	Travers
Cansico	Garcia	Janle	Ng	Turing
Chen	Gerson	Kaplan	Quinones	Ullmann
Cortez	Green	Kim	Rivera	Williams
Curzakis	Gupta	Lattimore	Roberts	Wong

3.33   Use Table B, starting at line 120, to do the randomization required by your design for the package liner experiment in Exercise 3.31.

3.34   You decide to use a completely randomized design in the two-factor experiment on response to advertising described in Example 3.9 (page 199). You have 36 students who will serve as subjects. Outline the design. Then use Table B at line 130 to randomly assign the subjects to the 6 treatments.

The logic of experimental design

The logic behind a randomized comparative design is as follows.

• Randomization produces groups of experimental units that should be similar in all respects before the treatments are applied.

• Comparative design ensures that influences other than the experimental treatments operate equally on all groups.

• Therefore, differences in the response variable must be due to the effects of the treatments. That is, the treatments not only are associated with the observed differences in the response but must actually cause them.

The great advantage of randomized comparative experiments is that they can produce data that give good evidence for a cause-and-effect relationship between the explanatory and response variables. We know that in general a strong association does not imply causation. A strong association in data from a well-designed experiment does imply causation.

Experimenters were slow to accept randomized comparative designs when they were first put forward in the 1920s. One reason is that the outcome of a randomized experiment depends on chance. If the researchers in Example 3.11 drew cage numbers out of the hat a second time, they would get a different 15 rats in the experimental group and no doubt a somewhat different result when they compare the average weight gains of the two groups.

The outcomes of a randomized experiment, like the result of a random sample, do depend on chance. But in both settings, the laws of probability describe how much chance variation is present. *"Random" in statistics does not mean "haphazard."* Feeding the new breakfast food to the first 15 rats that the technician could catch would be haphazard, and perhaps biased if these rats were smaller, slower, or friendlier than the others. Randomization, on the other hand, gives each rat an equal chance to be chosen.

The presence of chance variation does require us to look more closely at the logic of randomized comparative experiments. We cannot say that *any* difference in average weight gain between the two groups of rats, however small, must be due to the diets. Some differences will appear even if the diets are identical, because the rats are not exactly alike. Some rats will grow faster than others, and by chance more of the faster-growing rats may end up in one group than in the other. Even though there are no systematic differences between the groups, there will still be chance differences. It is the business of statistical inference, using the laws of probability, to say whether the observed difference is too large to occur plausibly as a result of chance alone.

STATISTICAL SIGNIFICANCE

An observed effect too large to attribute plausibly to chance is called **statistically significant**.

If we observe statistically significant differences among the groups after a comparative randomized experiment, we have good evidence that the treatments actually caused these differences. We will explore statistical significance in detail in Chapter 5. One important point should be made immediately, however: *experiments with many subjects are better able to detect differences among the effects of the treatments than similar experiments with fewer subjects*. You would not trust the results of an experiment that fed each diet to only one rat. The role of chance is too large if we use two rats and toss a coin to decide which is fed the new diet. The more rats we use, the more likely it is that randomization will create groups that are alike on the average. When differences among the rats are averaged out, only the effects of the different treatments remain. Here is a third principle of *replication* statistical design of experiments, called ***replication***: repeat each treatment on a large enough number of experimental units or subjects to allow the systematic effects of the treatments to be seen.

PRINCIPLES OF EXPERIMENTAL DESIGN

The basic principles of statistical design of experiments are:

1. **Control** the effects of lurking variables on the response, most simply by comparing several treatments.
2. **Randomization**, the use of impersonal chance to assign subjects to treatments.
3. **Replication** of the experiment on many subjects to reduce chance variation in the results.

3.35   Example 3.12 describes an experiment to learn whether providing house-holds with electronic indicators or charts will reduce their electricity consumption. An executive of the electric company objects to including a control group. He says, "It would be simpler to just compare electricity use last year (before the indicator or chart was provided) with consumption in the same period this year. If households use less electricity this year, the indicator or chart must be working." Explain clearly why this design is inferior to that in Example 3.12.

3.36   Does regular exercise reduce the risk of a heart attack? Here are two ways to study this question. Explain clearly why the second design will produce more trustworthy data.

1. A researcher finds 2000 men over 40 who exercise regularly and have not had heart attacks. She matches each with a similar man who does not exercise regularly, and she follows both groups for 5 years.

2. Another researcher finds 4000 men over 40 who have not had heart attacks and are willing to participate in a study. She assigns 2000 of the men to a regular program of supervised exercise. The other 2000 continue their usual habits. The researcher follows both groups for 5 years.

3.37   The financial aid office of a university asks a sample of students about their employment and earnings. The report says that "for academic year earnings, a significant difference was found between the sexes, with men earning more on the average. No significant difference was found between the earnings of black and white students." Explain the meaning of "a significant difference" and "no significant difference" in plain language.

Cautions about experimentation

Randomized comparative experiments are the best means of gaining knowledge about the effects of explanatory variables on a response. You should nonetheless examine even experimental evidence with a critical eye.

hidden bias   First, the way the experiment is conducted may produce **hidden bias** despite the use of comparison and randomization. Experimenters must take great care to deal with all experimental units or subjects in exactly the same way, so that the treatments are the only systematic differences present. Unequal conditions introduce bias.

EXAMPLE 3.13

> A study of the roasting of meat in large commercial ovens found mysteriously large weight losses in the roasts cooked at the right front corner of the oven. The reason: in every run of the experiment, a meat thermometer was thrust into the right front roast, allowing juice to escape. ◀

Less obvious violations of equal treatment can occur in medical and behavioral experiments. Suppose that in the second gastric freezing experiment of Example 3.10 the person interviewing the subjects knew which patients received the real freezing and which received "only" the placebo. This knowledge may subconsciously affect the interviewer's attitude toward the patient and recording of the patient's degree of pain relief. The experiment should therefore be *double-blind*.

DOUBLE-BLIND EXPERIMENT

In a double-blind experiment, neither the subjects nor the people who have contact with them know which treatment a subject received.

The gastric freezing experiment was double-blind, with attention to such details as ensuring that the tube in the mouth of each subject was cold, whether or not the fluid in the balloon was refrigerated. Careful planning and attention to detail are the keys to avoiding hidden bias.

lack of realism in experiments   The most serious potential weakness of experiments is **lack of realism**. The subjects or treatments or setting of an experiment may not realistically duplicate the conditions we really want to study. Here are some examples.

EXAMPLE 3.14

> The study of television advertising in Example 3.9 showed a 40-minute videotape to students who knew an experiment was going on. We can't be sure that the results apply to everyday television viewers. Many behavioral science experiments use as subjects students who know they are subjects in an experiment. That's not a realistic setting.

> An industrial experiment uses a small-scale pilot production process to find the choices of catalyst concentration and temperature that maximize yield. These may not be the best choices for the operation of a full-scale plant. ◀

Lack of realism can limit our ability to apply the conclusions of an experiment to the settings of greatest interest. Most experimenters want to generalize their conclusions to some setting wider than that of the actual experiment. Statistical analysis of the original experiment cannot tell us how far the results will generalize. Rather, the experimenters in Example 3.14 must argue based on an understanding of psychology or chemical engineering that the experimental results do describe the wider world. Other psychologists or engineers may disagree. This is one reason why a single experiment is rarely completely convincing, despite the compelling logic of experimental design. The true scope of a new finding must usually be explored by a number of experiments in various settings.

A convincing case that an experiment is sufficiently realistic to produce useful information is based not on statistics but on the experimenter's knowledge of the subject matter of the experiment. The attention to detail required to avoid hidden bias also rests on subject matter knowledge. Good experiments combine statistical principles with understanding of a specific field of study.

EXERCISES

3.38   An experiment that claimed to show that meditation lowers anxiety proceeded as follows. The experimenter interviewed the subjects and rated their level of anxiety. Then the subjects were randomly assigned to two groups. The experimenter taught one group how to meditate and they meditated daily for a month. The other group was simply told to relax more. At the end of the month, the experimenter interviewed all the subjects again and rated their anxiety level. The meditation group now had less anxiety. Psychologists said that the results were suspect because the ratings were not blind. Explain what this means and how lack of blindness could bias the reported results.

3.39   Fizz Laboratories, a pharmaceutical company, has developed a new pain-relief medication. Sixty patients suffering from arthritis and needing pain relief are available. Each patient will be treated and asked an hour later, "About what percentage of pain relief did you experience?"

(a) Why should Fizz not simply administer the new drug and record the patients' responses?

(b) Outline the design of an experiment to compare the drug's effectiveness with that of aspirin and of a placebo.

(c) Should patients be told which drug they are receiving? How would this knowledge probably affect their reactions?

(d) If patients are not told which treatment they are receiving, the experiment is single-blind. Should this experiment be double-blind also? Explain.

Other experimental designs*

Completely randomized designs are the simplest statistical designs for experiments. They are the analog of simple random samples. In fact, each treatment group is an SRS drawn from the available subjects. Completely randomized designs illustrate clearly the principles of control, randomization, and replication. However, just as in sampling, more elaborate statistical designs are often superior. In particular, matching the subjects in various ways can produce more precise results than simple randomization.

EXAMPLE 3.15

Women and men respond differently to advertising. An experiment to compare the effectiveness of three television commercials for the same product will want to look separately at the reactions of men and women, as well as assess the overall response to the ads.

A completely randomized design considers all subjects, both men and women, as a single pool. The randomization assigns subjects to three treatment groups without regard to their sex. This ignores the differences between men and women. A better design considers women and men separately. Randomly assign the women to three groups, one to view each commercial. Then separately assign the men at random to three groups. Figure 3.3 outlines this improved design. ◀

*This section expands the discussion of experimental designs. It is not needed in order to read the rest of the book.

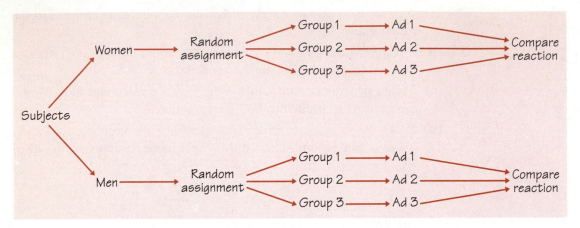

FIGURE 3.3   The outline of a block design to compare the effectiveness of three TV advertisements. Male and female subjects form two blocks.

The design of Figure 3.3 uses the principles of comparison, random-ization, and replication. However, the randomization is not complete (all subjects randomly assigned to treatment groups) but restricted to operate only within groups of similar subjects. The groups are called *blocks*, and the design is called a *block design*.

BLOCK DESIGN

A **block** is a group of experimental units or subjects that are similar in ways that are expected to affect the response to the treatments. In a **block design**, the random assignment of units to treatments is carried out separately within each block.

Blocks are another form of *control*. They control the effects of some lurking variables by bringing those variables into the experiment to form the blocks. Block designs are similar to stratified designs for sampling. Blocks and strata both group similar units. We use two different names only because the idea developed separately for sampling and experiments. Blocks allow us to draw separate conclusions about each block, for exam-ple, about men and women in Example 3.15. Blocking also allows more precise overall conclusions, because the systematic differences between men and women can be removed when we study the overall effects of the three commercials.

matched pairs design

A simple and common special type of block design is the **matched pairs design**. Matched pairs designs compare just two treatments. Each block consists of just two units, as closely matched as possible. These units are assigned at random to the treatments, by tossing a coin or reading odd and even digits from Table B. Alternatively, each block in a matched pairs design may consist of just one subject, who gets both treatments one after the other. Each subject serves as his or her own control. The *order* of the treatments can influence the subject's response, so the order is randomized for each subject, again by a coin toss.

EXAMPLE 3.16

Pepsi once wanted to demonstrate that Coke drinkers in fact prefer Pepsi when they taste both colas blind. The subjects, all people who said they were Coke drinkers, tasted both colas from glasses without brand markings and said which they liked better. This is a matched pairs design in which each subject compares the two colas. Because responses may depend on which cola is tasted first, the order of tasting should be chosen at random for each subject.

When more than half the Coke drinkers chose Pepsi, Coke claimed that the experiment was biased. The Pepsi glasses were marked M and Coke glasses were marked Q. Aha, said Coke, this just shows that people like the letter M better than the letter Q. A careful experiment would in fact take care to avoid any distinction other than the actual treatments.[13] ◀

EXERCISES

3.40   Twenty overweight females have agreed to participate in a study of the effectiveness of four weight-loss treatments: A, B, C, and D. The researcher first calculates how overweight each subject is by comparing the subject's actual weight with her "ideal" weight. The subjects and their excess weights in pounds are

Birnbaum	35	Hernandez	25	Moses	25	Smith	29
Brown	34	Jackson	33	Nevesky	39	Stall	33
Brunk	30	Kendall	28	Obrach	30	Tran	35
Cruz	34	Loren	32	Rodriguez	30	Wilansky	42
Deng	24	Mann	28	Santiago	27	Williams	22

The response variable is the weight lost after 8 weeks of treatment. Because a subject's excess weight will influence the response, a block design is appropriate.

(a) Arrange the subjects in order of increasing excess weight. Form 5 blocks of 4 subjects each by grouping the 4 least overweight, then the next 4, and so on.

(b) Use Table B to randomly assign the 4 subjects in each block to the 4 weight-loss treatments. Be sure to explain exactly how you used the table.

3.41   Return to the advertising experiment of Example 3.9 (page 199). You have 36 subjects: 24 women and 12 men. Men and women often react differently to advertising. You therefore decide to use a block design with the two genders as blocks. You must assign the 6 treatments at random within each block separately.

(a) Outline the design with a diagram.

(b) Use Table B, beginning at line 140, to do the randomization. Report your result in a table that lists the 24 women and 12 men and the treatment you assigned to each.

3.42   Is the right hand generally stronger than the left in right-handed people? You can crudely measure hand strength by placing a bathroom scale on a shelf with the end protruding, then squeezing the scale between the thumb below and the four fingers above it. The reading of the scale shows the force exerted. Describe the design of a matched pairs experiment to compare the strength of the right and left hands, using 10 right-handed people as subjects. (You need not actually do the randomization.)

3.43   Some investment advisors believe that charts of past trends in the prices of securities can help predict future prices. Most economists disagree. In an experiment to examine the effects of using charts, business students trade (hypothetically) a foreign currency at computer screens. There are 20 student subjects available, named for convenience A, B, C, ..., T. Their goal is to make as much money as possible, and the best performances are rewarded with small prizes. The student traders have the price history of the foreign currency in dollars in their computers. They may or may not also have software that highlights trends. Describe *two* designs for this experiment, a completely randomized design and a matched pairs design in which each student serves as his or her own control. In both cases, carry out the randomization required by the design.

3.44   An expert on worker performance is interested in the effect of room temperature on the performance of tasks requiring manual dexterity. She chooses temperatures of 70° F and 90° F as treatments. The response variable is the number of correct insertions, during a 30-minute period, in a peg-and-hole apparatus that requires the use of both hands simultaneously. Each subject is trained on the apparatus and then asked to make as many insertions as possible in 30 minutes of continuous effort.

(a) Outline a completely randomized design to compare dexterity at 70° and 90°. Twenty subjects are available.

(b) Because individuals differ greatly in dexterity, the wide variation in individual scores may hide the systematic effect of temperature unless there are many subjects in each group. Describe in detail the design of a matched pairs experiment in which each subject serves as his or her own control.

3.45 There are several psychological tests that measure the extent to which Mexican Americans are oriented toward Mexican/Spanish or Anglo/English culture. Two such tests are the Bicultural Inventory (BI) and the Acculturation Rating Scale for Mexican Americans (ARSMA). To study the correlation between the scores on these two tests, researchers will give both tests to a group of 22 Mexican Americans.

(a) Briefly describe a matched pairs design for this study. In particular, how will you use randomization in your design?

(b) You have an alphabetized list of the subjects (numbered 1 to 22). Carry out the randomization required by your design and report the result.

3.46 You are participating in the design of a medical experiment to investigate whether a calcium supplement in the diet will reduce the blood pressure of middle-aged men. Preliminary work suggests that calcium may be effective and that the effect may be greater for black men than for white men.

(a) Outline in graphical form the design of an appropriate experiment.

(b) Choosing the sizes of the treatment groups requires more statistical expertise. We will learn more about this aspect of design in later chapters. Explain in plain language the advantage of using larger groups of subjects.

SUMMARY

An **observational study** gathers data without trying to influence the responses. An **experiment** imposes some treatment in order to observe the response.

In an experiment, we impose one or more **treatments** on the experimental **units** or **subjects**. Each treatment is a combination of levels of the explanatory variables, which we call **factors**.

The **design** of an experiment refers to the choice of treatments and the manner in which the experimental units or subjects are assigned to the treatments.

The basic principles of statistical design of experiments are **control**, **randomization**, and **replication**.

The simplest form of control is **comparison**. Experiments should compare two or more treatments in order to avoid **confounding** of the effect of a treatment with other influences, such as lurking variables.

Randomization uses chance to assign subjects to the treatments. Randomization creates treatment groups that are similar (except for chance variation) before the treatments are applied. Randomization and comparison together prevent **bias**, or systematic favoritism, in experiments.

Randomization can be carried out by giving numerical labels to the experimental units and using a **table of random digits** to choose treatment groups.

Replication of the treatments on many units reduces the role of chance variation and makes the experiment more sensitive to differences among the treatments.

The validity of experimental results may be threatened by **hidden bias** or **lack of realism**. Statistics alone does not protect against these threats.

In addition to comparison, a second form of control is to restrict randomization by forming **blocks** of experimental units that are similar in some way that is important to the response. Randomization is then carried out separately within each block.

Matched pairs are a common form of blocking for comparing just two treatments. In some matched pairs designs, each subject receives both treatments in a random order. In others, the subjects are matched in pairs as closely as possible, and one subject in each pair receives each treatment.

SECTION 3.2 EXERCISES

3.47   (a) Exercise 2.48 (page 145) describes a study of the effects of herbal tea on the attitude of nursing home residents. Is this study an experiment? Why?

(b) Exercise 2.55 (page 149) describes a study of the effect of learning a foreign language on scores on an English test. Is this study an experiment? Why?

3.48   If children are given more choices within a class of products, will they tend to prefer that product to a competing product that offers fewer choices? Marketers want to know. An experiment prepared three "choice sets" of beverages. The first contained two milk drinks and two fruit drinks. The second had the same two fruit drinks but four milk drinks. The third contained four fruit drinks but only the original two milk drinks. The researchers divided 210 children aged 4 to 12 years into 3 groups at random.

They offered each group one of the choice sets. As each child chose a beverage to drink from the choice set presented, the researchers noted whether the choice was a milk drink or a fruit drink.

(a) What are the experimental units or subjects?

(b) What is the factor, and what are its levels?

(c) What is the response variable?

3.49 Can aspirin help prevent heart attacks? The Physicians' Health Study, a large medical experiment involving 22,000 male physicians, attempted to answer this question. One group of about 11,000 physicians took an aspirin every second day, while the rest took a placebo. After several years the study found that subjects in the aspirin group had significantly fewer heart attacks than subjects in the placebo group.

(a) Identify the experimental subjects, the factor and its levels, and the response variable in the Physicians' Health Study.

(b) Use a diagram to outline a completely randomized design for the Physicians' Health Study.

3.50 Use a diagram to outline a completely randomized design for the children's choice study of Exercise 3.48.

3.51 Some medical researchers suspect that added calcium in the diet reduces blood pressure. You have available 40 men with high blood pressure who are willing to serve as subjects.

(a) Outline an appropriate design for the experiment, taking the placebo effect into account.

(b) The names of the subjects appear below. Use Table B, beginning at line 119, to do the randomization required by your design, and list the subjects to whom you will give the drug.

Alomar	Denman	Han	Liang	Rosen
Asihiro	Durr	Howard	Maldonado	Solomon
Bennett	Edwards	Hruska	Marsden	Tompkins
Bikalis	Farouk	Imrani	Moore	Townsend
Chen	Fratianna	James	O'Brian	Tullock
Clemente	George	Kaplan	Ogle	Underwood
Cranston	Green	Krushchev	Plochman	Willis
Curtis	Guillen	Lawless	Rodriguez	Zhang

3.52 The children's choice experiment in Exercise 3.48 has 210 subjects. Explain how you would assign labels to the 210 children in the actual experiment. Then use Table B at line 125 to choose *only the first 5* children assigned to the first treatment.

3.53   A survey of physicians found that some doctors give a placebo to a patient who complains of pain for which the physician can find no cause. If the patient's pain improves, these doctors conclude that it had no physical basis. The medical school researchers who conducted the survey claimed that these doctors do not understand the placebo effect. Why?

CHAPTER REVIEW

Designs for producing data are essential parts of statistics in practice. Random sampling and randomized comparative experiments are perhaps the most important statistical inventions in this century. Both were slow to gain acceptance, and you will still see many voluntary response samples and uncontrolled experiments. This chapter has explained good techniques for producing data and has also explained why bad techniques often produce worthless data.

The deliberate use of chance in producing data is a central idea in statistics. It allows use of the laws of probability to analyze data, as we will see in the following chapters. Here are the major skills you should have now that you have studied this chapter.

A. SAMPLING

1. Identify the population in a sampling situation.
2. Recognize bias due to voluntary response samples and other inferior sampling methods.
3. Use Table B of random digits to select a simple random sample (SRS) from a population.
4. Recognize the presence of undercoverage and nonresponse as sources of error in a sample survey. Recognize the effect of the wording of questions on the responses.
5. Use random digits to select a stratified random sample from a population when the strata are identified.

B. EXPERIMENTS

1. Recognize whether a study is an observational study or an experiment.
2. Recognize bias due to confounding of explanatory variables with lurking variables in either an observational study or an experiment.
3. Identify the factors (explanatory variables), treatments, response variables, and experimental units or subjects in an experiment.

4. Outline the design of a completely randomized experiment using a diagram like those in Examples 3.11 and 3.12. The diagram in a specific case should show the sizes of the groups, the specific treatments, and the response variable.

5. Use Table B of random digits to carry out the random assignment of subjects to groups in a completely randomized experiment.

6. Recognize the placebo effect. Recognize when the double-blind technique should be used.

7. Explain why a randomized comparative experiment can give good evidence for cause-and-effect relationships.

CHAPTER 3 REVIEW EXERCISES

3.54 The Ministry of Health in the Canadian Province of Ontario wants to know whether the national health care system is achieving its goals in the province. Much information about health care comes from patient records, but that source doesn't allow us to compare people who use health services with those who don't. So the Ministry of Health conducted the Ontario Health Survey, which interviewed a random sample of 61,239 people who live in the Province of Ontario.[14]

(a) What is the population for this sample survey? What is the sample?

(b) The survey found that 76% of males and 86% of females in the sample had visited a general practitioner at least once in the past year. Do you think these estimates are close to the truth about the entire population? Why?

3.55 What is the preferred treatment for breast cancer that is detected in its early stages? The most common treatment was once removal of the breast. It is now usual to remove only the tumor and nearby lymph nodes, followed by radiation. To study whether these treatments differ in their effectiveness, a medical team examines the records of 25 large hospitals and compares the survival times after surgery of all women who have had either treatment.

(a) What are the explanatory and response variables?

(b) Explain carefully why this study is not an experiment.

(c) Explain why confounding will prevent this study from discovering which treatment is more effective. (The current treatment was in fact recommended after a large randomized comparative experiment.)

3.56 A study of the relationship between physical fitness and leadership uses as subjects middle-aged executives who have volunteered for an exercise program. The executives are divided into a low-fitness group and a high-fitness group on the basis of a physical examination. All subjects then take a psychological test designed to measure leadership, and the results for the two groups are compared. Is this study an experiment? Explain your answer.

3.57   A university's financial aid office wants to know how much it can expect students to earn from summer employment. This information will be used to set the level of financial aid. The population contains 3478 students who have completed at least one year of study but have not yet graduated. The university will send a questionnaire to an SRS of 100 of these students, drawn from an alphabetized list.

(a) Describe how you will label the students in order to select the sample.

(b) Use Table B, beginning at line 105, to select the *first* 5 students in the sample.

3.58   A labor organization wants to study the attitudes of college faculty members toward collective bargaining. These attitudes appear to be different depending on the type of college. The American Association of University Professors classifies colleges as follows:

Class I.   Offer doctorate degrees and award at least 15 per year.

Class IIA.   Award degrees above the bachelor's but are not in Class I.

Class IIB.   Award no degrees beyond the bachelor's.

Class III.   Two-year colleges.

Discuss the design of a sample of faculty from colleges in your state, with total sample size about 200.

3.59   You want to investigate the attitudes of students at your school about the school's policy on sexual harassment. You have a grant that will pay the costs of contacting about 500 students.

(a) Specify the exact population for your study. For example, will you include part-time students?

(b) Describe your sample design. Will you use a stratified sample?

(c) Briefly discuss the practical difficulties that you anticipate. For example, how will you contact the students in your sample?

3.60   New varieties of corn with altered amino acid content may have higher nutritional value than standard corn, which is low in the amino acid lysine. An experiment compares two new varieties, called opaque-2 and floury-2, with normal corn. The researchers mix corn-soybean meal diets using each type of corn at each of three protein levels, 12% protein, 16% protein, and 20% protein. They feed each diet to 10 one-day-old male chicks and record their weight gains after 21 days. The weight gain of the chicks is a measure of the nutritional value of their diet.

(a) What are the experimental units and the response variable in this experiment?

(b) How many factors are there? How many treatments? Use a diagram like Figure 3.2 to describe the treatments. How many experimental units does the experiment require?

(c) Use a diagram to describe a completely randomized design for this experiment. (You do not need to actually do the randomization.)

3.61 A chemical engineer is designing the production process for a new product. The chemical reaction that produces the product may have higher or lower yield, depending on the temperature and the stirring rate in the vessel in which the reaction takes place. The engineer decides to investigate the effects of all combinations of two temperatures (50° C and 60° C) and three stirring rates (60 rpm, 90 rpm, and 120 rpm) on the yield of the process. She will process two batches of the product at each combination of temperature and stirring rate. In Exercise 3.29 you identified the treatments.

(a) Outline in graphic form the design of an appropriate experiment.

(b) The randomization in this experiment determines the order in which batches of the product will be processed according to each treatment. Use Table B, starting at line 128, to carry out the randomization and state the result.

3.62 Is the number of days a letter takes to reach another city affected by the time of day it is mailed and whether or not the zip code is used? Describe briefly the design of a two-factor experiment to investigate this question. Be sure to specify the treatments exactly and to tell how you will handle lurking variables such as the day of the week on which the letter is mailed.

3.63 (Optional) Do consumers prefer the taste of a cheeseburger from McDonald's or from Wendy's in a blind test in which neither burger is identified? Describe briefly the design of a matched pairs experiment to investigate this question.

3.64 The previous two exercises illustrate the use of statistically designed experiments to answer questions that arise in everyday life. Select a question of interest to you that an experiment might answer and briefly discuss the design of an appropriate experiment.

3.65 A study on predicting job performance reports that there is a statistically significant correlation between the score on a screening test given to potential employees and a new employee's score on an evaluation made after a year on the job. What does "statistically significant" mean here?

NOTES AND DATA SOURCES

1. Based in part on Randall Rothenberger, "The trouble with mall interviewing," *New York Times*, August 16, 1989.

2. The information in this example is taken from *The Ascap Survey and Your Royalties*, ASCAP, New York, undated.

3. For more detail on the material of this section, along with references, see P. E. Converse and M. W. Traugott, "Assessing the accuracy of polls and surveys," *Science*, 234 (1986), pp. 1094–1098.

4. The estimates of the census undercount come from Howard Hogan, "The 1990 post-enumeration survey: operations and results," *Journal of the American Statistical Association*, 88 (1993), pp. 1047–1060. The information about nonresponse appears in Eugene P. Eriksen and Teresa K. DeFonso, "Beyond the net undercount: how to measure census error," *Chance*, 6, no. 4 (1993), pp. 38–43 and 14.

5. For more detail on the limits of memory in surveys, see N. M. Bradburn, L. J. Rips, and S. K. Shevell, "Answering autobiographical questions: the impact of memory and inference on surveys," *Science*, 236 (1987), pp. 157–161.

6. The Levi jeans and disposable diaper examples are taken from Cynthia Crossen, "Margin of error: studies galore support products and positions, but are they reliable?" *Wall Street Journal*, November 14, 1991.

7. Giuliana Coccia, "An overview of non-response in Italian telephone surveys," *Proceedings of the 99th Session of the International Statistical Institute, 1993*, Book 3, pp. 271–272.

8. The first question was asked by Ross Perot, and the second by a Time/CNN poll, both in March 1993. The example comes from W. Mitofsky, "Mr. Perot, you're no pollster," *New York Times*, March 27, 1993.

9. From the *New York Times* of August 21, 1989.

10. Simplified from Arno J. Rethans, John L. Swasy, and Lawrence J. Marks, "Effects of television commercial repetition, receiver knowledge, and commercial length: A test of the two-factor model," *Journal of Marketing Research*, 23 (February 1986), pp. 50–61.

11. L. L. Miao, "Gastric freezing: an example of the evaluation of medical therapy by randomized clinical trials," in J. P. Bunker, B. A. Barnes, and F. Mosteller (eds.), *Costs, Risks and Benefits of Surgery*, Oxford University Press, New York, 1977, pp. 198–211.

12. Based on Christopher Anderson, "Measuring what works in health care," *Science*, 263 (1994), pp. 1080–1082.

13. Taken from "Advertising: the cola war," *Newsweek*, August 30, 1976, p. 67.

14. Information from Warren McIsaac and Vivek Goel, "Is access to physician services in Ontario equitable?" Institute for Clinical Evaluative Sciences in Ontario, October 18, 1993.

The purpose of statistics is to gain understanding from data. It is not surprising that we can approach so broad a goal in different ways, depending on the circumstances. We have studied one approach to data, *exploratory data analysis,* in some detail. In examining the design of samples and experiments, we began to move from data analysis toward *statistical inference.* Both types of reasoning are essential to effective work with data. Here is a brief sketch of the differences between them.

Exploratory Data Analysis	Statistical Inference
Purpose is unrestricted exploration of the data, searching for interesting patterns.	Purpose is to answer specific questions, posed before the data were produced.
Conclusions apply only to the individuals and circumstances for which we have data in hand.	Conclusions apply to a larger group of individuals or a broader class of circumstances.
Conclusions are informal, based on what we see in the data.	Conclusions are formal, backed by a statement of our confidence in them.

These distinctions help us understand how inference differs from data analysis, but in practice the two approaches cooperate. Inference usually requires that the pattern of the data be reasonably regular. Data analysis can spot outliers or other deviations that would make inference unreliable. Data analysis, especially using graphs, is an essential first step when we want to do inference. Data production is also closely linked to inference. A good design for producing data is the best guarantee that inference makes sense.

In Part II of this book you will gain an understanding of the reasoning and basic methods of statistical inference. Chapters 4 and 5 provide the core concepts. Probability, especially the central idea of a sampling distribution, is the topic of Chapter 4. Chapter 5 presents the reasoning of statistical inference in detail. Chapters 6 and 7 discuss practical examples of inference. The link between inference, data analysis, and data production is clearest when we face real settings in Chapters 6 and 7.

PART II

UNDERSTANDING INFERENCE

"Oh, sure, he remembers everything. But show me one significant insight he's been able to draw from all that data."

Reprinted by permission of Harald Bakkan and Mischa Richter from the Chronicle of Higher Education.

DAVID BLACKWELL

Statistical practice rests in part on statistical theory. Statistics has been advanced not only by people concerned with practical problems, from Florence Nightingale to R. A. Fisher and John Tukey, but also by people whose first love is mathematics for its own sake. David Blackwell (1919–) is one of the major contemporary contributors to the mathematical study of statistics.

Blackwell grew up in Illinois, earned a doctorate in mathematics at the age of 22, and in 1944 joined the faculty of Howard University in Washington, D.C. "It was the ambition of every black scholar in those days to get a job at Howard University," he says. "That was the best job you could hope for." Society changed, and in 1954 Blackwell became professor of statistics at the University of California at Berkeley.

Washington, D.C., had an active statistical community, and the young mathematician Blackwell soon began to work on mathematical aspects of statistics. He explored the behavior of statistical procedures which, rather than working with a fixed sample, keep taking observations until there is enough information to reach a firm conclusion. He found insights into statistical inference by thinking of inference as a game in which nature plays against the statistician. Blackwell's work uses probability theory, the mathematics that describes chance behavior. We must travel the same route, though only a short distance. This chapter presents, in a rather informal fashion, the probabilistic ideas needed to understand the reasoning of inference.

CHAPTER **4**

Sampling Distributions and Probability

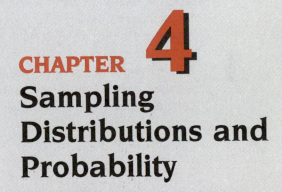

INTRODUCTION

The reasoning of statistical inference rests on asking, "How often would this method give a correct answer if I used it very many times?" If it doesn't make sense to imagine repeatedly producing your data in the same circumstances, statistical inference is not possible.[1] Exploratory data analysis makes sense for any data, but formal inference does not. Even experts can disagree about how widely statistical inference should be used. But all agree that inference is most secure when we produce data by random sampling or randomized comparative experiments. The reason is that when we use chance to choose respondents or assign subjects, the laws of probability answer the question "What would happen if we did this many times?" The purpose of this chapter is to see what the laws of probability tell us, but without going into the mathematics of probability theory. Chapter 5 then applies our new understanding to inference.

This chapter, more than the rest of the book, offers you choices. Sections 4.1, 4.3, and 4.5 present the basic ideas and facts about probability that you need to understand inference. You can go forward after reading only these sections. Sections 4.2 and 4.4 offer more facts about probability but remain focused on probability as background for statistical inference. Probability is a large and important subject in its own right, but it is not our subject in this book. Statistical process control (SPC) is the topic of Section 4.6. SPC ideas are very important in some, but not all, settings where statistical methods are used. What is more, they immediately apply probability facts to draw conclusions and so form a bridge between probability and inference.

4.1 SAMPLING DISTRIBUTIONS

"Are you afraid to go outside at night within a mile of your home because of crime?" When the Gallup Poll asked this question, 45% of the people in the sample said "Yes." That 45% describes the sample, but we use it as an estimate for the entire population. We must now take care to keep straight whether a number describes a sample or a population. Here is the vocabulary we use.

> ### PARAMETER, STATISTIC
>
> A **parameter** is a number that describes the population. In statistical practice, the value of a parameter is not known.

230

(continued on next page)

(continued from previous page)

A **statistic** is a number that can be computed from the sample data without making use of any unknown parameters. In practice, we often use a statistic to estimate an unknown parameter.

EXAMPLE 4.1

Are attitudes toward shopping changing? Sample surveys show that fewer people enjoy shopping than in the past. A recent survey asked a nationwide random sample of 2500 adults if they agreed or disagreed that "I like buying new clothes, but shopping is often frustrating and time-consuming." Of the respondents, 1650, or 66%, said they agreed.[2] The number 66% is a *statistic*. The population that the poll wants to draw conclusions about is all U.S. residents age 18 and over. The *parameter* of interest is the percent of all adult U.S. residents who would have said "Agree" if asked the same question. We don't know the value of this parameter. ◄

EXERCISES

State whether each boldface number in Exercises 4.1 to 4.4 is a *parameter* or a *statistic*.

4.1 The Bureau of Labor Statistics last month interviewed 60,000 members of the U.S. labor force, of whom **7.2%** were unemployed.

4.2 A carload lot of ball bearings has mean diameter **2.5003** centimeters (cm). This is within the specifications for acceptance of the lot by the purchaser. By chance, an inspector chooses 100 bearings from the lot that have mean diameter **2.5009** cm. Because this is outside the specified limits, the lot is mistakenly rejected.

4.3 A telemarketing firm in Los Angeles uses a device that dials residential telephone numbers in that city at random. Of the first 100 numbers dialed, **48%** are unlisted. This is not surprising because **52%** of all Los Angeles residential phones are unlisted.

4.4 A researcher carries out a randomized comparative experiment with young rats to investigate the effects of a toxic compound in food. She feeds the control group a normal diet. The experimental group receives a diet with 2500 parts per million of the toxic material. After 8 weeks, the mean weight gain is **335** grams for the control group and **289** grams for the experimental group.

Sampling variability

The sample survey of Example 4.1 produced information on attitudes about shopping. Let's examine it in more detail. We want to estimate the proportion of the population that find clothes shopping frustrating. Call this population proportion p. It is a parameter. The poll found that 1650 out of 2500 randomly selected adults agreed with the statement that shopping is often frustrating. The proportion of the sample who agreed was

$$\hat{p} = \frac{1650}{2500} = .66$$

The sample proportion $\hat{p}$ is a statistic. We use $\hat{p}$ to estimate the unknown parameter p.

How can $\hat{p}$, based on only 2500 of the more than 180 million American adults, be an accurate estimate of p? After all, a second random sample taken at the same time would choose different people and no doubt produce a different value of $\hat{p}$. This basic fact is called **sampling variability**: the value of a statistic varies in repeated random sampling.

sampling variability

To understand why sampling variability is not fatal, we ask, "What would happen if we took many samples?" Here's how to answer that question:

- Take a large number of samples from the same population.
- Calculate the sample proportion $\hat{p}$ for each sample.
- Make a histogram of the values of $\hat{p}$.
- Examine the distribution displayed in the histogram for overall pattern, center and spread, and outliers or other deviations.

In practice it is too expensive to take many samples from a population like all adult U.S. residents. But we can imitate many samples by using random digits. Using random digits from a table or computer software to imitate chance behavior is called **simulation**.

simulation

EXAMPLE 4.2

We will simulate drawing simple random samples (SRSs) of size 100 from the population of all adult U.S. residents. Suppose that in fact 60% of the population find clothes shopping time-consuming and frustrating. Then the true value of the parameter we want to estimate is $p = .6$. (Of course, we would not sample in practice if we already knew that $p = .6$. We are sampling here to find out how sampling behaves.)

We can imitate the population by a huge table of random digits, with each entry standing for a person. Six of the ten digits (say 0 to 5) stand for people who find shopping frustrating. The remaining four digits, 6 to 9, stand for those who do not. Because all digits in a random number table are equally likely, this assignment produces a population proportion of frustrated shoppers equal to $p = .6$. We then imitate an SRS of 100 people from the population by taking 100 consecutive digits from Table B. The statistic $\hat{p}$ is the proportion of 0s to 5s in the sample.

For example, the first 100 entries in Table B contain 63 digits between 0 and 5, so $\hat{p} = 63/100 = .63$. A second SRS based on the second 100 entries in Table B gives a different result, $\hat{p} = .56$. The two sample results are different, and neither is equal to the true population value $p = .6$. That's sampling variability.

◀

Simulation is a powerful tool for studying chance. It is much faster to use Table B than to actually draw repeated SRSs, and much faster yet to use a computer programmed to produce random digits. Figure 4.1 is the histogram of values of $\hat{p}$ from 1000 separate SRSs of size 100 drawn from a

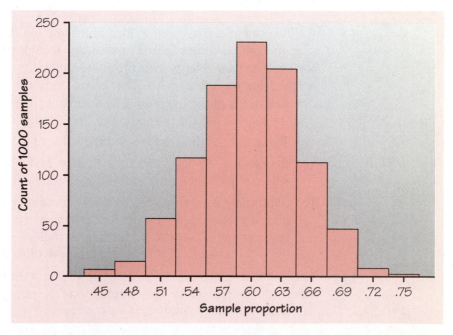

FIGURE 4.1   The sampling distribution of the sample proportion $\hat{p}$ from SRSs of size 100 drawn from a population with population proportion $p = .6$. The histogram shows the results of drawing 1000 SRSs.

population with $p = .6$. This histogram shows what would happen if we drew many samples. It displays the *sampling distribution* of $\hat{p}$.

SAMPLING DISTRIBUTION

The **sampling distribution** of a statistic is the distribution of values taken by the statistic in all possible samples of the same size from the same population.

Strictly speaking, the sampling distribution is the ideal pattern that would emerge if we looked at all possible samples of size 100 from our population. A distribution obtained from a fixed number of trials, like the 1000 trials in Figure 4.1, is only an approximation to the sampling distribution. One of the uses of probability theory in statistics is to obtain exact sampling distributions without simulation. The interpretation of a sampling distribution is the same, however, whether we obtain it by simulation or by the mathematics of probability.

EXERCISES

4.5   Coin tossing can illustrate the idea of a sampling distribution. The population is all outcomes (heads or tails) we would get if we tossed a coin forever. The parameter p is the proportion of heads in this population. We suspect that p is close to 0.5. That is, we think the coin will show about one-half heads in the long run. The sample is the outcomes of 20 tosses, and the statistic $\hat{p}$ is the proportion of heads in these 20 tosses.

(a) Toss a coin 20 times and record the value of $\hat{p}$.

(b) Repeat this sampling process 10 times. Make a histogram of the 10 values of $\hat{p}$. Is the center of this distribution close to 0.5?

(c) (**Optional**) Ten repetitions give a very crude approximation to the sampling distribution. Pool your work with that of other students to obtain several hundred repetitions. Make a histogram of all the values of $\hat{p}$. Is the center close to 0.5? Is the shape approximately normal?

4.6   Let us illustrate the idea of a sampling distribution in the case of a very small sample from a very small population. The population is the scores of 10 students on an exam:

Student	0	1	2	3	4	5	6	7	8	9
Score	82	62	80	58	72	73	65	66	74	62

The parameter of interest is the mean score in this population, which is 69.4. The sample is an SRS of size $n = 4$ drawn from the population. Because the students are labeled 0 to 9, a single random digit from Table B chooses one student for the sample.

(a) Use Table B to draw an SRS of size 4 from this population. Write the four scores in your sample and calculate the mean $\bar{x}$ of the sample scores. This statistic is an estimate of the population parameter.

(b) Repeat this process 10 times. Make a histogram of the 10 values of $\bar{x}$. You are constructing the sampling distribution of $\bar{x}$. Is the center of your histogram close to 69.4?

(c) (**Optional**) Ten repetitions give a very crude approximation to the sampling distribution. Pool your work with that of other students—using different parts of Table B—to obtain several hundred repetitions. Make a histogram of all the values of $\bar{x}$. Is the center close to 69.4? Is the shape approximately normal? This histogram is a better approximation to the sampling distribution.

Describing sampling distributions

We can use the tools of data analysis to describe any distribution. Let's apply those tools to Figure 4.1. Remember that the figure shows the values of the sample proportion $\hat{p}$ in a large number of samples from the same population.

- The *overall shape* of the distribution is symmetric and approximately normal.

- There are no *outliers* or other important deviations from the overall pattern.

- The *center* of the distribution is very close to the true value $p = .6$ for the population from which the samples were drawn. In fact, the mean of the 1000 $\hat{p}$s is 0.598 and their median is exactly 0.6.

- The values of $\hat{p}$ have a large *spread*. They range from 0.45 to 0.75. Because the distribution is close to normal, we can use the standard deviation to describe its spread. The standard deviation is about 0.05.

Figure 4.1 shows that a sample of 100 people often gives a $\hat{p}$ quite far from the population parameter $p = .6$. That is, a sample of 100 people does not produce a trustworthy estimate of the population proportion. That is why the sample survey of attitudes toward shopping interviewed not 100 but 2500 people. Let's repeat our simulation, this time taking 1000

SRSs of size 2500 from a population with proportion $p = .6$ who find shopping frustrating.

Figure 4.2 displays the sampling distribution of the 1000 values of $\hat{p}$ from these new samples. Figure 4.2 uses the same horizontal scale as Figure 4.1 to make comparison easy. Here's what we see:

- The *center* of the distribution is again close to 0.6. In fact, the mean is 0.6002 and the median is exactly 0.6.

- The *spread* of Figure 4.2 is much less than that of Figure 4.1. The range of the values of $\hat{p}$ from 1000 samples is only 0.5728 to 0.6296. The standard deviation is about 0.01. Almost all samples of 2500 people give a $\hat{p}$ close to the population parameter $p = .6$.

- Because the values of $\hat{p}$ cluster so tightly about 0.6, it is hard to see the *shape* of the distribution in Figure 4.2. Figure 4.3 displays the same 1000 values of $\hat{p}$ on an expanded scale that makes the shape clearer. The distribution is again approximately normal in shape.

The appearance of the sampling distributions in Figures 4.1 to 4.3 is a consequence of random sampling. Haphazard sampling does not give such regular and predictable results. When randomization is used in a design

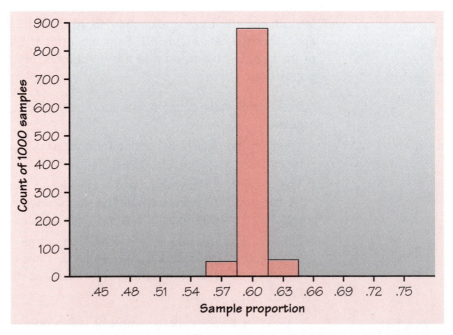

FIGURE 4.2 The sampling distribution of the sample proportion $\hat{p}$ from SRSs of size 2500 drawn from a population with population proportion $p = .6$. The histogram shows the results of drawing 1000 SRSs. The scale is the same as in Figure 4.1.

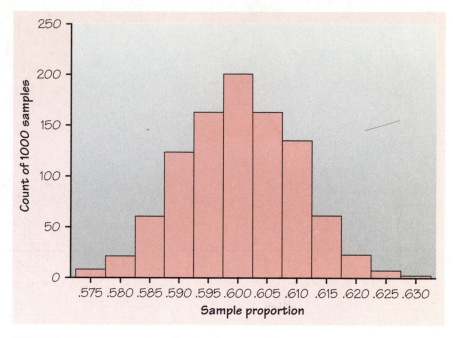

FIGURE 4.3   The sampling distribution from Figure 4.2, for samples of size 2500, redrawn on a different scale to better display the shape.

for producing data, statistics computed from the data have a definite pattern of behavior over many repetitions, even though the result of a single repetition is uncertain.

The bias of a statistic

The fact that statistics from random samples have definite sampling distributions allows a more careful answer to the question of how trustworthy a statistic is as an estimate of a parameter. Figure 4.4 shows the two sampling distributions of $\hat{p}$, for samples of 100 people and samples of 2500 people, side by side and drawn to the same scale. Both distributions are approximately normal, so we have also drawn normal curves for both. How trustworthy is the sample proportion $\hat{p}$ as an estimator of the population proportion p in each case?

First, we can describe *bias* more exactly by speaking of the bias of a statistic rather than bias in a sampling method. Bias concerns the center of the sampling distribution. The centers of the sampling distributions in Figure 4.4 are very close to the true value of the population parameter. Those distributions show the results of 1000 samples. In fact, the mean of the sampling distribution (think of taking all possible samples, not just 1000 samples) is *exactly* equal to 0.6, the parameter in the population.

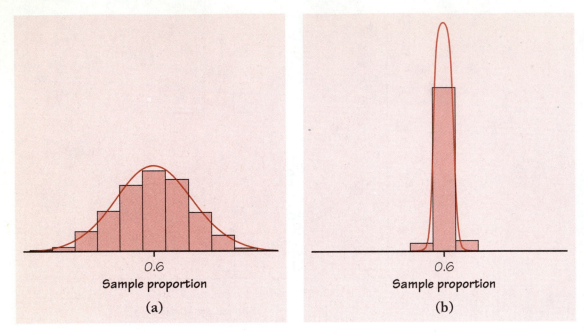

FIGURE 4.4 The sampling distributions for sample proportions $\hat{p}$ for SRSs of two sizes drawn from a population having population proportion $p = .6$. **(a)** Sample size 100. **(b)** Sample size 2500. Both statistics are unbiased because the mean of their distributions equals the true population value $p = .6$. The statistic from the larger sample is less variable.

UNBIASED STATISTIC

A statistic used to estimate a parameter is **unbiased** if the mean of its sampling distribution is equal to the true value of the parameter being estimated.

An unbiased statistic will sometimes fall above the true value of the parameter and sometimes below if we take many samples. Because its sampling distribution is centered at the true value, however, there is no systematic tendency to overestimate or underestimate the parameter. This makes the idea of lack of bias in the sense of "no favoritism" more precise. The sample proportion $\hat{p}$ from an SRS is an unbiased estimator of the population proportion p. If we draw an SRS from a population in which 60% find shopping frustrating, the mean of the sampling distribution of $\hat{p}$ is 0.6. If we draw an SRS from a population with 50% frustrated shoppers, the mean of $\hat{p}$ is then 0.5.

The variability of a statistic

The statistics whose sampling distributions appear in Figure 4.4 are both unbiased. That is, both distributions are centered at 0.6, the true population parameter. The sample proportion $\hat{p}$ from a random sample of any size is an unbiased estimate of the parameter p. Larger samples have a clear advantage, however. They are much more likely to produce an estimate close to the true value of the parameter because there is much less variability among large samples than among small samples.

EXAMPLE 4.3

The sampling distribution of $\hat{p}$ for samples of size 100, shown in Figure 4.4(a), is close to the normal distribution with mean 0.6 and standard deviation 0.05. Recall the 68–95–99.7 rule for normal distributions. It says that 95% of values of $\hat{p}$ fall within two standard deviations of the mean of the distribution. So 95% of all samples give an estimate $\hat{p}$ between

$$\text{mean} \pm (2 \times \text{standard deviation}) = .6 \pm (2 \times .05) = .6 \pm .1$$

If in fact 60% of the population find clothes shopping frustrating, the estimates from repeated SRSs of size 100 will usually fall between 50% and 70%. That's not very satisfactory.

For samples of size 2500, Figure 4.4(b) shows that the standard deviation is only about 0.01. So 95% of these samples will give an estimate within about 0.02 of the mean, that is, between 0.58 and 0.62. An SRS of size 2500 can be trusted to give sample estimates that are very close to the truth about the entire population. ◀

In Section 4.3 we will give the standard deviation of $\hat{p}$ for any size sample. We will then see Example 4.3 as part of a general rule that shows exactly how the variability of sample results decreases for larger samples. One important and surprising fact is that the spread of the sampling distribution does *not* depend very much on the size of the *population*.

VARIABILITY OF A STATISTIC

The variability of a statistic is described by the spread of its sampling distribution. This spread is determined by the sampling design and the size of the sample. Larger samples give smaller spread.

As long as the population is much larger than the sample (say, at least 10 times as large), the spread of the sampling distribution is approximately the same for any population size.

Why does the size of the population have little influence on the behavior of statistics from random samples? To see that this is plausible, imagine sampling harvested corn by thrusting a scoop into a lot of corn kernels. The scoop doesn't know whether it is surrounded by a bag of corn or by an entire truckload. As long as the corn is well mixed (so that the scoop selects a random sample), the variability of the result depends only on the size of the scoop.

The fact that the variability of sample results is controlled by the size of the sample has important consequences for sampling design. A statistic from an SRS of size 2500 from the more than 250,000,000 residents of the United States is just as precise as an SRS of size 2500 from the 740,000 inhabitants of San Francisco. This is good news for designers of national samples but bad news for those who want accurate information about the citizens of San Francisco. If both use an SRS, both must use the same size sample to obtain equally trustworthy results.

EXERCISES

4.7 The table below contains the results of simulating on a computer 100 repetitions of the drawing of an SRS of size 200 from a large lot of bearings. Ten percent of the bearings in the lot do not conform to the specifications. That is, $p = .10$ for this population. The numbers in the table are the counts of nonconforming bearings in each sample of 200.

17	23	18	27	15	17	18	13	16	18	20	15	18	16	21
17	18	19	16	23	20	18	18	17	19	13	27	22	23	26
17	13	16	14	24	22	16	21	24	21	30	24	17	14	16
16	17	24	21	16	17	23	18	23	22	24	23	23	20	19
20	18	20	25	16	24	24	24	15	22	22	16	28	15	22
9	19	16	19	19	25	24	20	15	21	25	24	19	19	20
28	18	17	17	25	17	17	18	19	18					

(a) Make a table that shows how often each count occurs. For each count in your table, give the corresponding value of the sample proportion

$$\hat{p} = \frac{\text{count}}{200}$$

Then draw a histogram for the values of the statistic $\hat{p}$.

(b) Is the shape of the distribution approximately normal?

(c) Find the mean of the 100 observations on $\hat{p}$. Mark the mean on your histogram to show its center. Does the statistic $\hat{p}$ appear to have large or small bias as an estimate of the population proportion p?

(d) The sampling distribution of $\hat{p}$ is the distribution of the values of $\hat{p}$ from all possible samples of size 200 from this population. What is the mean of this distribution?

(e) If we repeatedly selected SRSs of size 1000 instead of 200 from this same population, what would be the mean of the sampling distribution of the sample proportion $\hat{p}$? Would the spread be larger, smaller, or about the same when compared with the spread of your histogram in (a)?

4.8 Figure 4.5 shows histograms of four sampling distributions of statistics intended to estimate the same parameter. Label each distribution relative to the others as large or small bias and as large or small variability.

4.9 The Internal Revenue Service plans to examine an SRS of individual federal income tax returns from each state. One variable of interest is the proportion of returns claiming itemized deductions. The total number of tax returns in a state varies from almost 14 million in California to fewer than 210,000 in Wyoming.

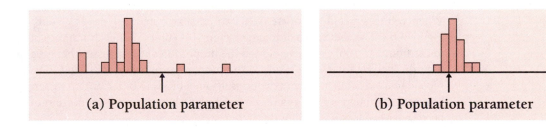

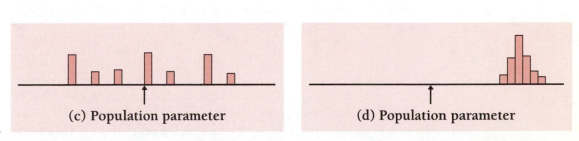

FIGURE 4.5 Which of these sampling distributions displays large or small bias and large or small variability? (Exercise 4.8)

(a) Will the sampling variability of the sample proportion change from state to state if an SRS of 2000 tax returns is selected in each state? Explain your answer.

(b) Will the sampling variability of the sample proportion change from state to state if an SRS of 1% of all tax returns is selected in each state? Explain your answer.

The language of probability

Toss a coin, or choose an SRS. The result can't be predicted in advance because the result will vary when you toss the coin or choose the sample repeatedly. But there is still a regular pattern in the results, a pattern that only emerges after many repetitions. The histograms in Figures 4.1 to 4.4 show the regular pattern in the values of the sample proportion $\hat{p}$ in many SRSs from the same population. Tossing a coin is simpler than choosing an SRS. Let's examine the long-term regularity of coin tossing.

EXAMPLE 4.4

When you toss a coin, there are only two possible outcomes, heads or tails. Figure 4.6 shows the results of tossing a coin 1000 times. For each number of tosses from 1 to 1000, we have plotted the proportion of those tosses that gave a head. The first toss was a head, so the proportion of heads starts at 1. The second toss was a tail, reducing the proportion of heads to 0.5 after two tosses. The next three tosses gave a tail followed by two heads, so the proportion of heads after five tosses is 3/5, or 0.6.

The proportion of tosses that produce heads is quite variable at first, but settles down as we make more and more tosses. Eventually this proportion gets close to 0.5 and stays there. We say that 0.5 is the *probability* of a head. The probability 0.5 appears as a horizontal line on the graph. ◀

"Random" in statistics does not mean haphazard but rather refers to a kind of order that emerges only in the long run. You can see that order emerging in Figure 4.6. In the very long run, the proportion of tosses that give a head is 0.5. This is the intuitive idea of probability. Probability 0.5 means "occurs half the time in a very large number of trials."

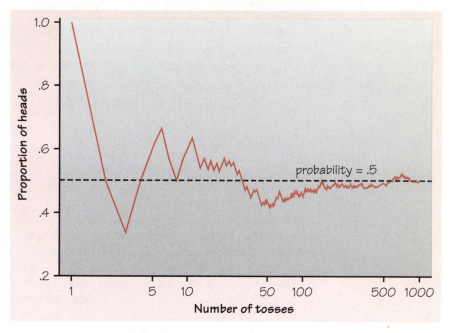

FIGURE 4.6 The behavior of the proportion of coin tosses that give a head, from 1 to 1000 tosses of a coin. In the long run, the proportion of heads approaches 0.5, the probability of a head.

RANDOMNESS AND PROBABILITY

We call a phenomenon **random** if individual outcomes are uncertain but there is nonetheless a regular distribution of outcomes in a large number of repetitions.

The **probability** of any outcome of a random phenomenon is the proportion of times the outcome would occur in a very long series of repetitions.

That some things are random is an observed fact about the world. The outcome of a coin toss, the time between emissions of particles by a radioactive source, and the sexes of the next litter of lab rats are all random. So is the value of a statistic from a random sample or a randomized experiment. Probability theory is the branch of mathematics that describes random behavior. Of course, we can never observe a probability exactly.

We could always continue tossing the coin, for example. Mathematical probability is an idealization based on imagining what would happen in an indefinitely long series of trials.

The sampling distributions in Figures 4.1 to 4.4 show the behavior of the statistic $\hat{p}$ in a large number of samples. We want to use the language of probability to describe this long-run behavior.

EXAMPLE 4.5

Choose an SRS of 2500 people from a population where $p = .6$ find shopping frustrating. Figure 4.3 is the sampling distribution produced by our simulation. It shows the values of $\hat{p}$ in 1000 such samples. Consider the outcome "$\hat{p}$ is greater than 0.61." In the simulation, this outcome occurred in 155 of the 1000 samples. So we estimate that the probability of this outcome is 155/1000, or 0.155. Only the language is new. Rather than saying "15.5% of all samples," we say "probability 0.155." Like the sampling distribution, probability really refers not to these specific 1000 samples but to an indefinitely large number of samples. Both are idealizations that we can approximate by looking at 1000 samples. *The sampling distribution of a statistic gives the probabilities that the statistic takes various values.* ◀

Although the mathematics of probability theory lies behind much that we do, we need only an intuitive understanding of probability and a few basic facts.

PROBABILITY FACTS

- Any probability is a number between 0 and 1.

- All possible outcomes together must have probability 1.

- The probability that an event does not occur is 1 minus the probability that the event does occur.

- If two events have no outcomes in common, the probability that one or the other occurs is the sum of their individual probabilities.

These facts follow from the idea of probability as "the long-run proportion of repetitions in which an event occurs." Any proportion is a number between 0 and 1, so any probability is also a number between 0 and 1. An event with probability 0 never occurs, and an event with probability 1 occurs on every trial. Because some outcome must occur on every trial, the sum of the probabilities for all possible outcomes must be exactly 1. If an event occurs in (say) 70% of all trials, it fails to occur in the other 30%. The probability that an event occurs and the probability that it does not occur always add to 100%, or 1. If one event occurs in 40% of all trials, a different event occurs in 25% of all trials, and the two can never occur together, then one or the other occurs on 65% of all trials because 40% + 25% = 65%.

EXAMPLE 4.6

Example 4.5 notes that 15.5% of the samples gave a $\hat{p}$ larger than 0.61. So the other 84.5% of the samples had a $\hat{p}$ less than or equal to 0.61. That commonsense statement translates into the language of probability as follows. The probability of an outcome greater than 0.61 is $P = .155$. So the probability of an outcome less than or equal to 0.61 is

$$1 - P = 1 - .155 = .845$$

◀

EXAMPLE 4.7

All human blood can be typed as one of O, A, B, or AB. The distribution of the types varies a bit with race. Choose a black American at random. Here are the probabilities that the person you choose will have blood type O, A, or B.

Blood type	O	A	B	AB
Probability	.49	.27	.20	?

What is the probability that the person chosen has either type A or type B blood? No person can have both blood types, so the probability of one or the other is the sum,

$$.27 + .20 = .47$$

What is the probability of type AB blood? The probability of all possible outcomes together must be 1. The outcomes O, A, and B together have probability

$$.49 + .27 + .20 = .96$$

So the probability of type AB blood must be $1 - .96 = .04$.   ◀

EXERCISES

4.10   Toss a thumbtack on a hard surface 100 times. How many times did it land with the point up? What is the approximate probability of landing point up?

4.11   The table of random digits (Table B) was produced by a random mechanism that gives each digit probability 0.1 of being a 0. What proportion of the first 200 digits in the table are 0s? This proportion is an estimate, based on 200 repetitions, of the true probability, which in this case is known to be 0.1.

4.12   Probability is a measure of how likely an event is to occur. Match one of the probabilities that follow with each statement of likelihood given. (The probability is usually a more exact measure of likelihood than is the verbal statement.)

0 0.01 0.3 0.6 0.99 1

(a) This event is impossible. It can never occur.

(b) This event is certain. It will occur on every trial of the random phenomenon.

(c) This event is very unlikely, but it will occur once in a while in a long sequence of trials.

(d) This event will occur more often than not.

4.13   If you draw an M&M candy at random from a bag of the candies, the candy you draw will have one of six colors. The probability of drawing each color depends on the proportion of each color among all candies made. The table below gives the probability that a randomly chosen M&M has each color.

Color	Brown	Red	Yellow	Green	Orange	Tan
Probability	.3	.2	.2	.1	.1	?

(a) What must be the probability of drawing a tan candy?

(b) What is the probability that the candy drawn is any of red, yellow, or orange?

4.14 Choose at random an American woman aged 25 to 29 years. The probability that you choose a single woman is equal to the proportion of single women in this age group. The Census Bureau tells us that the probabilities that we choose a woman who is single, widowed, or divorced are

Outcome	Single	Married	Widowed	Divorced
Probability	.288	?	.003	.076

(a) What is the probability that a woman in this age group is married?

(b) What is the probability that a woman in this age group is either widowed or divorced?

SUMMARY

A number that describes a population is called a **parameter**. A number that can be computed from the sample data is called a **statistic**. The purpose of sampling or experimentation is usually to use statistics to make statements about unknown parameters.

A statistic from a probability sample or randomized experiment has a **sampling distribution** that describes how the statistic varies in repeated data production. The sampling distribution answers the question, "What would happen if we repeated the sample or experiment many times?" Formal statistical inference is based on the sampling distributions of statistics.

A statistic as an estimator of a parameter may suffer from **bias** or from high **variability**. Bias means that the center of the sampling distribution is not equal to the true value of the parameter. The variability of the statistic is described by the spread of its sampling distribution.

Properly chosen statistics from randomized data production designs have no bias resulting from the way the sample is selected or the way the experimental units are assigned to treatments. The variability of the statistic is determined by the size of the sample or by the size of the experimental groups. Statistics from larger samples have less variability.

The **probability** of any outcome of a random phenomenon is the proportion of a very large number of repetitions in which the outcome will occur. Any probability is a number between 0 and 1.

The sampling distribution of a statistic gives the probabilities for all the values of the statistic.

SECTION 4.1. EXERCISES

4.15   An entomologist samples a field for egg masses of a harmful insect by placing a yard-square frame at random locations and examining the ground within the frame carefully. He wants to estimate the proportion of square yards in which egg masses are present. Suppose that in a large field egg masses are present in 20% of all possible yard-square areas. That is, $p = .2$ in this population.

(a) Use Table B to simulate the presence or absence of egg masses in each square yard of an SRS of 10 square yards from the field. Be sure to explain clearly which digits you used to represent the presence and the absence of egg masses. What proportion of your 10 sample areas had egg masses? This is the statistic $\hat{p}$.

(b) Repeat (a) with different lines from Table B, until you have simulated the results of 20 SRSs of size 10. What proportion of the square yards in each of your 20 samples had egg masses? Make a stemplot from these 20 values to display the distribution of your 20 observations on $\hat{p}$. What is the mean of this distribution? What is its shape?

(c) If you looked at all possible SRSs of size 10, rather than just 20 SRSs, what would be the mean of the values of $\hat{p}$? This is the mean of the sampling distribution of $\hat{p}$.

(d) In another field, 40% of all square-yard areas contain egg masses. What is the mean of the sampling distribution of $\hat{p}$ in samples from this field?

4.16   An opinion poll asks, "Are you afraid to go outside at night within a mile of your home because of crime?" Suppose that the proportion of all adults who would say "Yes" to this question is $p = .4$.

(a) Use Table B to simulate the result of an SRS of 20 adults. Be sure to explain clearly which digits you used to represent each of "Yes" and "No." What proportion of your 20 responses were "Yes"?

(b) Repeat (a) using different lines in Table B until you have simulated the results of 10 SRSs of size 20 from the same population. Compute the proportion of "Yes" responses in each sample. These are the values of the statistic $\hat{p}$ in 10 samples. Find the mean of your 10 values of $\hat{p}$. Is it close to p?

(c) The sampling distribution of $\hat{p}$ is the distribution of $\hat{p}$ from all possible SRSs of size 20 from this population. What would be the mean of this distribution?

(d) If the population proportion changed to $p = .5$, what would then be the mean of the sampling distribution of $\hat{p}$?

4.17   A national opinion poll recently estimated that 44% ($\hat{p} = .44$) of all adults agree that parents of school-age children should be given vouchers good for education at any public or private school of their choice. The polling organization used a probability sampling method for which the sample proportion $\hat{p}$ has a normal distribution with standard deviation about 0.015. If a sample were drawn by the same method from the state of New Jersey (population 7.8 million) instead of from the entire United States (population 250 million), would this standard deviation be larger, about the same, or smaller? Explain your answer.

4.18   Hold a penny on edge on a flat surface with the index finger of one hand and snap it with your other index finger so that it spins rapidly until finally falling with either heads or tails upward. Repeat this 50 times and record the number of heads. What is your estimate of the probability of a head? (Disregard any trial in which the penny spins for less than several seconds or in which it hits an obstacle.)

4.19   When we toss a penny, experience shows that the probability of a head is close to 1/2. Suppose now that we toss the penny repeatedly until we get a head. What is the probability that the first head comes up in an *odd* number of tosses (1, 3, 5, and so on)? To find out, toss a penny to the first head 50 times. Keep a record of the result of the first toss and of the number of tosses needed to get a head on each of your 50 trials.

(a) Based on the first tosses in your 50 trials, estimate the probability of getting a head on the first toss. We expect that very many tosses would give an estimate close to 0.5.

(b) Use your 50 trials to estimate the probability that the first head appears on an odd-numbered toss.

4.20   Government data assign a single cause for each death that occurs in the United States. The data show that the probability is 0.42 that a randomly chosen death was due to cardiovascular (mainly heart) diseases, and 0.24 that it was due to cancer. What is the probability that a death was due either to cardiovascular disease or to cancer? What is the probability that the death was due to some other cause?

4.21   Choose an acre of land in Canada at random. The probability is 0.35 that it is forest and 0.03 that it is pasture.

(a) What is the probability that the acre chosen is not forested?

(b) What is the probability that it is either forest or pasture?

(c) What is the probability that a randomly chosen acre in Canada is something other than forest or pasture?

4.22   Select a first-year college student at random and ask what his or her academic rank was in high school. Here are the probabilities, based on a large sample survey of students.

Outcome	Top 20%	Second 20%	Third 20%	Fourth 20%	Lowest 20%
Probability	.41	.23	.29	.06	.01

(a) What is the sum of these probabilities? Why do you expect the sum to have this value?

(b) What is the probability that a randomly chosen first-year college student was not in the top 20% of his or her high school class?

(c) What is the probability that a first-year student was in the top 40% in high school?

4.23   **(Optional)** A roulette wheel contains compartments numbered 1 through 36 plus 0 and 00. Of the 38 compartments, 0 and 00 are colored green, 18 of the others are red, and 18 are black. A ball is spun in the direction opposite to the wheel's motion, and bets are made on the number where the ball comes to rest. A simple wager is "red or black," in which you bet that the ball will stop in, say, a red compartment. If the wheel is fair, all 38 compartments are equally likely. What is the probability of a red?

4.2   PROBABILITY DISTRIBUTIONS*

The distribution of a variable tells us what values it takes and how often it takes each value. We can apply the same idea to describe the outcomes of a random phenomenon and their probabilities. We are most interested in statistics such as proportions and means from random samples and randomized comparative experiments. These statistics take numerical values that vary when we repeat the randomization, so we call them *random variables*.

RANDOM VARIABLE

A random variable is a variable whose value is a numerical outcome of a random phenomenon.

---

*This more advanced section gives more detail about probability. It is not needed to read the rest of the book.

It is usual to denote random variables by capital letters near the end of the alphabet, such as X or Y. Of course, we are most interested in statistics like $\hat{p}$ and $\overline{x}$, and we won't give them new names. The ***probability distribution*** of a random variable tells us what the possible values of the variable are and how probabilities are assigned to those values.

probability distribution

Discrete random variables

The probability distribution of a random variable is most straightforward when the variable has only a finite number of possible values.

EXAMPLE 4.8

The instructor of a large class has a rigid grading scheme. She gives 15% each of A's and D's, 30% each of B's and C's, and 10% F's. Choose a student at random from this class. To "choose at random" means to give every student the same chance to be chosen. The student's grade on a four-point scale (A = 4.0) is a random variable X.

The value of X changes when we repeatedly choose students at random, but it is always one of 0, 1, 2, 3, or 4. Here is the distribution of X.

Grade	0	1	2	3	4
Probability	.10	.15	.30	.30	.15

The probabilities add to exactly 1 because every student must get one of the five grades.

The probability that the student got a B or better is the sum of the probabilities of an A and a B,

$$P(\text{grade is 3 or 4}) = P(X = 3) + P(X = 4)$$
$$= .30 + .15 = .45$$

The capital P stands for "probability." Read an expression like $P(X = 3)$ as "the probability that X equals 3." ◀

Example 4.8 presents the distribution of X in the form of a table of the possible values and their probabilities. We can make such a table whenever X takes only finitely many values.

DISCRETE RANDOM VARIABLE

A **discrete random variable** X has a finite number of possible values. The **probability distribution** of X lists the values and their probabilities,

Value of X	x_1	x_2	x_3	$\cdots$	x_k
Probability	p_1	p_2	p_3	$\cdots$	p_k

The probabilities p_i must satisfy two requirements:

1. Every probability p_i is a number between 0 and 1.
2. $p_1 + p_2 + \cdots + p_k = 1$.

Find the probability of any event by adding the probabilities p_i of the particular values x_i that make up the event.

EXAMPLE 4.9

A household is a group of people living together, regardless of their relationship to each other. Many sample surveys such as the Current Population Survey select a random sample of households. Choose a household at random, and let the random variable X be the number of people living there. Here is the distribution of X.

Household size	1	2	3	4	5	6	7
Probability	.251	.321	.171	.154	.067	.022	.014

These probabilities are the proportions for all households in the country. They give the probabilities that a single household chosen at random will have each size. (The very few households with more than 7 members are placed in the 7 group.) Check that the probabilities add to exactly 1.

The probability that a randomly chosen household has more than two members is

$$P(X > 2) = P(X = 3) + P(X = 4) + P(X = 5) + P(X = 6) + P(X = 7)$$
$$= .171 + .154 + .067 + .022 + .014 = .428 \quad \blacktriangleleft$$

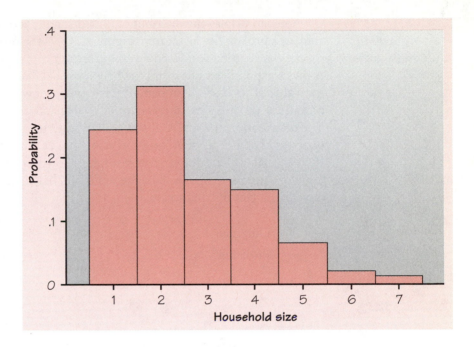

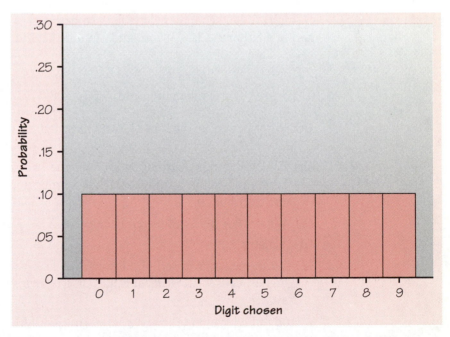

FIGURE 4.7 Probability histograms for two discrete random variables. (a) The size of a randomly chosen household (Example 4.9). (b) One digit drawn from the table of random digits.

probability
histogram

You can display the distribution of a discrete random variable graphically in a **probability histogram**. Figure 4.7(a) is the probability histogram for the distribution of household size in Example 4.9. The horizontal scale shows the possible values, and the height of each bar is the probability for the value at its base. A probability histogram shows quickly what values a random variable can take and which values are more or less probable. The distribution of household size is skewed to the right, and about three-quarters of all households consist of one, two, or three people. In contrast, Figure 4.7(b) is the probability histogram for drawing one digit from a table of random digits. Each of the ten digits has the same probability. In table form, this distribution is

Outcome	0	1	2	3	4	5	6	7	8	9
Probability	.1	.1	.1	.1	.1	.1	.1	.1	.1	.1

EXERCISES

4.24   A study of social mobility in England looked at the social class reached by the sons of lower-class fathers. Social classes are numbered from 1 (low) to 5 (high). Take the random variable X to be the class of a randomly chosen son of a father in Class 1. The study found that the distribution of X is

Son's class	1	2	3	4	5
Probability	.48	.38	.08	.05	.01

(a) What percent of the sons of lower-class fathers reach the highest class, Class 5?

(b) Check that this distribution satisfies the two requirements for a discrete probability distribution. Draw a probability histogram to display the distribution.

(c) What is $P(X \leq 3)$? (Be careful: the event "$X \leq 3$" includes the value 3.)

(d) What is $P(X < 3)$?

(e) Write the event "a son of a lower-class father reaches one of the two highest classes" in terms of values of X. What is the probability of this event?

4.25   A study of education followed a large group of fifth-grade children to see how many years of school they eventually completed. Let X be the highest year of school that a randomly chosen fifth grader completes. (Students who go on to college are included in the outcome $X = 12$.) The study found this probability distribution for X:

Years	4	5	6	7	8	9	10	11	12
Probability	.010	.007	.007	.013	.032	.068	.070	.041	.752

(a) What percent of fifth graders eventually finished twelfth grade?

(b) Check that this is a legitimate discrete probability distribution and draw a probability histogram to display it.

(c) Find $P(X \geq 6)$. (Be careful: the event "$X \geq 6$" includes the value 6.)

(d) Find $P(X > 6)$.

(e) What values of X make up the event "The student completed at least one year of high school?" (High school begins with the ninth grade.) What is the probability of this event?

4.26   If a carefully made die is rolled once, it is reasonable to assign probability 1/6 to each of the six faces. What is the probability of rolling a number less than 3?

Equally likely outcomes

When we look up an entry in a table of random digits, the 10 digits 0, 1, 2, 3, 4, 5, 6, 7, 8, and 9 are all equally likely. Because their probabilities must add to 1, each of the 10 outcomes must have probability 1/10. Five of the 10 possible outcomes are odd, so the probability of getting an odd number is 5/10. When we draw a card from a shuffled deck, all 52 cards in the deck are equally likely. So any one card has probability 1/52. Of the 52 cards, 13 are spades. The probability of drawing a spade is therefore 13/52. Here is the general rule for finding probabilities when all outcomes are equally likely.

EQUALLY LIKELY OUTCOMES

If a random phenomenon has k possible outcomes, all equally likely, then each individual outcome has probability $1/k$. The probability of any event A is

$$P(A) = \frac{\text{count of outcomes in } A}{\text{count of all possible outcomes}}$$
$$= \frac{\text{count of outcomes in } A}{k}$$

Probabilities in the equally likely case are found by counting outcomes. Here is an example.

EXAMPLE 4.10

Roll two dice and record the pips on each of the two up-faces. Figure 4.8 shows the 36 possible outcomes. If the dice are carefully made, all 36 outcomes are equally likely. So each has probability 1/36. Gamblers are often interested in the sum of the pips on the up faces.

What is the probability of rolling a 5? The event "roll a 5" contains the four outcomes

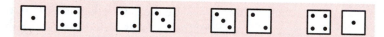

The probability is therefore 4/36, or about 0.111. What about the probability of rolling a 7? In Figure 4.8 you will find six outcomes for which the sum of the pips is 7. The probability is 6/36, or about 0.167.

The sum of the pips is a discrete random variable X. By continuing to count outcomes in Figure 4.8 you can obtain the distribution of X. Here it is.

Value of X	2	3	4	5	6	7	8	9	10	11	12
Probability	$\frac{1}{36}$	$\frac{2}{36}$	$\frac{3}{36}$	$\frac{4}{36}$	$\frac{5}{36}$	$\frac{6}{36}$	$\frac{5}{36}$	$\frac{4}{36}$	$\frac{3}{36}$	$\frac{2}{36}$	$\frac{1}{36}$

Notice that the values of X are *not* equally likely. ◀

FIGURE 4.8 The 36 possible outcomes in rolling two dice. These outcomes are equally likely to occur, so each has probability 1/36.

EXERCISES

4.27   A computer system assigns passwords to users at random. Each password begins with one of the letters A to Z. Because the assignment is random, it is reasonable to assign probability 1/26 to each of the 26 possible first letters. What is the probability that a password begins with a vowel (A, E, I, O, or U)?

4.28   A couple plans to have three children. There are 8 possible arrangements of girls and boys. For example, GGB means the first two children are girls and the third child is a boy. All 8 arrangements are (approximately) equally likely.

   (a) Write down all 8 arrangements of the sexes of three children. What is the probability of any one of these arrangements?

   (b) Let X be the number of girls the couple has. What is the probability that $X = 2$?

   (c) Starting from your work in (a), find the distribution of X. That is, what values can X take, and what are the probabilities for each value?

4.29   Automobile license plate numbers in Indiana consist of seven characters. The first three describe the county in which the car is licensed, while the last four are digits assigned at random. You are hoping for a plate on which these four digits are identical (like 7777). What is your probability of receiving such a plate? Answer this question in these steps:

   (a) How many different four-digit numbers 0000, 0001, ..., 9999 are there?

   (b) If the plate number is assigned at random, what is the probability of getting any specific four-digit number on your license plate?

   (c) How many four-digit numbers are there with all four digits identical? What is the probability that your plate has such a number?

The mean and standard deviation of a discrete random variable

Probability distributions of random variables, like distributions of data, can be described by numerical measures of center and spread. The most common measure of center is the mean and the most common measure of spread or variability is the standard deviation. The idea of the mean is that it is the average of the values of the variable. In the case of random variables, we need to take into account the fact that not all values need be equally probable.

EXAMPLE 4.11

The number of people living in a randomly chosen household is a random variable X. The distribution of X is

Household size	1	2	3	4	5	6	7
Probability	.251	.321	.171	.154	.067	.022	.014

The average size of a household is not the ordinary average of the sizes 1 to 7. After all, 32.1% of households have 2 people and only 1.4% have 7 people. The correct average is found by weighting each size by how common it is, and this is how we define the mean. The mean number of people in a household is

$$\mu = (1)(.251) + (2)(.321) + (3)(.171) + (4)(.154) + (5)(.067)$$
$$+ (6)(.022) + (7)(.014)$$
$$= 2.587$$

The Greek letter μ (mu) is the usual notation for the mean of a random variable.

◄

MEAN OF A DISCRETE RANDOM VARIABLE

Suppose that X is a discrete random variable whose distribution is

Value of X	x_1	x_2	x_3	$\cdots$	x_k
Probability	p_1	p_2	p_3	$\cdots$	p_k

Find the **mean** of X by multiplying each possible value by its probability and adding over all the values:

$$\mu = x_1 p_1 + x_2 p_2 + \cdots + x_k p_k$$

The mean of a random variable X is a single fixed number μ. It gives the average value of X in several senses:

- The mean μ is the average of the possible values of X, each weighted by how likely it is to occur. That's what the definition of μ says.
- The mean μ is the point at which the probability histogram of the distribution of X would balance if made of solid material. See Figure 4.9. Recall that the mean μ of a density curve has this same property.
- If we actually repeat the random phenomenon many times, record the value of X each time, and average these observed values, this average will get closer and closer to μ as we make more and more repetitions. This fact is called the **law of large numbers**.

law of large numbers

We will discuss the law of large numbers in more detail in Section 4.5. It says that we can think of μ as the ordinary average of the outcomes we get in an indefinitely large number of trials. As we make more and more repetitions of a random phenomenon, the proportion of trials on which each possible value x_i occurs approaches its probability p_i, and the average of all the observed values approaches the mean μ.

Probability distributions, like distributions of data, also have standard deviations. The idea of the standard deviation is the same in both settings:

- Start with the deviations of the possible values of the variable from their mean. The *variance* is an average of the squares of these deviations.
- The *standard deviation* is the square root of the variance.

Just as with the mean, the right average in dealing with a random variable weights each possible value by its probability. Here is the definition of the standard deviation.

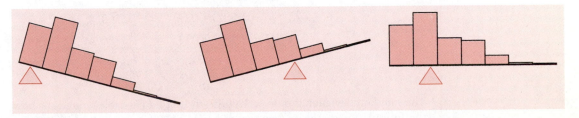

FIGURE 4.9   The mean μ of a discrete random variable is the point at which the probability histogram balances.

STANDARD DEVIATION OF A DISCRETE RANDOM VARIABLE

Suppose that X is a discrete random variable whose distribution is

Value of X	x_1	x_2	x_3	$\cdots$	x_k
Probability	p_1	p_2	p_3	$\cdots$	p_k

and that μ is the mean of X. The **variance** of X is

$$\sigma^2 = (x_1 - \mu)^2 p_1 + (x_2 - \mu)^2 p_2 + \cdots + (x_k - \mu)^2 p_k$$

The **standard deviation** σ is the square root of the variance.

EXAMPLE 4.12

Example 4.11 gives the distribution of the number of people X living in a randomly chosen household. The mean of X is $\mu = 2.587$. To find the standard deviation of X we can arrange the calculation in a table as follows.

x_i	p_i	$(x_i - \mu)^2 p_i$		
1	.251	$(1 - 2.587)^2 (.251)$	=	.6322
2	.321	$(2 - 2.587)^2 (.321)$	=	.1106
3	.171	$(3 - 2.587)^2 (.171)$	=	.0292
4	.154	$(4 - 2.587)^2 (.154)$	=	.3075
5	.067	$(5 - 2.587)^2 (.067)$	=	.3901
6	.022	$(6 - 2.587)^2 (.022)$	=	.2563
7	.014	$(7 - 2.587)^2 (.014)$	=	.2726
		σ_X^2	=	1.9985

The standard deviation is $\sigma = \sqrt{1.9985} = 1.414$. (The table reminds us how the formula for σ^2 works. You can reduce roundoff error by carrying all the steps on your calculator rather than writing down the seven intermediate results.) ◀

The standard deviation of a random variable shares the properties of the standard deviation of a distribution of data. For example, it is always 0 or positive and $\sigma = 0$ only when there is no variability (that is, when the random variable takes only one value).

EXERCISES

4.30 Example 4.10 (page 256) gives the distribution of the sum X of the pips when rolling two dice.

(a) Make a probability histogram of the distribution of X.

(b) Find the mean μ of X. Then explain why the result you get could be seen from the histogram in (a) without calculation.

4.31 One digit from the table of random digits can take the values 0, 1, 2, 3, 4, 5, 6, 7, 8, and 9. All 10 values are equally likely. Figure 4.7(b) (page 253) is a probability histogram of the distribution. Find the mean μ. Then explain why the result you get could be seen from the histogram without calculation.

4.32 Find the standard deviation σ of the sum X of the pips when rolling two dice. Start from the mean μ that you found in Exercise 4.30. Mark on your histogram from Exercise 4.30 the points one standard deviation above and below the mean. What is the probability of getting an outcome in this range?

4.33 Find the standard deviation σ of a digit drawn from the table of random digits. Start from the mean μ that you found in Exercise 4.31. Make a copy of the probability histogram (Figure 4.7(b)) and mark on it the mean and the points one standard deviation above and below the mean. What is the probability of getting an outcome in this range?

Continuous random variables

Some random variables have values that are not isolated numbers but an entire interval of numbers. When you measure the height of a randomly chosen young woman, for example, you might get any number between about 3 feet and about 7 feet. In practice, you may measure only to the nearest tenth of an inch or hundredth of an inch. It is simpler not to worry about just how precise the measurements are and to think of possible heights as continuously distributed between 3 feet and 7 feet. When we described the distributions of data, we sometimes replaced histograms by density curves. We do the same thing to describe probabilities when we want to allow an entire interval of possible values.

CONTINUOUS RANDOM VARIABLE

A **continuous random variable** X takes all values in an interval of numbers. The **probability distribution** of X is described by a density curve. The probability of any event is the area under the density curve and above the values of X that make up the event.

The distribution of a continuous random variable assigns probabilities as areas under a density curve. Figure 4.10 illustrates the idea. The total area under any density curve is 1, so once again any probability is a number between 0 and 1, and the total probability assigned to all possible values of X is 1. Discrete distributions assign a probability to each individual value. Continuous distributions assign probabilities to intervals of values rather than to individual values.

A continuous random variable, like a discrete random variable, has a mean and a standard deviation. You can interpret the mean μ of a continuous random variable in the same three ways as the mean of a discrete random variable.

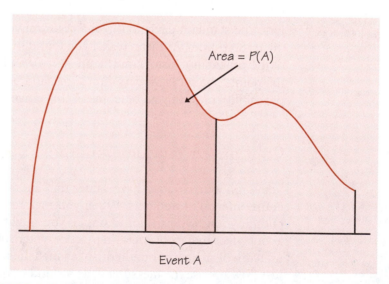

FIGURE 4.10 The probability distribution of a continuous random variable assigns probabilities as areas under a density curve.

- μ is an average of the possible values of the variable, weighted by their probabilities. Because a continuous random variable has infinitely many possible values, we can no longer express this idea by writing a sum.

- μ is the mean of the density curve, the point at which the curve would balance if made of solid material.

- The law of large numbers remains true: as we make more and more observations on X, the average of the outcomes we observe will get ever closer to μ.

The variance σ^2 is again the average of the squared deviations of the possible values of the variable from their mean. The standard deviation σ is the square root of the variance. As in the case of distributions of data, the standard deviation of a density curve isn't easy to interpret except for normal distributions.

Normal distributions

Any density curve gives the distribution of a continuous random variable. The density curves that are most familiar to us are the normal curves. So *normal distributions are probability distributions*. There is a close connection between a normal distribution as an idealized description for data and a normal probability distribution. If we look at the heights of all young women, we find that they closely follow the normal distribution with mean $\mu = 64.5$ inches and standard deviation $\sigma = 2.5$ inches. That is a distribution for a large set of data. Imagine choosing one young woman at random. Her height is a random variable X. If we repeat the random choice very many times, the distribution of values of X is the same normal distribution.

EXAMPLE 4.13

What is the probability that a randomly chosen young woman has height between 68 and 70 inches?

The height is a random variable X with the $N(64.5, 2.5)$ distribution. Find the probability by standardizing and using Table A, the table of standard normal probabilities. We will reserve capital Z for a standard normal variable.

$$P(68 \leq X \leq 70) = P\left(\frac{68 - 64.5}{2.5} \leq \frac{X - 64.5}{2.5} \leq \frac{70 - 64.5}{2.5}\right)$$

$$= P(1.4 \leq Z \leq 2.2)$$

$$= .9861 - .9192 = .0669$$

Figure 4.11 shows the areas under the standard normal curve. The calculation is the same as those we did in Chapter 1. Only the language of probability is new.   ◀

EXERCISES

4.34   Many computer systems have "random number generators" that produce numbers that are distributed uniformly between 0 and 1. The number produced on one trial is a continuous random variable X. Figure 4.12 graphs the density curve of X.

(a) Check that the area under this curve is 1.

(b) What is the probability that the number is less than 0.25? (Sketch the density curve, shade the area that represents the probability, then find that area. Do this for (c) and (d) also.)

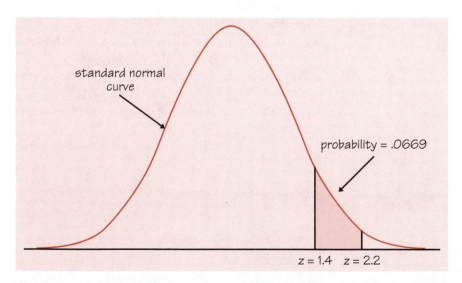

FIGURE 4.11   The probability in Example 4.13 as an area under the standard normal curve.

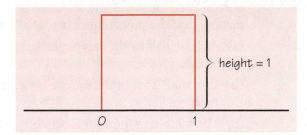

FIGURE 4.12 The density curve for the output of a random number generator. The values of this continuous random variable are distributed uniformly between 0 and 1.

(c) What is the probability that the number lies between 0.1 and 0.9?

(d) What is the probability that the number is either less than 0.1 or greater than 0.9?

4.35 Generate two random numbers between 0 and 1 and take Y to be their sum. Then Y is a continuous random variable that can take any value between 0 and 2. The density curve of Y is the triangle shown in Figure 4.13.

(a) Check that the area under this curve is 1.

(b) What is the probability that Y is less than 1? (Sketch the density curve, shade the area that represents the probability, then find that area. Do this for (c) also.)

(c) What is the probability that Y is less than 0.5?

4.36 An SRS of 400 American adults is asked, "What do you think is the most serious problem facing our schools?" Suppose that in fact 30% of all adults would answer "drugs" if asked this question. That is, the population proportion is $p = .3$. The sample proportion $\hat{p}$ who answer "drugs" will vary in repeated sampling. The sampling distribution will be approximately

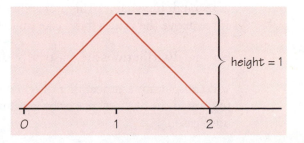

FIGURE 4.13 The density curve for the sum of two random numbers. This continuous random variable takes values between 0 and 2.

normal with mean 0.3 and standard deviation 0.023. Using this approximation, find the probabilities of the following events:

(a) At least half of the sample believes that drugs are the schools' most serious problem.

(b) Less than 25% of the sample believes that drugs are the most serious problem.

(c) The sample proportion is between 0.25 and 0.35.

SUMMARY

A **random variable** is a variable taking numerical values determined by the outcome of a random phenomenon. The **probability distribution** of a random variable X tells us what the possible values of X are and how probabilities are assigned to those values.

A random variable X and its distribution can be **discrete** or **continuous**.

A **discrete random variable** has finitely many possible values. The probability distribution assigns each of these values a probability between 0 and 1 such that the sum of all the probabilities is exactly 1. The probability of any event is the sum of the probabilities of all the values that make up the event.

If all the possible outcomes of a random phenomenon are **equally likely**, we find the probability of any event by counting the outcomes in the event and dividing by the total number of possible outcomes.

A **continuous random variable** takes all values in some interval of numbers. A **density curve** describes the probability distribution of a continuous random variable. The probability of any event is the area under the curve above the values that make up the event.

Normal distributions are one type of continuous probability distribution.

You can picture a probability distribution by drawing a **probability histogram** in the discrete case or by graphing the density curve in the continuous case.

The distribution of a random variable X, like a distribution of data, has a **mean μ** and a **standard deviation σ**.

The **mean** μ is the balance point of the probability histogram or density curve. The **law of large numbers** says that the average of the values of X observed in many trials must approach μ. If X is discrete with possible values x_i having probabilities p_i, you can find the mean from the distribution as the average of the values, each weighted by its probability:

$$\mu = x_1 p_1 + x_2 p_2 + \cdots + x_k p_k$$

The **variance** σ^2 is the average squared deviation of the values of the variable from their mean. For a discrete random variable,

$$\sigma^2 = (x_1 - \mu)^2 p_1 + (x_2 - \mu)^2 p_2 + \cdots + (x_k - \mu)^2 p_k$$

The **standard deviation** σ is the square root of the variance. The standard deviation measures the variability of the distribution about the mean. It is easiest to interpret for normal distributions.

SECTION 4.2 EXERCISES

4.37   Example 4.8 (page 251) gives the distribution of grades (A = 4, B = 3, and so on) in a large course as

Outcome	0	1	2	3	4
Probability	.10	.15	.30	.30	.15

Find the mean grade in this course. Is the mean grade higher or lower than a B? Find the standard deviation of the grades.

4.38   Keno is a favorite game in casinos. Balls numbered 1 to 80 are tumbled in a machine as the bets are placed, then 20 of the balls are chosen at random. Players select numbers by marking a card. The simplest of the many wagers available is "Mark 1 Number." The payoff is $3 on a $1 bet if the number you select is one of those chosen. Because 20 of 80 numbers are chosen, the probability of winning is 20/80, or 0.25. The winnings X on one play have two possible values: $2 with probability 0.25 and $-$1 (that is, the player loses his dollar) with probability 0.75.

(a) What is the mean amount μ of a player's winnings?

(b) The casino wins what a player loses. According to the law of large numbers, what are the casino's long-run average winnings from each dollar bet?

(c) What is the standard deviation σ of a player's winnings on one play of keno?

4.39   Exercise 4.25 (page 254) gives the distribution of the number of years of school eventually completed by children who have reached the fifth grade. What is the mean number of years of school completed by these children? What is the standard deviation?

4.40   Most software random number generators allow users to specify the range of the random numbers to be produced. You can, for example, generate a random number X between 0 and 2. The distribution of X spreads its probability uniformly between 0 and 2. The density curve of X should have constant height between 0 and 2, and height 0 elsewhere.

(a) What is the height of the density curve between 0 and 2? Draw a graph of the density curve of X.

(b) What is the mean μ (the balance point) of this distribution?

(c) Use your graph from (a) and the fact that probability is area under the curve to find $P(X \leq 1)$.

(d) Find $P(.5 < X < 1.3)$.

(e) Find $P(X \geq .8)$.

4.41   An opinion poll asks an SRS of 1500 adults, "Do you happen to jog?" Suppose (as is approximately correct) that the population proportion who jog is $p = .15$. Then the proportion $\hat{p}$ in the sample who answer "Yes" will be approximately normally distributed with mean $\mu = .15$ and standard deviation $\sigma = .0092$. Find the following probabilities:

(a) $P(\hat{p} \geq .16)$

(b) $P(.14 \leq \hat{p} \leq .16)$

4.3   SAMPLE PROPORTIONS

population proportion p

sample proportion $\hat{p}$

Retailers would like to know what proportion of all adults find clothes shopping frustrating and time-consuming. This unknown **population proportion** is a parameter p. A random sample of 2500 people found 1650 frustrated by clothes shopping. The **sample proportion**

$$\hat{p} = \frac{1650}{2500} = .66$$

is a statistic that we use to gain information about the parameter p. In everyday language we often express proportions as percents. We may say, "66% of the sample found clothes shopping frustrating." Statistical recipes work with proportions as decimal fractions, so 66% becomes 0.66.

The sampling distribution of $\hat{p}$

How good is the statistic $\hat{p}$ as an estimate of the parameter p? To find out, we ask, "What would happen if we took many samples?" The sampling distribution of $\hat{p}$ answers this question. In the simulation examples in Section 4.1, we found:

- The sampling distribution of the sample proportion $\hat{p}$ has a shape that is close to normal.
- Its mean is close to the population proportion p.
- Its standard deviation gets smaller as the size of the sample gets larger.

We looked at the results of 1000 samples. The mathematics of probability describes the actual sampling distribution from all possible samples. Here are the facts.

SAMPLING DISTRIBUTION OF A SAMPLE PROPORTION

Choose an SRS of size n from a large population with population proportion p having some characteristic of interest. Let $\hat{p}$ be the proportion of the sample having that characteristic. Then:

- The sampling distribution of $\hat{p}$ is **approximately normal** and is closer to a normal distribution when the sample size n is large.

- The **mean** of the sampling distribution is exactly p.

- The **standard deviation** of the sampling distribution is

$$\sqrt{\frac{p(1-p)}{n}}$$

Because the mean of the sampling distribution of $\hat{p}$ is always equal to the parameter p, the sample proportion $\hat{p}$ is an unbiased estimator of p. The standard deviation of $\hat{p}$ gets smaller as the sample size n increases because n appears in the denominator of the formula for the standard deviation. That is, $\hat{p}$ is less variable in larger samples. What is more, the formula shows just how quickly the standard deviation decreases as n increases. The sample size n is under the square root sign, so to cut the standard deviation in half, we must take a sample four times as large, not just twice as large.

The formula for the standard deviation of $\hat{p}$ doesn't apply when the sample is a large part of the population. You can't use this recipe if you choose an SRS of 50 of the 100 people in a class, for example. In practice, we usually take a sample only when the population is large. Otherwise, we could examine the entire population. Here is a practical guide.[3]

RULE OF THUMB 1

Use the recipe for the standard deviation of $\hat{p}$ only when the population is at least 10 times as large as the sample.

EXAMPLE 4.14

You ask an SRS of 1500 first-year college students whether they applied for admission to any other college. There are over 1.7 million first-year college students, so the rule of thumb is easily satisfied. In fact, 35% of all first-year students applied to colleges besides the one they are attending. What is the probability that your sample will give a result within 2 percentage points of this true value?

We have an SRS of size $n = 1500$ drawn from a population in which the proportion $p = .35$ applied to other colleges. The sample proportion $\hat{p}$ has mean 0.35 and standard deviation

$$\sqrt{\frac{p(1-p)}{n}} = \sqrt{\frac{(.35)(.65)}{1500}} = \sqrt{.00015167} = .0123$$

We want the probability that $\hat{p}$ falls between 0.33 and 0.37 (within 2 percentage points, or 0.02, of 0.35). This is a normal distribution calculation.

Standardize $\hat{p}$ by subtracting its mean 0.35 and dividing by its standard deviation 0.0123. That produces a new statistic that has the standard normal distribution. It is usual to call such a statistic Z:

$$Z = \frac{\hat{p} - .35}{.0123}$$

Then draw a picture of the areas under the standard normal curve (Figure 4.14), and use Table A to find them. Here is the calculation.

$$P(.33 \leq \hat{p} \leq .37) = P\left(\frac{.33 - .35}{.0123} \leq \frac{\hat{p} - .35}{.0123} \leq \frac{.37 - .35}{.0123}\right)$$

$$= P(-1.63 \leq Z \leq 1.63)$$

$$= .9484 - .0516 = .8968$$

We see that almost 90% of all samples will give a result within 2 percentage points of the truth about the population. ◀

The outline of the calculation in Example 4.14 is familiar from Chapter 1, but the language of probability is new. The sampling distribution of

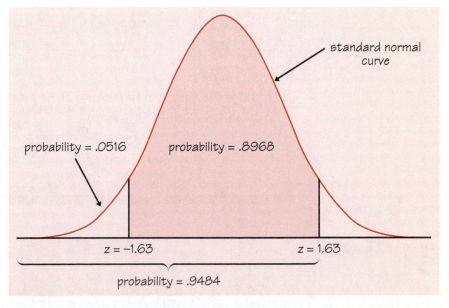

FIGURE 4.14   Probabilities in Example 4.14 as areas under the standard normal curve.

$\hat{p}$ gives probabilities for its values, so the entries in Table A are now probabilities. We used a brief notation that is common in statistics. The capital P in

$$P(.33 \le \hat{p} \le .37)$$

stands for "probability." The expression inside the parentheses tells us what event we are finding the probability of. This entire expression is a short way of writing "the probability that $\hat{p}$ lies between 0.33 and 0.37."

Using the normal approximation for $\hat{p}$

The sampling distribution of $\hat{p}$ is only approximately normal. For example, if we sample 100 individuals, the only possible values of $\hat{p}$ are 0, 1/100, 2/100, and so on. The statistic has only 101 possible values, so its distribution cannot be exactly normal. The accuracy of the normal approximation improves as the sample size n increases. For a fixed sample size n, the normal approximation is most accurate when p is close to 1/2, and least accurate when p is near 0 or 1. If $p = 1$, for example, then $\hat{p} = 1$ in every sample because every individual in the population has the characteristic we are counting. The normal approximation is no good at all when $p = 1$ or $p = 0$. Here is a rule of thumb that ensures that normal calculations are accurate enough for most statistical purposes. Unlike the first rule of thumb, this one rules out some settings of practical interest.

> **RULE OF THUMB 2**
>
> We will use the normal approximation to the sampling distribution of $\hat{p}$ for values of n and p that satisfy
>
> $$np \ge 10 \quad \text{and} \quad n(1 - p) \ge 10$$

EXAMPLE 4.15

One way of checking the effect of undercoverage, nonresponse, and other sources of error in a sample survey is to compare the sample with known facts about the population. About 11% of American adults are black. The proportion $\hat{p}$ of blacks in an SRS of 1500 adults should therefore be close to 11%. It is unlikely to be exactly 11% because of sampling variability. If a national sample contains only 9.2% blacks, should we suspect that the sampling procedure is somehow underrepresenting blacks?

We will find the probability that a sample contains no more than 9.2% blacks when the population is 11% black. First, check our rule of thumb for using the normal approximation to the sampling distribution of $\hat{p}$:

$$np = (1500)(.11) = 165$$

and

$$n(1 - p) = (1500)(.89) = 1335$$

Both are much larger than 10, so the approximation will be quite accurate.

The mean of $\hat{p}$ is $p = .11$. The standard deviation is

$$\sqrt{\frac{p(1 - p)}{n}} = \sqrt{\frac{(.11)(.89)}{1500}}$$

$$= \sqrt{.00006527} = .00808$$

Now do the normal probability calculation, illustrated by Figure 4.15.

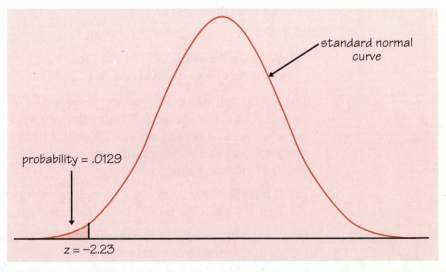

probability = .0129

z = −2.23

standard normal curve

FIGURE 4.15 The probability in Example 4.15 as an area under the standard normal curve.

$$P(\hat{p} \leq .092) = P\left(\frac{\hat{p} - .11}{.00808} \leq \frac{.092 - .11}{.00808}\right)$$

$$= P(Z \leq -2.23)$$

$$= .0129$$

Only 1.29% of all samples would have so few blacks. Because it is unlikely that a sample would include so few blacks, we have good reason to suspect that the sampling procedure underrepresents blacks. ◀

EXERCISES

4.42   The Gallup Poll once asked a random sample of 1540 adults, "Do you happen to jog?" Suppose that in fact 15% of all adults jog.

   (a) Find the mean and standard deviation of the proportion $\hat{p}$ of the sample who jog. (Assume the sample is an SRS.)

   (b) Explain why you can use the formula for the standard deviation of $\hat{p}$ in this setting (rule of thumb 1).

   (c) Check that you can use the normal approximation for the distribution of $\hat{p}$ (rule of thumb 2).

   (d) Find the probability that between 13% and 17% of the sample jog.

   (e) What sample size would be required to reduce the standard deviation of the sample proportion to one-half the value you found in (a)?

4.43   The Gallup Poll asked a probability sample of 1785 adults whether they attended church or synagogue during the past week. Suppose that 40% of the adult population did attend. We would like to know the probability that an SRS of size 1785 would come within plus or minus 3 percentage points of this true value.

   (a) If $\hat{p}$ is the proportion of the sample who did attend church or synagogue, what is the mean of $\hat{p}$? What is its standard deviation?

   (b) Explain why you can use the formula for the standard deviation of $\hat{p}$ in this setting (rule of thumb 1).

   (c) Check that you can use the normal approximation for the distribution of $\hat{p}$ (rule of thumb 2).

(d) Find the probability that $\hat{p}$ takes a value between 0.37 and 0.43. Will an SRS of size 1785 usually give a result $\hat{p}$ within plus or minus 3 percentage points of the true population proportion?

4.44 Suppose that 15% of all adults jog. Exercise 4.42 asks the probability that the sample proportion $\hat{p}$ from an SRS estimates $p = .15$ within ± 2 percentage points. Find this probability for SRSs of sizes 200, 800, and 3200. What general conclusion can you draw from your calculations?

4.45 Suppose that 40% of the adult population attended church or synagogue last week. Exercise 4.43 asks the probability that $\hat{p}$ from an SRS estimates $p = .4$ within ± 3 percentage points. Find this probability for SRSs of sizes 300, 1200, and 4800. What general fact do your results illustrate?

4.46 According to a market research firm, 52% of all residential telephone numbers in Los Angeles are unlisted. A telephone sales firm uses random digit dialing equipment that dials residential numbers at random, whether or not they are listed in the telephone directory. The firm calls 500 numbers in Los Angeles.

(a) What are the mean and standard deviation of the proportion of unlisted numbers in the sample?

(b) What is the probability that at least half the numbers dialed are unlisted? (Remember to check that you can use the normal approximation.)

4.47 Your mail-order company advertises that it ships 90% of its orders within three working days. You select an SRS of 100 of the 5000 orders received in the past week for an audit. The audit reveals that 86 of these orders were shipped on time.

(a) What is the sample proportion of orders shipped on time?

(b) If the company really ships 90% of its orders on time, what is the probability that the proportion in an SRS of 100 orders is as small as the proportion in your sample or smaller?

(c) A critic says, "Aha! You claim 90%, but in your sample the on-time percentage is lower than that. So the 90% claim is wrong." Explain in simple language why your probability calculation in (b) shows that the result of the sample does not refute the 90% claim.

4.48 Exercise 4.46 asks for a probability calculation about telephone calls to a random sample of Los Angeles telephone numbers. Exercise 4.47 asks for a similar calculation about a random sample of mail orders. For which calculation does the normal approximation to the sampling distribution of $\hat{p}$ give a more accurate answer? Why? (You need not actually do either calculation.)

Sample counts

counts   Sometimes we are interested in the **count** of special individuals in a sample rather than the *proportion* of such individuals. To deal with these problems, just restate them in terms of proportions.

EXAMPLE 4.16

A quality engineer selects an SRS of 100 switches from a large shipment for detailed inspection. She will accept the shipment if she finds 9 or fewer bad switches in the sample. Unknown to the engineer, 10% of the switches in the shipment fail to meet the specifications. What is the probability that 9 or fewer of the 100 switches in the sample fail inspection?

The *count* of failures among 100 switches is ≤ 9 exactly when the *proportion* $\hat{p}$ of failures is $\leq 9/100 = .09$. So the probability we want is $P(\hat{p} \leq .09)$.

First check that the normal approximation can be used:

$$np = (100)(.1) = 10 \quad \text{and} \quad n(1 - p) = (100)(.9) = 90$$

This setting just does satisfy the rule of thumb.

The mean of $\hat{p}$ is $p = .1$ and the standard deviation is

$$\sqrt{\frac{p(1 - p)}{n}} = \sqrt{\frac{(.1)(.9)}{100}} = .03$$

The probability of 9 or fewer bad switches in the sample is then

$$P(\hat{p} \leq .09) = P\left(\frac{\hat{p} - .1}{.03} \leq \frac{.09 - .1}{.03}\right)$$

$$= P(Z \leq -.33)$$

$$= .3707$$

The mathematics of probability makes it possible to find the exact value of this probability. It is 0.4513. The normal approximation 0.3707 is accurate to only one decimal place. Because $np = 10$, this combination of n and p is on the border of the values for which we are willing to use the approximation. The approximation gets better as np and $n(1 - p)$ get larger. ◄

We might also ask the *mean* number of switches that will fail inspection in the setting of Example 4.16. We know that the mean of the *proportion* of switches that fail in repeated samples is equal to the population proportion $p = .1$. So the mean *count* of defective switches in 100 samples is $100p = (100)(.1) = 10$. In general, the mean count of special individuals in an SRS of size n from a population with proportion p of special individuals is np.

EXERCISES

4.49 A selective college would like to have an entering class of 1200 students. Because not all students who are offered admission accept, the college admits 1700 students. Past experience shows that about 70% of the students admitted will accept. Treat the students admitted as an SRS of size 1700 from the population of all qualified students, 70% of whom would accept admission.

(a) What is the mean number of students who will accept admission?

(b) What is the probability that fewer than 1200 students accept?

(c) What is the probability that between 1150 and 1250 accept?

4.50 A study by a federal agency found that lie detector tests given to truthful persons have probability about 0.2 of suggesting that the person is deceptive.[4] A company gives lie detector tests to 500 job applicants each year. Suppose that all of the applicants are truthful. We can think of them as an SRS from the large population of truthful people, of which the test will find a proportion $p = .2$ to be deceptive.

(a) What is the mean number of applicants that the test finds deceptive in a year?

(b) What is the probability that the test finds at least 90 applicants to be deceptive?

SUMMARY

When we want information about the **population proportion** p of individuals with some special characteristic, we often take an SRS and use the **sample proportion** $\hat{p}$ to estimate the unknown parameter p.

The **sampling distribution** of $\hat{p}$ describes how the statistic varies in all possible samples from the population.

The **mean** of the sampling distribution is equal to the population proportion p. That is, $\hat{p}$ is an unbiased estimator of p.

The **standard deviation** of the sampling distribution is $\sqrt{p(1-p)/n}$ for an SRS of size n. This recipe can be used in practice if the population is at least 10 times as large as the sample.

The standard deviation of $\hat{p}$ gets smaller as the sample size n gets larger. Because of the square root, a sample four times larger is needed to cut the standard deviation in half.

When the sample size n is large, the sampling distribution of $\hat{p}$ is close to a normal distribution with mean p and standard deviation $\sqrt{p(1-p)/n}$. In practice, use this **normal approximation** when both $np \geq 10$ and $n(1-p) \geq 10$.

To find a probability involving a **sample count**, restate the problem in terms of a sample proportion. The mean of a sample count is np.

SECTION 4.3 EXERCISES

4.51   According to government data, 22% of American children under the age of 6 live in households with incomes less than the official poverty level. A study of learning in early childhood chooses an SRS of 300 children.

(a) What is the probability that more than 20% of the sample are from poverty households? (Remember to check that you can use the normal approximation.)

(b) What is the probability that more than 30% of the sample are from poverty households?

4.52   Here is a simple probability model for multiple-choice tests. Suppose that a student has probability p of correctly answering a question chosen at random from a universe of possible questions. (A strong student has a higher p than a weak student.) The correctness of an answer to any specific question doesn't depend on other questions. A test contains n questions. Then the proportion of correct answers that a student gives is a sample proportion $\hat{p}$ from an SRS of size n drawn from a population with population proportion p.

(a) Julie is a good student for whom $p = .75$. Find the probability that Julie scores 70% or lower on a 100-question test.

(b) If the test contains 250 questions, what is the probability that Julie will score 70% or lower?

(c) How many questions must the test contain in order to reduce the standard deviation of Julie's proportion of correct answers to half its value for a 100-item test?

(d) Laura is a weaker student for whom $p = .6$. Does the answer you gave in (c) for the standard deviation of Julie's score apply to Laura's standard deviation also?

4.53   The Helsinki Heart Study asks whether the anticholesterol drug gemfibrozil will reduce heart attacks. In planning such an experiment, the researchers must be confident that the sample sizes are large enough to enable them to observe enough heart attacks. The Helsinki study plans to give gemfibrozil to 2000 men and a placebo to another 2000. The probability of a heart attack during the 5-year period of the study for men this age is about 0.04. We can think of the study participants as an SRS from a large population, of which the proportion $p = .04$ will have heart attacks.

(a) What is the mean number of heart attacks that the study will find in one group of 2000 men if the treatment does not change the probability 0.04?

(b) What is the probability that the group will suffer at least 75 heart attacks?

4.54   Explain why you *cannot* use the methods of this section to find the following probabilities.

(a) A factory employs 3000 unionized workers, of whom 30% are Hispanic. The 15-member union executive committee contains 3 Hispanics. What would be the probability of 3 or fewer Hispanics if the executive committee were chosen at random from all the workers?

(b) A university is concerned about the academic standing of its intercollegiate athletes. A study committee chooses an SRS of 50 of the 316 athletes to interview in detail. Suppose that in fact 40% of the athletes have been told by coaches to neglect their studies on at least one occasion. What is the probability that at least 15 in the sample are among this group?

4.4   THE BINOMIAL DISTRIBUTIONS*

The distribution of a sample proportion or sample count is approximately normal when the sample size n and population proportion p satisfy the

---

*This more advanced section continues the optional discussion of probability from Section 4.2. It is not needed in order to read the rest of the book.

rule of thumb that both np and $n(1 - p)$ be 10 or greater. We will use this normal approximation for statistical inference about p. Of course, sample counts and sample proportions are not exactly normal. They are discrete random variables. In Example 4.16, an engineer inspected a sample of 100 switches. The count of bad switches can only take the values 0, 1, 2, ..., 99, 100. The proportion of bad switches must be one of 0, 1/100, 2/100, ..., 99/100, 1. What is the discrete probability distribution of a count or proportion?

The binomial setting

We will concentrate on the distribution of sample counts. These are a bit simpler than proportions because counts always take whole-number values. You can apply our results to proportions by restating any problem involving a proportion in terms of the count that the proportion comes from. The distribution of a count X depends on how the data are produced. Here is one common situation.

THE BINOMIAL SETTING

1. There are a fixed number n of observations.
2. The n observations are all **independent**. That is, knowing the result of one observation tells you nothing about the other observations.
3. Each observation falls into one of just two categories, which for convenience we call "success" and "failure."
4. The probability of a success, call it p, is the same for each observation.

Think of tossing a coin n times as an example of the binomial setting. Each toss gives either heads or tails. Knowing the outcome of one toss doesn't tell us anything about other tosses, so the n tosses are independent. If we call heads a success, then p is the probability of a head and remains the same as long as we toss the same coin. The number of heads we count is a random variable X. The distribution of X is called a *binomial distribution*.

BINOMIAL DISTRIBUTION

The distribution of the count X of successes in the binomial setting is the **binomial distribution** with parameters n and p. The parameter n is the number of observations, and p is the probability of a success on any one observation. The possible values of X are the whole numbers from 0 to n.

The binomial distributions are an important class of discrete probability distributions. Pay attention to the binomial setting, because not all counts have binomial distributions.

EXAMPLE 4.17

Blood type is inherited. If both parents carry genes for the O and A blood types, each child has probability 0.25 of getting two O genes and so of having blood type O. Different children inherit independently of each other. The number of O blood types among 5 children of these parents is the count X of successes in 5 independent observations with probability 0.25 of a success on each observation. So X has the binomial distribution with $n = 5$ and $p = .25$. ◄

EXAMPLE 4.18

Deal 10 cards from a shuffled deck and count the number X of red cards. There are 10 observations, and each gives either a red or a black card. A "success" is a red card. But the observations are *not* independent. If the first card is black, the second is more likely to be red because there are more red cards than black cards left in the deck. The count X does *not* have a binomial distribution. ◄

EXAMPLE 4.19

An engineer chooses an SRS of 10 switches from a shipment of 10,000 switches. Suppose that (unknown to the engineer) 10% of the switches in the shipment are bad. The engineer counts the number X of bad switches in the sample.

This is not quite a binomial setting. Just as removing one card in Example 4.18 changed the makeup of the deck, removing one switch changes the proportion of bad switches remaining in the shipment. So the state of the second switch chosen is not independent of the first. But removing one switch from a shipment of 10,000 changes the makeup of the remaining 9999 switches very little. In practice, the distribution of X is very close to the binomial distribution with $n = 10$ and $p = .1$. ◀

Example 4.19 shows how we can use the binomial distributions in the statistical setting of selecting an SRS. When the population is much larger than the sample, a count of successes in an SRS of size n has approximately the binomial distribution with n equal to the sample size and p equal to the proportion of successes in the population.

EXERCISES

In each of Exercises 4.55 to 4.58, X is a count. Does X have a binomial distribution? Give your reasons in each case.

4.55   You observe the sex of the next 20 children born at a local hospital; X is the number of girls among them.

4.56   A couple decides to continue to have children until their first girl is born; X is the total number of children the couple has.

4.57   A student studies binomial distributions using computer-assisted instruction. After the lesson, the computer presents 10 problems. The student solves each problem and enters her answer. The computer gives additional instruction between problems if the answer is wrong. The count X is the number of problems that the student gets right.

4.58   Joe buys a state lottery ticket every week. The count X is the number of times in a year that he wins a prize.

Binomial probabilities

We can find a formula for the probability that a binomial random variable takes any value by adding probabilities for the different ways of getting exactly that many successes in n observations. Here is the example we will use to show the idea.

EXAMPLE 4.20

Each child born to a particular set of parents has probability 0.25 of having blood type O. If these parents have 5 children, what is the probability that exactly 2 of them have type O blood?

The count of children with type O blood is a binomial random variable X with $n = 5$ tries and probability $p = .25$ of a success on each try. We want $P(X = 2)$. ◀

Because the method doesn't depend on the specific example, let's use "S" for success and "F" for failure for short. Do the work in two steps.

Step 1: Find the probability that a specific 2 of the 5 tries give successes, say the first and the third. This is the outcome SFSFF. Here's how to find the probability of this outcome:

- The probability that the first try is a success is 0.25. That is, in many repetitions, we succeed on the first try 25% of the time.
- Out of all the repetitions with a success on the first try, 75% have a failure on the second try. So the proportion of repetitions on which the first two tries are SF is (.25)(.75). We can multiply here because the tries are *independent*. That is, the first try has no influence on the second.
- Keep going: of these repetitions, the proportion 0.25 have S on the third try. So the probability of SFS is (.25)(.75)(.25). After two more tries, the probability of SFSFF is the product of the try-by-try probabilities:

$$(.25)(.75)(.25)(.75)(.75) = (.25)^2(.75)^3$$

Step 2: Observe that the probability of *any one* arrangement of 2 S's and 3 F's has this same probability. That's true because we multiply together 0.25 twice and 0.75 three times whenever we have 2 S's and 3 F's. The probability that $X = 2$ is the probability of getting 2 S's and 3 F's in any arrangement whatsoever. Here are all the possible arrangements:

SSFFF   SFSFF   SFFSF   SFFFS   FSSFF
FSFSF   FSFFS   FFSSF   FFSFS   FFFSS

There are 10 of them, all with the same probability. The overall probability of 2 successes is therefore

$$P(X = 2) = 10(.25)^2(.75)^3 = .2637$$

The pattern of this calculation works for any binomial probability. To use it, we need to be able to count the number of arrangements of k successes in n observations without actually listing them. We use the following fact to do the counting.

BINOMIAL COEFFICIENT

The number of ways of arranging k successes among n observations is given by the **binomial coefficient**

$$\binom{n}{k} = \frac{n!}{k!\,(n - k)!}$$

for $k = 0, 1, 2, \ldots, n$.

---

factorial   The formula for binomial coefficients uses the *factorial* notation. For any positive whole number n, its factorial $n!$ is

$$n! = n \times (n - 1) \times (n - 2) \times \cdots \times 3 \times 2 \times 1$$

Also, $0! = 1$.

Notice that the larger of the two factorials in the denominator of a binomial coefficient will cancel much of the $n!$ in the numerator. For example, the binomial coefficient we need for Example 4.20 is

$$\binom{5}{2} = \frac{5!}{2!\,3!}$$

$$= \frac{(5)(4)(3)(2)(1)}{(2)(1) \times (3)(2)(1)}$$

$$= \frac{(5)(4)}{(2)(1)}$$

$$= \frac{20}{2} = 10$$

The notation $\binom{n}{k}$ is *not* related to the fraction $\frac{n}{k}$. A helpful way to remember its meaning is to read it as "binomial coefficient *n* choose *k*." Binomial coefficients have many uses in mathematics, but we are interested in them only as an aid to finding binomial probabilities. The binomial coefficient $\binom{n}{k}$ counts the number of ways in which *k* successes can be distributed among *n* observations. The binomial probability $P(X = k)$ is this count multiplied by the probability of any specific arrangement of the *k* successes. Here is the formula we seek.

BINOMIAL PROBABILITY

If *X* has the binomial distribution with *n* observations and probability *p* of success on each observation, the possible values of *X* are 0, 1, 2, ..., *n*. If *k* is any one of these values,

$$P(X = k) = \binom{n}{k} p^k (1 - p)^{n-k}$$

EXAMPLE 4.21

The number *X* of switches that fail inspection in Example 4.19 has approximately the binomial distribution with $n = 10$ and $p = .1$. The probability that no more than 1 switch fails is

$$P(X \leq 1) = P(X = 1) + P(X = 0)$$

$$= \binom{10}{1}(.1)^1(.9)^9 + \binom{10}{0}(.1)^0(.9)^{10}$$

$$= \frac{10!}{1!\,9!}(.1)(.3874) + \frac{10!}{0!\,10!}(1)(.3487)$$

$$= (10)(.1)(.3874) + (1)(1)(.3487)$$

$$= .3874 + .3487$$

$$= .7361$$

Notice that the calculation uses the facts that $0! = 1$ and that $a^0 = 1$ for any number *a* other than 0. ◀

EXERCISES

4.59   If the parents in Example 4.20 have 5 children, the number who have type O blood is a random variable X that has the binomial distribution with $n = 5$ and $p = .25$.

(a) What are the possible values of X?

(b) Find the probability of each value of X. Draw a probability histogram for this distribution.

4.60   A factory employs several thousand workers, of whom 30% are Hispanic. If the 15 members of the union executive committee were chosen from the workers at random, the number of Hispanics on the committee would have the binomial distribution with $n = 15$ and $p = .3$.

(a) What is the probability that exactly 3 members of the committee are Hispanic?

(b) What is the probability that 3 or fewer members of the committee are Hispanic?

4.61   A university claims that 80% of its basketball players get degrees. An investigation examines the fate of all 20 players who entered the program over a period of several years that ended six years ago. Of these players, 10 graduated and the remaining 10 are no longer in school. If the university's claim is true, the number of players among the 20 who graduate should have the binomial distribution with $n = 20$ and $p = .8$. What is the probability that exactly 10 out of 20 players graduate?

4.62   Among employed women, 25% have never been married. Select 10 employed women at random.

(a) The number in your sample who have never been married has a binomial distribution. What are n and p?

(b) What is the probability that exactly 2 of the 10 women in your sample have never been married?

(c) What is the probability that 2 or fewer have never been married?

Binomial mean and standard deviation

If a count X has the binomial distribution based on n observations with probability p of success, what is its mean μ? We can guess the answer. If a basketball player makes 80% of her free throws, the mean number made in 10 tries should be 80% of 10, or 8. In general, the mean of a binomial distribution should be $\mu = np$. With some hard work, we could use the definitions of the mean and standard deviation of a discrete random variable in Section 4.2 to verify that this is true and also find a formula for the standard deviation. Here are the facts.

MEAN AND STANDARD DEVIATION OF A BINOMIAL RANDOM VARIABLE

If a count X has the binomial distribution with number of observations n and probability of success p, the **mean** and **standard deviation** of X are

$$\mu = np$$

$$\sigma = \sqrt{np(1-p)}$$

These short formulas are good only for binomial distributions. They can't be used for other discrete random variables.

EXAMPLE 4.22

Continuing Example 4.21, the count X of bad switches is binomial with $n = 10$ and $p = .1$. This is the sampling distribution the engineer would see if she drew all possible SRSs of 10 switches from the shipment and recorded the value of X for each sample. You can use the binomial probability formula to obtain the probability of each value of X. Figure 4.16 is a probability histogram of the distribution. The possible counts of bad switches in a sample of 10 switches are the whole numbers from 0 to 10. The counts from 5 to 10 have probabilities so small that they can't be seen in the graph.

The mean and standard deviation of the binomial distribution in Figure 4.16 are

$$\mu = np$$
$$= (10)(.1) = 1$$
$$\sigma = \sqrt{np(1-p)}$$
$$= \sqrt{(10)(.1)(.9)} = \sqrt{.9} = .9487$$

The mean is marked on the probability histogram in Figure 4.16.   ◀

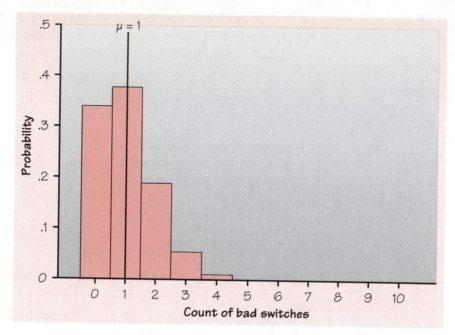

FIGURE 4.16 Probability histogram for the binomial distribution with $n = 10$ and $p = 0.1$.

EXERCISES

4.63 What are the mean and standard deviation of the number of children with type O blood in Exercise 4.59? Mark the location of the mean on the probability histogram you made in that exercise.

4.64 (a) What is the mean number of Hispanics on randomly chosen committees of 15 workers in Exercise 4.60?

(b) What is the standard deviation σ of the count X of Hispanic members?

(c) Suppose that 10% of the factory workers were Hispanic. Then $p = .1$. What is σ in this case? What is σ if $p = .01$? What does your work show about the behavior of the standard deviation of a binomial distribution as the probability of a success gets closer to 0?

4.65 (a) Find the mean number of graduates out of 20 players in the setting of Exercise 4.61.

(b) Find the standard deviation σ of the count X.

(c) Suppose that the 20 players came from a population of which $p = .9$ graduated. What is the standard deviation σ of the count of graduates?

If $p = .99$, what is σ? What does your work show about the behavior of the standard deviation of a binomial distribution as the probability p of success gets closer to 1?

4.66   You choose 10 employed women at random, as in Exercise 4.62. What is the mean number of women in such a sample who have never been married? What is the standard deviation?

SUMMARY

A count X of successes has a binomial distribution in the **binomial setting**: there are n observations; the observations are **independent** of each other; each observation results in a success or a failure; and each observation has the same probability p of a success.

The **binomial distribution** with n observations and probability p of success gives a good approximation to the sampling distribution of the count of successes in an SRS of size n from a large population containing proportion p of successes.

If X has the binomial distribution with parameters n and p, the possible values of X are the whole numbers $0, 1, 2, \ldots, n$. The **binomial probability** that X takes any value is

$$P(X = k) = \binom{n}{k} p^k (1 - p)^{n-k}$$

The **binomial coefficient**

$$\binom{n}{k} = \frac{n!}{k!\,(n - k)!}$$

counts the number of ways k successes can be arranged among n observations. Here the **factorial $n!$** is

$$n! = n \times (n - 1) \times (n - 2) \times \cdots \times 3 \times 2 \times 1$$

for positive whole numbers n, and $0! = 1$.

The **mean** and **standard deviation** of a binomial count X are

$$\mu = np$$

$$\sigma = \sqrt{np(1 - p)}$$

SECTION 4.4 EXERCISES

4.67 In a test for ESP (extrasensory perception), a subject is told that cards the experimenter can see but he cannot contain either a star, a circle, a wave, or a square. As the experimenter looks at each of 20 cards in turn, the subject names the shape on the card. A subject who is just guessing has probability 0.25 of guessing correctly on each card.

  (a) The count of correct guesses in 20 cards has a binomial distribution. What are n and p?

  (b) What is the mean number of correct guesses?

  (c) What is the probability of exactly 5 correct guesses?

4.68 A believer in the "random walk" theory of stock markets thinks that an index of stock prices has probability 0.65 of increasing in any year. Moreover, the change in the index in any given year is not influenced by whether it rose or fell in earlier years. Let X be the number of years among the next 6 years in which the index rises.

  (a) X has a binomial distribution. What are n and p?

  (b) What are the possible values that X can take?

  (c) Find the probability of each value X. Draw a probability histogram for the distribution of X.

  (d) What are the mean and standard deviation of this distribution? Mark the location of the mean on the histogram.

4.69 A federal report finds that lie detector tests given to truthful persons have probability about 0.2 of suggesting that the person is deceptive.

  (a) A company asks 12 job applicants about thefts from previous employers, using a lie detector to assess their truthfulness. Suppose that all 12 answer truthfully. What is the probability that the lie detector says all 12 are truthful? What is the probability that the lie detector says at least one is deceptive?

  (b) What is the mean number among 12 truthful persons who will be classified as deceptive? What is the standard deviation of this number?

4.70   A test for the presence of antibodies to the AIDS virus in blood has probability 0.99 of detecting the antibodies when they are present. Suppose that during a year 20 units of blood with AIDS antibodies pass through a blood bank.

(a) Take X to be the number of these 20 units that the test detects. What is the distribution of X?

(b) What is the probability that the test detects all 20 contaminated units? What is the probability that at least one unit is not detected?

(c) What is the mean number of units among the 20 that will be detected? What is the standard deviation of the number detected?

4.71   (Optional) A deck of 52 cards contains 26 red cards and 26 black cards. If the deck is well shuffled, all the cards are equally likely to be drawn.

(a) Draw one card. What is the probability that it is black? What is the probability that it is red?

(b) Suppose that the first card you drew was black. How many cards remain in the deck? How many of them are black? You now draw a second card, without returning the first card to the deck. What is the probability that this card is black? That it is red?

(c) Answer the questions in (b) again, this time supposing that the first card you drew was red.

Comment: Your work in (b) and (c) shows that knowing the color of the first card changes the probabilities for the second card. That is, the colors of the first and second cards are not independent. The probabilities in (b) and (c) are called *conditional probabilities* because they depend on a specific condition, the color of the first card.

4.72   (Optional) A shipment contains 10,000 switches. Of these, 1000 are bad. An inspector draws switches at random, so that each switch has the same chance to be drawn.

(a) Draw one switch. What is the probability that the switch you draw is bad? What is the probability that it is not bad?

(b) Suppose the first switch drawn is bad. How many switches remain? How many of them are bad? Draw a second switch at random. What is the probability that this switch is bad?

(c) Answer the questions in (b) again, but now suppose that the first switch drawn is not bad.

Comment: As in the previous problem, knowing the result of the first trial changes the probabilities for the second trial. But because the shipment is large, the probabilities change very little. The trials are almost independent.

4.5   SAMPLE MEANS

Sample proportions arise most often when we are interested in categorical variables. We then ask questions like "What proportion of the shoppers sampled feel that clothes shopping is frustrating?" or "What percent of the sample respondents are black?" When we record quantitative variables— the income of a household, the diameter of a bearing, the blood pressure of a patient—we are interested in other statistics, such as the median or mean or standard deviation of the variable. Because sample means are just averages of observations, they are among the most common statistics. This section describes the sampling distribution of the mean of the responses in an SRS.

EXAMPLE 4.23

A basic principle of investment is that diversification reduces risk. That is, buying several securities rather than just one reduces the variability of the return on an investment. Figure 4.17 illustrates this principle in the case of common stocks listed on the New York Stock Exchange. Figure 4.17(a) shows the distribution of returns for all 1815 stocks listed on the Exchange for the entire year 1987.[5] This was a year of extreme swings in stock prices, including a record loss of over 20% in a single day. The mean return for all 1815 stocks was −3.5% and the distribution shows a very wide spread.   ◄

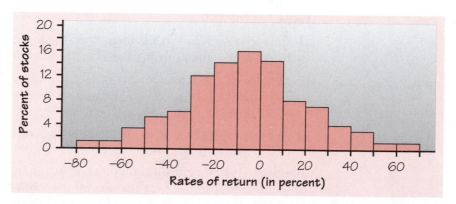

FIGURE 4.17(a)   The distribution of returns for New York Stock Exchange common stocks in 1987.

Figure 4.17(b) shows the distribution of returns for all possible portfolios that invested equal amounts in each of 5 stocks. A portfolio is just a sample of 5 stocks and its return is the average return for the 5 stocks chosen. The mean return for all portfolios is still −3.5%, but the variation among portfolios is much less than the variation among individual stocks. For example, 11% of all individual stocks had a loss of more than 40%, but only 1% of the portfolios had a loss that large.   ◄

The histograms in Figure 4.17 emphasize a principle that we will make precise in this section:

• Averages are less variable than individual observations.

More detailed examination of the distributions would point to a second principle:

• Averages are more normal than individual observations.

These two facts contribute to the popularity of sample means in statistical inference.

For means just as for proportions, we must distinguish carefully between a population parameter and a sample statistic. We write μ (the Greek

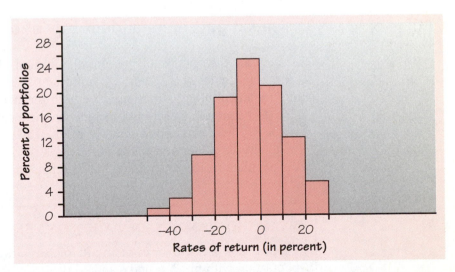

FIGURE 4.17(b)   The distribution of returns for portfolios of five stocks in 1987. Figure 4.17 is taken with permission from John K. Ford, "A method for grading 1987 stock recommendations," *American Association of Individual Investors Journal*, March 1988, pp. 16–17.

letter mu) for the mean of a population. This is a fixed *parameter* that is unknown when we use a sample for inference. The mean of the sample is the familiar $\overline{x}$, the average of the observations in the sample. This is a *statistic* that would almost certainly take a different value if we chose another sample from the same population. The sample mean $\overline{x}$ from a sample or an experiment is an estimate of the mean μ of the underlying population, just as a sample proportion $\hat{p}$ is an estimate of a population proportion p. In the same way, we write σ (the Greek letter sigma) for the standard deviation of a population and the familiar s for the standard deviation of a sample. As long as we were just doing data analysis, the distinction between population and sample was not important. Now, however, it is essential.

PARAMETERS AND STATISTICS

The mean and standard deviation of a population are **parameters**. We use Greek letters to write these parameters: μ for the mean and σ for the standard deviation.

The mean and standard deviation calculated from sample data are **statistics**. We write the sample mean as $\overline{x}$ and the sample standard deviation as s.

The mean and the standard deviation of $\overline{x}$

The sampling distribution of $\overline{x}$ is the distribution of the values of $\overline{x}$ in all possible samples of the same size from the population. Figure 4.17(a) shows the distribution of a population, with mean $\mu = -3.5\%$. Figure 4.17(b) is the sampling distribution of the sample mean $\overline{x}$ from all samples of size $n = 5$ from this population. The mean of all the values of $\overline{x}$ is again -3.5%, but the values of $\overline{x}$ are less spread out than the individual values in the population. This is an example of a general fact.

MEAN AND STANDARD DEVIATION OF A SAMPLE MEAN

Suppose that $\overline{x}$ is the mean of an SRS of size n drawn from a large population with mean μ and standard deviation σ. Then the **mean** of the sampling distribution of $\overline{x}$ is μ and its **standard deviation** is $\sigma/\sqrt{n}$.

The behavior of $\bar{x}$ in repeated samples is much like that of the sample proportion $\hat{p}$:

- The sample mean $\bar{x}$ is an unbiased estimator of the population mean μ.
- The values of $\bar{x}$ are less spread out for larger samples. Their standard deviation decreases at the rate $\sqrt{n}$, so you must take a sample four times as large to cut the standard deviation of $\bar{x}$ in half.
- You should only use the recipe $\sigma/\sqrt{n}$ for the standard deviation of $\bar{x}$ when the population is at least 10 times as large as the sample. This is almost always the case in practice.

Notice that these facts about the mean and standard deviation of $\bar{x}$ are true no matter what the shape of the population distribution is.

EXAMPLE 4.24

The height of young women varies approximately according to the $N(64.5, 2.5)$ distribution. This is a population distribution with $\mu = 64.5$ and $\sigma = 2.5$. If we choose one young woman at random, the heights we get in repeated choices follow this distribution. That is, the distribution of the population is also the distribution of one observation chosen at random. So we can think of the population distribution as a distribution of probabilities, just like a sampling distribution.

Now measure the height of an SRS of 10 young women. The sampling distribution of their sample mean height $\bar{x}$ will have mean $\mu = 64.5$ inches and standard deviation

$$\frac{\sigma}{\sqrt{n}} = \frac{2.5}{\sqrt{10}} = .79 \text{ inch}$$

The heights of individual women vary widely about the population mean, but the average height of a sample of 10 women is less variable. Figure 4.18 compares the two distributions. ◀

The fact that averages of several observations are less variable than individual observations is important in many settings. For example, it is common practice to repeat a careful measurement several times and report

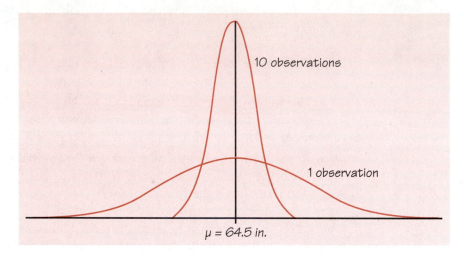

FIGURE 4.18   The sampling distribution of the mean height $\bar{x}$ for samples of 10 young women compared with the distribution of the height of a single woman chosen at random.

the average of the results. Think of the results of n repeated measurements as an SRS from the population of outcomes we would get if we repeated the measurement forever. The average of the n results (the sample mean $\bar{x}$) is less variable than a single measurement.

EXERCISES

4.73   Investors remember 1987 as the year stocks lost 20% of their value in a single day. For 1987 as a whole, the mean return of all common stocks on the New York Stock Exchange was $\mu = -3.5\%$. (That is, these stocks lost an average of 3.5% of their value in 1987.) The standard deviation of the returns was about $\sigma = 26\%$. Figure 4.17(a) shows the distribution of returns. Figure 4.17(b) is the sampling distribution of the mean returns $\bar{x}$ for all possible samples of 5 stocks. What are the mean and the standard deviation of the distribution in Figure 4.17(b)?

4.74   The scores of individual students on the American College Testing (ACT) composite college entrance examination have a normal distribution with mean 18.6 and standard deviation 5.9. At Northside High, 76 seniors take the test. If the scores at this school have the same distribution as national scores, what are the mean and standard deviation of the average (sample mean) score for the 76 students? Do your results depend on the fact that individual scores have a normal distribution?

4.75   An automatic grinding machine in an auto parts plant prepares axles with a target diameter $\mu = 40.125$ millimeters (mm). The machine has some variability, so the standard deviation of the diameters is $\sigma = .002$ mm. The machine operator inspects a sample of 4 axles each hour for quality control purposes and records the sample mean diameter. What will be the mean and standard deviation of the numbers recorded? Do your results depend on whether or not the axle diameters have a normal distribution?

4.76   The law requires coal mine operators to test the amount of dust in the atmosphere of the mine. A laboratory carries out the test by weighing filters that have been exposed to the air in the mine. The test has a standard deviation of $\sigma = .08$ milligram in repeated weighings of the same filter. The laboratory weighs each filter 3 times and reports the mean result. What is the standard deviation of the reported results?

The central limit theorem

We have described the mean and standard deviation of the sampling distribution of a sample mean $\overline{x}$, but not the distribution itself. The shape of the distribution of $\overline{x}$ depends on the shape of the population distribution. In particular, if the population distribution is normal, then so is the distribution of the sample mean.

> **SAMPLING DISTRIBUTION OF A SAMPLE MEAN**
>
> Draw an SRS of size n from a population that has the normal distribution with mean μ and standard deviation σ. Then the sample mean $\overline{x}$ has the normal distribution $N(\mu, \sigma/\sqrt{n})$ with mean μ and standard deviation $\sigma/\sqrt{n}$.

We already knew the mean and standard deviation of the sampling distribution. All that we have added now is the normal shape. Although many populations have roughly normal distributions, very few indeed are exactly normal. What happens to $\overline{x}$ when the population distribution is not normal? It turns out that as the sample size increases, the distribution of $\overline{x}$ gets closer to a normal distribution. This is true no matter what shape the population distribution has, as long as the population has a finite standard deviation σ. This famous fact of probability theory is called the *central limit theorem*. It is much more useful than the fact that the distribution of $\bar{x}$ is exactly normal if the population is exactly normal.

CENTRAL LIMIT THEOREM

Draw an SRS of size n from any population whatsoever with mean μ and finite standard deviation σ. When n is large, the sampling distribution of the sample mean $\bar{x}$ is close to the normal distribution $N(\mu, \sigma/\sqrt{n})$ with mean μ and standard deviation $\sigma/\sqrt{n}$.

How large a sample size n is needed for $\bar{x}$ to be close to normal depends on the population distribution. More observations are required if the shape of the population distribution is far from normal.

EXAMPLE 4.25

Figure 4.19 shows the central limit theorem in action for a very nonnormal population. Figure 4.19(a) displays the density curve for the distribution of the population. The distribution is strongly right skewed, and the most probable outcomes are near 0 at one end of the range of possible values. The mean μ of this distribution is 1 and its standard deviation σ is also 1. This particular distribution is called an *exponential distribution* from the shape of its density curve. Exponential distributions are used to describe the lifetime in service of electronic components and the time required to serve a customer or repair a machine.

Figures 4.19(b), (c), and (d) are the density curves of the sample means of 2, 10, and 25 observations from this population. As n increases, the shape becomes more normal. The mean remains at $\mu = 1$ and the standard deviation decreases, taking the value $1/\sqrt{n}$. The density curve for 10 observations is still somewhat skewed to the right but already resembles a normal curve with $\mu = 1$ and $\sigma = 1/\sqrt{10} = .32$. The density curve for $n = 25$ is yet more normal. The contrast between the shape of the population distribution and the distribution of the mean of 10 or 25 observations is striking. ◄

The central limit theorem allows us to use normal probability calculations to answer questions about sample means from many observations even when the population distribution is not normal.

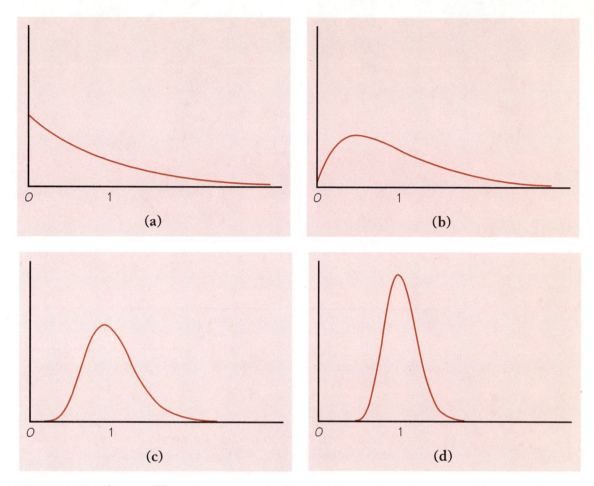

FIGURE 4.19 The central limit theorem in action: the distribution of sample means $\bar{x}$ from a strongly nonnormal population become more normal as the sample size increases. **(a)** The distribution of one observation. **(b)** The distribution of $\bar{x}$ for 2 observations. **(c)** The distribution of $\bar{x}$ for 10 observations. **(d)** The distribution of $\bar{x}$ for 25 observations.

EXAMPLE 4.26

The time that a technician requires to perform preventive maintenance on an air-conditioning unit is governed by the exponential distribution whose density curve appears in Figure 4.19(a). The mean time is $\mu = 1$ hour and the standard deviation is $\sigma = 1$ hour. Your company operates 70 of these units. What is the probability that their average maintenance time exceeds 50 minutes?

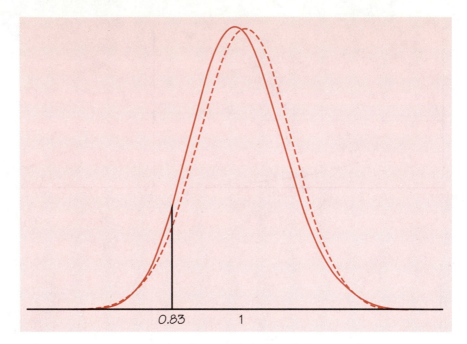

FIGURE 4.20   The exact distribution (*dashed*) and the normal approximation from the central limit theorem (*solid*) for the average time needed to maintain an air conditioner, for Example 4.26.

The central limit theorem says that the sample mean time $\bar{x}$ (in hours) spent working on 70 units has approximately the normal distribution with mean equal to the population mean $\mu = 1$ hour and standard deviation

$$\frac{\sigma}{\sqrt{70}} = \frac{1}{\sqrt{70}} = .12 \text{ hour}$$

The distribution of $\bar{x}$ is therefore approximately $N(1, 0.12)$. Figure 4.20 shows this normal curve (solid) and also the actual density curve of $\bar{x}$ (dashed).

Because 50 minutes is 50/60 of an hour, or 0.83 hour, the probability we want is

$$P(\bar{x} > .83) = P\left(\frac{\bar{x} - 1}{.12} > \frac{.83 - 1}{.12}\right)$$

$$= P(Z > -1.42) = .9222$$

This is the area to the right of 0.83 under the solid normal curve in Figure 4.20. The exactly correct probability is the area under the dashed density curve in the figure. It is 0.9294. The central limit theorem normal approximation is off by only about 0.007. ◀

EXERCISES

4.77   The scores of students on the ACT college entrance examination in a recent year had the normal distribution with mean $\mu = 18.6$ and standard deviation $\sigma = 5.9$.

(a) What is the probability that a single student randomly chosen from all those taking the test scores 21 or higher?

(b) Now take an SRS of 50 students who took the test. What is the probability that the mean score $\bar{x}$ of these students is 21 or higher?

4.78   A bottling company uses a filling machine to fill plastic bottles with cola. The bottles are supposed to contain 300 milliliters (ml). In fact, the contents vary according to a normal distribution with mean $\mu = 298$ ml and standard deviation $\sigma = 3$ ml.

(a) What is the probability that an individual bottle contains less than 295 ml?

(b) What is the probability that the mean contents of the bottles in a six-pack is less than 295 ml?

4.79   A laboratory weighs filters from a coal mine to measure the amount of dust in the mine atmosphere. Repeated measurements of the weight of dust on the same filter vary normally with standard deviation $\sigma = .08$ milligrams (mg) because the weighing is not perfectly precise. The dust on a particular filter actually weighs 123 mg. Repeated weighings will then have the normal distribution with mean 123 mg and standard deviation 0.08 mg.

(a) The laboratory reports the mean of 3 weighings. What is the distribution of this mean?

(b) What is the probability that the laboratory reports a weight of 124 mg or higher for this filter?

4.80   The number of flaws per square yard in a type of carpet material varies with mean 1.6 flaws per square yard and standard deviation 1.2 flaws per square yard. The population distribution cannot be normal, because a count takes only whole-number values. An inspector studies 200 square yards of the material, records the number of flaws found in each square yard, and calculates $\bar{x}$, the mean number of flaws per square yard inspected. Use the central limit theorem to find the approximate probability that the mean number of flaws exceeds 2 per square yard.

4.81   The distribution of annual returns on common stocks is roughly symmetric, but extreme observations are more frequent than in a normal distribution. Because the distribution is not strongly nonnormal, the mean return over even a moderate number of years is close to normal. In the long run, annual real returns on common stocks have varied with mean about 9% and standard deviation about 28%. Andrew plans to retire in 45 years and is considering investing in stocks. What is the probability (assuming that the past pattern of variation continues) that the mean annual return on common stocks over the next 45 years will exceed 15%? What is the probability that the mean return will be less than 5%?

The law of large numbers

The central limit theorem says that in large samples the sample mean $\bar{x}$ must be close to the population mean μ. This is true because the values of $\bar{x}$ in all possible samples closely follow a normal distribution with mean μ and standard deviation that gets smaller and smaller as the sample size n increases. We can turn this fact around to help us better understand the population mean μ.

For any population, μ is the average value of the variable we are measuring for all the individuals in the population. If we use a density curve to describe the distribution of the population, the mean μ is the "balance point" of the curve. (See page 58 in Chapter 1.) Now we can add that μ is the value approached by the mean $\bar{x}$ of observations drawn at random from the population as we draw more and more observations. This is true for any population whatever with a finite mean μ, even though the full central limit theorem requires that the population also have a finite variance.

LAW OF LARGE NUMBERS

Draw observations at random from any population with finite mean μ. As the number of observations drawn increases, the mean $\bar{x}$ of the observed values gets closer and closer to μ.

The law of large numbers can be proved mathematically starting from the basic laws of probability. The behavior of $\bar{x}$ is similar to the idea of probability. In the long run, the proportion of outcomes taking any value gets close to the probability of that value, and the average outcome gets close to the population mean.

The law of large numbers is the foundation of such business enterprises as gambling casinos and insurance companies. The winnings (or losses) of a gambler on a single play are uncertain—that's why gambling is exciting.

The house can calculate the distribution of winnings—what the possible amounts are and how probable each amount is—from the laws of probability. The average winnings of the house on tens of thousands of plays will be very close to the mean of the distribution of winnings. That's why gambling can be a business, for this distribution guarantees the house a profit.

EXERCISE

4.82   It would be quite risky for you to insure the life of a 21-year-old friend for $100,000. There is a high probability that your friend would live and you would gain the few dollars you charged him in insurance premiums. But if he were to die, you would lose almost $100,000. Explain carefully why selling insurance is not risky for an insurance company that insures many thousands of 21-year-old men.

SUMMARY

When we want information about the **population mean μ** for some variable, we often take an SRS and use the **sample mean $\overline{x}$** to estimate the unknown parameter μ.

The **sampling distribution of $\overline{x}$** describes how the statistic $\overline{x}$ varies in all possible samples from the population.

The **mean** of the sampling distribution is μ, so that $\overline{x}$ is an unbiased estimator of μ.

The **standard deviation** of the sampling distribution of $\overline{x}$ is $\sigma/\sqrt{n}$ for an SRS of size n if the population has standard deviation σ. This recipe can be used in practice if the population is at least 10 times as large as the sample.

If the population has a normal distribution, so does $\overline{x}$.

The **central limit theorem** states that for large n the sampling distribution of $\overline{x}$ is approximately normal for any population with finite standard deviation σ. The mean and standard deviation of the normal distribution are the mean μ and standard deviation $\sigma/\sqrt{n}$ of $\overline{x}$ itself.

The **law of large numbers** states that the actually observed mean outcome $\overline{x}$ of a large number of observations must approach the mean μ of the population.

SECTION 4.5. EXERCISES

4.83   A company that owns and services a fleet of cars for its sales force has found that the service lifetime of disc brake pads varies from car to car according to a normal distribution with mean $\mu = 55,000$ miles and standard deviation $\sigma = 4500$ miles. The company installs a new brand of brake pads on 8 cars.

(a) If the new brand has the same lifetime distribution as the previous type, what is the distribution of the sample mean lifetime for the 8 cars?

(b) The average life of the pads on these 8 cars turns out to be $\bar{x} = 51,800$ miles. What is the probability that the sample mean lifetime is 51,800 miles or less if the lifetime distribution is unchanged? The company takes this probability as evidence that the average lifetime of the new brand of pads is less than 55,000 miles.

4.84   The number of traffic accidents per week at an intersection varies with mean 2.2 and standard deviation 1.4. The number of accidents in a week must be a whole number, so the population distribution is not normal.

(a) Let $\bar{x}$ be the mean number of accidents per week at the intersection during a year (52 weeks). What is the approximate distribution of $\bar{x}$ according to the central limit theorem?

(b) What is the approximate probability that $\bar{x}$ is less than 2?

(c) What is the approximate probability that there are fewer than 100 accidents at the intersection in a year? (Hint: Restate this event in terms of $\bar{x}$.)

4.85   Judy's doctor is concerned that she may suffer from hypokalemia (low potassium in the blood). There is variation both in the actual potassium level and in the blood test that measures the level. Judy's measured potassium level varies according to the normal distribution with $\mu = 3.8$ and $\sigma = .2$. A patient is classified as hypokalemic if the potassium level is below 3.5.

(a) If a single potassium measurement is made, what is the probability that Judy is diagnosed as hypokalemic?

(b) If measurements are made instead on 4 separate days and the mean result is compared with the criterion 3.5, what is the probability that Judy is diagnosed as hypokalemic?

4.86   The level of nitrogen oxide (NOX) in the exhaust of a particular car model varies with mean 1.4 grams per mile (g/mi) and standard deviation 0.3 g/mi. A company has 125 cars of this model in its fleet. If $\bar{x}$ is the mean NOX emission level for these cars, what is the level L such that the probability that $\bar{x}$ is greater than L is only 0.01? (Hint: This requires a backward normal calculation. See page 71 of Chapter 1 if you need to review.)

4.87   Children in kindergarten are sometimes given the Ravin Progressive Matrices Test (RPMT) to assess their readiness for learning. Experience at Southwark Elementary School suggests that the RPMT scores for its kindergarten pupils have mean 13.6 and standard deviation 3.1. The distribution is close to normal. Mr. Lavin has 22 children in his kindergarten class this year. He suspects that their RPMT scores will be unusually low because the test was interrupted by a fire drill. To check this suspicion, he wants to find the level L such that there is probability only 0.05 that the mean score of 22 children falls below L when the usual Southwark distribution remains true. What is the value of L? (Hint: This requires a backward normal calculation. See page 71 of Chapter 1 if you need to review.)

4.6   CONTROL CHARTS[*]

There are many situations in which our goal is to hold a variable constant over time. You may monitor your weight or blood pressure and plan to modify your behavior if either changes. Manufacturers watch the results of regular measurements made during production and plan to take action if quality deteriorates. Statistics plays a central role in these situations because of the presence of variation. *All processes have variation.* Your weight fluctuates from day to day; the critical dimension of a machined part varies a bit from item to item. Variation occurs in even the most precisely made product due to small changes in the raw material, the adjustment of the machine, the behavior of the operator, and even the temperature in the plant. Because variation is always present, we can't expect to hold a variable exactly constant over time. The statistical description of stability over time requires that the pattern of variation remain stable, not that there be no variation in the variable measured.

STATISTICAL CONTROL

A variable that continues to be described by the same distribution when observed over time is said to be in statistical control, or simply **in control**.

Control charts are statistical tools that monitor the control of a process and alert us when the process has been disturbed.

---

[*]Control charts are important in industry and also illustrate the use of sampling distributions in inference. Nonetheless, this section is not required for an understanding of later material.

Control charts expect that variation will be present. They work by distinguishing the natural variation in the process from the additional variation that suggests that the process has changed. A control chart sounds an alarm when it sees too much variation. The most common application of control charts is to monitor the performance of an industrial process. The same methods, however, can be used to check the stability of quantities as varied as the ratings of a television show and the level of ozone in the atmosphere. Control charts combine graphical and numerical descriptions of data with use of sampling distributions. They therefore provide a natural bridge between exploratory data analysis and formal statistical inference.

$\bar{x}$ charts

The population in the control chart setting is all items that would be produced by the process if it ran on forever in its present state. The items actually produced form samples from this population. We generally speak of the process rather than the population. Choose a quantitative variable, such as a diameter or a voltage, that is an important measure of the quality of an item. The process mean μ is the long-term average value of this variable; μ describes the center or aim of the process. The sample mean $\bar{x}$ of several items estimates μ and helps us judge whether the center of the process has moved away from its proper value. The most common control chart plots the means $\bar{x}$ of small samples taken from the process at regular intervals over time.

EXAMPLE 4.27

A manufacturer of video screens for computer monitors must control the tension on the mesh of fine wires that lies behind the surface of the screen. Too much tension will tear the mesh, and too little will allow wrinkles. Tension is measured by an electrical device with output readings in millivolts (mV). The proper tension is 275 mV. Some variation is always present in the production process. Careful study shows that when the process is operating properly, the standard deviation of the tension readings is $\sigma = 43$ mV.

The operator measures the tension on a sample of 4 screens each hour. The mean $\bar{x}$ of each sample estimates the mean tension μ for the process at the time of the sample. Table 4.1 shows the observed $\bar{x}$'s for 20 consecutive hours of production. How can we use these data to keep the process in control? ◀

TABLE 4.1 $\bar{x}$ from 20 samples of size 4			
Sample	$\bar{x}$	Sample	$\bar{x}$
1	269.5	11	264.7
2	297.0	12	307.7
3	269.6	13	310.0
4	283.3	14	343.3
5	304.8	15	328.1
6	280.4	16	342.6
7	233.5	17	338.8
8	257.4	18	340.1
9	317.5	19	374.6
10	327.4	20	336.1

A time plot helps us see whether or not the process is stable. Figure 4.21 is a plot of the successive sample means against the order in which the samples were taken. Because the target value for the process mean is $\mu = 275$ mV, we draw a *centerline* at that level across the plot. The means from the later samples fall above this line and are consistently higher than those from earlier samples. This suggests that the process mean μ may

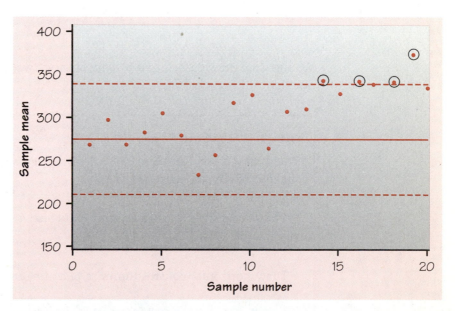

FIGURE 4.21   An $\bar{x}$ control chart for the data in Table 4.1. The points plotted are mean tension measurements $\bar{x}$ for samples of 4 video screens taken hourly during production. The solid center line and dashed control limits help determine whether the process has been disturbed.

have shifted upward, away from its target value of 275 mV. But perhaps the drift in $\bar{x}$ simply reflects the natural variation in the process. We need to back up our graph by calculation.

We expect $\bar{x}$ to have a distribution that is close to normal. Not only are the tension measurements roughly normal, but also the central limit theorem effect implies that sample means will be closer to normal than individual measurements. Because a control chart is a warning device, it is not necessary that our probability calculations be exactly correct. Approximate normality is good enough. In that same spirit, control charts use the approximate normal probabilities given by the 68–95–99.7 rule rather than more exact calculations using Table A.

If the standard deviation of the individual screens remains at $\sigma = 43$ mV, the standard deviation of $\bar{x}$ from 4 screens is

$$\frac{\sigma}{\sqrt{n}} = \frac{43}{\sqrt{4}} = 21.5 \text{ mV}$$

As long as the mean remains at its target value $\mu = 275$ mV, the 99.7 part of the 68–95–99.7 rule says that almost all values of $\bar{x}$ will lie between

$$\mu - 3\frac{\sigma}{\sqrt{n}} = 275 - (3)(21.5) = 210.5$$

$$\mu + 3\frac{\sigma}{\sqrt{n}} = 275 + (3)(21.5) = 339.5$$

We therefore draw dashed *control limits* at these two levels on the plot. We now have an $\bar{x}$ control chart.

$\bar{X}$ CONTROL CHART

To evaluate the control of a process with given standards μ and σ, make an $\bar{x}$ **control chart** as follows:

- Plot the means $\bar{x}$ of regular samples of size n against time.

- Draw a horizontal **centerline** at μ.

- Draw horizontal **control limits** at $\mu \pm 3\sigma/\sqrt{n}$.

Any $\bar{x}$ that does not fall between the control limits is evidence that the process is out of control.

Four points, which are circled in Figure 4.21, lie above the upper control limit of the control chart. The 99.7 part of the 68–95–99.7 rule says that the probability is only 0.003 that a particular point would fall outside the control limits if μ and σ remain at their target values. These points are therefore good evidence that the distribution of mesh tension has changed. It appears that the process mean moved up at about sample number 12. In practice, the operators search for a disturbance in the process as soon as they notice the first out-of-control point, that is, after sample number 14. Lack of control might be caused by a new operator, a new batch of mesh, or a breakdown in the tensioning apparatus. The out-of-control signal alerts us to the change before a large number of defective screens are produced.

$\overline{x}$ chart   An $\overline{x}$ control chart is often called simply an *$\overline{x}$ chart*. Points $\overline{x}$ that vary between the control limits of an $\overline{x}$ chart represent the chance variation that is present in a normally operating process. Points that are out of control suggest that some source of additional variability has disturbed the stable operation of the process. Such a disturbance makes out-of-control points probable rather than unlikely. For example, if the process mean μ in Example 4.27 shifts from 275 mV to 339.5 mV (which is the value of the upper control limit), the probability that the next point falls above the upper control limit increases from about 0.0015 to 0.5.

EXERCISES

4.88   A maker of auto air conditioners checks a sample of 4 thermostats from each hour's production. The thermostats are set at 75° F and then placed in a chamber where the temperature rises gradually. The tester records the temperature at which the thermostat turns on the air conditioner. The target for the process mean is $\mu = 75°$. Past experience indicates that the response temperature of properly adjusted thermostats varies with $\sigma = .5°$. The mean response temperature $\overline{x}$ for each hour's sample is plotted on an $\overline{x}$ control chart. Calculate the centerline and control limits for this chart.

4.89   The width of a slot cut by a milling machine is important to the proper functioning of an aircraft hydraulic system. The manufacturer checks the control of the milling process by measuring a sample of 5 consecutive items during each hour's production. The mean slot width for each sample is plotted on an $\overline{x}$ control chart. The target width for the slot is $\mu = .8750$ inch. When properly adjusted, the milling machine should produce slots with mean width equal to the target value and standard deviation $\sigma = .0012$ inch. What centerline and control limits should you draw on the $\overline{x}$ chart?

Using control charts

The basic signal for lack of control in an $\bar{x}$ chart is a single point beyond the control limits. In practice, however, other signals are used as well. In particular, a *run signal* is almost always combined with the basic one-point-out signal.

OUT-OF-CONTROL SIGNALS

The most common signals for lack of control in an $\bar{x}$ chart are:

- One point falling outside the control limits.

- A **run** of 9 points in a row on the same side of the centerline.

Begin a search for the cause as soon as a chart shows either signal.

Nine consecutive points on the same side of the centerline are unlikely to occur unless the process aim has moved away from the target. Think of 9 straight heads or 9 straight tails in tossing a coin. The run signal often responds to a gradual drift in the process mean before the one-point-out signal, while the one-point-out signal often catches a sudden shift in the process mean more quickly. The two signals together make a good team. In the $\bar{x}$ chart of Figure 4.21, the run signal does not give an out-of-control signal until sample number 20. The one-point-out signal alerts us at sample 14.

$\hat{p}$ chart   There are many types of control charts. A control chart gets its name from the statistic it plots against time. An $\bar{x}$ chart plots $\bar{x}$, while a *$\hat{p}$ chart* plots the sample proportion $\hat{p}$. (By an accident of history, a $\hat{p}$ chart is called a p chart in most writing on quality control.) We find the centerline and control limits in any control chart from the sampling distribution of the statistic. You know, for example, that $\hat{p}$ from a sample of size n has approximately the $N(p, \sqrt{p(1-p)/n})$ distribution. You can use this fact and the 68–95–99.7 rule to give the centerline and control limits for a $\hat{p}$ chart to control a population proportion at a target value p. Such $\hat{p}$ charts are most often used to monitor the proportion p of items produced that fail to conform to a set of specifications.

There are many variations on even the $\bar{x}$ chart. In practice, we rarely know the process mean μ and the process standard deviation σ. We must then base control limits on estimates of μ and σ from past samples. There are some fine points in using control limits based on past data. For example, we must first check that the process was already in control when we gathered the data. We will content ourselves with the case in which values μ and σ are given.

The purpose of a control chart is not to ensure good quality by inspecting many of the items produced. Control charts focus on the manufacturing process itself rather than on the products. By checking the process at regular intervals, we can detect disturbances and correct them before they become serious. This is called **statistical process control**. Process control achieves high quality at a lower cost than inspecting all of the products. Small samples of 4 or 5 items are usually adequate for $\overline{x}$ charts.

statistical process control

A process that is in control is stable over time, but stability alone does not guarantee good quality. The natural variation in the process may be so large that many of the products are unsatisfactory. Nonetheless, establishing control brings a number of advantages.

- In order to assess whether the process quality is satisfactory, we must observe the process operating in control free of breakdowns and other disturbances.

- A process in control is predictable. We can predict both the quantity and the quality of items produced.

- When a process is in control we can easily see the effects of attempts to improve the process, which are not hidden by the unpredictable variation that characterizes lack of statistical control.

A process in control is doing as well as it can. If the process is not capable of producing adequate quality even when undisturbed, we must make some major change in the process, such as installing new machines or retraining the operators.

EXERCISES

4.90   A pharmaceutical manufacturer forms tablets by compressing together the active ingredient and various fillers. The operators measure the hardness of a sample from each lot of tablets in order to control the compression process. The target values for the hardness are $\mu = 11.5$ and $\sigma = .2$. Table 4.2 gives three sets of data, each representing $\overline{x}$ for 20 successive samples of $n = 4$ tablets. One set remains in control at the target value. In a second set, the process mean μ shifts suddenly to a new value. In a third, the process mean drifts gradually.

TABLE 4.2	Three sets of $\bar{x}$ from 20 samples of size 4		
Sample	Data Set A	Data Set B	Data Set C
1	11.602	11.627	11.495
2	11.547	11.613	11.475
3	11.312	11.493	11.465
4	11.449	11.602	11.497
5	11.401	11.360	11.573
6	11.608	11.374	11.563
7	11.471	11.592	11.321
8	11.453	11.458	11.533
9	11.446	11.552	11.486
10	11.522	11.463	11.502
11	11.664	11.383	11.534
12	11.823	11.715	11.624
13	11.629	11.485	11.629
14	11.602	11.509	11.575
15	11.756	11.429	11.730
16	11.707	11.477	11.680
17	11.612	11.570	11.729
18	11.628	11.623	11.704
19	11.603	11.472	12.052
20	11.816	11.531	11.905

(a) What are the centerline and control limits for an $\bar{x}$ chart for hardness?

(b) Draw a separate $\bar{x}$ chart for each of the three data sets. Circle any points that are beyond the control limits. Also, check for runs of 9 points above or below the centerline and mark the ninth point of any run as being out of control.

(c) Based on your work in (b) and the appearance of the control charts, which set of data comes from a process that is in control? In which case does the process mean shift suddenly and at about which sample do you think that the mean changed? Finally, in which case does the mean drift gradually?

4.91   The diameter of a bearing deflector in an electric motor is supposed to be 2.205 cm. Experience shows that when the manufacturing process is properly adjusted, it produces items with mean 2.2050 cm and standard deviation 0.0010 cm. A sample of 5 consecutive items is measured once each hour. The sample means $\bar{x}$ for the past 12 hours are

Hour	1	2	3	4	5	6
$\bar{x}$	2.2047	2.2047	2.2050	2.2049	2.2053	2.2043

Hour	7	8	9	10	11	12
$\bar{x}$	2.2036	2.2042	2.2038	2.2045	2.2026	2.2040

Make an $\bar{x}$ control chart for the deflector diameter. Use both the one-point-out and run-of-nine signals to assess the control of the process. At what point should action have been taken to correct the process as the hourly point was added to the chart?

4.92 A manager who knows no statistics asks you, "What does it mean to say that a process is in control? Does being in control guarantee that the quality of our product is good?" Answer these questions in plain language that the manager can understand.

SUMMARY

A process that continues over time is **in control** if any variable measured on the process has the same distribution at any time. A process that is in control is operating under stable conditions.

An $\bar{x}$ **control chart** is a graph of sample means plotted against the time order of the samples, with a solid **centerline** at the target value μ of the process mean and dashed **control limits** at $\mu \pm 3\sigma/\sqrt{n}$. An $\bar{x}$ chart helps us decide if a process is in control with mean μ and standard deviation σ.

The probability that the next point lies outside the control limits on an $\bar{x}$ chart is about 0.003 if the process is in control. Such a point is evidence that the process is **out of control**. That is, some disturbance has changed the distribution of the process. A cause for the change in the process should be sought.

There are other common signals for lack of control, such as a **run** of 9 consecutive points on the same side of the centerline.

There are other types of control charts, such as $\hat{p}$ **charts**. A control chart is named for the statistic that it plots.

The purpose of **statistical process control** using control charts is to monitor a process so that any changes can be detected and corrected quickly. This is an economical method of maintaining good product quality.

SECTION 4.6 EXERCISES

4.93   Ceramic insulators are baked in lots in a large oven. After the baking, the process operator chooses 3 insulators at random from each lot, tests them for breaking strength, and plots the mean breaking strength for each sample on a control chart. The specifications call for a mean breaking strength of at least 10 pounds per square inch (psi). Past experience suggests that if the ceramic is properly formed and baked, the standard deviation in the breaking strength is about 1.2 psi. Here are the sample means from the last 15 lots.

Lot	1	2	3	4	5	6	7	8
$\bar{x}$	12.94	11.45	11.78	13.11	12.69	11.77	11.66	12.60

Lot	9	10	11	12	13	14	15
$\bar{x}$	11.23	12.02	10.93	12.38	7.59	13.17	12.14

(a) Find the centerline and control limits for an $\bar{x}$ chart with standards $\mu = 10$ and $\sigma = 1.2$. Make a control chart with these lines drawn on it and plot the $\bar{x}$ points on your chart.

(b) A process mean breaking strength greater than 10 psi is acceptable, so points out of control in the high direction do not call for remedial action. With this in mind, use both the one-point-out and run-of-nine signals to assess the control of the process and recommend action.

4.94   A process molds plastic screw-on caps for containers of motor oil. The strength of properly made caps (the torque that would break the cap when it is screwed tight) has the normal distribution with mean 10 inch-pounds and standard deviation 1.2 inch-pounds. You monitor the molding process by testing a sample of 6 caps every 20 minutes. You measure the breaking strength of the sample caps and keep an $\bar{x}$ chart. Find the centerline and control limits for this $\bar{x}$ chart.

4.95   An important step in the manufacture of integrated circuit chips is etching the lines that conduct current between components on the chip. The chips contain a line width test pattern for process control measurements. The target width is 3.0 micrometers (μm). Past data show that the line width varies in production according to a normal distribution with mean 2.829 μm and standard deviation 0.1516 μm.

(a) The acceptable range of line widths is $3.0 \pm .2$ μm. What percent of chips have line widths outside this range?

(b) What are the control limits for an $\bar{x}$ chart for line width if samples of size 5 are measured at regular intervals during production? (Use the target value 3.0 as your centerline.)

4.96 The control limits of an $\bar{x}$ chart describe the normal range of variation of sample means. Don't confuse this with the range of values taken by individual items. Example 4.27 describes tension measurements for video screens, which vary with mean $\mu = 275$ and standard deviation $\sigma = 43$. The control limits, within which 99.7% of $\bar{x}$'s from samples of size 4 lie, are 210.5 and 339.5. A manager notices that many individual screens have tension outside this range. He is upset about this. Explain to the manager why many individuals fall outside the control limits. Then (assuming normality) find the range of values that contains 99.7% of individual tension measurements.

4.97 **(Optional)** The usual American and Japanese practice in making $\bar{x}$ charts is to place the control limits 3 standard deviations of $\bar{x}$ out from the centerline. That is, the control limits are $\bar{x} \pm 3\sigma/\sqrt{n}$. The probability that a particular $\bar{x}$ falls outside these limits when the process is in control is about 0.003, using the 99.7 part of the 68–95–99.7 rule. European practice, on the other hand, places the control limits c standard deviations out, where the number c is chosen to give exactly probability 0.001 of a point $\bar{x}$ falling above $\mu + c\sigma/\sqrt{n}$ when the target μ and σ remain true. (The probability that $\bar{x}$ falls below $\mu - c\sigma/\sqrt{n}$ is also 0.001 because of the symmetry of the normal distributions.) Use Table A to find the value of c.

4.98 **(Optional)** A manufacturer of compact disc players uses statistical process control to monitor the quality of the circuit board that contains most of the player's electronic components. Inspectors test every finished circuit board for proper function using a computerized test bed. The plant produces 400 circuit boards per shift. The proportion $\hat{p}$ of the 400 boards that fail the test is recorded each shift. Company standards call for a failure rate of no more than 10%, or $p = .1$.

(a) If the process is in control with proportion $p = .1$ of defective boards, what are the mean and standard deviation of the sample proportion $\hat{p}$?

(b) According to the normal approximation, what is the distribution of $\hat{p}$ when the process is in control?

(c) Give the centerline (at the mean of $\hat{p}$) and control limits (at three standard deviations of $\hat{p}$ above and below the mean).

(d) The table below lists the proportions of failures in 16 consecutive shifts' production. Make a $\hat{p}$ control chart with centerline and control limits from (c). Plot these values of $\hat{p}$ on the chart. Are any points out of control? Are there any runs of 9 or more points?

Day	1	2	3	4	5	6	7	8
$\hat{p}$	.1150	.1600	.1300	.1225	.1000	.1225	.1900	.1150

Day	9	10	11	12	13	14	15	16
$\hat{p}$	.1000	.1600	.1675	.1225	.1375	.1975	.1525	.1675

4.99 (Optional) A manufacturer inspects successive samples of n items for process control purposes. Each item is classified as conforming to specifications or as failing to conform. The inspectors record the proportion $\hat{p}$ of nonconforming items in each sample and plot the values of $\hat{p}$ against the order of the samples. There is a target value p for the population proportion of nonconforming items. Generalize the results of the previous exercise to give recipes in terms of p for the centerline and control limits for a $\hat{p}$ control chart. Use the normal approximation for $\hat{p}$ and follow the model of the $\bar{x}$ chart.

4.100 (Optional) A producer of agricultural machinery purchases bolts from a supplier in lots of several thousand. The producer submits 80 bolts from each lot to a shear test and accepts the lot if no more than 3 of them fail the test. Past records show that an average of 1.8 bolts per lot have failed the test. This is an opportunity to keep a control chart to monitor the quality level of the supplier over time. Because $1.8/80 = .0225$, set the target proportion of nonconforming bolts at $p = .0225$. The proportion of nonconforming bolts among the 80 tested from each incoming lot will be plotted on a $\hat{p}$ control chart. Give the centerline and control limits for this chart. (Because of the informal nature of process control procedures, the chart is based on the normal approximation to the distribution of the sample proportion $\hat{p}$ even though the approximation is not very accurate for this n and p.)

CHAPTER REVIEW

This chapter lays the foundations for the study of statistical inference. Statistical inference uses data to draw conclusions about the population or process from which the data come. What is special about inference is that the conclusions include a statement, in the language of probability, about how reliable they are. The statement gives a probability that answers the question "What would happen if I used this method very many times?"

This chapter introduced the language of probability, especially sampling distributions of statistics. A sampling distribution describes the values a statistic would take in very many repetitions of a sample or an experiment under the same conditions. Understanding that idea is the key to understanding statistical inference. The chapter gave details about the sampling distributions of two important statistics: a sample proportion $\hat{p}$ and a sample mean $\overline{x}$. These statistics behave much the same. In particular, their sampling distributions are approximately normal if the sample is large. This is a main reason why normal distributions are so important in statistics. We can use everything we know about normal distributions to study the sampling distributions of proportions and means.

Here is a review list of the most important things you should be able to do after studying this chapter. The list doesn't include the optional material on probability distributions, binomial distributions, or control charts.

A. SAMPLING DISTRIBUTIONS

1. Identify parameters and statistics in a sample or experiment.
2. Recognize the fact of sampling variability: a statistic will take different values when you repeat a sample or experiment.
3. Interpret a sampling distribution as describing the values taken by a statistic in all possible repetitions of a sample or experiment under the same conditions.
4. Describe the bias and variability of a statistic in terms of the mean and spread of its sampling distribution.

5. Understand that the variability of a statistic is controlled by the size of the sample. Statistics from larger samples are less variable.

B. PROBABILITY

1. Recognize that some phenomena are random. Probability describes the long-run order of a random phenomenon.

2. Know that any probability is a number between 0 and 1. Understand that the probability of an event is the proportion of times the event occurs out of very many repetitions of a random phenomenon.

3. Interpret the sampling distribution of a statistic as describing the probabilities of its possible values.

C. SAMPLE PROPORTIONS

1. Recognize when a problem involves a sample proportion $\hat{p}$.

2. Find the mean and standard deviation of a sample proportion $\hat{p}$ for an SRS of size n from a population having population proportion p.

3. Know that the standard deviation (spread) of the sampling distribution of $\hat{p}$ gets smaller at the rate $\sqrt{n}$ as the sample size n gets larger.

4. Recognize when you can use the normal approximation to the sampling distribution of $\hat{p}$. Use the normal approximation to calculate probabilities that concern $\hat{p}$.

D. SAMPLE MEANS

1. Recognize when a problem involves the mean $\bar{x}$ of a sample.

2. Find the mean and standard deviation of a sample mean $\bar{x}$ from an SRS of size n when the mean μ and standard deviation σ of the population are known.

3. Know that the standard deviation (spread) of the sampling distribution of $\bar{x}$ gets smaller at the rate $\sqrt{n}$ as the sample size n gets larger.

4. Understand that $\bar{x}$ has approximately a normal distribution when the sample is large (central limit theorem). Use this normal distribution to calculate probabilities that concern $\bar{x}$.

5. Use the law of large numbers to interpret the population mean μ as the average of an indefinitely large number of observations drawn from the population.

4.101 An opinion poll asks a sample of 500 adults whether they favor giving parents of school-age children vouchers that can be exchanged for education at any public or private school of their choice. Each school would be paid by the government on the basis of how many vouchers it collected. Suppose that in fact 45% of the population favor this idea. What is the probability that more than half of the sample are in favor? (Assume that the sample is an SRS.)

4.102 High school dropouts make up 14.1% of all Americans aged 18 to 24. A vocational school that wants to attract dropouts mails an advertising flyer to 25,000 persons between the ages of 18 and 24.

(a) If the mailing list can be considered a random sample of the population, what is the mean number of high school dropouts who will receive the flyer?

(b) What is the probability that at least 3500 dropouts will receive the flyer?

4.103 The Wechsler Adult Intelligence Scale (WAIS) is a common "IQ test" for adults. The distribution of WAIS scores for persons over 16 years of age is approximately normal with mean 100 and standard deviation 15.

(a) What is the probability that a randomly chosen individual has a WAIS score of 105 or higher?

(b) What are the mean and standard deviation of the average WAIS score $\bar{x}$ for an SRS of 60 people?

(c) What is the probability that the average WAIS score of an SRS of 60 people is 105 or higher?

(d) Would your answers to any of (a), (b), or (c) be affected if the distribution of WAIS scores in the adult population were distinctly nonnormal?

4.104 The weight of the eggs produced by a certain breed of hen is normally distributed with mean 65 grams (g) and standard deviation 5 g. Think of cartons of such eggs as SRSs of size 12 from the population of all eggs. What is the probability that the weight of a carton falls between 750 g and 825 g?

4.105 A study of rush-hour traffic in San Francisco counts the number of people in each car entering a freeway at a suburban interchange. Suppose that this count has mean 1.5 and standard deviation 0.75 in the population of all cars that enter at this interchange during rush hours.

(a) Could the exact distribution of the count be normal? Why or why not?

(b) Traffic engineers estimate that the capacity of the interchange is 700 cars per hour. According to the central limit theorem, what is the approximate distribution of the mean number of persons $\bar{x}$ in 700 randomly selected cars at this interchange?

(c) What is the probability that 700 cars will carry more than 1075 people? (Hint: Restate this event in terms of the mean number of people $\bar{x}$ per car.)

4.106   Power companies trim trees growing near their lines to avoid power failures due to falling limbs in storms. Applying a chemical to slow the growth of the trees is cheaper than trimming, but the chemical kills some of the trees. Suppose that one such chemical would kill 20% of sycamore trees. The power company tests this chemical on 250 sycamores. Consider these as an SRS from the population of all sycamore trees.

(a) What are the mean and standard deviation of the number of trees in the sample that are killed?

(b) What is the probability that at least 60 trees (24% of the sample) are killed?

4.107   While he was a prisoner of the Germans during World War II, the British mathematician John Kerrich tossed a coin 10,000 times. He got 5067 heads. Take Kerrich's tosses to be an SRS from the population of all possible tosses of his coin. If the coin is perfectly balanced, $p = .5$. Is there reason to think that Kerrich's coin gave too many heads to be balanced? To answer this question, find the probability that a balanced coin would give 5067 or more heads in 10,000 tosses. What do you conclude?

NOTES AND DATA SOURCES

1. In this book we discuss only the most widely used kind of statistical inference. This is sometimes called *frequentist* inference because it is based on answering the question "What would happen in many repetitions?" Another approach to inference, called *Bayesian*, can be used even for one-time situations. Bayesian inference is important but is conceptually complex and much less widely used in practice.

2. The survey question and result are reported in Trish Hall, "Shop? Many say 'Only if I must,' " *New York Times*, November 28, 1990.

3. Strictly speaking, the recipes we give for the standard deviations of $\hat{p}$ and $\bar{x}$ assume that an SRS of size n is drawn from an *infinite* population. If the population has finite size N, the standard deviations in the recipes are multiplied by

$\sqrt{1 - n/N}$. This "finite population correction" approaches 1 as N increases. When $n/N \leq .1$, it is $\geq .948$.

4. Office of Technology Assessment, *Scientific Validity of Polygraph Testing: A Research Review and Evaluation*, Government Printing Office, Washington, D.C., 1983.

5. From John K. Ford, "A method for grading 1987 stock recommendations," *American Association of Individual Investors Journal*, March 1988, pp. 16–17.

JERZY NEYMAN

The most-used methods of statistical inference are confidence intervals and tests of significance. Both are products of the twentieth century. From complex and sometimes confusing origins, statistical tests took their current form in the writings of R. A. Fisher, whom we met at the beginning of Chapter 3. Confidence intervals appeared in 1934, the brainchild of Jerzy Neyman (1894–1981).

Neyman was trained in Poland and, like Fisher, worked at an agricultural research institute. He moved to London in 1934 and in 1938 joined the University of California at Berkeley. He founded Berkeley's Statistical Laboratory and remained its head even after his official retirement as a professor in 1961. Retirement did not slow Neyman's work—he remained active until the end of his long life and almost doubled his list of publications after "retiring." Statistical problems arising from astronomy, biology, and attempts to modify the weather attracted his attention.

Neyman ranks with Fisher as a founder of modern statistical practice. In addition to introducing confidence intervals, he helped systematize the theory of sample surveys and reworked significance tests from a new point of view. Fisher, who was very argumentative, disliked Neyman's approach to tests and said so. Neyman, who wasn't shy, replied vigorously.

Tests and confidence intervals are our topic in this chapter. Like most users of statistics, we will stay close to Fisher's approach to tests. You can find some of Neyman's ideas in the optional final section.

CHAPTER **5**

Introduction to Inference

INTRODUCTION

To infer means to draw a conclusion. Statistical inference provides us with methods for drawing conclusions from data. We have, of course, been drawing conclusions from data all along. What is new in formal inference is that we use probability to express the strength of our conclusions. Probability allows us to take chance variation into account and so to correct our judgment by calculation. Here are two examples of how probability can correct our judgment.

EXAMPLE 5.1

In the Vietnam War years, a lottery determined the order in which men were drafted for army service. The lottery assigned draft numbers by choosing birth dates in random order. We expect a correlation near zero between birth dates and draft numbers if the draft numbers come from random choice. The actual correlation between birth date and draft number in the first draft lottery was $r = -.226$. That is, men born later in the year tended to get lower draft numbers. Is this small correlation evidence that the lottery was biased? Our unaided judgment can't tell, because any two variables will have some association in practice, just by chance. So we calculate that a correlation this far from zero has probability less than 0.001 in a truly random lottery. Because a correlation as strong as that observed would almost never occur in a random lottery, there is strong evidence that the lottery was unfair.

Probability calculations can also protect us from jumping to a conclusion when only chance variation is at work. Investigate 20 new businesses started by women and another 20 started by men. After two years, 12 of those headed by women have failed, but only 8 of the businesses started by men have failed. Can we conclude from this sample that failures are more frequent among all new businesses started by women? A difference this large or larger between the results for two groups of 20 businesses would occur about one time in five simply because of chance variation. An effect that could so easily be just chance is not convincing. ◀

In this chapter we will meet the two most common types of formal statistical inference. Section 5.1 concerns *confidence intervals* for estimating the value of a population parameter. Section 5.2 presents *tests of significance*, which assess the evidence for a claim about a population. Both types of inference are based on the sampling distributions of statistics. That is, both report probabilities that state *what would happen if we used the inference method many times*. This kind of probability statement is characteristic of statistical inference. Users of statistics must understand the meaning of the probability statements that appear, for example, on computer output for statistical procedures.

The methods of formal inference require the long-run regular behavior that probability describes. Inference is most reliable when the data are produced by a properly randomized design. *When you use statistical inference you are acting as if the data are a random sample or come from a randomized experiment.* If this is not true, your conclusions may be open to challenge. Do not be overly impressed by the complex details of formal inference. This elaborate machinery cannot remedy basic flaws in producing the data such as voluntary response samples and uncontrolled experiments. Use the common sense developed in your study of the first three chapters of this book, and proceed to formal inference only when you are satisfied that the data deserve such analysis.

The purpose of this chapter is to describe the reasoning used in statistical inference. We will illustrate the reasoning by a few specific inference techniques, but these are oversimplified so that they are not very useful in practice. Later chapters will first show how to modify these techniques to make them practically useful and will then introduce inference methods for use in most of the settings we met in learning to explore data. There are libraries—both of books and of computer software—full of more elaborate statistical techniques. Informed use of any of these methods requires an understanding of the underlying reasoning. A computer will do the arithmetic, but you must still exercise judgment based on understanding.

5.1 ESTIMATING WITH CONFIDENCE

Young people have a better chance of full-time employment and good wages if they are good with numbers. How strong are the quantitative skills of young Americans of working age? One source of data is the National Assessment of Educational Progress (NAEP) Young Adult Literacy Assessment Survey, which is based on a nationwide probability sample of households.

EXAMPLE 5.2

The NAEP survey includes a short test of quantitative skills, covering mainly basic arithmetic and the ability to apply it to realistic problems. Scores on the test range from 0 to 500. For example, a person who scores 233 can add the amounts of two checks appearing on a bank deposit slip; someone scoring 325 can determine the price of a meal from a menu; a person scoring 375 can transform a price in cents per ounce into dollars per pound.

In a recent year, 840 men 21 to 25 years of age were in the NAEP sample. Their mean quantitative score was $\bar{x} = 272$. These 840 men are an SRS from the population of all young men. On the basis of this sample, what can we say about the mean score μ in the population of all 9.5 million young men of these ages?[1]   ◄

The sample mean $\bar{x}$ is an unbiased estimator of the unknown population mean μ. Because $\bar{x} = 272$, we guess that μ is "somewhere around 272." To make "somewhere around 272" more precise, we ask: *How would the sample mean $\bar{x}$ vary if we took many samples of 840 young men from this same population?* Recall the essential facts about the sampling distribution of $\bar{x}$:

- $\bar{x}$ has a normal distribution. (The central limit theorem tells us that the average of 840 scores has a distribution that is very close to normal.)
- The mean of this normal sampling distribution is the same as the unknown population mean μ.
- The standard deviation of $\bar{x}$ for an SRS of 840 men is $\sigma/\sqrt{840}$, where σ is the standard deviation of individual NAEP scores among all young men.

Let us suppose that we know from long experience that the standard deviation of scores in the population of all young men is $\sigma = 60$. The standard deviation of $\bar{x}$ is then

$$\frac{\sigma}{\sqrt{n}} = \frac{60}{\sqrt{840}} \doteq 2.1$$

(It is not realistic to assume we know σ. We will see in the next chapter how to proceed when σ is not known. For now, we are more interested in statistical reasoning than in details of realistic methods.)

In many repeated samples of size 840, the sample mean score $\bar{x}$ would vary according to the normal distribution with mean equal to the unknown μ and standard deviation 2.1. Inference about the unknown μ starts from this sampling distribution. Figure 5.1 displays the distribution. The differ-

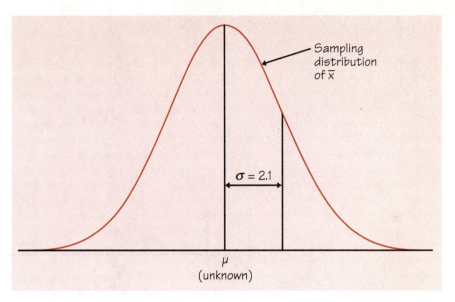

FIGURE 5.1   The sampling distribution of the mean score $\bar{x}$ of an SRS of 840 young men on the NAEP quantitative test.

ent values of $\bar{x}$ appear along the axis in the figure, and the normal curve shows how probable these values are.

Statistical confidence

Figure 5.2 is another picture of the same sampling distribution. It illustrates the following line of thought:

- The 68–95–99.7 rule says that in about 95% of all samples, $\bar{x}$ will be within two standard deviations of the population mean score μ. That is, $\bar{x}$ will be within 4.2 points of μ in 95% of all samples.
- Whenever $\bar{x}$ is within 4.2 points of the unknown μ, then of course μ is within 4.2 points of the observed $\bar{x}$. This happens in 95% of all samples.
- So in 95% of all samples the unknown μ lies between $\bar{x} - 4.2$ and $\bar{x} + 4.2$.

    This conclusion just restates a fact about the sampling distribution of $\bar{x}$. The language of statistical inference uses this fact about what would happen in the long run to express our confidence in the results of any

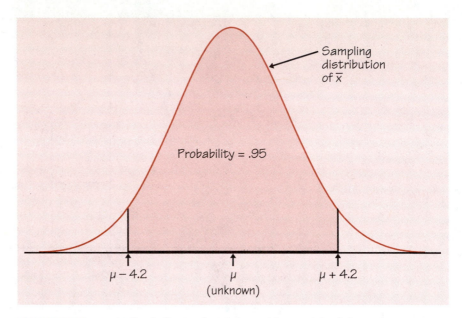

FIGURE 5.2   In 95% of all samples, $\bar{x}$ lies within ± 4.2 of the unknown population mean μ. So μ also lies within ± 4.2 of $\bar{x}$ in those samples.

one sample. Our sample gave $\bar{x} = 272$. We say that we are *95% confident* that the unknown mean NAEP quantitative score for all young men lies between

$$\bar{x} - 4.2 = 272 - 4.2 = 267.8$$

and

$$\bar{x} + 4.2 = 272 + 4.2 = 276.2$$

Be sure you understand the grounds for our confidence. There are only two possibilities:

1. The interval between 267.8 and 276.2 contains the true μ.
2. Our SRS was one of the few samples for which $\bar{x}$ is not within 4.2 points of the true μ. Only 5% of all samples give such inaccurate results.

We cannot know whether our sample is one of the 95% for which the interval $\bar{x} \pm 4.2$ catches μ, or one of the unlucky 5%. The statement that we are 95% confident that the unknown μ lies between 267.8 and 276.2 is shorthand for saying, "We got these numbers by a method that gives correct results 95% of the time."

confidence
interval

The interval of numbers between the values $\bar{x} \pm 4.2$ is called a **95% confidence interval** for μ. Like most confidence intervals we will meet, this one has the form

$$\text{estimate} \pm \text{margin of error}$$

margin of error

The estimate ($\bar{x}$ in this case) is our guess for the value of the unknown parameter. The **margin of error** ± 4.2 shows how accurate we believe our guess is, based on the variability of the estimate. This is a 95% confidence interval because it catches the unknown μ in 95% of all possible samples.

Figure 5.3 illustrates the behavior of 95% confidence intervals in repeated sampling. The center of each interval is at $\bar{x}$ and therefore varies from sample to sample. The sampling distribution of $\bar{x}$ appears at the top of the figure to show the long-term pattern of this variation. The 95% confidence intervals $\bar{x} \pm 4.2$ from 25 SRSs appear below. The center $\bar{x}$ of each interval is marked by a dot. The arrows on either side of the dot span the confidence interval. All except one of these 25 intervals cover the true value of μ. In a very large number of samples, 95% of the confidence intervals would contain μ.

EXERCISES

5.1 A *New York Times* poll on women's issues interviewed 1025 women randomly selected from the United States, excluding Alaska and Hawaii. The poll found that 47% of the women said they do not get enough time for themselves.

(a) The poll announced a margin of error of ± 3 percentage points for 95% confidence in its conclusions. What is the 95% confidence interval for the percent of all adult women who think they do not get enough time for themselves?

(b) Explain to someone who knows no statistics why we can't just say that 47% of all adult women do not get enough time for themselves.

(c) Then explain clearly what "95% confidence" means.

5.2 A student reads that a 95% confidence interval for the mean NAEP quantitative score for men of ages 21 to 25 is 267.8 to 276.2. Asked to explain

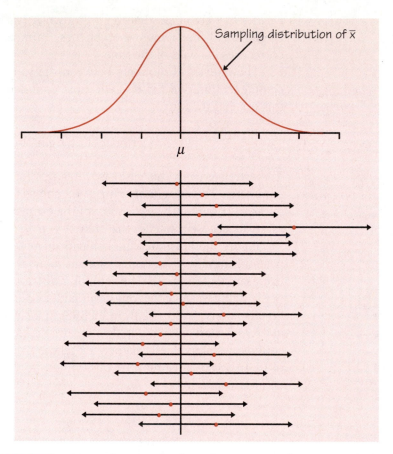

FIGURE 5.3   Twenty-five samples from the same population gave these 95% confidence intervals. In the long run, 95% of all samples give an interval that contains the population mean μ.

the meaning of this interval, the student says, "95% of all young men have scores between 267.8 and 276.2." Is the student right? Justify your answer.

5.3   Suppose that you give the NAEP test to an SRS of 1000 people from a large population in which the scores have mean 280 and standard deviation $\sigma = 60$. The mean $\bar{x}$ of the 1000 scores will vary if you take repeated samples.

(a) The sampling distribution of $\bar{x}$ is approximately normal. What are its mean and standard deviation?

(b) Sketch the normal curve that describes how $\bar{x}$ varies in many samples from this population. Mark its mean and the values one, two, and three standard deviations on either side of the mean.

(c) According to the 68–95–99.7 rule, about 95% of all the values of $\bar{x}$ fall within _____ of the mean of this curve. What is the missing number? Call it m for "margin of error." Shade the region from the mean minus m to the mean plus m on the axis of your sketch with a heavy line, as in Figure 5.2.

(d) Whenever $\bar{x}$ falls in the region you shaded, the true value of the population mean, $\mu = 280$, lies in the confidence interval between $\bar{x} - m$ and $\bar{x} + m$. Draw the confidence interval below your sketch for one value of $\bar{x}$ inside the shaded region and one value of $\bar{x}$ outside the shaded region. (Use Figure 5.3 as a model for the drawing.)

(e) In what percent of all samples will the true mean $\mu = 280$ be covered by the confidence interval $\bar{x} \pm m$?

5.4   Oxides of nitrogen (called NOX for short) emitted by cars and trucks are important contributors to air pollution. The amount of NOX emitted by a particular model varies from vehicle to vehicle. For one light truck model, NOX emissions vary with mean μ that is unknown and standard deviation $\sigma = .4$ grams per mile. You test an SRS of 50 of these trucks. The sample mean NOX level $\bar{x}$ estimates the unknown μ. You will get different values of $\bar{x}$ if you repeat your sampling.

(a) The sampling distribution of $\bar{x}$ is approximately normal. What are its mean and standard deviation?

(b) Sketch the normal curve for the sampling distribution of $\bar{x}$. Mark its mean and the values one, two, and three standard deviations on either side of the mean.

(c) According to the 68–95–99.7 rule, about 95% of all values of $\bar{x}$ lie within a distance m of the mean of the sampling distribution. What is m? Shade the region on the axis of your sketch that is within m of the mean, as in Figure 5.2.

(d) Whenever $\bar{x}$ falls in the region you shaded, the unknown population mean μ lies in the confidence interval $\bar{x} \pm m$. For what percent of all possible samples does this happen?

(e) Following the style of Figure 5.3, draw the confidence intervals below your sketch for two values of $\bar{x}$, one that falls within the shaded region and one that falls outside it.

Confidence intervals

Statisticians have constructed confidence intervals for many different parameters based on a variety of designs for producing data. We will meet a

confidence level

number of these in later chapters. Any confidence interval has two parts: an *interval* computed from the data and a **confidence level** giving the probability that the method produces an interval that covers the parameter. Users can choose the confidence level, most often 90% or higher because we most often want to be quite sure of our conclusions. We will use C to stand for the confidence level in decimal form. For example, a 95% confidence level corresponds to $C = .95$. Here is the general definition of a confidence interval for an unknown parameter. In our examples, the parameter is the mean μ of the population, but it might be a population proportion p, the population standard deviation σ, or any other parameter.

CONFIDENCE INTERVAL

A **level C confidence interval** for a parameter is an interval computed from sample data by a method that has probability C of producing an interval containing the true value of the parameter.

We can now give the recipe for a level C confidence interval for the mean μ of a population when the data are an SRS of size n. The interval is based on the fact that the sampling distribution of the sample mean $\bar{x}$ is at least approximately normal. To get confidence level C we want to catch the central probability C under a normal curve. To do that, we must go out z^* standard deviations on either side of the mean. The number z^* is the same for any normal distribution, so we use the standard normal table. Here is an example of how to find z^*.

EXAMPLE 5.3

To find an 80% confidence interval, we must catch the central 80% of the normal sampling distribution of $\bar{x}$. In catching the central 80% we leave out 20%, or 10% in each tail. So z^* is the point with area 0.1 to its right (and 0.9 to its left) under the standard normal curve. Search the body of Table A to find the point with area 0.9 to its left. The closest entry is $z^* = 1.28$. There is area 0.8 under the standard normal curve between -1.28 and 1.28. Figure 5.4 shows how z^* is related to areas under the curve. ◀

Figure 5.5 shows the general situation for any confidence level C. If we catch the central area C, the leftover tail area is $1 - C$, or $(1 - C)/2$

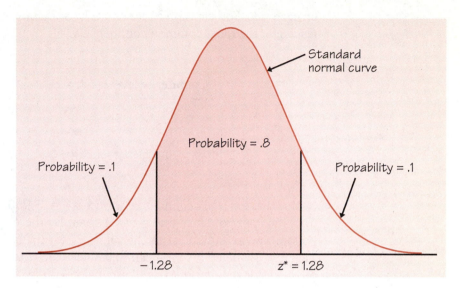

FIGURE 5.4 The central probability 0.8 under a standard normal curve lies between -1.28 and 1.28. That is, there is area 0.1 to the right of 1.28 under the curve.

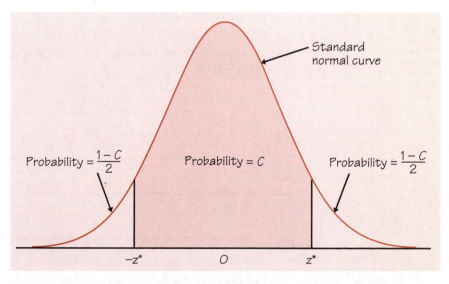

FIGURE 5.5 In general, the central probability C under a standard normal curve lies between $-z^*$ and z^*. Because z^* has area $(1 - C)/2$ to its right under the curve, we call it the upper $(1 - C)/2$ critical value.

on each side. You can find z^* for any C by searching Table A. Here are the results for the most common confidence levels:

Confidence level	Tail area	z^*
90%	.05	1.645
95%	.025	1.960
99%	.005	2.576

Notice that for 95% confidence we use $z^* = 1.960$. This is more exact than the approximate value $z^* = 2$ given by the 68–95–99.7 rule. The bottom row in Table C gives the values z^* for many confidence levels C. This row is labeled z^*. (You can find Table C in the back of the book and on the inside rear cover. We will use the other rows of the table in the next chapter.) Although we can find z^* from Table C by simply looking above the confidence level C, it is usual to describe the point z^* in terms of the probability to its right. For example, we call 1.960 the upper 0.025 *critical value* of the standard normal distribution.

CRITICAL VALUES

The number z^* with probability p lying to its right under the standard normal curve is called the **upper p critical value** of the standard normal distribution.

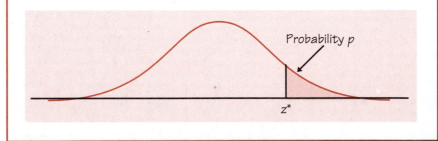

Here's how to find the level C confidence interval:

- Any normal curve has probability C between the point z^* standard deviations below its mean and the point z^* standard deviations above its mean.

- The standard deviation of the sampling distribution of $\bar{x}$ is $\sigma/\sqrt{n}$, and its mean is the population mean μ. So there is probability C that the observed sample mean $\bar{x}$ takes a value between

$$\mu - z^* \frac{\sigma}{\sqrt{n}} \quad \text{and} \quad \mu + z^* \frac{\sigma}{\sqrt{n}}$$

- Whenever this happens, the population mean μ is contained between

$$\bar{x} - z^* \frac{\sigma}{\sqrt{n}} \quad \text{and} \quad \bar{x} + z^* \frac{\sigma}{\sqrt{n}}$$

That is our confidence interval. The estimate of the unknown μ is $\bar{x}$, and the margin of error is $z^*\sigma/\sqrt{n}$.

CONFIDENCE INTERVAL FOR A POPULATION MEAN

Draw an SRS of size n from a population having unknown mean μ and known standard deviation σ. A level C confidence interval for μ is

$$\bar{x} \pm z^* \frac{\sigma}{\sqrt{n}}$$

Here z^* is the upper $(1 - C)/2$ critical value for the standard normal distribution, found in Table C. This interval is exact when the population distribution is normal and is approximately correct for large n in other cases.

EXAMPLE 5.4

A manufacturer of pharmaceutical products analyzes a specimen from each batch of a product to verify the concentration of the active ingredient. The chemical analysis is not perfectly precise. Repeated measurements on the same specimen give slightly different results. The results of repeated measurements follow a normal distribution quite closely. The analysis procedure has no bias, so the mean μ of the population of all measurements is the true concentration in the specimen. The standard deviation of this distribution is known to be

$\sigma = .0068$ grams per liter. The laboratory analyzes each specimen three times and reports the mean result.

Three analyses of one specimen give concentrations

$$0.8403 \qquad 0.8363 \qquad 0.8447$$

We want a 99% confidence interval for the true concentration μ.

The sample mean of the three readings is

$$\bar{x} = \frac{.8403 + .8363 + .8447}{3} = .8404$$

For 99% confidence, we see from Table C that $z^* = 2.576$. A 99% confidence interval for μ is therefore

$$\bar{x} \pm z^* \frac{\sigma}{\sqrt{n}} = .8404 \pm 2.576 \frac{.0068}{\sqrt{3}}$$
$$= .8404 \pm .0101$$
$$= (.8303, .8505)$$

We are 99% confident that the true concentration lies between 0.8303 and 0.8505 grams per liter. ◄

Suppose that a single measurement gave $x = .8404$, the same value that the sample mean took in Example 5.4. Repeating the calculation with $n = 1$ shows that the 99% confidence interval based on a single measurement is

$$\bar{x} \pm z^* \frac{\sigma}{\sqrt{1}} = .8404 \pm (2.576)(.0068)$$
$$= .8404 \pm .0175$$
$$= (.8229, .8579)$$

The mean of three measurements gives a smaller margin of error and therefore a shorter interval than a single measurement. Figure 5.6 illustrates the gain from using three observations.

The form of confidence intervals for the population mean μ rests on the fact that the statistic $\bar{x}$ used to estimate μ has a normal distribution.

FIGURE 5.6   Confidence intervals for $n = 3$ and $n = 1$ for Example 5.4. Larger samples give shorter intervals.

Because many sample statistics have normal distributions (at least approximately), it is useful to notice that the confidence interval has the form

$$\text{estimate} \pm z^* \sigma_{\text{estimate}}$$

The estimate based on the sample is the center of the confidence interval. The margin of error is $z^* \sigma_{\text{estimate}}$. The desired confidence level determines z^* from Table C. The standard deviation of the estimate, σ_{estimate}, depends on the particular estimate we use. When the estimate is $\bar{x}$ from an SRS, the standard deviation of the estimate is $\sigma/\sqrt{n}$.

EXERCISES

5.5   A study of the career paths of hotel general managers sent questionnaires to an SRS of 160 hotels belonging to major U.S. hotel chains. There were 114 responses. The average time these 114 general managers had spent with their current company was 11.78 years. Give a 99% confidence interval for the mean number of years general managers of major-chain hotels have spent with their current company. (Take it as known that the standard deviation of time with the company for all general managers is 3.2 years.)

5.6   The Degree of Reading Power (DRP) is a test of the reading ability of children. Here are DRP scores for a sample of 44 third-grade students in a suburban school district.[2]

40	26	39	14	42	18	25	43	46	27	19
47	19	26	35	34	15	44	40	38	31	46
52	25	35	35	33	29	34	41	49	28	52
47	35	48	22	33	41	51	27	14	54	45

(a) We expect the distribution of DRP scores to be close to normal. Make a stemplot or histogram of the distribution of these 44 scores and describe its shape.

(b) Suppose that the standard deviation of the population of DRP scores is known to be $\sigma = 11$. Give a 99% confidence interval for the mean score in the school district.

(c) Would you trust your conclusion from (b) if these scores came from a single class in one school in the district? Why?

5.7   Here are measurements (in millimeters) of a critical dimension on a sample of auto engine crankshafts.

224.120   224.001   224.017   223.982   223.989   223.961

223.960   224.089   223.987   223.976   223.902   223.980

224.098   224.057   223.913   223.999

The data come from a production process that is known to have standard deviation $\sigma = .060$ mm. The process mean is supposed to be $\mu = 224$ mm but can drift away from this target during production.

(a) We expect the distribution of the dimension to be close to normal. Make a stemplot or histogram of these data and describe the shape of the distribution.

(b) Give a 95% confidence interval for the process mean at the time these crankshafts were produced.

5.8   A test for the level of potassium in the blood is not perfectly precise. Moreover, the actual level of potassium in a person's blood varies slightly from day to day. Suppose that repeated measurements for the same person on different days vary normally with $\sigma = .2$.

(a) Julie's potassium level is measured once. The result is $x = 3.2$. Give a 90% confidence interval for her mean potassium level.

(b) If three measurements were taken on different days and the mean result is $\bar{x} = 3.2$, what is a 90% confidence interval for Julie's mean blood potassium level?

How confidence intervals behave

The confidence interval $\bar{x} \pm z^* \sigma / \sqrt{n}$ for the mean of a normal population illustrates several important properties that are shared by all confidence intervals in common use. The user chooses the confidence level, and the margin of error follows from this choice. We would like high confidence and also a small margin of error. High confidence says that our method almost always gives correct answers. A small margin of error says that we have pinned down the parameter quite precisely. The margin of error is

$$\text{margin of error} = z^* \frac{\sigma}{\sqrt{n}}$$

This expression has z^* and σ in the numerator and $\sqrt{n}$ in the denominator. So the margin of error gets smaller when

- z^* gets smaller. Smaller z^* is the same as smaller confidence level C (look at Figure 5.5 again). There is a trade-off between the confidence level and the margin of error. To obtain a smaller margin of error from the same data, you must be willing to accept lower confidence.

- σ gets smaller. The standard deviation σ measures the variation in the population. You can think of the variation among individuals in the population as noise that obscures the average value μ. It is easier to pin down μ when σ is small.

- n gets larger. Increasing the sample size n reduces the margin of error for any fixed confidence level. Because n appears under a square root sign, we must take four times as many observations in order to cut the margin of error in half.

EXAMPLE 5.5

Suppose that the pharmaceutical manufacturer in Example 5.4 is content with 90% confidence rather than 99%. Table C gives the critical value for 90% confidence as $z^* = 1.645$. The 90% confidence interval for μ based on three repeated measurements with mean $\bar{x} = .8404$ is

$$\bar{x} \pm z^* \frac{\sigma}{\sqrt{n}} = .8404 \pm 1.645 \frac{.0068}{\sqrt{3}}$$
$$= .8404 \pm .0065$$
$$= (.8339, .8469)$$

Settling for 90%, rather than 99%, confidence has reduced the margin of error from ± 0.0101 to ± 0.0065. Figure 5.7 compares these two intervals.

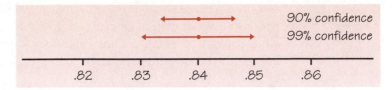

FIGURE 5.7   90% and 99% confidence intervals for Example 5.5. Higher confidence requires a wider interval.

Increasing the number of measurements from 3 to 12 will also reduce the width of the 99% confidence interval in Example 5.4. Check that replacing $\sqrt{3}$ by $\sqrt{12}$ cuts the ± 0.0101 margin of error in half, because we now have four times as many observations.   ◄

EXERCISES

5.9   Examples 5.4 and 5.5 give confidence intervals for the concentration μ based on 3 measurements with $\bar{x} = .8404$ and $\sigma = .0068$. The 99% confidence interval is 0.8303 to 0.8505 and the 90% confidence interval is 0.8339 to 0.8469.

(a) Find the 80% confidence interval for μ.

(b) Find the 99.9% confidence interval for μ.

(c) Make a sketch like Figure 5.7 to compare all four intervals. How does increasing the confidence level affect the length of the confidence interval?

5.10   Find the margin of error for 99% confidence in Example 5.4 if the laboratory measures the concentration of each specimen 12 times. Check that your result is half as large as the margin of error based on 3 measurements in Example 5.4.

5.11   The National Assessment of Educational Progress (NAEP) test (Example 5.2) was also given to a sample of 1077 women of ages 21 to 25 years. Their mean quantitative score was 275. Take it as known that the standard deviation of all individual scores is $\sigma = 60$.

(a) Give a 95% confidence interval for the mean score μ in the population of all young women.

(b) Give the 90% and 99% confidence intervals for μ.

(c) What are the margins of error for 90%, 95%, and 99% confidence? How does increasing the confidence level affect the margin of error of a confidence interval?

5.12   The NAEP sample of 1077 young women had mean quantitative score $\overline{x} = 275$. Take it as known that the standard deviation of all individual scores is $\sigma = 60$.

(a) Give a 95% confidence interval for the mean score μ in the population of all young women.

(b) Suppose that the same result, $\overline{x} = 275$, had come from a sample of 250 women. Give the 95% confidence interval for the population mean μ in this case.

(c) Then suppose that a sample of 4000 women had produced the sample mean $\overline{x} = 275$, and again give the 95% confidence interval for μ.

(d) What are the margins of error for samples of size 250, 1077, and 4000? How does increasing the sample size affect the margin of error of a confidence interval?

Choosing the sample size

A wise user of statistics never plans data collection without at the same time planning the inference. You can arrange to have both high confidence and a small margin of error by taking enough observations. The margin of error of the confidence interval for the mean of a normally distributed population is $m = z^*\sigma/\sqrt{n}$. To obtain a desired margin of error m, substitute the value of z^* for your desired confidence level, and solve for the sample size n. Here is the result.

SAMPLE SIZE FOR DESIRED MARGIN OF ERROR

The confidence interval for a population mean will have a specified margin of error m when the sample size is

$$n = \left(\frac{z^*\sigma}{m}\right)^2$$

This formula is not the proverbial free lunch. In practice, taking observations costs time and money. The required sample size may be impossibly expensive. Do notice once again that it is the size of the *sample* that determines the margin of error. The size of the *population* (as long as the population is much larger than the sample) does not influence the sample size we need.

EXAMPLE 5.6

Management asks the laboratory of Example 5.4 to produce results accurate to within ±0.005 with 95% confidence. How many measurements must be averaged to comply with this request?

The desired margin of error is $m = .005$. For 95% confidence, Table C gives $z^* = 1.960$. We know that $\sigma = .0068$. Therefore,

$$n = \left(\frac{z^*\sigma}{m}\right)^2 = \left(\frac{1.96 \times .0068}{.005}\right)^2 = 7.1$$

Because 7 measurements will give a slightly larger margin of error than desired, and 8 measurements a slightly smaller margin of error, the lab must take 8 measurements on each specimen to meet management's demand. (Always round *up* to the next higher whole number when finding *n*.) On learning the cost of this many measurements, management may reconsider its request.   ◄

EXERCISES

5.13   To assess the accuracy of a laboratory scale, a standard weight known to weigh 10 grams is weighed repeatedly. The scale readings are normally distributed with unknown mean (this mean is 10 grams if the scale has no bias). The standard deviation of the scale readings is known to be 0.0002 gram.

(a) The weight is weighed five times. The mean result is 10.0023 grams. Give a 98% confidence interval for the mean of repeated measurements of the weight.

(b) How many measurements must be averaged to get a margin of error of ±0.0001 with 98% confidence?

5.14   How large a sample of the hotel managers in Exercise 5.5 would be needed to estimate the mean μ within ±1 year with 99% confidence?

5.15   How large a sample of the crankshafts in Exercise 5.7 would be needed to estimate the mean μ within ±0.020 mm with 95% confidence?

Some cautions

Any formula for inference is correct only in specific circumstances. If statistical procedures carried warning labels like those on drugs, most inference methods would have long labels indeed. Our handy formula $\bar{x} \pm z^*\sigma/\sqrt{n}$ for estimating a normal mean comes with the following list of warnings for the user.

- The data must be an SRS from the population. We are completely safe if we actually carried out the random selection of an SRS. We are not in great danger if the data can plausibly be thought of as observations taken at random from a population. That is the case in Examples 5.4 to 5.6, where we have in mind the population resulting from a very large number of repeated analyses of the same specimen.

- The formula is not correct for probability sampling designs more complex than an SRS. Correct methods for other designs are available. We will not discuss confidence intervals based on multistage or stratified samples. If you plan such samples, be sure that you (or your statistical consultant) know how to carry out the inference you desire.

- There is no correct method for inference from data haphazardly collected with bias of unknown size. Fancy formulas cannot rescue badly produced data.

- Because $\bar{x}$ is strongly influenced by a few extreme observations, outliers can have a large effect on the confidence interval. You should search for outliers and try to correct them or justify their removal before computing the interval. If the outliers cannot be removed, ask your statistical consultant about procedures that are not sensitive to outliers.

- If the sample size is small and the population is not normal, the true confidence level will be different from the value C used in computing the interval. Examine your data carefully for skewness and other signs of nonnormality. The interval relies only on the distribution of $\bar{x}$, which even for quite small sample sizes is much closer to normal than the individual observations. When $n \geq 15$, the confidence level is not greatly disturbed by nonnormal populations unless extreme outliers or quite strong skewness are present. We will discuss this issue in more detail in the next chapter.

- You must know the standard deviation σ of the population. This unrealistic requirement renders the interval $\bar{x} \pm z^*\sigma/\sqrt{n}$ of little use in statistical practice. We will learn in the next chapter what to do when σ is unknown. However, if the sample is large, the sample standard deviation s will be close to the unknown σ. Then $\bar{x} \pm z^*s/\sqrt{n}$ is an approximate confidence interval for μ.

The most important caution concerning confidence intervals is a consequence of the first of these warnings. *The margin of error in a confidence interval covers only random sampling errors.* The margin of error is obtained from the sampling distribution and indicates how much error can be expected because of chance variation in randomized data production. Practical difficulties such as undercoverage and nonresponse in a sample survey can cause additional errors that may be larger than the random sampling error. Remember this unpleasant fact when reading the results of an opinion poll or other sample survey. The practical conduct of the survey influences the trustworthiness of its results in ways that are not included in the announced margin of error.

Every inference procedure that we will meet has its own list of warnings. Because many of the warnings are similar to those above, we will not print the full warning label each time. It is easy to state (from the mathematics of probability) conditions under which a method of inference is exactly correct. These conditions are *never* fully met in practice. For example, no population is exactly normal. Deciding when a statistical procedure should be used in practice often requires judgment assisted by exploratory analysis of the data.

Finally, you should understand what statistical confidence does not say. We are 95% confident that the mean NAEP quantitative score for all men aged 21 to 25 years lies between 267.8 and 276.2. That is, these numbers were calculated by a method that gives correct results in 95% of all possible samples. We *cannot* say that the probability is 95% that the true mean falls between 267.8 and 276.2. No randomness remains after we draw one particular sample and get from it one particular interval. The true mean either is or is not between 267.8 and 276.2. The probability calculations of standard statistical inference describe how often the *method* gives correct answers.

EXERCISES

5.16   A radio talk show invites listeners to enter a dispute about a proposed pay increase for city council members. "What yearly pay do you think council members should get? Call us with your number." In all, 958 people call. The mean pay they suggest is $\bar{x} = \$8740$ per year, and the standard deviation of the responses is $s = \$1125$. For a large sample such as this, s is very close to the unknown population σ. The station calculates the 95% confidence interval for the mean pay μ that all citizens would propose for council members to be $8669 to $8811.

(a) Is the station's calculation correct?

(b) Does their conclusion describe the population of all the city's citizens? Explain your answer.

5.17 The last closely contested presidential election pitted Jimmy Carter against Gerald Ford in 1976. A poll taken immediately before the 1976 election showed that 51% of the sample intended to vote for Carter. The polling organization announced that they were 95% confident that the sample result was within ±2 points of the true percent of all voters who favored Carter.

(a) Explain in plain language to someone who knows no statistics what "95% confident" means in this announcement.

(b) The poll showed Carter leading. Yet the polling organization said the election was too close to call. Explain why.

(c) On hearing of the poll, a nervous politician asked, "What is the probability that over half the voters prefer Carter?" A statistician replied that this question can't be answered from the poll results, and that it doesn't even make sense to talk about such a probability. Explain why.

5.18 The New York Times/CBS News Poll recently asked the question, "Do you favor an amendment to the Constitution that would permit organized prayer in public schools?" Sixty-six percent of the sample answered "Yes." The article describing the poll says that it "is based on telephone interviews conducted from Sept. 13 to Sept. 18 with 1,664 adults around the United States, excluding Alaska and Hawaii. . . . the telephone numbers were formed by random digits, thus permitting access to both listed and unlisted residential numbers."

(a) The article gives the margin of error as 3 percentage points. Make a confidence statement about the percent of all adults who favor a school prayer amendment.

(b) The news article goes on to say: "The theoretical errors do not take into account a margin of additional error resulting from the various practical difficulties in taking any survey of public opinion." List some of the "practical difficulties" that may cause errors in addition to the ±3% margin of error. Pay particular attention to the news article's description of the sampling method.

SUMMARY

The purpose of a **confidence interval** is to estimate an unknown parameter with an indication of how accurate the estimate is and of how confident we are that the result is correct.

Any confidence interval has two parts: an interval computed from the data and a confidence level. The **interval** often has the form

$$\text{estimate} \pm \text{margin of error}$$

The **confidence level** states the probability that the method will give a correct answer. That is, if you use 95% confidence intervals often, in the long run 95% of your intervals will contain the true parameter value. You do not know whether a 95% confidence interval calculated from a particular set of data contains the true parameter value.

A level C **confidence interval for the mean** μ of a normal population with known standard deviation σ, based on an SRS of size n, is given by

$$\bar{x} \pm z^* \frac{\sigma}{\sqrt{n}}$$

Here z^* is chosen so that the standard normal curve has area C between $-z^*$ and z^*. Because of the central limit theorem, this interval is approximately correct for large samples when the population is not normal.

The number z^* is called the **upper p critical value** of the standard normal distribution for $p = (1 - C)/2$. Critical values for many confidence levels appear in Table C.

Other things being equal, the **margin of error** of a confidence interval gets smaller as

- the confidence level C decreases,
- the population standard deviation σ decreases, and
- the sample size n increases.

The sample size required to obtain a confidence interval with specified margin of error m for a normal mean is

$$n = \left(\frac{z^* \sigma}{m}\right)^2$$

where z^* is the critical value for the desired level of confidence. Always round n up when you use this formula.

A specific confidence interval recipe is correct only under specific conditions. The most important conditions concern the method used to produce the data. Other factors such as the form of the population distribution may also be important.

SECTION 5.1 EXERCISES

5.19 The Acculturation Rating Scale for Mexican Americans (ARSMA) is a psychological test that measures the degree to which Mexican Americans have adopted Mexican/Spanish versus Anglo/English culture. The distribution of ARSMA scores in a population used to develop the test was approximately normal, with mean 3.0 and standard deviation 0.8. A further study gave ARSMA to 42 first-generation Mexican Americans. The mean of their scores was $\bar{x} = 2.13$. Assuming the standard deviation for the first-generation population is also $\sigma = .8$, give a 95% confidence interval for the mean ARSMA score for first-generation Mexican Americans.

5.20 How satisfied are hotel managers with the computer systems their hotels use? A survey was sent to 560 managers in hotels of size 200 to 500 rooms in Chicago and Detroit.[3] In all, 135 managers returned the survey. Two questions concerned their degree of satisfaction with the ease of use of their computer systems and with the level of computer training they had received. The managers responded using a seven-point scale, with 1 meaning "not satisfied," 4 meaning "moderately satisfied," and 7 meaning "very satisfied."

(a) What do you think is the population for this study? There are some major shortcomings in the data production. What are they? These shortcomings reduce the value of the formal inference you are about to do.

(b) The mean response for satisfaction with ease of use was $\bar{x} = 5.396$. Give a 95% confidence interval for the mean in the entire population. (Assume that the population standard deviation is $\sigma = 1.75$.)

(c) For satisfaction with training, the mean response was $\bar{x} = 4.398$. Taking $\sigma = 1.75$, give a 99% confidence interval for the population mean.

(d) The measurements of satisfaction are certainly not normally distributed, because they take only whole-number values from 1 to 7. Nonetheless, the use of confidence intervals based on the normal distribution is justified for this study. Why?

5.21 Consumers can purchase nonprescription medications at food stores, mass merchandise stores such as Kmart and Wal-Mart, or pharmacies. About 45% of consumers make such purchases at pharmacies. What accounts for the popularity of pharmacies, which often charge higher prices?

A study examined consumers' perceptions of overall performance of the three types of stores, using a long questionnaire that asked about such things as "neat and attractive store," "knowledgeable staff," and "assistance in choosing among various types of nonprescription medication." A performance score was based on 27 such questions. The subjects were 201 people chosen at random from the Indianapolis telephone directory. Here are the means and standard deviations of the performance scores for the sample.[4]

Store type	$\overline{x}$	s
Food stores	18.67	24.95
Mass merchandisers	32.38	33.37
Pharmacies	48.60	35.62

We do not know the population standard deviations, but a sample standard deviation s from so large a sample is usually close to σ. Use s in place of the unknown σ in this exercise.

(a) What population do you think the authors of the study want to draw conclusions about? What population are you certain they can draw conclusions about?

(b) Give 95% confidence intervals for the mean performance for each type of store in the population.

(c) Based on these confidence intervals, are you convinced that consumers think that pharmacies offer higher performance than the other types of stores?

5.22   The 1990 census "long form" was sent to a random sample of 17% of the nation's households. One question asked the total 1989 income of the householder. (The householder is the person in whose name the dwelling unit is owned or rented.) Suppose that the households that returned the long form are an SRS of the population of all households in each district. In Middletown, a city of 40,000 persons, 2621 householders reported their income. The mean of the responses was $\overline{x} = \$23,453$, and the standard deviation was $s = \$8721$. The sample standard deviation for so large a sample will be very close to the population standard deviation σ. Use these facts to give an approximate 99% confidence interval for the 1989 mean income of Middletown householders.

5.23   How large a sample of households would enable you to estimate the mean income of Middletown householders (see the previous exercise) within a margin of error of $1000 with 99% confidence?

5.24   A *New York Times* poll on women's issues interviewed 1025 women and 472 men randomly selected from the United States, excluding Alaska and Hawaii. The poll announced a margin of error of ±3 percentage points for 95% confidence in conclusions about women. The margin of error for results concerning men was ±4 percentage points. Why is this larger than the margin of error for women?

5.25   When the statistic that estimates an unknown parameter has a normal distribution, a confidence interval for the parameter has the form

$$\text{estimate} \pm z^* \sigma_{\text{estimate}}$$

In a complex sample survey design, the estimate of the population mean and the standard deviation of this estimate require elaborate computations. But when we are given the estimate and its standard deviation, we can calculate a confidence interval for μ without knowing the formulas that led to the numbers given.

A report based on the Current Population Survey estimates the median weekly earnings of families of wage or salary workers as $664 and also estimates that the standard deviation of this estimate is $3.50. The Current Population Survey uses an elaborate multistage sampling design to select a sample of about 60,000 households. The sampling distribution of the estimated median income is approximately normal. Give a 95% confidence interval for the median weekly earnings of all families of wage and salary workers.

5.2   TESTS OF SIGNIFICANCE

Confidence intervals are one of the two most common types of formal statistical inference. Use them when your goal is to estimate a population parameter. The second common type of inference has a different goal: to assess the evidence provided by the data in favor of some claim about the population. The reasoning of statistical tests, like that of confidence intervals, is based on asking what would happen if we repeated the sample or experiment many times. Here is the first example we will explore.

EXAMPLE 5.7

Diet colas use artificial sweeteners to avoid sugar. Colas with artificial sweeteners gradually lose their sweetness over time. Manufacturers therefore test new colas for loss of sweetness before marketing them. Trained tasters sip the cola along with drinks of standard sweetness and score the cola on a "sweetness score" of 1 to 10. The cola is then stored for a month at high temperature to imitate the effect of four months' storage at room temperature. After a month, each taster scores the stored cola. This is a matched pairs experiment. Our data are the differences (score before storage minus score after storage) in the tasters' scores. The bigger these differences, the bigger the loss of sweetness.

Here are the sweetness losses for a new cola, as measured by 10 trained tasters:

$$2.0 \quad 0.4 \quad 0.7 \quad 2.0 \quad -0.4 \quad 2.2 \quad -1.3 \quad 1.2 \quad 1.1 \quad 2.3$$

Most are positive. That is, most tasters found a loss of sweetness. But the losses are small, and two tasters (the negative scores) thought the cola gained sweetness. *Are these data good evidence that the cola lost sweetness in storage?* ◀

The reasoning of a significance test

The average sweetness loss for our cola is given by the sample mean,

$$\overline{x} = \frac{2.0 + 0.4 + \cdots + 2.3}{10} = 1.02$$

That's not a large loss. Ten different tasters would almost surely give a different result. Maybe it's just chance that produced this result. A ***test of*** ***significance*** asks:

significance test

Does the sample result $\overline{x} = 1.02$ reflect a real loss of sweetness?

OR

Could we easily get the outcome $\overline{x} = 1.02$ just by chance?

The significance test starts with a careful statement of these alternatives. First, we always draw conclusions about some parameter of the population, so we must identify this parameter. In this case, it's the population mean μ. The mean μ is the average loss in sweetness that a very large number of tasters would detect in the cola. Our 10 tasters are a sample from this population.

null hypothesis

Next, state the **null hypothesis**. The null hypothesis says that there is no effect or no change in the population. If the null hypothesis is true, the sample result is just chance at work. Here, the null hypothesis says that the cola does not lose sweetness (no change). We can write that in terms of the mean sweetness loss μ in the population as

$$H_0 : \mu = 0$$

We write H_0, read as "H-nought," to indicate the null hypothesis.

alternative hypothesis

The effect we suspect is true, the alternative to "no effect" or "no change," is described by the **alternative hypothesis**. We suspect that the cola does lose sweetness. In terms of the mean sweetness loss μ, the alternative hypothesis is

$$H_a : \mu > 0$$

The reasoning of a significance test goes like this.

- Suppose for the sake of argument that the null hypothesis is true, that on the average there is no loss of sweetness.
- *Is the sample outcome $\bar{x} = 1.02$ surprisingly large under that supposition?* If it is, that's evidence against H_0 and in favor of H_a.

To answer the question, we use our knowledge of how the sample mean $\bar{x}$ would vary in repeated samples if H_0 really were true. That's the sampling distribution of $\bar{x}$ once again.

From long experience we know that individual tasters' scores vary according to a normal distribution. The mean of this distribution is the parameter μ. We're asking what would happen if there is really no change in sweetness on the average, so μ is 0. That's just what the null hypothesis says. From long experience we also know that the standard deviation for all individual tasters is $\sigma = 1$. (It is not realistic to suppose that we know the population standard deviation σ. We will eliminate this assumption in the next chapter.) The sampling distribution of $\bar{x}$ from 10 tasters is then normal with mean $\mu = 0$ and standard deviation

$$\frac{\sigma}{\sqrt{n}} = \frac{1}{\sqrt{10}} = .316$$

We can judge whether any observed $\bar{x}$ is surprising by locating it on this distribution. Figure 5.8 shows the sampling distribution with the observed values of $\bar{x}$ for two types of cola.

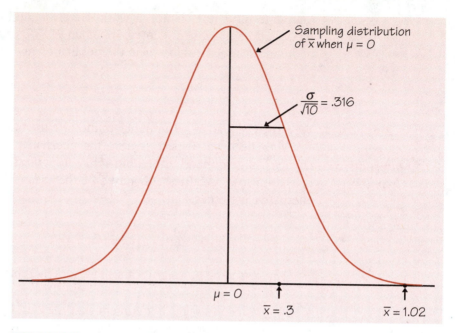

FIGURE 5.8 If a cola does not lose sweetness in storage, the mean score $\bar{x}$ for 10 tasters will have this sampling distribution. The actual result for one cola was $\bar{x} = .3$. That could easily happen just by chance. Another cola had $\bar{x} = 1.02$. That's so far out on the normal curve that it is good evidence that this cola did lose sweetness.

- One cola had $\bar{x} = .3$ for a sample of 10 tasters. It is clear from Figure 5.8 that an $\bar{x}$ this large could easily occur just by chance when the population mean is $\mu = 0$. That 10 tasters find $\bar{x} = .3$ is not evidence of a sweetness loss.

- The taste test for our cola produced $\bar{x} = 1.02$. That's way out on the normal curve in Figure 5.8, so far out that an observed value this large would almost never occur just by chance if the true μ were 0. This observed value is good evidence that in fact the true μ is greater than 0, that is, that the cola lost sweetness. The manufacturer must reformulate the cola and try again.

A significance test works by asking how unlikely the observed outcome would be if the null hypothesis were really true. The final step in our test is to assign a number to measure how unlikely our observed $\bar{x}$ is if H_0 is true. The less likely this outcome is, the stronger is the evidence against H_0.

Look again at Figure 5.8. If the alternative hypothesis is true, there is a sweetness loss and we expect the mean loss $\overline{x}$ found by the tasters to be positive. The farther out $\overline{x}$ is in the positive direction, the more convinced we are that the population mean μ is not zero but positive. We measure the strength of the evidence against H_0 by the probability under the normal curve in Figure 5.8 to the right of the observed $\overline{x}$. This probability is called the ***P-value***. It is the probability of a result at least as far out as the result we actually got. The lower this probability, the more surprising our result, and the stronger the evidence against the null hypothesis.

P-value

- For one new cola, our 10 tasters gave $\overline{x} = .3$. Figure 5.9 shows the P-value for this outcome. It is the probability to the right of 0.3. This probability is about 0.17. That is, 17% of all samples would give a mean score as large or larger than 0.3 just by chance when the true population mean is 0. An outcome this likely to occur just by chance is not good evidence against the null hypothesis.

- Our cola showed a larger sweetness loss, $\overline{x} = 1.02$. The probability of a result this large or larger is only 0.0006. This probability is

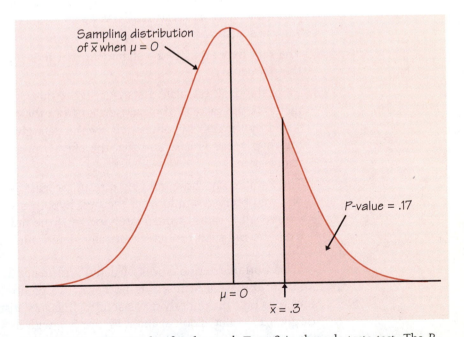

FIGURE 5.9 The P-value for the result $\overline{x} = .3$ in the cola taste test. The P-value is the probability (when H_0 is true) that $\overline{x}$ takes a value as large or larger than the actually observed value.

the *P*-value. Ten tasters would have an average score as large as 1.02 only 6 times in 10,000 tries if the true mean sweetness change were 0. An outcome this unlikely convinces us that the true mean is really greater than 0.

Small *P*-values are evidence against H_0, because they say that the observed result is unlikely to occur just by chance. Large *P*-values fail to give evidence against H_0. How small must a *P*-value be in order to persuade us? There's no fixed rule. But the level 0.05 (a result that would occur no more than once in 20 tries just by chance) is a common rule of thumb. A result with a small *P*-value, say less than 0.05, is called **statistically significant**. That's just a way of saying that chance alone would rarely produce so extreme a result.

statistically significant

Outline of a test

Here is the reasoning of a significance test in outline form:

1. Describe the effect you are searching for in terms of a population parameter like the mean μ. (Never state a hypothesis in terms of a sample statistic like $\bar{x}$.)

2. The null hypothesis is the statement that this effect is *not* present in the population.

3. From the data, calculate a statistic like $\bar{x}$ that estimates the parameter. Is the value of this statistic far from the parameter value stated by the null hypothesis? If so, the data give evidence that the null hypothesis is false and that the effect you are looking for is really there.

4. The *P*-value says how unlikely a result at least as extreme as the one we observed would be if the null hypothesis were true. Results with small *P*-values would rarely occur if the null hypothesis were true. We call such results statistically significant.

This outline overlooks lots of detail and many fine points. But it is important to have it firmly in mind before we go on. Recipes for significance tests hide the underlying reasoning. In fact, statistical software often just gives a *P*-value. Look again at Figure 5.8, with its $\bar{x}$-values from taste tests of two colas. It should be clear that one result is not surprising if the true mean score in the population is 0, and that the other is surprising. A significance test simply says that more precisely.

EXERCISES

5.26 The Survey of Study Habits and Attitudes (SSHA) is a psychological test that measures the attitude toward school and study habits of students. Scores range from 0 to 200. The mean score for U.S. college students is about 115, and the standard deviation is about 30. A teacher suspects that older students have better attitudes toward school. She gives the SSHA to 25 students who are at least 30 years of age. Assume that scores in the population of older students are normally distributed with standard deviation $\sigma = 30$. The teacher wants to test the hypotheses

$$H_0 : \mu = 115$$

$$H_a : \mu > 115$$

(a) What is the sampling distribution of the mean score $\bar{x}$ of a sample of 25 older students if the null hypothesis is true? Sketch the density curve of this distribution. (Hint: Sketch a normal curve first, then mark the axis using what you know about locating μ and σ on a normal curve.)

(b) Suppose that the sample data give $\bar{x} = 118.6$. Mark this point on the axis of your sketch. In fact, the result was $\bar{x} = 125.7$. Mark this point on your sketch. Using your sketch, explain in simple language why one result is good evidence that the mean score of all older students is greater than 115 and why the other outcome is not.

(c) Shade the area under the curve that is the P-value for the sample result $\bar{x} = 118.6$.

5.27 The Census Bureau reports that households spend an average of 31% of their total spending on housing. A homebuilders association in Cleveland believes that this average is lower in their area. They interview a sample of 40 households in the Cleveland metropolitan area to learn what percent of their spending goes toward housing. Take μ to be the mean percent of spending devoted to housing among all Cleveland households. We want to test the hypotheses

$$H_0 : \mu = 31\%$$

$$H_a : \mu < 31\%$$

The population standard deviation is $\sigma = 9.6\%$.

(a) What is the sampling distribution of the mean percent $\bar{x}$ that the sample spends on housing if the null hypothesis is true? Sketch the density curve of the sampling distribution. (Hint: Sketch a normal curve first, then mark the axis using what you know about locating μ and σ on a normal curve.)

(b) Suppose that the study finds $\bar{x} = 30.2\%$ for the 40 households in the sample. Mark this point on the axis in your sketch. Then suppose that the study result is $\bar{x} = 27.6\%$. Mark this point on your sketch. Referring to your sketch, explain in simple language why one result is good evidence that average Cleveland spending on housing is less than 31% and the other result is not.

(c) Shade the area under the curve that gives the P-value for the result $\bar{x} = 30.2\%$. (Note that we are looking for evidence that spending is *less* than the null hypothesis states.)

More detail: stating hypotheses

We will now look in more detail at some aspects of significance tests. The first step in a test of significance is to state a claim that we will try to find evidence *against*.

NULL HYPOTHESIS H_0

The statement being tested in a test of significance is called the **null hypothesis**. The test of significance is designed to assess the strength of the evidence against the null hypothesis. Usually the null hypothesis is a statement of "no effect" or "no difference."

The alternative hypothesis H_a is the claim about the population that we are trying to find evidence *for*. In Example 5.7, we were seeking evidence of a loss in sweetness. The null hypothesis says "no loss" on the average in a large population of tasters. The alternative hypothesis says "there is a loss." So the hypotheses are

$$H_0 : \mu = 0$$

$$H_a : \mu > 0$$

*one-sided
alternative*

This alternative hypothesis is **one-sided** because we are interested only in deviations from the null hypothesis in one direction.

EXAMPLE 5.8

Does the job satisfaction of assembly workers differ when their work is machine-paced rather than self-paced? One study chose 28 subjects at random from a group of women who worked at assembling electronic devices. Half of the subjects were assigned at random to each of two groups. Both groups did similar assembly work, but one work setup allowed workers to pace themselves and the other featured an assembly line that moved at fixed time intervals so that the workers were paced by machine. After two weeks, all subjects took the Job Diagnosis Survey (JDS), a test of job satisfaction. Then they switched work setups, and took the JDS again after two more weeks. This is another matched pairs design. The response variable is the difference in JDS scores, self-paced minus machine-paced.[5]

The parameter of interest is the mean μ of the differences in JDS scores in the population of all female assembly workers. The null hypothesis says that there is no difference between self-paced and machine-paced work, that is,

$$H_0 : \mu = 0$$

*two-sided
alternative*

The authors of the study wanted to know if the two work conditions have different levels of job satisfaction. They did not specify the direction of the difference. The alternative hypothesis is therefore **two-sided**,

$$H_a : \mu \neq 0 \qquad \blacktriangleleft$$

Hypotheses always refer to some population, not to a particular outcome. For this reason, always state H_0 and H_a in terms of population parameters. Because H_a expresses the effect that we hope to find evidence *for*, it is often easier to begin by stating H_a and then set up H_0 as the statement that the hoped-for effect is not present.

It is not always easy to decide whether H_a should be one-sided or two-sided. In Example 5.8, the alternative $H_a : \mu \neq 0$ is two-sided. That is, it simply says there is a difference in job satisfaction without specifying the direction of the difference. The alternative $H_a : \mu > 0$ in the taste test example is one-sided. Because colas can only lose sweetness in storage, we are interested only in detecting an upward shift in μ. The alternative hypothesis should express the hopes or suspicions we bring to the data.

It is cheating to first look at the data and then frame H_a to fit what the data show. Thus the fact that the workers in the study of Example 5.8 were more satisfied with self-paced work should not influence our choice of H_a. If you do not have a specific direction firmly in mind in advance, use a two-sided alternative.

The choice of the hypotheses in Example 5.7 as

$$H_0 : \mu = 0$$

$$H_a : \mu > 0$$

deserves a final comment. The cola maker is not concerned with the possibility that the tasters may detect a gain in sweetness, indicated by a negative mean loss μ. However, we can allow for the possibility that μ is less than zero by including this case in the null hypothesis. Then we would write

$$H_0 : \mu \leq 0$$

$$H_a : \mu > 0$$

This statement is logically satisfying because the hypotheses account for all possible values of μ. However, only the parameter value in H_0 that is closest to H_a influences the form of the test in all common significance testing situations. We will therefore take H_0 to be the simpler statement that the parameter has a specific value, in this case $H_0 : \mu = 0$.

EXERCISES

Each of the following situations calls for a significance test for a population mean μ. State the null hypothesis H_0 and the alternative hypothesis H_a in each case.

5.28   The diameter of a spindle in a small motor is supposed to be 5 mm. If the spindle is either too small or too large, the motor will not work properly. The manufacturer measures the diameter in a sample of motors to determine whether the mean diameter has moved away from the target.

5.29   Census Bureau data show that the mean household income in the area served by a shopping mall is $42,500 per year. A market research firm questions shoppers at the mall. The researchers suspect the mean household income of mall shoppers is higher than that of the general population.

5.30   The examinations in a large accounting class are scaled after grading so that the mean score is 50. The professor thinks that one teaching assistant is a poor teacher and suspects that his students have a lower mean score than the class as a whole. The TA's students this semester can be considered a sample from the population of all students in the course, so the professor compares their mean score with 50.

5.31   Last year, your company's service technicians took an average of 2.6 hours to respond to trouble calls from business customers who had purchased service contracts. Do this year's data show a different average response time?

More detail: *P*-values and statistical significance

test statistic   A significance test uses data in the form of a ***test statistic***. The test statistic is usually based on a statistic that estimates the parameter that appears in the hypotheses. In our examples, the parameter is μ and the test statistic is the sample mean $\overline{x}$.

A test of significance assesses the evidence against the null hypothesis in terms of probability. If the test statistic falls far from the value suggested by the null hypothesis in the direction specified by the alternative hypothesis, it is good evidence against H_0 and in favor of H_a. To describe how strong the evidence is, find the probability of getting an outcome *as extreme or more extreme than the actually observed outcome*. "Extreme" means "far from what we would expect if H_0 were true." The direction or directions that count as "far from what we would expect" are determined by the alternative hypothesis H_a.

P-VALUE

The probability, computed assuming that H_0 is true, that the test statistic would take a value as extreme or more extreme than that actually observed is called the **P-value** of the test. The smaller the P-value is, the stronger is the evidence against H_0 provided by the data.

Computer software that carries out tests of significance usually calculates the P-value for us. In some cases we can find P-values from our knowledge of sampling distributions.

EXAMPLE 5.9

In Example 5.7 the observations are an SRS of size $n = 10$ from a normal population with $\sigma = 1$. The observed mean sweetness loss for one cola was $\bar{x} = .3$. The P-value for testing

$$H_0 : \mu = 0$$

$$H_a : \mu > 0$$

is therefore

$$P(\bar{x} \geq .3)$$

calculated assuming that H_0 is true. When H_0 is true, $\bar{x}$ has the normal distribution with mean 0 and standard deviation

$$\frac{\sigma}{\sqrt{n}} = \frac{1}{\sqrt{10}} = .316$$

Find the P-value by a normal probability calculation. Start by drawing a picture that shows the P-value as an area under a normal curve. Figure 5.10 is the picture for this example. Then standardize $\bar{x}$ to get a standard normal Z and use Table A,

$$P(\bar{x} \geq .3) = P\left(\frac{\bar{x} - 0}{.316} \geq \frac{.3 - 0}{.316}\right)$$

$$= P(Z \geq .95)$$

$$= 1 - .8289 = .1711$$

This is the value that was reported on page 353.   ◀

We sometimes take one final step to assess the evidence against H_0. We can compare the P-value with a fixed value that we regard as decisive. This amounts to announcing in advance how much evidence against H_0 we will *significance level* insist on. The decisive value of P is called the **significance level**. We write it as α, the Greek letter alpha. If we choose $\alpha = .05$, we are requiring that the data give evidence against H_0 so strong that it would happen no more than 5% of the time (1 time in 20) when H_0 is true. If we choose $\alpha = .01$, we are insisting on stronger evidence against H_0, evidence so strong that it would appear only 1% of the time (1 time in 100) if H_0 is in fact true.

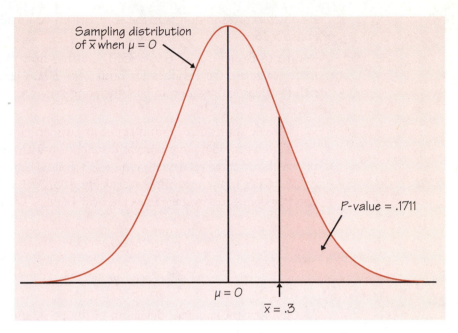

FIGURE 5.10   The *P*-value for the one-sided test in Example 5.9.

STATISTICAL SIGNIFICANCE

If the *P*-value is as small or smaller than α, we say that the data are **statistically significant at level α**.

"Significant" in the statistical sense does not mean "important." It means simply "not likely to happen just by chance." The significance level α makes "not likely" more exact. Significance at level 0.01 is often expressed by the statement "The results were significant ($P < .01$)." Here *P* stands for the *P*-value. The *P*-value is more informative than a statement of significance, because we can then assess significance at any level we choose. For example, a result with $P = .03$ is significant at the $\alpha = .05$ level, but not significant at the $\alpha = .01$ level.

EXERCISES

5.32   Return to Exercise 5.26 (page 355). Starting from the picture you drew there, calculate the *P*-values for both $\bar{x} = 118.6$ and $\bar{x} = 125.7$. The two

P-values express in numbers the comparison you made informally in Exercise 5.26.

5.33   Return to Exercise 5.27 (page 355). Starting from the picture you drew there, calculate the P-values for both $\bar{x} = 30.2\%$ and $\bar{x} = 27.6\%$. The two P-values express in numbers the comparison you made informally in Exercise 5.27.

5.34   Weekly sales of regular ground coffee at a supermarket have in the recent past varied according to a normal distribution with mean $\mu = 354$ units per week and standard deviation $\sigma = 33$ units. The store reduces the price by 5%. Sales in the next three weeks are 405, 378, and 411 units. Is this good evidence that average sales are now higher? The hypotheses are

$$H_0 : \mu = 354$$

$$H_a : \mu > 354$$

Assume that the standard deviation of the population of weekly sales remains $\sigma = 33$.

(a) Find the value of the test statistic $\bar{x}$.

(b) Sketch the normal curve for the sampling distribution of $\bar{x}$ when H_0 is true. Shade the area that represents the P-value for the observed outcome.

(c) Calculate the P-value.

(d) Is the result statistically significant at the $\alpha = .05$ level? Is it significant at the $\alpha = .01$ level? Do you think there is convincing evidence that mean sales are higher?

5.35   A study of the pay of corporate chief executive officers (CEOs) examined the increase in cash compensation of the CEOs of 104 companies, adjusted for inflation, in a recent year. The mean increase in real compensation was $\bar{x} = 6.9\%$ and the standard deviation of the increases was $s = 55\%$. Is this good evidence that the mean real compensation μ of all CEOs increased that year? The hypotheses are

$$H_0 : \mu = 0 \quad \text{(no increase)}$$

$$H_a : \mu > 0 \quad \text{(an increase)}$$

Because the sample size is large, the sample s is close to the population σ, so take $\sigma = 55\%$.

(a) Sketch the normal curve for the sampling distribution of $\bar{x}$ when H_0 is true. Shade the area that represents the P-value for the observed outcome $\bar{x} = 6.9\%$.

(b) Calculate the P-value.

(c) Is the result significant at the $\alpha = .05$ level? Do you think the study gives strong evidence that the mean compensation of all CEOs went up?

5.36   A social psychologist reports that "in our sample, ethnocentrism was significantly higher $(P < .05)$ among church attenders than among non-attenders." Explain what this means in language understandable to someone who knows no statistics. Do not use the word "significance" in your answer.

5.37   The financial aid office of a university asks a sample of students about their employment and earnings. The report says that "for academic year earnings, a significant difference $(P = .038)$ was found between the sexes, with men earning more on the average. No difference $(P = .476)$ was found between the earnings of black and white students." Explain both of these conclusions, for the effects of sex and of race on mean earnings, in language understandable to someone who knows no statistics.[6]

Tests for a population mean

Although the reasoning of significance testing isn't simple, carrying out a test is. There are three steps:

1. State the hypotheses.
2. Calculate the test statistic.
3. Find the P-value.

Once you have stated your hypotheses and identified the proper test, you or your computer can do Steps 2 and 3 by following a recipe. We now develop the recipe for one significance test, the one we have used in our examples.

We have an SRS of size n drawn from a normal population with unknown mean μ. We want to test the hypothesis that μ has a specified value. Call the specified value μ_0. The null hypothesis is

$$H_0 : \mu = \mu_0$$

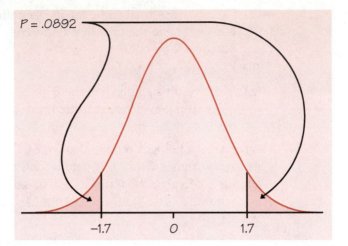

FIGURE 5.11   The P-value for the two-sided test in Example 5.10 is the sum of the area above 1.7 and below -1.7.

The test is based on the sample mean $\bar{x}$. Because normal calculations require standardized variables, we will use as our test statistic the *standardized sample mean*

$$z = \frac{\bar{x} - \mu_0}{\sigma/\sqrt{n}}$$

z test statistic   This **z test statistic** has the standard normal distribution when H_0 is true. If the alternative is one-sided on the high side

$$H_a : \mu > \mu_0$$

then the P-value is the probability that a standard normal variable Z takes a value at least as large as the observed z. That is,

$$P = P(Z \geq z)$$

Example 5.9 calculates this P-value for the cola taste test. There, $\mu_0 = 0$, the standardized sample mean was $z = .95$, and the P-value was $P(Z \geq .95) = .1711$. Similar reasoning applies when the alternative hypothesis states that the true μ lies below the hypothesized μ_0 (one-sided).

When H_a states that μ is simply unequal to μ_0 (two-sided), values of z away from zero in either direction count against the null hypothesis. The P-value is the probability that a standard normal Z is at least as far from zero *in either direction* as the observed z.

EXAMPLE 5.10

Suppose that the z test statistic for a two-sided test is $z = 1.7$. The two-sided P-value is the probability that $Z \leq -1.7$ or $Z \geq 1.7$. Figure 5.11 shows this probability as areas under the standard normal curve. Because the standard normal distribution is symmetric, we can calculate this probability by finding $P(Z \geq 1.7)$ and *doubling* it.

$$P(Z \leq -1.7 \text{ or } Z \geq 1.7) = 2P(Z \geq 1.7) = 2(1 - .9554) = .0892$$

We would make exactly the same calculation if we observed $z = -1.7$. It is the absolute value $|z|$ that matters, not whether z is positive or negative. ◀

---

z TEST FOR A POPULATION MEAN

To test the hypothesis $H_0 : \mu = \mu_0$ based on an SRS of size n from a population with unknown mean μ and known standard deviation σ, compute the **z test statistic**

$$z = \frac{\bar{x} - \mu_0}{\sigma/\sqrt{n}}$$

In terms of a variable Z having the standard normal distribution, the P-value for a test of H_0 against

$H_a : \mu > \mu_0$ is $P(Z \geq z)$

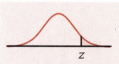

$H_a : \mu < \mu_0$ is $P(Z \leq z)$

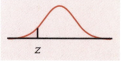

$H_a : \mu \neq \mu_0$ is $2P(Z \geq |z|)$

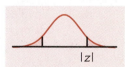

These P-values are exact if the population distribution is normal and are approximately correct for large n in other cases.

EXAMPLE 5.11

The National Center for Health Statistics reports that the mean systolic blood pressure for males 35 to 44 years of age is 128 and the standard deviation in this population is 15. The medical director of a large company looks at the medical records of 72 executives in this age group and finds that the mean systolic blood pressure in this sample is $\bar{x} = 126.07$. Is this evidence that the company's executives have a different mean blood pressure from the general population? As usual in this chapter, we make the unrealistic assumption that we know the population standard deviation. Assume that executives have the same $\sigma = 15$ as the general population of middle-aged males.

Step 1: Hypotheses. The null hypothesis is "no difference" from the national mean $\mu_0 = 128$. The alternative is two-sided, because the medical director did not have a particular direction in mind before examining the data. So the hypotheses about the unknown mean μ of the executive population are

$$H_0 : \mu = 128$$

$$H_a : \mu \neq 128$$

Step 2: Test statistic. The z test statistic is

$$z = \frac{\bar{x} - \mu_0}{\sigma/\sqrt{n}}$$

$$= \frac{126.07 - 128}{15/\sqrt{72}}$$

$$= -1.09$$

Step 3: P-value. You should still draw a picture to help find the P-value, but now you can sketch the standard normal curve with the observed value of z. Figure 5.12 shows that the P-value is the probability that a standard normal variable Z takes a value at least 1.09 away from zero. From Table A we find that this probability is

$$P = 2P(Z \geq 1.09) = 2(1 - .8621) = .2758$$

Conclusion: More than 27% of the time, an SRS of size 72 from the general male population would have a mean blood pressure at least as far from 128 as

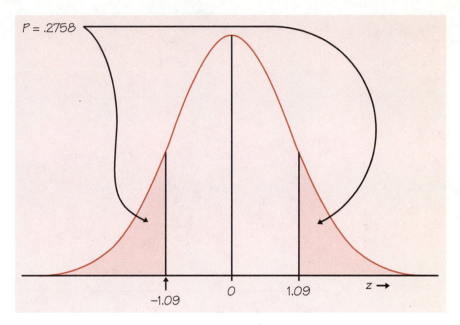

FIGURE 5.12   The P-value for the two-sided test in Example 5.11.

that of the executive sample. The observed $\bar{x}$ = 126.07 is therefore not good evidence that executives differ from other men. ◄

The z test assumes that the 72 executives in the sample are an SRS from the population of all middle-aged male executives in the company. We should check this assumption by asking how the data were produced. If medical records are available only for executives with recent medical problems, for example, the data are of little value for our purpose. It turns out that all executives are given a free annual medical exam, and that the medical director selected 72 exam results at random.

The data in Example 5.11 do *not* establish that the mean blood pressure μ for this company's executives is 128. We sought evidence that μ differed from 128 and failed to find convincing evidence. That is all we can say. No doubt the mean blood pressure of the entire executive population is not exactly equal to 128. A large enough sample would give evidence of the difference, even if it is very small. Tests of significance assess the evidence *against* H_0. If the evidence is strong, we can confidently reject H_0 in favor of the alternative. Failing to find evidence against H_0 means only that the data are consistent with H_0, not that we have clear evidence that H_0 is true.

(326.0pt x 208.0pt)

FIGURE 5.13   The *P*-value for the one-sided test in Example 5.12.

EXAMPLE 5.12

In a discussion of the education level of the American workforce, someone says, "The average young person can't even balance a checkbook." The National Assessment of Educational Progress (NAEP) says that a score of 275 or higher on its quantitative test (see Example 5.2 on page 325) reflects the skill needed to balance a checkbook. The NAEP random sample of 840 young men had a mean score of $\overline{x} = 272$, a bit below the checkbook-balancing level. Is this sample result good evidence that the mean for *all* young men is less than 275? As in Example 5.2, assume that $\sigma = 60$.

Step 1: Hypotheses.   The hypotheses are

$$H_0 : \mu = 275$$

$$H_a : \mu < 275$$

Step 2: Test statistic.   The z statistic is

$$z = \frac{\overline{x} - \mu_0}{\sigma/\sqrt{n}} = \frac{272 - 275}{60/\sqrt{840}}$$

$$= -1.45$$

Step 3: P-value.   Because H_a is one-sided on the low side, small values of z count against H_0. Figure 5.13 illustrates the P-value. Using Table A, we find that

$$P = P(Z \le -1.45)$$
$$= .0735$$

Conclusion: A mean score as low as 272 would occur about 7 times in 100 samples if the population mean were 275. This is modest evidence that the mean NAEP score for all young men is less than 275.   ◀

EXERCISES

5.38   Here are measurements (in millimeters) of a critical dimension on a sample of automobile engine crankshafts.

224.120	224.001	224.017	223.982
223.960	224.089	223.987	223.976
224.098	224.057	223.913	223.999
223.989	223.902	223.961	223.980

The manufacturing process is known to vary normally with standard deviation $\sigma = .060$ mm. The process mean is supposed to be 224 mm. Do these data give evidence that the process mean is not equal to the target value 224 mm?

(a) State the H_0 and H_a that you will test.

(b) Calculate the test statistic z.

(c) Give the P-value of the test. Are you convinced that the process mean is not 224 mm?

5.39   Bottles of a popular cola are supposed to contain 300 milliliters (ml) of cola. There is some variation from bottle to bottle because the filling machinery is not perfectly precise. The distribution of the contents is normal with standard deviation $\sigma = 3$ ml. An inspector who suspects that the bottler is underfilling measures the contents of six bottles. The results are

299.4	297.7	301.0
298.9	300.2	297.0

Is this convincing evidence that the mean contents of cola bottles is less than the advertised 300 ml?

(a) State the hypotheses that you will test.

(b) Calculate the test statistic.

(c) Find the *P*-value and state your conclusion.

Tests with fixed significance level

Sometimes we demand a specific degree of evidence in order to reject the null hypothesis. A level of significance α says how much evidence we re-quire. In terms of the *P*-value, the outcome of a test is significant at level α if $P \le \alpha$. Significance at any level is easy to assess once you have the *P*-value. When you do not use statistical software, the *P*-value can be diffi-cult to calculate. Fortunately, you can decide whether a result is statistically significant without calculating *P*. The following example illustrates how to assess significance at a fixed level α by using a table of critical values, the same table used to obtain confidence intervals.

EXAMPLE 5.13

In Example 5.12, we examined whether the mean NAEP quantitative score of young men is less than 275. The hypotheses are

$$H_0 : \mu = 275$$

$$H_a : \mu < 275$$

The *z* statistic takes the value $z = -1.45$. Is the evidence against H_0 statistically significant at the 5% level?

To determine significance, we need only compare the observed $z = -1.45$ with the 5% critical value $z^* = 1.645$ from Table C. Because $z = -1.45$ is *not* farther from 0 than -1.645, it is *not* significant at level $\alpha = .05$.

Here is why. The *P*-value is the area to the left of -1.45 under the standard normal curve, shown in Figure 5.13. The result $z = -1.45$ is significant at the 5% level exactly when this area is no more than 5%. The area to the left of the critical value -1.645 is exactly 5%. So -1.645 separates values of *z* that are significant from those that are not. Figure 5.14 illustrates the procedure. ◀

FIXED SIGNIFICANCE LEVEL z TESTS FOR A POPULATION MEAN

To test the hypothesis $H_0 : \mu = \mu_0$ based on an SRS of size n from a population with unknown mean μ and known standard deviation σ, compute the z test statistic

$$z = \frac{\overline{x} - \mu_0}{\sigma / \sqrt{n}}$$

Reject H_0 at significance level α against a one-sided alternative

$$H_a : \mu > \mu_0 \text{ if } z \geq z^*$$

$$H_a : \mu < \mu_0 \text{ if } z \leq -z^*$$

where z^* is the upper α critical value from Table C. Reject H_0 at significance level α against a two-sided alternative

$$H_a : \mu \neq \mu_0 \text{ if } |z| \geq z^*$$

where z^* is the upper $\alpha/2$ critical value from Table C.

EXAMPLE 5.14

The analytical laboratory of Example 5.4 (page 335) is asked to evaluate the claim that the concentration of the active ingredient in a specimen is 0.86%. The lab makes 3 repeated analyses of the specimen. The mean result is $\overline{x} = .8404$. The true concentration is the mean μ of the population of all analyses of the specimen. The standard deviation of the analysis process is known to be $\sigma = .0068$. Is there significant evidence at the 1% level that $\mu \neq .86$?

Step 1: Hypotheses. The hypotheses are

$$H_0 : \mu = .86$$

$$H_a : \mu \neq .86$$

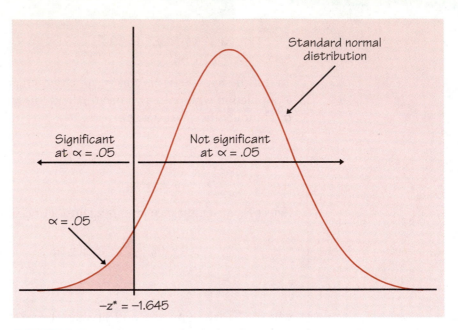

FIGURE 5.14 Deciding whether a z statistic is significant at the $\alpha = .05$ level in the one-sided test of Example 5.13.

Step 2: Test statistic.   The z statistic is

$$z = \frac{.8404 - .86}{.0068/\sqrt{3}}$$

$$= -4.99$$

Step 3: Significance.   Because the alternative is two-sided, we compare $|z| = 4.99$ with the $\alpha/2 = .005$ critical value from Table C. This critical value is $z^* = 2.576$. Figure 5.15 illustrates the values of z that are statistically significant. Because $|z| > 2.576$, we reject the null hypothesis and conclude (at the 1% significance level) that the concentration is not as claimed.   ◀

The observed result in Example 5.14 was $z = -4.99$. The conclusion that this result is significant at the 1% level does not tell the whole story. The observed z is far beyond the 1% critical value, and the evidence against H_0 is far stronger than 1% significance suggests. The P-value

$$2P(Z \geq 4.99) = .0000006$$

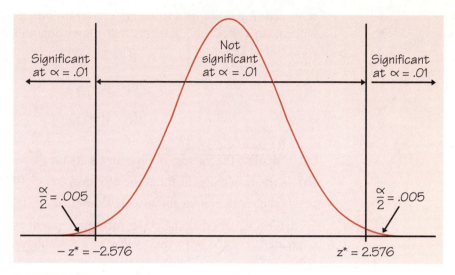

FIGURE 5.15   Deciding whether a z statistic is significant at the $\alpha = .01$ level in the two-sided test of Example 5.14.

gives a better sense of how strong the evidence is. *The P-value is the smallest level α at which the data are significant.* Knowing the P-value allows us to assess significance at any level.

Tables of critical values such as Table C allow us to estimate P-values without a probability calculation. In Example 5.14, compare the observed $z = -4.99$ with *all* of the normal critical values in the bottom row of the table. It is beyond even 3.291, the critical value for $P = .0005$. So we know that for the two-sided test, $P < .001$. In Example 5.13, $z = -1.45$ lies between the 0.05 and 0.10 entries in the table. So the P-value for the one-sided test lies between 0.05 and 0.10. This approximation is accurate enough for most purposes.

Because the practice of statistics almost always employs computer software that calculates P-values automatically, the use of tables of critical values is becoming outdated. The usual tables of critical values (such as Table C) appear in this book for learning purposes and to rescue students without good computing facilities.

EXERCISES

5.40   A computer has a random number generator designed to produce random numbers that are uniformly distributed on the interval from 0 to 1. If this is true, the numbers generated come from a population with $\mu = .5$ and

$\sigma = .2887$. A command to generate 100 random numbers gives outcomes with mean $\bar{x} = .4365$. Assume that the population σ remains fixed. We want to test

$$H_0 : \mu = .5$$

$$H_a : \mu \neq .5$$

(a) Calculate the value z of the z test statistic.

(b) Is the result significant at the 5% level ($\alpha = .05$)?

(c) Is the result significant at the 1% level ($\alpha = .01$)?

5.41   To determine whether the mean nicotine content of a brand of cigarettes is greater than the advertised value of 1.4 milligrams, a health advocacy group tests

$$H_0 : \mu = 1.4$$

$$H_a : \mu > 1.4$$

The calculated value of the test statistic is $z = 2.42$.

(a) Is the result significant at the 5% level?

(b) Is the result significant at the 1% level?

Tests from confidence intervals

The calculation in Example 5.14 for a 1% significance test is very similar to that in Example 5.4 for a 99% confidence interval. In fact, a two-sided test at significance level α can be carried out directly from a confidence interval with confidence level $C = 1 - \alpha$.

CONFIDENCE INTERVALS AND TWO-SIDED TESTS

A level α two-sided significance test rejects a hypothesis $H_0 : \mu = \mu_0$ exactly when the value μ_0 falls outside a level $1 - \alpha$ confidence interval for μ.

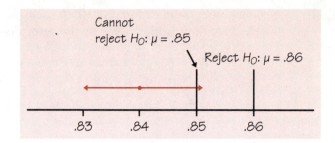

FIGURE 5.16 Values of μ falling outside a 99% confidence interval can be rejected at the 1% significance level. Values falling inside the interval cannot be rejected.

EXAMPLE 5.15

The 99% confidence interval for μ in Example 5.4 is

$$\bar{x} \pm z^* \frac{\sigma}{\sqrt{n}} = .8404 \pm .0101$$

$$= (.8303, .8505)$$

The hypothesized value $\mu_0 = 0.86$ in Example 5.14 falls outside this confidence interval, so we reject

$$H_0 : \mu = .86$$

at the 1% significance level. On the other hand, we cannot reject

$$H_0 : \mu = .85$$

at the 1% level in favor of the two-sided alternative $H_a : \mu \neq .85$, because 0.85 lies inside the 99% confidence interval for μ. Figure 5.16 illustrates both cases. ◀

EXERCISE

5.42 Radon is a colorless, odorless gas that is naturally released by rocks and soils and may concentrate in tightly closed houses. Because radon is slightly radioactive, there is some concern that it may be a health hazard. Radon detectors are sold to homeowners worried about this risk, but the detectors may be inaccurate. University researchers placed 12 detectors in a chamber where they were exposed to 105 picocuries per liter of radon over 3 days. Here are the readings given by the detectors.[7]

$$91.9 \quad 97.8 \quad 111.4 \quad 122.3 \quad 105.4 \quad 95.0$$
$$103.8 \quad 99.6 \quad 96.6 \quad 119.3 \quad 104.8 \quad 101.7$$

Assume (unrealistically) that you know that the standard deviation of readings for all detectors of this type is $\sigma = 9$.

(a) Give a 90% confidence interval for the mean reading μ for this type of detector.

(b) Is there significant evidence at the 10% level that the mean reading differs from the true value 105? State hypotheses and base a test on your confidence interval from (a).

SUMMARY

A **test of significance** is intended to assess the evidence provided by data against a **null hypothesis** H_0 in favor of an **alternative hypothesis** H_a.

The hypotheses are stated in terms of population parameters. Usually H_0 is a statement that no effect is present, and H_a says that a parameter differs from its null value in a specific direction (**one-sided alternative**) or in either direction (**two-sided alternative**).

The essential reasoning of a significance test is as follows. Suppose for the sake of argument that the null hypothesis is true. If we repeated our data production many times, would we often get data as inconsistent with H_0 as the data we actually have? If the data are unlikely when H_0 is true, they provide evidence against H_0.

A test is based on a **test statistic**. The **P-value** is the probability, computed supposing H_0 to be true, that the test statistic will take a value at least as extreme as that actually observed. Small P-values indicate strong evidence against H_0. Calculating P-values requires knowledge of the sampling distribution of the test statistic when H_0 is true.

If the P-value is as small or smaller than a specified value α, the data are **statistically significant** at significance level α.

Significance tests for the hypothesis $H_0 : \mu = \mu_0$ concerning the un-known mean μ of a population are based on the z **statistic**

$$z = \frac{\overline{x} - \mu_0}{\sigma/\sqrt{n}}$$

The z test assumes an SRS of size n, known population standard deviation σ, and either a normal population or a large sample. P-values are com-puted from the normal distribution (Table A). Fixed α tests use the table of standard normal **critical values** (bottom row of Table C).

SECTION 5.2 EXERCISES

5.43   The job satisfaction study of Example 5.8 (page 357) measured the JDS job satisfaction score of 28 female assemblers doing both self-paced and machine-paced work. The parameter μ is the mean amount by which the self-paced score exceeds the machine-paced score in the population of all such workers. Scores are normally distributed. The population standard deviation is $\sigma = .60$. The hypotheses are

$$H_0 : \mu = 0$$

$$H_a : \mu \neq 0$$

(a) What is the sampling distribution of the mean JDS score $\overline{x}$ for 28 workers if the null hypothesis is true? Sketch the density curve of this distribution. (Hint: Sketch a normal curve first, then mark the axis using what you know about locating μ and σ on a normal curve.)

(b) Suppose that the study had found $\overline{x} = .09$. Mark this point on the axis in your sketch. In fact, the study found $\overline{x} = .27$ for these 28 workers. Mark this point on your sketch. Referring to your sketch, explain in simple language why one result is good evidence that H_0 is not true, and why the other is not.

(c) Make another copy of your sketch. Shade the area under the curve that gives the P-value for the result $\overline{x} = .09$. Then calculate this P-value. (Note that H_a is two-sided.)

(d) Calculate the P-value for the result $\overline{x} = .27$ also. The two P-values express your explanation in (b) in numbers.

5.44   The mean area of the several thousand apartments in a new development is advertised to be 1250 square feet. A tenant group thinks that the apartments are smaller than advertised. They hire an engineer to measure a sample of apartments to test their suspicion. What are the null hypothesis H_0 and alternative hypothesis H_a?

5.45   Experiments on learning in animals sometimes measure how long it takes mice to find their way through a maze. The mean time is 18 seconds for one particular maze. A researcher thinks that a loud noise will cause the mice to complete the maze faster. She measures how long each of 10 mice takes with a noise as stimulus. What are the null hypothesis H_0 and alternative hypothesis H_a?

5.46   Cobra Cheese Company buys milk from several suppliers. Cobra suspects that some producers are adding water to their milk to increase their profits. Excess water can be detected by measuring the freezing point of the milk. The freezing temperature of natural milk varies normally, with mean $\mu = -.545°$ Celsius (C) and standard deviation $\sigma = .008°$ C. Added water raises the freezing temperature toward $0°$ C, the freezing point of water. Cobra's laboratory manager measures the freezing temperature of five consecutive lots of milk from one producer. The mean measurement is $\bar{x} = -.538°$ C. Is this good evidence that the producer is adding water to the milk? State hypotheses, carry out the test, give the P-value, and state your conclusion.

5.47   There are other z statistics that we have not yet studied. You can use Table C to assess the significance of any z statistic. A study compares American-Japanese joint ventures in which the U.S. company is larger than its Japanese partner with joint ventures in which the U.S. company is smaller. One variable measured is the excess returns earned by shareholders in the American company. The null hypothesis is "no difference" between the means for the two populations. The alternative hypothesis is two-sided. The value of the test statistic is $z = -1.37$.

(a) Is this result significant at the 5% level?

(b) Is the result significant at the 10% level?

5.48   Use Table C to find the approximate P-value for the test in Exercise 5.40 without doing a probability calculation. That is, find from the table two numbers that contain the P-value between them.

5.49   Use Table C to find the approximate P-value for the test in Exercise 5.41. That is, between what two numbers obtained from the table must the P-value lie?

5.50   Between what values from Table C does the P-value for the outcome $z = -1.37$ in Exercise 5.47 lie? (Remember that H_a is two-sided.) Calculate the

P-value using Table A, and verify that it lies between the values you found from Table C.

5.51   Market pioneers, companies that are among the first to develop a new product or service, tend to have higher market shares than latecomers to the market. What accounts for this advantage? Here is an excerpt from the conclusions of a study of a sample of 1209 manufacturers of industrial goods:

> *Can patent protection explain pioneer share advantages? Only 21% of the pioneers claim a significant benefit from either a product patent or a trade secret. Though their average share is two points higher than that of pioneers without this benefit, the increase is not statistically significant ($z = 1.13$). Thus, at least in mature industrial markets, product patents and trade secrets have little connection to pioneer share advantages.*[8]

Find the *P*-value for the given z. Then explain to someone who knows no statistics what "not statistically significant" in the study's conclusion means. Why does the author conclude that patents and trade secrets don't help, even though they contributed 2 percentage points to average market share?

5.52   The cigarette industry has adopted a voluntary code requiring that models appearing in its advertising must appear to be at least 25 years old. Studies have shown, however, that consumers think many of the models are younger. Here is a quote from a study that asked whether different brands of cigarettes use models that appear to be of different ages.[9]

> *The ANCOVA revealed that the brand variable is highly significant ($P < .001$), indicating that the average perceived age of the models is not equal across the 12 brands. As discussed previously, certain brands such as Lucky Strike Lights, Kool Milds, and Virginia Slims tended to have younger models . . .*

ANCOVA is an advanced statistical technique, but significance and *P*-values have their usual meaning. Explain to someone who knows no statistics what "highly significant ($P < .001$)" means and why this is good evidence of differences among all advertisements of these brands even though the subjects saw only a sample of ads.

5.53   Explain in plain language why a significance test that is significant at the 1% level must always be significant at the 5% level.

5.54   Asked to explain the meaning of "statistically significant at the $\alpha = .05$ level," a student says: "This means that the probability that the null hypothesis is true is less than .05." Is this explanation correct? Why or why not?

5.3   USING SIGNIFICANCE TESTS

Carrying out a test of significance is often quite simple, especially if you use a fixed significance level α or get the P-value effortlessly from a computer. Using tests wisely is not so simple. Some hesitation about the unthinking use of significance tests is a sign of statistical maturity.

Significance tests are widely used in reporting the results of research in many fields of applied science and in industry. Some products (such as pharmaceuticals) require significant evidence of effectiveness and safety. Courts inquire about statistical significance in hearing class action discrimination cases and in other legal proceedings. Marketers want to know whether a new ad campaign significantly outperforms the old one, and medical researchers want to know whether a new therapy performs significantly better. In all these uses, statistical significance is valued because it points to an effect that is unlikely to occur simply by chance. Here are some points to keep in mind when using or interpreting significance tests.

Choosing a level of significance

The purpose of a test of significance is to give a clear statement of the degree of evidence provided by the sample against the null hypothesis. The P-value does this. But sometimes you will make some decision or take some action if your evidence reaches a certain standard. A level of significance α sets such a standard. Perhaps you will publish a research finding if the effect is significant at the $\alpha = .01$ level. Or perhaps your company will lose a lawsuit alleging racial discrimination if the percent of blacks hired is significantly below the percent of blacks in the pool of potential employees at the $\alpha = .05$ level. Courts have in fact tended to accept this standard in discrimination cases.[10]

Making a decision is different in spirit from testing significance, though the two are often mixed in practice. Choosing a level α in advance makes sense if you must make a decision, but not if you wish only to describe the strength of your evidence. Using tests with fixed α for decision making is discussed at greater length in Section 5.4.

If you do use a fixed α significance test to make a decision, choose α by asking how much evidence is required to reject H_0. This depends mainly on two circumstances:

- *How plausible is H_0?* If H_0 represents an assumption that the people you must convince have believed for years, strong evidence (small α) will be needed to persuade them.

- *What are the consequences of rejecting H_0?* If rejecting H_0 in favor of H_a means making an expensive changeover from one type of product packaging to another, you need strong evidence that the new packaging will boost sales.

Both the plausibility of H_0 and H_a and the consequences of any action that rejection may lead to are somewhat subjective. Different people may want to use different levels of significance. It is better to report the P-value, which allows each of us to decide individually if the evidence is sufficiently strong.

Users of statistics have often emphasized certain standard levels of significance, such as 10%, 5%, and 1%. This emphasis reflects the time when tables of critical values rather than computer programs dominated statistical practice. The 5% level ($\alpha = .05$) is particularly common. *There is no sharp border between "significant" and "insignificant," only increasingly strong evidence as the P-value decreases.* There is no practical distinction between the P-values 0.049 and 0.051. It makes no sense to treat $\alpha = .05$ as a universal rule for what is significant.

EXERCISE

5.55 Suppose that in the absence of special preparation Scholastic Assessment Test mathematics (SATM) scores vary normally with mean $\mu = 475$ and $\sigma = 100$. One hundred students go through a rigorous training program designed to raise their SATM scores by improving their mathematics skills. Carry out a test of

$$H_0 : \mu = 475$$

$$H_a : \mu > 475$$

in each of the following situations:

(a) The students' average score is $\bar{x} = 491.4$. Is this result significant at the 5% level?

(b) The average score is $\bar{x} = 491.5$. Is this result significant at the 5% level?

The difference between the two outcomes in (a) and (b) is of no importance. Beware attempts to treat $\alpha = .05$ as sacred.

Statistical significance and practical significance

When a null hypothesis ("no effect" or "no difference") can be rejected at the usual levels, $\alpha = .05$ or $\alpha = .01$, there is good evidence that an effect is present. But that effect may be very small. When large samples are available, even tiny deviations from the null hypothesis will be significant.

EXAMPLE 5.16

We are testing the hypothesis of no correlation between two variables. With 1000 observations, an observed correlation of only $r = .08$ is significant evidence at the $\alpha = .01$ level that the correlation in the population is not zero but positive. The low significance level does not mean there is a strong association, only that there is strong evidence of some association. The true population correlation is probably quite close to the observed sample value, $r = .08$. We might well conclude that for practical purposes we can ignore the association between these variables, even though we are confident (at the 1% level) that the correlation is positive. ◄

Remember the wise saying: *Statistical significance is not the same thing as practical significance.* Exercise 5.56 demonstrates in detail the effect on P of increasing the sample size.

The remedy for attaching too much importance to statistical significance is to pay attention to the actual data as well as to the P-value. Plot your data and examine them carefully. Are there outliers or other deviations from a consistent pattern? A few outlying observations can produce highly significant results if you blindly apply common tests of significance. Outliers can also destroy the significance of otherwise convincing data. The foolish user of statistics who feeds the data to a computer without exploratory analysis will often be embarrassed. Is the effect you are seeking visible in your plots? If not, ask yourself if the effect is large enough to be practically important. It is usually wise to give a confidence interval for the parameter in which you are interested. A confidence interval actually estimates the size of an effect, rather than simply asking if it is too large to reasonably occur by chance alone. Confidence intervals are not used as often as they should be, while tests of significance are perhaps overused.

EXERCISES

5.56   Let us suppose that SATM scores in the absence of coaching vary normally with mean $\mu = 475$ and $\sigma = 100$. Suppose also that coaching may change μ but does not change σ. An increase in the SATM score from 475 to 478 is of no importance in seeking admission to college, but this unimportant change can be statistically very significant. To see this, calculate the P-value for the test of

$$H_0 : \mu = 475$$

$$H_a : \mu > 475$$

in each of the following situations:

(a) A coaching service coaches 100 students. Their SATM scores average $\bar{x} = 478$.

(b) By the next year, the service has coached 1000 students. Their SATM scores average $\bar{x} = 478$.

(c) An advertising campaign brings the number of students coached to 10,000. Their average score is still $\bar{x} = 478$.

5.57 Give a 99% confidence interval for the mean SATM score μ after coaching in each part of the previous exercise. For large samples, the confidence interval tells us, "Yes, the mean score is higher than 475 after coaching, but only by a small amount."

Statistical inference is not valid for all sets of data

We emphasize again that badly designed surveys or experiments often produce invalid results. Formal statistical inference cannot correct basic flaws in the design. Each test is valid only in certain circumstances, with properly produced data being particularly important. The z test, for example, should bear the same warning label that we attached on page 343 to the corresponding confidence interval. Similar warnings accompany the other tests that we will learn.

EXAMPLE 5.17

Hawthorne effect

You wonder whether background music would improve the productivity of the staff who process mail orders in your business. After discussing the idea with the workers, you add music and find a significant increase. You should not be impressed. In fact, almost any change in the work environment together with knowledge that a study is under way will produce a short-term productivity increase. This is the **Hawthorne effect**, named after the Western Electric manufacturing plant where it was first noted.

The significance test correctly informs you that an increase has occurred that is larger than would often arise by chance alone. It does not tell you *what* other than chance caused the increase. The most plausible explanation is that workers change their behavior when they know they are being studied. Your experiment was uncontrolled, so the significant result cannot be interpreted. A randomized comparative experiment would isolate the actual effect of background music and so make significance meaningful.   ◀

Tests of significance and confidence intervals are based on the laws of probability. Randomization in sampling or experimentation ensures that these laws apply. Yet we must often analyze data that do not arise from randomized samples or experiments. To apply statistical inference to such data, we must have confidence in the use of probability to describe the data. The diameters of successive holes bored in auto engine blocks during production, for example, may behave like a random sample from a normal distribution. We can check this probability model by examining the data. If the model appears correct, we can apply the recipes of this chapter to do inference about the process mean diameter μ. Always ask how the data were produced, and don't be too impressed by P-values on a printout until you are confident that the data deserve a formal analysis.

EXERCISE

5.58   A local television station announces a question for a call-in opinion poll on the six o'clock news, then gives the response on the eleven o'clock news. Today's question concerns a proposed gun-control ordinance. Of the 2372 calls received, 1921 oppose the new law. The station, following standard statistical practice, makes a confidence statement: "81% of the Channel 13 Pulse Poll sample oppose gun control. We can be 95% confident that the proportion of all viewers who oppose the law is within 1.6% of the sample result." Is the station's conclusion justified? Explain your answer.

Beware of multiple analyses

Statistical significance is a commodity much sought after. It ought to mean that you have found an effect that you were looking for. The reasoning behind statistical significance works well if you decide what effect you are seeking, design a study to search for it, and use a test of significance to weigh the evidence you get. In other settings, significance may have little meaning.

EXAMPLE 5.18

You want to learn what distinguishes managerial trainees who eventually become executives from those who, after expensive training, don't succeed and leave the company. You have abundant data on past trainees—data on their personalities and goals, their college preparation and performance, even their family backgrounds and their hobbies. Statistical software makes it easy to perform dozens of significance tests on these dozens of variables to see which ones best predict later success. Aha! You find that future executives are significantly more likely than washouts to have an urban or suburban upbringing and an undergraduate degree in a technical field.

Before basing future recruiting on these findings, pause for a moment of reflection. When you make dozens of tests at the 5% level, you expect a few of them to be significant by chance alone. After all, results significant at the 5% level do occur 5 times in 100 in the long run even when H_0 is true. Running one test and reaching the $\alpha = .05$ level is reasonably good evidence that you have found something. Running several dozen tests and reaching that level once or twice is not.   ◀

There are methods for testing many hypotheses simultaneously while controlling the risk of false findings of significance. But if you carry out many individual tests without these special methods, finding a few small P-values is only suggestive, not conclusive. The same is true of less formal analyses. Searching the trainee data for the variable with the biggest difference between future washouts and future executives, then testing whether that difference is significant, is bad statistics. The P-value assumes you had that specific difference in mind before you looked at the data. It is very misleading when applied to the largest of many differences.

Searching data for suggestive patterns is certainly legitimate. Exploratory data analysis is an important aspect of statistics. But the reasoning of formal inference does not apply when your search for a striking effect in the data is successful. The remedy is clear. Once you have a hypothesis, design a study to search specifically for the effect you now think is there. If the result of this study is statistically significant, you have real evidence.

EXERCISE

5.59   A researcher looking for evidence of extrasensory perception (ESP) tests 500 subjects. Four of these subjects do significantly better ($P < .01$) than random guessing.

(a) Is it proper to conclude that these four people have ESP? Explain your answer.

(b) What should the researcher now do to test whether any of these four subjects have ESP?

SUMMARY

P-values are more informative than the reject-or-not result of a fixed level α test. Beware of placing too much weight on traditional values of α, such as $\alpha = .05$.

Very small effects can be highly significant (small P), especially when a test is based on a large sample. A statistically significant effect need not be practically important. Plot the data to display the effect you are seeking, and use confidence intervals to estimate the actual value of parameters.

On the other hand, lack of significance does not imply that H_0 is true, especially when the test is based on just a few observations.

Significance tests are not always valid. Faulty data collection, outliers in the data, and testing a hypothesis on the same data that suggested the hypothesis can invalidate a test.

Many tests run at once will probably produce some significant results by chance alone, even if all the null hypotheses are true.

SECTION 5.3 EXERCISES

5.60   Which of the following questions does a test of significance answer?

(a) Is the sample or experiment properly designed?

(b) Is the observed effect due to chance?

(c) Is the observed effect important?

5.61   A company compares two package designs for a laundry detergent by placing bottles with both designs on the shelves of several markets. Checkout scanner data on more than 5000 bottles bought show that more shoppers bought Design A than Design B. The difference is statistically significant ($P = .02$). Can we conclude that consumers strongly prefer Design A? Explain your answer.

5.62   A group of psychologists once measured 77 variables on a sample of schizophrenic people and a sample of people who were not schizophrenic. They compared the two samples using 77 separate significance tests. Two of these tests were significant at the 5% level. Suppose that there is in fact no difference on any of the 77 variables between people who are and people who are not schizophrenic in the adult population. Then all 77 null hypotheses are true.

(a) What is the probability that one specific test shows a difference significant at the 5% level?

(b) Why is it not surprising that 2 of the 77 tests were significant at the 5% level?

5.4   INFERENCE AS DECISION*

Tests of significance assess the strength of evidence against the null hypothesis. We measure evidence by the P-value, which is a probability computed under the assumption that H_0 is true. The alternative hypothesis (the statement we seek evidence for) enters the test only to help us see what outcomes count against the null hypothesis.

Using significance tests with fixed level α, however, suggests another way of thinking. A level of significance α chosen in advance points to the outcome of the test as a *decision*. If our result is significant at level α, we reject H_0 in favor of H_a. Otherwise, we fail to reject H_0. The transition from measuring the strength of evidence to making a decision is not a small step. Many statisticians feel that making decisions should be left to the user rather than built into the statistical test. A test result is only one among many factors that influence a decision.

acceptance sampling

Yet there are circumstances that call for a decision or action as the end result of inference. ***Acceptance sampling*** is one such circumstance. A producer of bearings and the consumer of the bearings agree that each carload lot must meet certain quality standards. When a carload arrives, the consumer inspects a sample of the bearings. On the basis of the sample outcome, the consumer either accepts or rejects the carload. We will use acceptance sampling to show how a different concept—inference as decision—changes the reasoning used in tests of significance.

---

*The purpose of this more advanced section is to clarify the reasoning of significance tests by contrast with a related type of reasoning. This section is not needed to read the rest of the book.

Type I and Type II errors

Tests of significance concentrate on H_0, the null hypothesis. If a decision is called for, however, there is no reason to single out H_0. There are simply two hypotheses, and we must accept one and reject the other. It is convenient to continue to call the two hypotheses H_0 and H_a, but H_0 no longer has the special status (the statement we try to find evidence against) that it had in tests of significance. In the acceptance sampling problem, we must decide between

H_0: the lot of bearings meets standards

H_a: the lot does not meet standards

on the basis of a sample of bearings.

We hope that our decision will be correct, but sometimes it will be wrong. There are two types of incorrect decisions. We can accept a bad lot of bearings, or we can reject a good lot. Accepting a bad lot injures the consumer, while rejecting a good lot hurts the producer. To distinguish these two types of error, we give them specific names.

> **TYPE I AND TYPE II ERRORS**
>
> If we reject H_0 (accept H_a) when in fact H_0 is true, this is a **Type I error**.
>
> If we accept H_0 (reject H_a) when in fact H_a is true, this is a **Type II error**.

The possibilities are summed up in Figure 5.17. If H_0 is true, our decision is either correct (if we accept H_0) or is a Type I error. If H_a is true, our decision is either correct or is a Type II error. Only one error is possible at one time.

Error probabilities

We assess any rule for making decisions by looking at the probabilities of the two types of error. This is in keeping with the idea that statistical inference is based on asking, "What would happen if I used this procedure many times?"

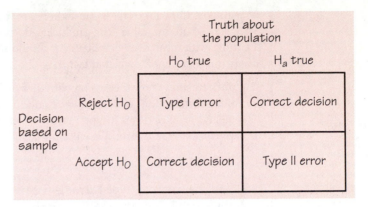

FIGURE 5.17   The two types of error in testing hypotheses.

Significance tests with fixed level α give a rule for making decisions, because the test either rejects H_0 or fails to reject it. If we adopt the decision-making way of thought, failing to reject H_0 means deciding that H_0 is true. We can then describe the performance of a test by the probabilities of Type I and Type II errors.

EXAMPLE 5.19

The mean diameter of a type of bearing is supposed to be 2.000 centimeters (cm). The bearing diameters vary normally with standard deviation $\sigma = .010$ cm. When a lot of the bearings arrives, the consumer takes an SRS of 5 bearings from the lot and measures their diameters. The consumer rejects the bearings if the sample mean diameter is significantly different from 2 at the 5% significance level.

This is a test of the hypotheses

$$H_0 : \mu = 2$$

$$H_a : \mu \neq 2$$

To carry out the test, the consumer computes the z statistic

$$z = \frac{\bar{x} - 2}{.01/\sqrt{5}}$$

and rejects H_0 if $z < -1.96$ or $z > 1.96$. A Type I error is to reject H_0 when in fact $\mu = 2$.

What about Type II errors? Because there are many values of μ in H_a, we will concentrate on one value. The producer and the consumer agree that a lot of bearings with mean diameter 2.015 cm should be rejected. So a particular Type II error is to accept H_0 when in fact $\mu = 2.015$.

Figure 5.18 shows how the two probabilities of error are obtained from the *two* sampling distributions of $\overline{x}$, for $\mu = 2$ and for $\mu = 2.015$. When $\mu = 2$, H_0 is true and to reject H_0 is a Type I error. When $\mu = 2.015$, H_a is true and to accept H_0 is a Type II error. We will now calculate these error probabilities. ◀

The probability of a Type I error is the probability of rejecting H_0 when it is really true. This is the probability that $|z| \geq 1.96$ when $\mu = 2$. But this is exactly the significance level of the test. The critical value 1.96 was chosen to make this probability 0.05, so we do not have to compute it again. The definition of "significance level 0.05" is that values of z this extreme will occur with probability 0.05 when H_0 is true.

SIGNIFICANCE AND TYPE I ERROR

The significance level α of any fixed level test is the probability of a Type I error. That is, α is the probability that the test will reject the null hypothesis H_0 when H_0 is in fact true.

The probability of a Type II error for the particular alternative $\mu = 2.015$ in Example 5.19 is the probability that the test will accept H_0 when μ has this alternative value. This is the probability that the test statistic z falls between -1.96 and 1.96, calculated assuming that $\mu = 2.015$. This probability is *not* $1 - .05$, because the probability 0.05 was found assuming that $\mu = 2$. Here is the calculation for Type II error.

EXAMPLE 5.20

To calculate the probability of a Type II error:

Step 1   *Write the rule for accepting H_0 in terms of $\overline{x}$.* The test accepts H_0 when

$$-1.96 \leq \frac{\overline{x} - 2}{.01/\sqrt{5}} \leq 1.96$$

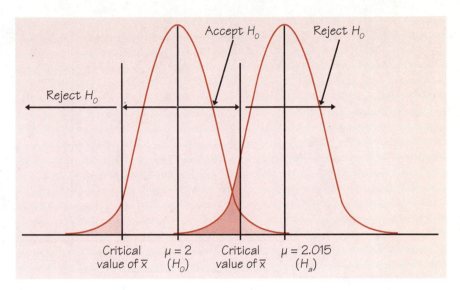

FIGURE 5.18 The two error probabilities for Example 5.19. The probability of a Type I error (*light shaded area*) is the probability of rejecting $H_0 : \mu = 2$ when in fact $\mu = 2$. The probability of a Type II error (*dark shaded area*) is the probability of accepting H_0 when in fact $\mu = 2.015$.

This is the same as

$$2 - 1.96\left(\frac{.01}{\sqrt{5}}\right) \le \bar{x} \le 2 + 1.96\left(\frac{.01}{\sqrt{5}}\right)$$

or, doing the arithmetic,

$$1.9912 \le \bar{x} \le 2.0088$$

This step does not involve the particular alternative $\mu = 2.015$.

Step 2   *Find the probability of accepting H_0 assuming that the alternative is true.* Take $\mu = 2.015$ and standardize to find the probability.

$$P(\text{Type II error}) = P(1.9912 \le \bar{x} \le 2.0088)$$

$$= P\left(\frac{1.9912 - 2.015}{.01/\sqrt{5}} \le \frac{\bar{x} - 2.015}{.01/\sqrt{5}} \le \frac{2.0088 - 2.015}{.01/\sqrt{5}}\right)$$

$$= P(-5.32 \le Z \le -1.39)$$

$$= .0823$$

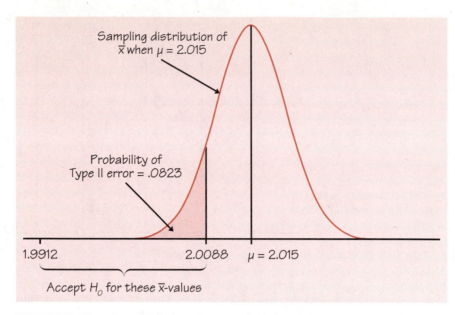

FIGURE 5.19 The probability of a Type II error for Example 5.20. This is the probability that the test accepts H_0 when the alternative hypothesis is true.

Figure 5.19 illustrates this error probability in terms of the sampling distribution of $\bar{x}$ when $\mu = 2.015$. The test will wrongly accept the hypothesis that $\mu = 2$ in about 8% of all samples when in fact $\mu = 2.015$. ◀

This test will reject 5% of all good lots of bearings (for which $\mu = 2$). It will accept 8% of lots so bad that $\mu = 2.015$. Calculations of error probabilities help the producer and consumer decide whether the test is satisfactory.

EXERCISES

5.63   Your company markets a computerized medical diagnostic program. The program scans the results of routine medical tests (pulse rate, blood tests, etc.) and either clears the patient or refers the case to a doctor. The program is used to screen thousands of people who do not have specific medical complaints. The program makes a decision about each person.

(a) What are the two hypotheses and the two types of error that the program can make? Describe the two types of error in terms of "false positive" and "false negative" test results.

(b) The program can be adjusted to decrease one error probability, at the cost of an increase in the other error probability. Which error probability would you choose to make smaller, and why? (This is a matter of judgment. There is no single correct answer.)

5.64   You have the NAEP quantitative scores for an SRS of 840 young men. You plan to test hypotheses about the population mean score,

$$H_0 : \mu = 275$$

$$H_a : \mu < 275$$

at the 1% level of significance. The population standard deviation is known to be $\sigma = 60$. The z test statistic is

$$z = \frac{\bar{x} - 275}{60/\sqrt{840}}$$

(a) What is the rule for rejecting H_0 in terms of z?

(b) What is the probability of a Type I error?

(c) You want to know whether this test will usually reject H_0 when the true population mean is 270, 5 points lower than the null hypothesis claims. Answer this question by calculating the probability of a Type II error when $\mu = 270$.

5.65   You have an SRS of size $n = 9$ from a normal distribution with $\sigma = 1$. You wish to test

$$H_0 : \mu = 0$$

$$H_a : \mu > 0$$

You decide to reject H_0 if $\bar{x} > 0$ and to accept H_0 otherwise.

(a) Find the probability of a Type I error. That is, find the probability that the test rejects H_0 when in fact $\mu = 0$.

(b) Find the probability of a Type II error when $\mu = .3$. This is the probability that the test accepts H_0 when in fact $\mu = .3$.

(c) Find the probability of a Type II error when $\mu = 1$.

Power

A test makes a Type II error when it fails to reject a null hypothesis that really is false. A high probability of a Type II error for a particular alternative means that the test is not sensitive enough to usually detect that alternative. Calculations of the probability of Type II errors are therefore useful even if you don't think that a statistical test should be viewed as making a decision. The language used is a bit different in the significance test setting. It is usual to report the probability that a test *does* reject H_0 when an alternative is true. The higher this probability is, the more sensitive the test is.

> ### POWER
>
> The probability that a fixed level α significance test will reject H_0 when a particular alternative value of the parameter is true is called the **power** of the test against that alternative.
>
> The power of a test against any alternative is 1 minus the probability of a Type II error for that alternative.

Calculations of power are essentially the same as calculations of the probability of Type II error. In Example 5.20, the power is the probability of *rejecting* H_0 in Step 2 of the calculation. It is $1 - .0823$, or 0.9177.

Calculations of P-values and calculations of power both say what would happen if we repeated the test many times. A P-value describes what would happen supposing that the null hypothesis is true. Power describes what would happen supposing that a particular alternative is true.

In planning an investigation that will include a test of significance, a careful user of statistics decides what alternatives the test should detect and checks that the power is adequate. The power depends on which particular parameter value in H_a we are interested in. Values of the mean μ that are in H_a but lie close to the hypothesized value μ_0 are harder to detect (lower power) than values of μ that are far from μ_0. If the power is too low, a larger sample size will increase the power for the same significance level α. In order to calculate power, we must fix an α so that there is a fixed rule for rejecting H_0. We prefer to report P-values rather than to use a fixed significance level. The usual practice is to calculate the power at a common significance level such as $\alpha = .05$ even though you intend to report a P-value.

EXERCISES

5.66   The cola maker of Example 5.7 (page 349) determines that a sweetness loss is too large to accept if the mean response for all tasters is $\mu = 1.1$. Will a 5% significance test of the hypotheses

$$H_0 : \mu = 0$$

$$H_a : \mu > 0$$

based on a sample of 10 tasters usually detect a change this great?

We want the power of the test against the alternative $\mu = 1.1$. This is the probability that the test rejects H_0 when $\mu = 1.1$ is true. The calculation method is similar to that for Type II error.

(a) Step 1   *Write the rule for rejecting H_0 in terms of $\bar{x}$.* We know that $\sigma = 1$, so the test rejects H_0 at the $\alpha = .05$ level when

$$z = \frac{\bar{x} - 0}{1/\sqrt{10}} \geq 1.645$$

Restate this in terms of $\bar{x}$.

(b) Step 2   *The power is the probability of this event supposing that the alternative is true.* Standardize using $\mu = 1.1$ to find the probability that $\bar{x}$ takes a value that leads to rejection of H_0.

5.67   Exercise 5.39 (page 369) concerns a test about the mean contents of cola bottles. The hypotheses are

$$H_0 : \mu = 300$$

$$H_a : \mu < 300$$

The sample size is $n = 6$, and the population is assumed to have a normal distribution with $\sigma = 3$. A 5% significance test rejects H_0 if $z \leq -1.645$, where the test statistic z is

$$z = \frac{\bar{x} - 300}{3/\sqrt{6}}$$

Power calculations help us see how large a shortfall in the bottle contents the test can be expected to detect.

(a) Find the power of this test against the alternative $\mu = 299$.

(b) Find the power against the alternative $\mu = 295$.

(c) Is the power against $\mu = 290$ higher or lower than the value you found in (b)? (Don't actually calculate that power.) Explain your answer.

5.68   Increasing the sample size increases the power of a test when the level α is unchanged. Suppose that in the previous exercise a sample of n bottles had been measured. In that exercise, $n = 6$. The 5% significance test still rejects H_0 when $z \le -1.645$, but the z statistic is now

$$z = \frac{\bar{x} - 300}{3/\sqrt{n}}$$

(a) Find the power of this test against the alternative $\mu = 299$ when $n = 25$.

(b) Find the power against $\mu = 299$ when $n = 100$.

Different views of statistical tests

The distinction between tests of significance and tests as rules for deciding between two hypotheses does not lie in the calculations but in the reasoning that motivates the calculations. In a test of significance we focus on a single hypothesis (H_0) and a single probability (the P-value). The goal is to measure the strength of the sample evidence against H_0. Calculations of power are done to check the sensitivity of the test. If we cannot reject H_0, we conclude only that there is not sufficient evidence against H_0, not that H_0 is actually true. If the same inference problem is thought of as a decision problem, we focus on two hypotheses and give a rule for deciding between them based on the sample evidence. We therefore must focus equally on two probabilities, the probabilities of the two types of error. We must choose one or the other hypothesis and cannot abstain on grounds of insufficient evidence.

There are clear distinctions between the two ways of thinking about statistical tests. But sometimes the two approaches merge. Jerzy Neyman advocated an approach called ***testing hypotheses*** that mixes the reasoning of significance tests and decision rules as follows:

testing
hypotheses

1. State H_0 and H_a just as in a test of significance. In particular, we are seeking evidence against H_0.

2. Think of the problem as a decision problem, so that the probabilities of Type I and Type II errors are relevant.

3. Because of Step 1, Type I errors are more serious. So choose an α (significance level) and consider only tests with probability of Type I error no greater than α.

4. Among these tests, select one that makes the probability of a Type II error as small as possible (that is, power as large as possible). If this probability is too large, you will have to take a larger sample to reduce the chance of an error.

Hypothesis testing is often emphasized in mathematical presentations of statistics because Neyman developed an impressive mathematical theory. In simple settings, this theory shows how to find the test that has the smallest possible probability of a Type II error among all tests with a given probability (like 0.05) of a Type I error. In part because such pleasing results aren't available for many practical settings, the significance test way of thinking prevails in statistical practice.

SUMMARY

An alternative to significance testing regards H_0 and H_a as two statements of equal status that we must decide between. This **decision analysis** point of view regards statistical inference in general as giving rules for making decisions in the presence of uncertainty.

In the case of testing H_0 versus H_a, decision analysis chooses a decision rule on the basis of the probabilities of two types of error. A **Type I error** occurs if we reject H_0 when it is in fact true. A **Type II error** occurs if we accept H_0 when in fact H_a is true.

The **power** of a significance test measures its ability to detect an alternative hypothesis. The power against a specific alternative is the probability that the test will reject H_0 when the alternative is true.

In a fixed level α significance test, the significance level α is the probability of a Type I error, and the power against a specific alternative is 1 minus the probability of a Type II error for that alternative.

Increasing the size of the sample increases the power (reduces the probability of a Type II error) when the significance level remains fixed.

SECTION 5.4 EXERCISES

5.69   Power calculations for two-sided tests follow the same outline as for one-sided tests. Example 5.14 (page 371) presents a test of

$$H_0 : \mu = .86$$

$$H_a : \mu \neq .86$$

at the 1% level of significance. The sample size is $n = 3$ and $\sigma = .0068$. We will find the power of this test against the alternative $\mu = .845$.

(a) The test in Example 5.14 rejects H_0 when $|z| \geq 2.576$. The test statistic z is

$$z = \frac{\bar{x} - .86}{.0068/\sqrt{3}}$$

Write the rule for rejecting H_0 in terms of the values of $\bar{x}$. (Because the test is two-sided, it rejects when $\bar{x}$ is either too large or too small.)

(b) Now find the probability that $\bar{x}$ takes values that lead to rejecting H_0 if the true mean is $\mu = .845$. This probability is the power.

(c) What is the probability that this test makes a Type II error when $\mu = .845$?

5.70   In Example 5.11 (page 366), a company medical director failed to find significant evidence that the mean blood pressure of a population of executives differed from the national mean $\mu = 128$. The medical director now wonders if the test used would detect an important difference if one were present. For the SRS of size 72 from a population with standard deviation $\sigma = 15$, the z statistic is

$$z = \frac{\bar{x} - 128}{15/\sqrt{72}}$$

The two-sided test rejects $H_0 : \mu = 128$ at the 5% level of significance when $|z| \geq 1.96$.

(a) Find the power of the test against the alternative $\mu = 134$.

(b) Find the power of the test against $\mu = 122$. Can the test be relied on to detect a mean that differs from 128 by 6?

(c) If the alternative were farther from H_0, say $\mu = 136$, would the power be higher or lower than the values calculated in (a) and (b)?

5.71 In Exercise 5.67 you found the power of a test against the alternative $\mu = 295$. Use the result of that exercise to find the probabilities of Type I and Type II errors for that test and that alternative.

5.72 In Exercise 5.64 you found the probabilities of the two types of error for a test of $H_0 : \mu = 275$, with the specific alternative that $\mu = 270$. Use the result of that exercise to give the power of the test against the alternative $\mu = 270$.

5.73 You are reading an article in a business journal that discusses the "efficient market hypothesis" for the behavior of securities prices. The author admits that most tests of this hypothesis have failed to find significant evidence against it. But he says this failure is a result of the fact that the tests used have low power. "The widespread impression that there is strong evidence for market efficiency may be due just to a lack of appreciation of the low power of many statistical tests."[11]

Explain in simple language why tests having low power often fail to give evidence against a hypothesis even when the hypothesis is really false.

CHAPTER REVIEW

Statistical inference draws conclusions about a population on the basis of sample data and uses probability to indicate how reliable the conclusions are. A confidence interval estimates an unknown parameter. A significance test shows how strong the evidence is for some claim about a parameter.

The probabilities in both confidence intervals and tests tell us what would happen if we used the recipe for the interval or test very many times. A confidence level is the probability that the recipe for a confidence interval actually produces an interval that contains the unknown parameter. A 99%

confidence interval gives a correct result 99% of the time when we use it repeatedly. A P-value is the probability that the test would produce a result at least as extreme as the observed result if the null hypothesis really were true. That is, a P-value tells us how surprising the observed outcome is. Very surprising outcomes (small P-values) are good evidence that the null hypothesis is not true.

The rest of this book presents confidence intervals and tests for use in many specific settings. We will have a good deal to say about practical aspects of using statistical methods. But in every case, the basic reasoning of confidence intervals and significance tests remains the same. Sections 5.1 to 5.3 in this chapter are the foundation for your understanding of statistical inference. Here are the most important things you should be able to do after studying this chapter.

A. CONFIDENCE INTERVALS

1. State in nontechnical language what is meant by "95% confidence" or other statements of confidence in statistical reports.

2. Calculate a confidence interval for the mean μ of a normal population with known standard deviation σ, using the recipe

$$\bar{x} \pm z^* \sigma/\sqrt{n}$$

3. Recognize when you can safely use this confidence interval recipe and when the sample design or a small sample from a skewed population makes it inaccurate.

4. Understand how the margin of error of a confidence interval changes with the sample size and the level of confidence C.

5. Find the sample size required to obtain a confidence interval of specified margin of error m when the confidence level and other information are given.

B. SIGNIFICANCE TESTS

1. State the null and alternative hypotheses in a testing situation when the parameter in question is a population mean μ.

2. Explain in nontechnical language the meaning of the P-value when you are given the numerical value of P for a test.

3. Calculate the z statistic and the P-value for both one-sided and two-sided tests about the mean μ of a normal population.

4. Assess statistical significance at standard levels α, either by comparing P to α or by comparing z to standard normal critical values.

5. Recognize that significance testing does not measure the size or importance of an effect.

6. Recognize when you can use the z test and when the data collection design or a small sample from a skewed population makes it inappropriate.

CHAPTER 5 REVIEW EXERCISES

5.74   Sulfur compounds cause "off-odors" in wine, so winemakers want to know the odor threshold, the lowest concentration of a compound that the human nose can detect. The odor threshold for dimethyl sulfide (DMS) in trained wine tasters is about 25 micrograms per liter of wine (μg/l). The untrained noses of consumers may be less sensitive, however. Here are the DMS odor thresholds for 10 untrained students:

$$31 \quad 31 \quad 43 \quad 36 \quad 23 \quad 34 \quad 32 \quad 30 \quad 20 \quad 24$$

Assume that the standard deviation of the odor threshold for untrained noses is known to be $\sigma = 7 \ \mu$g/l.

(a) Make a stemplot to verify that the distribution is roughly symmetric with no outliers. (More data confirm that there are no systematic departures from normality.)

(b) Give a 95% confidence interval for the mean DMS odor threshold among all students.

(c) Are you convinced that the mean odor threshold for students is higher than the published threshold, 25 μg/l? Carry out a significance test to justify your answer.

5.75   An agronomist examines the cellulose content of a variety of alfalfa hay. Suppose that the cellulose content in the population has standard deviation $\sigma = 8$ mg/g. A sample of 15 cuttings has mean cellulose content $\bar{x} = 145$ mg/g.

(a) Give a 90% confidence interval for the mean cellulose content in the population.

(b) A previous study claimed that the mean cellulose content was $\mu = 140$ mg/g, but the agronomist believes that the mean is higher than that figure. State H_0 and H_a and carry out a significance test to see if the new data support this belief.

(c) The statistical procedures used in (a) and (b) are valid when several assumptions are met. What are these assumptions?

5.76   Researchers studying iron deficiency in infants examined infants who were following different feeding patterns. One group of 26 infants was being breast-fed. At 6 months of age, these children had mean hemoglobin level $\bar{x} = 12.9$ grams per 100 milliliters of blood. Assume that the population standard deviation is $\sigma = 1.6$. Give a 95% confidence interval for the mean hemoglobin level of breast-fed infants. What assumptions (other than the unrealistic assumption that we know σ) does the method you used to get the confidence interval require? •

5.77   Here are the Degree of Reading Power (DRP) scores for an SRS of 44 third-grade students from a suburban school district.

40	26	39	14	42	18	25	43	46	27	19
47	19	26	35	34	15	44	40	38	31	46
52	25	35	35	33	29	34	41	49	28	52
47	35	48	22	33	41	51	27	14	54	45

DRP scores are approximately normal. Suppose that the standard deviation of scores in this school district is known to be $\sigma = 11$. The researcher believes that the mean score μ of all third graders in this district is higher than the national mean, which is 32.

(a) State H_0 and H_a to test this suspicion.

(b) Carry out the test. Give the P-value, and then interpret the result in plain language.

5.78   A government report gives a 99% confidence interval for the 1990 median family income as $29,943 ± $397. This result was calculated by advanced methods from the Current Population Survey, a multistage random sample of about 60,000 households.

   (a) Would a 95% confidence interval be wider or narrower? Explain your answer.

   (b) Would the null hypothesis that the 1990 median family income was $32,000 be rejected at the 1% significance level in favor of the two-sided alternative?

5.79   Statisticians prefer large samples. Describe briefly the effect of increasing the size of a sample (or the number of subjects in an experiment) on each of the following:

   (a) The margin of error of a 95% confidence interval.

   (b) The P-value of a test, when H_0 is false and all facts about the population remain unchanged as n increases.

   (c) **(Optional)** The power of a fixed level α test, when α, the alternative hypothesis, and all facts about the population remain unchanged.

5.80   A roulette wheel has 18 red slots among its 38 slots. You observe many spins and record the number of times that red occurs. Now you want to use these data to test whether the probability p of a red has the value that is correct for a fair roulette wheel. State the hypotheses H_0 and H_a that you will test. (We will describe the test for this situation in Chapter 7.)

5.81   When asked to explain the meaning of "the P-value was $P = .03$," a student says, "This means there is only probability 0.03 that the null hypothesis is true." Is this an essentially correct explanation? Explain your answer.

5.82   Another student, when asked why statistical significance appears so often in research reports, says, "Because saying that results are significant tells us that they cannot easily be explained by chance variation alone." Do you think that this statement is essentially correct? Explain your answer.

5.83   A study compares two groups of mothers with young children who were on welfare two years ago. One group attended a voluntary training program offered free of charge at a local vocational school and advertised in the local news media. The other group did not choose to attend the training program. The study finds a significant difference ($P < .01$) between the proportions of the mothers in the two groups who are still on welfare. The difference is not only significant but quite large. The report says that with 95% confidence the percent of the nonattending group still on welfare is

21% ± 4% higher than that of the group who attended the program. You are on the staff of a member of Congress who is interested in the plight of welfare mothers, and who asks you about the report.

(a) Explain in simple language what "a significant difference ($P < .01$)" means.

(b) Explain clearly and briefly what "95% confidence" means.

(c) Is this study good evidence that requiring job training of all welfare mothers would greatly reduce the percent who remain on welfare for several years?

NOTES AND DATA SOURCES

1. Information from Francisco L. Rivera-Batiz, "Quantitative literacy and the likelihood of employment among young adults," *Journal of Human Resources*, 27 (1992), pp. 313–328.

2. Data provided by Maribeth Cassidy Schmitt, from her Ph.D. dissertation, *The Effects of an Elaborated Directed Reading Activity on the Metacomprehension Skills of Third Graders*, Purdue University, 1987.

3. Data provided by John Rousselle and Huei-Ru Shieh, Department of Restaurant, Hotel, and Institutional Management, Purdue University.

4. Data provided by Mugdha Gore and Joseph Thomas, Purdue University School of Pharmacy.

5. Based on G. Salvendy, G. P. McCabe, S. G. Sanders, J. L. Knight, and E. J. McCormick, "Impact of personality and intelligence on job satisfaction of assembly line and bench work—an industrial study," *Applied Ergonomics*, 13 (1982), pp. 293–299.

6. From a study by M. R. Schlatter et al., Division of Financial Aid, Purdue University.

7. Data provided by Diana Schellenberg, Purdue University School of Health Sciences.

8. William T. Robinson, "Sources of market pioneer advantages: the case of industrial goods industries," *Journal of Marketing Research*, 25 (February 1988), pp. 87–94.

9. Michael B. Maziz et al., "Perceived age and attractiveness of models in cigarette advertisements," *Journal of Marketing*, 56 (January 1992), pp. 22–37.

10. For a discussion of statistical significance in the legal setting, see D. H. Kaye, "Is proof of statistical significance relevant?" *Washington Law Review*,

61 (1986), pp. 1333–1365. Kaye argues that "Presenting the *P*-value without characterizing the evidence by a significance test is a step in the right direction. Interval estimation, in turn, is an improvement over *P*-values."

11. Robert J. Schiller, "The volatility of stock market prices," *Science*, 235 (1987), pp. 33–36.

WILLIAM S. GOSSET

What would cause the head brewer of the famous Guinness brewery in Dublin, Ireland, not only to use statistics but to invent new statistical methods? The search for better beer, of course.

William S. Gosset (1876–1937), fresh from Oxford University, joined Guinness as a brewer in 1899. He soon became involved in experiments and in statistics to understand the data from these experiments. What are the best varieties of barley and hops for brewing? How should they be grown, dried, and stored? The results of the field experiments, as you can guess, varied. Statistical inference can uncover the pattern behind the variation. The statistical methods available at the turn of the century ended with a version of the z test for means—even confidence intervals were not yet available.

Gosset faced in his job the problem we noted in using the z test to introduce the reasoning of statistical tests: he didn't know the population standard deviation σ. What is more, field experiments give only small numbers of observations. Just replacing σ by s in the z statistic and calling the result roughly normal wasn't accurate enough. So Gosset asked the key question, What is the exact sampling distribution of the statistic $(\bar{x} - \mu)/s$?

By 1907 Gosset was brewer-in-charge of Guiness's experimental brewery. He also had the answer to his question and had calculated a table of critical values for his new distribution. We call it the t distribution. The new t test identified the best barley variety, and Guinness promptly bought up all the available seed. Guinness allowed Gosset to publish his discoveries, but not under his own name. He used the name "Student," and the t test is sometimes called "Student's t" in his honor. Gosset's statistical work helped him become head brewer, a more interesting title than professor of statistics.

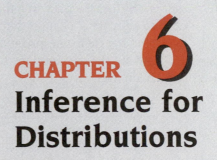

CHAPTER **6**

Inference for Distributions

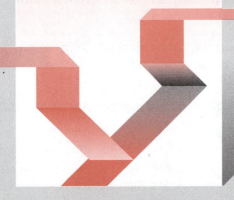

INTRODUCTION

With the principles in hand, we proceed to practice. This chapter describes confidence intervals and significance tests for the mean of a single population and for comparing the means of two populations. An optional section discusses a test for comparing the standard deviations of two populations. Later chapters present procedures for inference about population proportions, comparing the means of more than two populations, and for studying relationships among variables.

6.1   INFERENCE FOR THE MEAN OF A POPULATION

Confidence intervals and tests of significance for the mean μ of a normal population are based on the sample mean $\overline{x}$. The sampling distribution of $\overline{x}$ has μ as its mean. (That is, $\overline{x}$ is an unbiased estimator of the unknown μ.) Its spread depends on the sample size and also on the population standard deviation σ. In the previous chapter we made the unrealistic assumption that we knew the value of σ. In practice, σ is unknown. We must then estimate σ from the data even though we are primarily interested in μ. The need to estimate σ changes some details of tests and confidence intervals for μ, but not their interpretation.

Here are the assumptions we make in order to do inference about a population mean:

ASSUMPTIONS FOR INFERENCE ABOUT A MEAN

- Our data are a **simple random sample** (SRS) of size n from the population.

- Observations from the population have a **normal distribution** with mean μ and standard deviation σ. Both μ and σ are unknown parameters.

In this setting, the sample mean $\overline{x}$ has the normal distribution with mean μ and standard deviation $\sigma/\sqrt{n}$. Because we don't know σ, we estimate it by the sample standard deviation s. We then estimate the standard deviation of $\overline{x}$ by $s/\sqrt{n}$. This quantity is called the *standard error* of the sample mean $\overline{x}$.

> STANDARD ERROR
>
> When the standard deviation of a statistic is estimated from the data, the result is called the **standard error** of the statistic. The standard error of the sample mean $\bar{x}$ is $s/\sqrt{n}$.

The t distributions

When we know the value of σ, we base confidence intervals and tests for μ on the standardized sample mean

$$z = \frac{\bar{x} - \mu}{\sigma/\sqrt{n}}$$

This z statistic has the standard normal distribution $N(0, 1)$. When we do not know σ, we substitute the standard error $s/\sqrt{n}$ of $\bar{x}$ for its standard deviation $\sigma/\sqrt{n}$. The statistic that results does *not* have a normal distribution. It has a distribution that is new to us, called a t *distribution*.

> THE ONE-SAMPLE t STATISTIC AND THE t DISTRIBUTIONS
>
> Draw an SRS of size n from a population that has the normal distribution with mean μ and standard deviation σ. The **one-sample t statistic**
>
> $$t = \frac{\bar{x} - \mu}{s/\sqrt{n}}$$
>
> has the t **distribution** with $n - 1$ degrees of freedom.

degrees of freedom

The t statistic has the same interpretation as any standardized statistic: it says how far $\bar{x}$ is from its mean μ in standard deviation units. There is a different t distribution for each sample size. We specify a particular t distribution by giving its *degrees of freedom*. The degrees of freedom for the one-sample t statistic come from the sample standard deviation s

in the denominator of t. We saw in Chapter 1 (page 49) that s has $n - 1$ degrees of freedom. There are other t statistics with different degrees of freedom, some of which we will meet later in this chapter. We will write the t distribution with k degrees of freedom as $t(k)$ for short.

Figure 6.1 compares the density curves of the standard normal distribution and the t distributions with 2 and 9 degrees of freedom. The figure illustrates these facts about the t distributions:

- The density curves of the t distributions are similar in shape to the standard normal curve. They are symmetric about zero and are bell-shaped.

- The spread of the t distributions is a bit greater than that of the standard normal distribution. The t distributions in Figure 6.1 have more probability in the tails and less in the center than does the standard normal. This is true because substituting the estimate s for the fixed parameter σ introduces more variation into the statistic.

- As the degrees of freedom k increase, the $t(k)$ density curve approaches the $N(0, 1)$ curve ever more closely. This happens because s estimates σ more accurately as the sample size increases. So using s in place of σ causes little extra variation when the sample is large.

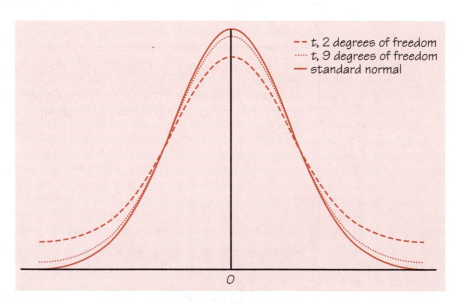

FIGURE 6.1   Density curves for the t distributions with 2 and 9 degrees of freedom and the standard normal distribution. All are symmetric with center 0. The t distributions have more probability in the tails than does the standard normal.

Table C in the back of the book gives critical values for the t distributions. Table C also appears inside the rear cover. Each row in the table contains critical values for one of the t distributions; the degrees of freedom appear at the left of the row. For convenience, we label the table entries both by p, the upper tail probability needed for significance tests, and by the confidence level C (in percent) required for confidence intervals. You have already used the standard normal critical values in the bottom row of Table C. By looking down any column, you can check that the t critical values approach the normal values as the degrees of freedom increase. As in the case of the normal table, computer software often makes Table C unnecessary.

EXERCISES

6.1   The scores of four roommates on the Law School Aptitude Test have mean $\bar{x} = 589$ and standard deviation $s = 37$. What is the standard error of the mean?

6.2   What critical value t^* from Table C satisfies each of the following conditions:

(a) The t distribution with 5 degrees of freedom has probability 0.05 to the right of t^*.

(b) The t distribution with 21 degrees of freedom has probability 0.99 to the left of t^*.

6.3   What critical value t^* from Table C satisfies each of the following conditions:

(a) The one-sample t statistic from a sample of 15 observations has probability 0.025 to the right of t^*.

(b) The one-sample t statistic from an SRS of 20 observations has probability 0.75 to the left of t^*.

The t confidence intervals and tests

To analyze samples from normal populations with unknown σ, just replace the standard deviation $\sigma/\sqrt{n}$ of $\bar{x}$ by its standard error $s/\sqrt{n}$ in the z procedures of Chapter 5. The z procedures then become *one-sample t procedures*. Use P-values or critical values from the t distribution with $n - 1$ degrees of freedom in place of the normal values. The one-sample t procedures are similar in both reasoning and computational detail to the z

procedures of Chapter 5. So we will now pay more attention to questions about using these methods in practice.

THE ONE-SAMPLE t PROCEDURES

Draw an SRS of size n from a population having unknown mean μ. A level C confidence interval for μ is

$$\bar{x} \pm t^* \frac{s}{\sqrt{n}}$$

where t^* is the upper $(1 - C)/2$ critical value for the $t(n - 1)$ distribution. This interval is exact when the population distribution is normal and is approximately correct for large n in other cases.

To test the hypothesis $H_0 : \mu = \mu_0$ based on an SRS of size n, compute the one-sample t statistic

$$t = \frac{\bar{x} - \mu_0}{s/\sqrt{n}}$$

In terms of a variable T having the $t(n - 1)$ distribution, the P-value for a test of H_0 against

$H_a : \mu > \mu_0$ is $P(T \geq t)$

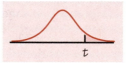

$H_a : \mu < \mu_0$ is $P(T \leq t)$

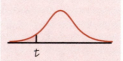

$H_a : \mu \neq \mu_0$ is $2P(T \geq |t|)$

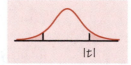

These P-values are exact if the population distribution is normal and are approximately correct for large n in other cases.

EXAMPLE 6.1

To study the metabolism of insects, researchers fed cockroaches measured amounts of a sugar solution. After 2, 5, and 10 hours, they dissected some of the cockroaches and measured the amount of sugar in various tissues.[1] Five roaches fed the sugar D-glucose and dissected after 10 hours had the following amounts (in micrograms) of D-glucose in their hindguts:

$$55.95 \quad 68.24 \quad 52.73 \quad 21.50 \quad 23.78$$

The researchers gave a 95% confidence interval for the mean amount of D-glucose in cockroach hindguts under these conditions.

First calculate that

$$\bar{x} = 44.44 \quad \text{and} \quad s = 20.741$$

The degrees of freedom are $n - 1 = 4$. From Table C we find that for 95% confidence $t^* = 2.776$. The confidence interval is

$$\bar{x} \pm t^* \frac{s}{\sqrt{n}} = 44.44 \pm 2.776 \frac{20.741}{\sqrt{5}}$$

$$= 44.44 \pm 25.75$$

$$= (18.69, \ 70.19)$$

Comparing this estimate with those for other body tissues and different times before dissection led to new insight into cockroach metabolism and to new ways of eliminating roaches from homes and restaurants. The large margin of error is due to the small sample size and the rather large variation among the cockroaches, reflected in the large value of s. ◀

The one-sample t confidence interval has the form

$$\text{estimate} \pm t^* \, \text{SE}_{\text{estimate}}$$

where "SE" stands for "standard error." We will meet a number of confidence intervals that have this common form. Like the confidence interval, t tests are close in form to the z tests we met earlier. Here is an example. In Chapter 5 we used the z test on these data. That required the unrealistic assumption that we knew the population standard deviation σ. Now we can do a realistic analysis.

EXAMPLE 6.2

Cola makers test new recipes for loss of sweetness during storage. Trained tasters rate the sweetness before and after storage. Here are the sweetness losses (sweetness before storage minus sweetness after storage) found by 10 tasters for one new cola recipe.

$$2.0 \quad 0.4 \quad 0.7 \quad 2.0 \quad -0.4 \quad 2.2 \quad -1.3 \quad 1.2 \quad 1.1 \quad 2.3$$

Are these data good evidence that the cola lost sweetness?

Step 1: Hypotheses.   Tasters vary in their perception of sweetness loss. So we ask the question in terms of the mean loss μ for a large population of tasters. The null hypothesis is "no loss," and the alternative hypothesis says "there is a loss."

$$H_0 : \mu = 0$$

$$H_a : \mu > 0$$

Step 2: Test statistic.   The basic statistics are

$$\bar{x} = 1.02 \quad \text{and} \quad s = 1.196$$

The one-sample t test statistic is

$$t = \frac{\bar{x} - \mu_0}{s/\sqrt{n}} = \frac{1.02 - 0}{1.196/\sqrt{10}}$$

$$= 2.70$$

Step 3: P-value.   The P-value for $t = 2.70$ is the area to the right of 2.70 under the t distribution curve with degrees of freedom $n - 1 = 9$. Figure 6.2 shows this area. We can't find the exact value of P without software. But we can pin P between two values by using Table C. Search the df = 9 row of Table C for entries that bracket $t = 2.70$. Because the observed t lies between the critical values for 0.02 and 0.01, the P-value lies between 0.01 and 0.02. Computer software gives the more exact result $P = .012$. There is quite strong evidence for a loss of sweetness. ◄

df = 9

p	.02	.01
t^*	2.398	2.821

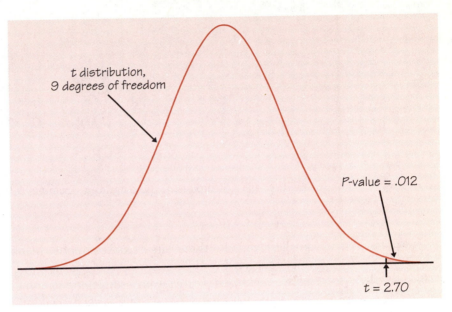

t distribution,
9 degrees of freedom

P-value = .012

t = 2.70

FIGURE 6.2   The P-value for the one-sided t test in Example 6.2.

The t confidence interval in Example 6.1 and the t test in Example 6.2 rest on assumptions that are reasonable but are not all easy to check. Both examples are experiments. In both cases the researchers took pains to avoid bias. The cockroaches were assigned at random to different sugars and different times before dissection and treated identically in every other way. The taste testers worked in isolation booths to avoid influence by other tasters. So we trust the results for these particular cockroaches and tasters.

The statistical analysis rests on two assumptions: random sampling and normal population distributions. We must be willing to treat the cockroaches and the tasters as SRSs from larger populations if we want to draw conclusions about cockroaches in general or tasters in general. The roaches were chosen at random from a population grown in the laboratory for research purposes. The tasters all have the same training. Even though we don't actually have SRSs from the populations we are interested in, we are willing to act as if we did. This is a matter of judgment.

The assumption that the population distribution is normal cannot be effectively checked with only 5 or 10 observations. In part, the researchers rely on experience with similar variables. They also look at the data. Stemplots of both distributions appear in Figure 6.3 (we rounded the cockroach data). The distribution of the 10 taste scores doesn't have a regular shape, but there are no gaps or outliers or other signs of nonnormal behavior. The cockroach data, on the other hand, have a wide gap between the two

(a)		(b)	
2	2 4	−1	3
3		−0	4
4		0	4 7
5	3 6	1	1 2
6	8	2	0 0 2 3
7		3	
8			

FIGURE 6.3   Stemplots of the data in (a) Example 6.1 and (b) Example 6.2.

smallest and the three largest observations. In observational data, this might suggest two different species of roaches. In this case we know that all five cockroaches came from a single population grown in the laboratory. The gap is just chance variation in a very small sample.

Because the t procedures are so common, all statistical software systems will do the calculations for you. Figure 6.4 shows the output from three statistical packages: Data Desk, Minitab, and S-PLUS. In each case, we entered the 10 sweetness losses as values of a variable called "cola" and asked for the one-sample t test of $H_0 : \mu = 0$ against $H_a : \mu > 0$. The three outputs report slightly different information, but all include the basic facts: $\bar{x} = 1.02$, $t = 2.70$, $P = .012$. These are the results we found in Example 6.2.

EXERCISES

6.4   What critical value t^* from Table C would you use for a confidence interval for the mean of the population in each of the following situations?

(a) A 95% confidence interval based on $n = 10$ observations.

(b) A 99% confidence interval from an SRS of 20 observations.

(c) An 80% confidence interval from a sample of size 7.

6.5   The one-sample t statistic for testing

$$H_0 : \mu = 0$$

$$H_a : \mu > 0$$

from a sample of $n = 15$ observations has the value $t = 1.82$.

Data Desk

```
cola :
Test Ho: mu(cola) = 0 vs Ha: mu(cola) > 0
Sample Mean = 1.02000 t-Statistic = 2.697 w/9 df
Reject Ho at Alpha = 0.0500
p = 0.0123
```

Minitab

```
TEST OF MU = 0.000 VS MU G.T. 0.000
```

	N	MEAN	STDEV	SE MEAN	T	P VALUE
cola	10	1.020	1.196	0.378	2.70	0.012

S-PLUS

```
data:  cola
t = 2.6967, df = 9, p-value = 0.0123
alternative hypothesis: true mean is greater than 0
sample estimates:
 mean of x
     1.02
```

FIGURE 6.4   Output for the one-sample t test of Example 6.2 from three statistical software packages. You can easily locate the basic results in output from any statistical software.

(a) What are the degrees of freedom for this statistic?

(b) Give the two critical values t^* from Table C that bracket t. What are the right-tail probabilities p for these two entries?

(c) Between what two values does the P-value of the test fall?

(d) Is the value $t = 1.82$ significant at the 5% level? Is it significant at the 1% level?

6.6   The one-sample t statistic from a sample of $n = 25$ observations for the two-sided test of

$$H_0 : \mu = 64$$

$$H_a : \mu \neq 64$$

has the value $t = 1.12$.

(a) What are the degrees of freedom for t?

(b) Locate the two critical values t^* from Table C that bracket t. What are the right-tail probabilities p for these two values?

(c) Between what two values does the P-value of the test fall? (Note that H_a is two-sided.)

(d) Is the value $t = 1.12$ statistically significant at the 10% level? At the 5% level?

6.7   Poisoning by the pesticide DDT causes tremors and convulsions. In a study of DDT poisoning, researchers fed several rats a measured amount of DDT. They then made measurements on the rats' nervous systems that might explain how DDT poisoning causes tremors. One important variable was the "absolutely refractory period," the time required for a nerve to recover after a stimulus. This period varies normally. Measurements on four rats gave the data below (in milliseconds).[2]

$$1.6 \quad 1.7 \quad 1.8 \quad 1.9$$

(a) Find the mean refractory period $\bar{x}$ and the standard error of the mean.

(b) Give a 90% confidence interval for the mean absolutely refractory period for all rats of this strain when subjected to the same treatment.

6.8   The level of various substances in the blood of kidney dialysis patients is of concern because kidney failure and dialysis can lead to nutritional problems. A researcher did blood tests on several dialysis patients on six consecutive clinic visits. One variable she measured was the level of phosphate in the blood. An individual's phosphate levels tend to vary normally over time. The data on one patient, in milligrams of phosphate per deciliter of blood, are[3]

$$5.6 \quad 5.1 \quad 4.6 \quad 4.8 \quad 5.7 \quad 6.4$$

(a) Calculate the sample mean $\bar{x}$ and also its standard error.

(b) Use the t procedures to give a 90% confidence interval for this patient's mean phosphate level.

6.9   Suppose that the mean absolutely refractory period for unpoisoned rats is known to be 1.3 milliseconds. DDT poisoning should slow nerve recovery and so increase this period. Do the data in Exercise 6.7 give good evidence for this supposition? State H_0 and H_a and do a t test. Between what levels from Table C does the P-value lie? What do you conclude from the test?

Matched pairs *t* procedures

The cockroach study in Example 6.1 estimated the mean amount of sugar in the hindgut, but the researchers then compared results for several body tissues and several times before dissection to get a bigger picture. The taste test in Example 6.2 was a matched pairs study in which the same 10 tasters rated before-and-after sweetness. Comparative studies are more convincing than single-sample investigations. For that reason, one-sample inference is less common than comparative inference. One common design to com-

matched pairs design

pare two treatments makes use of one-sample procedures. In a **matched pairs design**, subjects are matched in pairs and each treatment is given to one subject in each pair. The experimenter can toss a coin to assign two treatments to the two subjects in each pair. Another situation calling for matched pairs is before-and-after observations on the same subjects, as in the taste test of Example 6.2.

MATCHED PAIRS *t* PROCEDURES

To compare the responses to the two treatments in a matched pairs design, apply the one-sample *t* procedures to the observed differences.

The parameter μ in a matched pairs *t* procedure is the mean difference in the responses to the two treatments within matched pairs of subjects in the entire population.

EXAMPLE 6.3

The National Endowment for the Humanities sponsors summer institutes to improve the skills of high school language teachers. One institute hosted 20 French teachers for four weeks. At the beginning of the period, the teachers took the Modern Language Association's listening test of understanding of spoken French. After four weeks of immersion in French in and out of class, they took the listening test again. (The actual spoken French in the two tests was different, so that simply taking the first test should not improve the score on the second test.) Table 6.1 gives the pretest and posttest scores. The maximum possible score on the test is 36.[4]

To analyze these data, subtract the pretest score from the posttest score to obtain the improvement for each teacher. These 20 differences form a single sample.

TABLE 6.1 MLA listening scores for 20 French teachers							
Teacher	Pretest	Posttest	Gain	Teacher	Pretest	Posttest	Gain
1	32	34	2	11	30	36	6
2	31	31	0	12	20	26	6
3	29	35	6	13	24	27	3
4	10	16	6	14	24	24	0
5	30	33	3	15	31	32	1
6	33	36	3	16	30	31	1
7	22	24	2	17	15	15	0
8	25	28	3	18	32	34	2
9	32	26	−6	19	23	26	3
10	20	26	6	20	23	26	3

They appear in the "Gain" column in Table 6.1. The first teacher, for example, improved from 32 to 34, so the gain is $34 - 32 = 2$.

Step 1: Hypotheses. To assess whether the institute significantly improved the teachers' comprehension of spoken French, we test

$$H_0 : \mu = 0$$

$$H_a : \mu > 0$$

Here μ is the mean improvement that would be achieved if the entire population of French teachers attended a summer institute. The null hypothesis says that no improvement occurs, and H_a says that posttest scores are higher on the average.

Step 2: Test statistic. The 20 differences have

$$\bar{x} = 2.5 \quad \text{and} \quad s = 2.893$$

The one-sample t statistic is therefore

$$t = \frac{\bar{x} - 0}{s/\sqrt{n}} = \frac{2.5 - 0}{2.893/\sqrt{20}} = 3.86$$

df = 19

p	.001	.0005
t^*	3.579	3.883

Step 3: P-value. Find the P-value from the $t(19)$ distribution. (Remember that the degrees of freedom are 1 less than the sample size.) Table C shows that 3.86 lies between the upper 0.001 and 0.0005 critical values of the $t(19)$ distribution. The P-value therefore lies between these values. A computer statistical package gives the value $P = .00053$. The improvement in listening scores is

very unlikely to be due to chance alone. We have strong evidence that the institute was effective in raising scores. In scholarly publications, the details of routine statistical procedures are usually omitted. This test would be reported in the form "The improvement in scores was significant ($t = 3.86$, df = 19, $P = .00053$)."

A 90% confidence interval for the mean improvement in the entire population requires the critical value $t^* = 1.729$ from Table C. The confidence interval is

$$\bar{x} \pm t^* \frac{s}{\sqrt{n}} = 2.5 \pm 1.729 \frac{2.893}{\sqrt{20}} = 2.5 \pm 1.12$$

$$= (1.38, \ 3.62)$$

The estimated average improvement is 2.5 points, with margin of error 1.12 for 90% confidence. Though statistically significant, the effect of attending the institute was rather small. ◀

Example 6.3 illustrates how to restate matched pairs data as single-sample data by taking differences within each pair. We are in fact making inferences about a single population, the population of all differences within matched pairs. It is incorrect to ignore the pairs and analyze the data as if we had two samples, one from teachers who attended an institute and a second from teachers who did not. Inference procedures for comparing two samples assume that the samples are selected independently of each other. This assumption does not hold when the same subjects are measured twice. The proper analysis depends on the design used to produce the data.

What about the assumptions of simple random sampling and normality? The use of the t procedures in Example 6.3 is a bit questionable. First, the teachers are not an SRS from the population of high school French teachers. There is some selection bias in favor of energetic, committed teachers who are willing to give up four weeks of their summer vacation. It is therefore not clear to exactly what population the results apply. This vagueness is common when we don't actually take an SRS from a population.

Second, a look at the data shows that several of the teachers had pretest scores close to the maximum of 36. They could not improve their scores very much even if their mastery of French increased substantially. This is a weakness in the listening test that is the measuring instrument in this study. The differences in scores may not adequately indicate the effectiveness of the institute. This is one reason why the average increase was small.

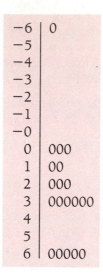

```
−6 | 0
−5 |
−4 |
−3 |
−2 |
−1 |
−0 |
 0 | 000
 1 | 00
 2 | 000
 3 | 000000
 4 |
 5 |
 6 | 00000
```

FIGURE 6.5   Stemplot of the gains in French comprehension for the 20 teachers in Example 6.3. All the leaves are 0 because all scores are whole numbers that are used as stems.

A final difficulty facing the *t* procedures in Example 6.3 is that the data show departures from normality. In a matched pairs analysis, the population of *differences* must have a normal distribution because the *t* procedures are applied to the differences. Figure 6.5 is a stemplot of the 20 differences. One teacher actually lost 6 points between the pretest and the posttest. This one subject lowered the sample mean from 2.95 for the other 19 subjects to 2.5 for all 20. The stemplot shows this outlier, and a gap between 3 and 6 that may be due to chance. The fact that all the leaves in the stemplot are 0 reminds us that only whole-number scores are possible. The distribution is not normal even if we remove the outlier. Does this nonnormality forbid use of the *t* test? The behavior of the *t* procedures when the population does not have a normal distribution is one of their most important properties. We will turn next to this issue.

EXERCISES

Many exercises from this point on ask you to give the *P*-value of a *t* test. If you have a suitable calculator or computer software, give the exact *P*-value. Otherwise, use Table C to give two values between which *P* lies.

6.10   An agricultural field trial compares the yield of two varieties of tomatoes for commercial use. The researchers divide in half each of 10 small plots of land in different locations and plant each tomato variety on one half of each plot. After harvest, they compare the yields in pounds per plant at each location. The 10 differences (Variety A − Variety B) give $\bar{x} = .34$ and $s = .83$. Is there convincing evidence that Variety A has the higher mean yield?

(a) Describe in words what the parameter μ is in this setting.

(b) State H_0 and H_a.

(c) Find the t statistic and give a P-value. What do you conclude?

6.11 The design of controls and instruments affects how easily people can use them. A student project investigated this effect by asking 25 right-handed students to turn a knob (with their right hands) that moved an indicator by screw action. There were two identical instruments, one with a right-hand thread (the knob turns clockwise) and the other with a left-hand thread (the knob must be turned counterclockwise). The table below gives the times in seconds each subject took to move the indicator a fixed distance.[5]

Subject	Right thread	Left thread	Subject	Right thread	Left thread
1	113	137	14	107	87
2	105	105	15	118	166
3	130	133	16	103	146
4	101	108	17	111	123
5	138	115	18	104	135
6	118	170	19	111	112
7	87	103	20	89	93
8	116	145	21	78	76
9	75	78	22	100	116
10	96	107	23	89	78
11	122	84	24	85	101
12	103	148	25	88	123
13	116	147			

(a) Each of the 25 students used both instruments. Discuss briefly how you would use randomization in arranging the experiment.

(b) The project hoped to show that right-handed people find right-hand threads easier to use. What is the parameter μ for a matched pairs t test? State H_0 and H_a in terms of μ.

(c) Carry out a test of your hypotheses. Give the P-value and report your conclusions.

6.12   Give a 90% confidence interval for the mean time advantage of right-hand over left-hand threads in the setting of Exercise 6.11. Do you think that the time saved would be of practical importance if the task were performed many times—for example, by an assembly line worker? To help answer this question, find the mean time for right-hand threads as a percent of the mean time for left-hand threads.

Robustness of t procedures

The one-sample t procedures are exactly correct only when the population is normal. Real populations are never exactly normal. The usefulness of the t procedures in practice therefore depends on how strongly they are affected by lack of normality.

ROBUST PROCEDURES

A confidence interval or significance test is called **robust** if the confidence level or P-value does not change very much when the assumptions of the procedure are violated.

Because the tails of normal curves drop off quickly, samples from normal distributions will have very few outliers. Outliers suggest that your data are not a sample from a normal population. Like $\bar{x}$ and s, *the t procedures are strongly influenced by outliers*. If we dropped the single outlier in Example 6.3, the test statistic would change from $t = 3.86$ to $t = 5.98$ and the P-value would be much smaller. In this case, the outlier makes the test result *less* significant and the margin of error of the confidence interval *larger* than they would otherwise be. The results of the t procedures in Example 6.3 are conservative in the sense that the conclusions show a smaller effect than would be the case if the outlier were not present.

Fortunately, the t procedures are quite robust against nonnormality of the population when there are no outliers, especially when the distribution is roughly symmetric. Larger samples improve the accuracy of P-values and critical values from the t distributions when the population is not normal. The main reason for this is the central limit theorem. The t statistic uses the sample mean $\bar{x}$, which becomes more nearly normal as the sample size gets larger even when the population does not have a normal distribution.

Always make a plot to check for skewness and outliers before you use the t procedures for small samples. For most purposes, you can safely use the one-sample t procedures when $n \geq 15$ unless an outlier or quite strong skewness is present. If we can justify removing the outlier in Example 6.3 (perhaps that teacher was ill when she took the posttest), we can use the t procedures on the 19 remaining observations. Here are practical guidelines for inference on a single mean.[6]

USING THE t PROCEDURES

- Except in the case of small samples, the assumption that the data are an SRS from the population of interest is more important than the assumption that the population distribution is normal.

- *Sample size less than 15.* Use t procedures if the data are close to normal. If the data are clearly nonnormal or if outliers are present, do not use t.

- *Sample size at least 15.* The t procedures can be used except in the presence of outliers or strong skewness.

- *Large samples.* The t procedures can be used even for clearly skewed distributions when the sample is large, roughly $n \geq 40$.

EXAMPLE 6.4

Consider several of the data sets we graphed in Chapter 1. Figure 6.6 shows the histograms.

- Figure 6.6(a) is a histogram of the percent of each state's residents who are over 65 years of age. *We have data on the entire population of 50 states, so formal inference makes no sense.* We can calculate the exact mean for the population. There is no uncertainty due to having only a sample from the population, and no need for a confidence interval or test.

- Figure 6.6(b) shows the time of the first lightning strike each day in a mountain region in Colorado. The data contain more than 70

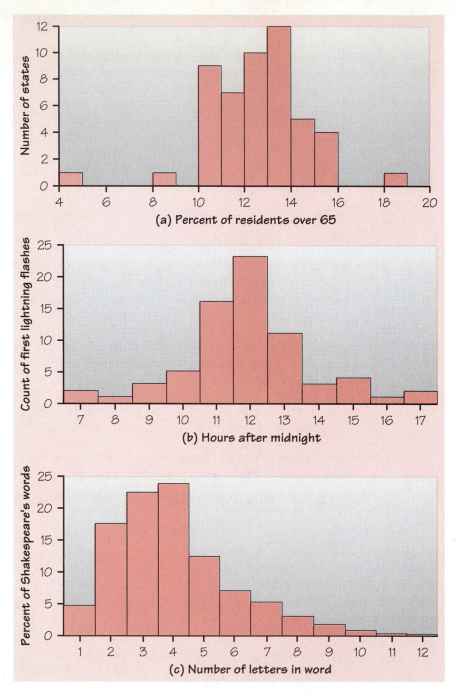

FIGURE 6.6 Can we use *t* procedures for these data? **(a)** Percent of residents over 65 years of age in the states. *No*: this is an entire population, not a sample. **(b)** Times of first lightning strikes each day at a site in Colorado. *Yes*: there are over 70 observations with a symmetric distribution. **(c)** Word lengths in Shakespeare's plays. *Yes, if the sample is large enough* to overcome the right skewness.

observations that have a symmetric distribution. You can use the t procedures to draw conclusions about the mean time of a day's first lightning strike with complete confidence.

- Figure 6.6(c) shows that the distribution of word lengths in Shakespeare's plays is skewed to the right. We aren't told how large the sample is. You can use the t procedures for a distribution like this if the sample size is roughly 40 or larger. ◀

EXERCISES

6.13 The Acculturation Rating Scale for Mexican Americans (ARSMA) measures the extent to which Mexican Americans have adopted Anglo/English culture. During the development of ARSMA, the test was given to a group of 17 Mexicans. Their scores, from a possible range of 1.00 to 5.00, had a symmetric distribution with $\bar{x} = 1.67$ and $s = .25$. Because low scores should indicate a Mexican cultural orientation, these results helped to establish the validity of the test.[7]

  (a) Give a 95% confidence interval for the mean ARSMA score of Mexicans.

  (b) What assumptions does your confidence interval require? Which of these assumptions is most important in this case?

6.14 A bank wonders whether omitting the annual credit card fee for customers who charge at least $2400 in a year would increase the amount charged on its credit cards. The bank makes this offer to an SRS of 200 of its credit card customers. It then compares how much these customers charge this year with the amount that they charged last year. The mean increase is $332, and the standard deviation is $108.

  (a) Is there significant evidence at the 1% level that the mean amount charged increases under the no-fee offer? State H_0 and H_a and carry out a t test.

  (b) Give a 99% confidence interval for the mean amount of the increase.

  (c) The distribution of the amount charged is skewed to the right, but outliers are prevented by the credit limit that the bank enforces on each card. Use of the t procedures is justified in this case even though the population distribution is not normal. Explain why.

  (d) A critic points out that the customers would probably have charged more this year than last even without the new offer, because the economy is more prosperous and interest rates are lower. Briefly describe the design of an experiment to study the effect of the no-fee offer that would avoid this criticism.

6.15   Here are measurements (in millimeters) of a critical dimension for 16 auto engine crankshafts.

224.120   224.001   224.017   223.982   223.989   223.961
223.960   224.089   223.987   223.976   223.902   223.980
224.098   224.057   223.913   223.999

The dimension is supposed to be 224 mm and the variability of the manufacturing process is unknown. Is there evidence that the mean dimension is not 224 mm?

(a) Check the data graphically for outliers or strong skewness that might threaten the validity of the t procedures. What do you conclude?

(b) State H_0 and H_a and carry out a t test. Give the P-value (from Table C or software). What do you conclude?

6.16   Many homeowners buy detectors to check for the invisible gas radon in their homes. How accurate are these detectors? To answer this question, university researchers placed 12 radon detectors in a chamber that exposed them to 105 picocuries per liter of radon. The detector readings were as follows.[8]

91.9   97.8   111.4   122.3   105.4   95.0
103.8   99.6   96.6   119.3   104.8   101.7

(a) Make a stemplot of the data. The distribution is somewhat skewed to the right, but not strongly enough to forbid use of the t procedures.

(b) Is there convincing evidence that the mean reading of all detectors of this type differs from the true value 105? Carry out a test in detail, then write a brief conclusion.

The power of the t test*

The power of a statistical test measures its ability to detect deviations from the null hypothesis. In practice we carry out the test in the hope of showing that the null hypothesis is false, so high power is important. The power of

---

*This more advanced section is not needed to read the rest of the book. It requires material in the optional Section 5.4.

the one-sample t test against a specific alternative value of the population mean μ is the probability that the test will reject the null hypothesis when the alternative value of the mean is true. To calculate the power, we assume a fixed level of significance, usually $\alpha = .05$.

Calculation of the exact power of the t test takes into account the estimation of σ by s and is a bit complex. But an approximate calculation that acts as if σ were known is usually adequate for planning a study. This calculation is very much like that for the power of the z test, presented on pages 394–395.

EXAMPLE 6.5

It is the winter before the summer language institute of Example 6.3. The director, thinking ahead to the report he must write, hopes that enrolling 20 teachers will enable him to be quite certain of detecting an average improvement of 2 points in the mean listening score. Is this realistic?

We wish to compute the power of the t test for

$$H_0 : \mu = 0$$

$$H_a : \mu > 0$$

against the alternative $\mu = 2$ when $n = 20$. We must have a rough guess of the size of σ in order to compute the power. People planning a large study often run a small pilot study for this and other purposes. In this case, listening-score improvements in past summer language institutes have had sample standard deviations of about 3. We therefore take both $\sigma = 3$ and $s = 3$ in our approximate calculation.

Step 1 *Write the rule for rejecting H_0 in terms of $\bar{x}$.* The t test with 20 observations rejects H_0 at the 5% significance level if the t statistic

$$t = \frac{\bar{x} - 0}{s/\sqrt{20}}$$

exceeds the upper 5% point of $t(19)$, which is 1.729. Taking $s = 3$, the test rejects H_0 when

$$t = \frac{\bar{x}}{3/\sqrt{20}} \geq 1.729$$

$$\bar{x} \geq 1.729\frac{3}{\sqrt{20}}$$

$$\bar{x} \geq 1.160$$

Step 2 *The power is the probability of rejecting H_0 assuming that the alternative is true.* We want the probability that $\bar{x} \geq 1.160$ when $\mu = 2$. Taking $\sigma = 3$, standardize $\bar{x}$ to find this probability:

$$P(\bar{x} \geq 1.160) = P\left(\frac{\bar{x} - 2}{3/\sqrt{20}} \geq \frac{1.160 - 2}{3/\sqrt{20}}\right)$$

$$= P(Z \geq -1.252)$$

$$= 1 - .1056 = .8944$$

A true difference of 2 points in the population mean scores will produce significance at the 5% level in 89% of all possible samples. The director can be reasonably confident of detecting a difference this large.   ◀

EXERCISES

6.17 The bank in Exercise 6.14 tested a new idea on a sample of 200 customers. The bank wants to be quite certain of detecting a mean increase of $\mu = \$100$ in the amount charged, at the $\alpha = .01$ significance level. Perhaps a sample of only $n = 50$ customers would accomplish this. Find the approximate power of the test with $n = 50$ against the alternative $\mu = \$100$ as follows:

(a) What is the critical value t^* for the one-sided test with $\alpha = .01$ and $n = 50$?

(b) Write the rule for rejecting $H_0 : \mu = 0$ in terms of the t statistic. Then take $s = 108$ (an estimate based on the data in Exercise 6.14) and state the rejection rule in terms of $\bar{x}$.

(c) Assume that $\mu = 100$ (the given alternative) and that $\sigma = 108$ (an estimate from the data in Exercise 6.14). The approximate power is the probability of the event you found in (b), calculated under these assumptions. Find the power. Would you recommend that the bank do a test on 50 customers, or should more customers be included?

6.18 The tomato experts who carried out the field trial described in Exercise 6.10 suspect that the large P-value there is due to low power. They would like to be able to detect a mean difference in yields of 0.5 pound per plant at the 0.05 significance level. Based on the previous study, use 0.83 as an estimate of both the population σ and the value of s in future samples.

(a) What is the power of the test from Exercise 6.10 with $n = 10$ against the alternative $\mu = .5$?

(b) If the sample size is increased to $n = 25$ plots of land, what will be the power against the same alternative?

SUMMARY

Tests and confidence intervals for the mean μ of a normal population are based on the sample mean $\bar{x}$ of an SRS. Because of the central limit theorem, the resulting procedures are approximately correct for other population distributions when the sample is large.

The standardized sample mean is the **one-sample z statistic,**

$$z = \frac{\bar{x} - \mu}{\sigma/\sqrt{n}}$$

When we know σ, we use the z statistic and the standard normal distribution.

In practice, we do not know σ. Replace the standard deviation $\sigma/\sqrt{n}$ of $\bar{x}$ by the **standard error** $s/\sqrt{n}$ to get the **one-sample t statistic**

$$t = \frac{\bar{x} - \mu}{s/\sqrt{n}}$$

The t statistic has the **t distribution** with $n - 1$ degrees of freedom.

There is a t distribution for every positive **degrees of freedom** k. All are symmetric distributions similar in shape to the standard normal distribution. The $t(k)$ distribution approaches the $N(0, 1)$ distribution as k increases.

An exact level C **confidence interval** for the mean μ of a normal population is

$$\bar{x} \pm t^* \frac{s}{\sqrt{n}}$$

where t^* is the upper $(1 - C)/2$ critical value of the $t(n - 1)$ distribution.

Significance tests for $H_0 : \mu = \mu_0$ are based on the t statistic. Use P-values or fixed significance levels from the $t(n - 1)$ distribution.

Use these one-sample procedures to analyze **matched pairs** data by first taking the difference within each matched pair to produce a single sample.

The t procedures are relatively **robust** when the population is nonnormal, especially for larger sample sizes. The t procedures are useful for nonnormal data when $n \geq 15$ unless the data show outliers or strong skewness.

SECTION 6.1 EXERCISES

When an exercise asks for a P-value, give P exactly if you have a suitable calculator or computer software. Otherwise, use Table C to give two values between which P lies.

6.19   The one-sample t statistic for a test of

$$H_0 : \mu = 10$$

$$H_a : \mu < 10$$

based on $n = 10$ observations has the value $t = -2.25$.

(a) What are the degrees of freedom for this statistic?

(b) Between what two probabilities p from Table C does the P-value of the test fall?

6.20   A manufacturer of small appliances employs a market research firm to estimate retail sales of its products by gathering information from a sample of retail stores. This month an SRS of 75 stores in the Midwest sales region finds that these stores sold an average of 24 of the manufacturer's hand mixers, with standard deviation 11.

(a) Give a 95% confidence interval for the mean number of mixers sold by all stores in the region.

(b) The distribution of sales is strongly right skewed, because there are many smaller stores and a few very large stores. The use of t in (a) is reasonably safe despite this violation of the normality assumption. Why?

6.21   In a randomized comparative experiment on the effect of calcium in the diet on blood pressure, researchers divided 54 healthy white males at random into two groups. One group received calcium; the other, a placebo.

At the beginning of the study, the researchers measured many variables on the subjects. The paper reporting the study gives $\bar{x} = 114.9$ and $s = 9.3$ for the seated systolic blood pressure of the 27 members of the placebo group.

(a) Give a 95% confidence interval for the mean blood pressure in the population from which the subjects were recruited.

(b) What assumptions about the population and the study design are required by the procedure you used in (a)? Which of these assumptions are important for the validity of the procedure in this case?

6.22 Gas chromatography is a sensitive technique used to measure small amounts of compounds. The response of a gas chromatograph is calibrated by repeatedly testing specimens containing a known amount of the compound to be measured. A calibration study for a specimen containing 1 nanogram (that's 10^{-9} gram) of a compound gave the following response readings.[9]

$$21.6 \quad 20.0 \quad 25.0 \quad 21.9$$

The response is known from experience to vary according to a normal distribution unless an outlier indicates an error in the analysis. Estimate the mean response to 1 nanogram of this substance, and give the margin of error for your choice of confidence level. Then explain to a chemist who knows no statistics what your margin of error means.

6.23 The embryos of brine shrimp can enter a dormant phase in which metabolic activity drops to a low level. Researchers studying this dormant phase measured the level of several compounds important to normal metabolism. They reported their results in a table, with the note, "Values are means ± SEM for three independent samples." The table entry for the compound ATP was $0.84 \pm .01$. Biologists reading the article must be able to decipher this.[10]

(a) What does the abbreviation SEM stand for?

(b) The researchers made three measurements of ATP, which had $\bar{x} = .84$. What was the sample standard deviation s for these measurements?

(c) Give a 90% confidence interval for the mean ATP level in dormant brine shrimp embryos.

6.24 The table below gives the pretest and posttest scores on the Modern Language Association's listening test in Spanish for 20 high school Spanish teachers who attended an intensive summer course in Spanish. The setting is identical to the French institute described in Example 6.3.[11]

Subject	Pretest	Posttest	Subject	Pretest	Posttest
1	30	29	11	30	32
2	28	30	12	29	28
3	31	32	13	31	34
4	26	30	14	29	32
5	20	16	15	34	32
6	30	25	16	20	27
7	34	31	17	26	28
8	15	18	18	25	29
9	28	33	19	31	32
10	20	25	20	29	32

(a) We hope to show that attending the institute improves listening skills. State an appropriate H_0 and H_a. Be sure to identify the parameter appearing in the hypotheses.

(b) Make a graphical check for outliers or strong skewness in the data that you will use in your statistical test, and report your conclusions on the validity of the test.

(c) Carry out a test. Can you reject H_0 at the 5% significance level? At the 1% significance level?

(d) Give a 90% confidence interval for the mean increase in listening score due to attending the summer institute.

6.25   The ARSMA test (Exercise 6.13) was compared with a similar test, the Bicultural Inventory (BI), by administering both tests to 22 Mexican Americans. Both tests have the same range of scores (1.00 to 5.00) and are scaled to have similar means for the groups used to develop them. There was a high correlation between the two scores, giving evidence that both are measuring the same characteristics. The researchers wanted to know whether the population mean scores for the two tests are the same. The differences in scores (ARSMA − BI) for the 22 subjects had $\bar{x} = .2519$ and $s = .2767$.

(a) Describe briefly how to arrange the administration of the two tests to the subjects, including randomization.

(b) Carry out a significance test for the hypothesis that the two tests have the same population mean. Give the P-value and state your conclusion.

(c) Give a 95% confidence interval for the difference between the two population mean scores.

6.26   A study of the pay of corporate CEOs (chief executive officers) examined the cash compensation, adjusted for inflation, of the CEOs of 104 corporations over the period 1977 to 1988. Among the data are the average annual pay increases for each of the 104 CEOs. The mean percent increase in pay was 6.9%. The data showed great variation, with a standard deviation of 17.4%. The distribution was strongly skewed to the right.[12]

(a) Despite the skewness of the distribution, there were no extreme outliers. Explain why we can use t procedures for these data.

(b) What are the degrees of freedom? When the exact degrees of freedom do not appear in Table C, use the next lower degrees of freedom in the table.

(c) Give a 99% confidence interval for the mean increase in pay for all corporate CEOs. What essential condition must the data satisfy if we are to trust your result?

6.27   Table 1.3 (page 35) gives the ages of U.S. presidents when they took office. It does not make sense to use the t procedures (or any other statistical procedures) to give a 95% confidence interval for the mean age of the presidents. Explain why not.

6.28   **(Optional)** Exercise 6.25 reports a small study comparing ARSMA and BI, two tests of the acculturation of Mexican Americans. Would this study usually detect a difference in mean scores of 0.2? To answer this question, calculate the approximate power of the test (with $n = 22$ subjects and $\alpha = .05$) of

$$H_0 : \mu = 0$$

$$H_a : \mu \neq 0$$

against the alternative $\mu = .2$. Note that this is a two-sided test.

(a) From Table C, what is the critical value for $\alpha = .05$?

(b) Write the rule for rejecting H_0 at the $\alpha = .05$ level. Then take $s = .3$, the approximate value observed in Exercise 6.25, and restate the rejection criterion in terms of $\bar{x}$.

(c) Find the probability of this event when $\mu = .2$ (the alternative given) and $\sigma = .3$ (estimated from the data in Exercise 6.25) by a normal probability calculation. This is the approximate power.

6.2   COMPARING TWO MEANS

Comparing two populations or two treatments is one of the most common situations encountered in statistical practice. We call such situations *two-sample problems*.

> **TWO-SAMPLE PROBLEMS**
>
> - The goal of inference is to compare the responses to two treatments or to compare the characteristics of two populations.
>
> - We have a separate sample from each treatment or each population.

Two-sample problems

A two-sample problem can arise from a randomized comparative experiment that randomly divides subjects into two groups and exposes each group to a different treatment. Comparing random samples separately selected from two populations is also a two-sample problem. Unlike the matched pairs designs studied earlier, there is no matching of the units in the two samples and the two samples can be of different sizes. Inference procedures for two-sample data differ from those for matched pairs. Here are some typical two-sample problems.

EXAMPLE 6.6

(a) A medical researcher is interested in the effect on blood pressure of added calcium in our diet. She conducts a randomized comparative experiment in which one group of subjects receives a calcium supplement and a control group gets a placebo.

(b) A psychologist develops a test that measures social insight. He compares the social insight of male college students with that of female college students by giving the test to a large group of students of each sex.

(c) A bank wants to know which of two incentive plans will most increase the use of its credit cards. It offers each incentive to a random sample of credit card customers and compares the amount charged during the following six months.

◀

EXERCISES

6.29   The following situations require inference about a mean or means. Identify each as (1) single sample, (2) matched pairs, or (3) two samples. The procedures of Section 6.1 apply to cases (1) and (2). We are about to learn procedures for (3).

(a) An education researcher wants to learn whether it is more effective to put questions before or after introducing a new concept in an elementary school mathematics text. He prepares two text segments that teach the concept, one with motivating questions before and the other with review questions after. He uses each text segment to teach a separate group of children. The researcher compares the scores of the groups on a test over the material.

(b) Another researcher approaches the same issue differently. She prepares text segments on two unrelated topics. Each segment comes in two versions, one with questions before and the other with questions after. The subjects are a single group of children. Each child studies both topics, one (chosen at random) with questions before and the other with questions after. The researcher compares test scores for each child on the two topics to see which topic he or she learned better.

6.30 The following situations require inference about a mean or means. Identify each as (1) single sample, (2) matched pairs, or (3) two samples. The procedures of Section 6.1 apply to cases (1) and (2). We are about to learn procedures for (3).

(a) To check a new analytical method, a chemist obtains a reference specimen of known concentration from the National Institute of Standards and Technology. She then makes 20 measurements of the concentration of this specimen with the new method and checks for bias by comparing the mean result with the known concentration.

(b) Another chemist is checking the same new method. He has no reference specimen, but a familiar analytic method is available. He wants to know if the new and old methods agree. He takes a specimen of unknown concentration and measures the concentration 10 times with the new method and 10 times with the old method.

Comparing two population means

We can examine two-sample data graphically by comparing stemplots (for small samples) or histograms or boxplots (for larger samples). Now we will apply the ideas of formal inference in this setting. When both population distributions are symmetric, and especially when they are at least approximately normal, a comparison of the mean responses in the two populations is the most common goal of inference. Here are the assumptions we will make.

ASSUMPTIONS FOR COMPARING TWO MEANS

• We have **two SRSs**, from two distinct populations. The samples are **independent**. That is, one sample has no influence on the other. Matching violates independence, for example. We measure the same variable for both samples.

• Both populations are **normally distributed**. The means and standard deviations of the populations are unknown.

Call the variable we measure x_1 in the first population and x_2 in the second because the variable may have different distributions in the two populations. Here is the notation we will use to describe the two populations:

Population	Variable	Mean	Standard deviation
1	x_1	μ_1	σ_1
2	x_2	μ_2	σ_2

There are four unknown parameters, the two means and the two standard deviations. The subscripts remind us which population a parameter describes. We want to compare the two population means, either by giving a confidence interval for their difference $\mu_1 - \mu_2$ or by testing the hypothesis of no difference, $H_0 : \mu_1 = \mu_2$.

We use the sample means and standard deviations to estimate the unknown parameters. Again, subscripts remind us which sample a statistic comes from. Here is the notation that describes the samples:

Population	Sample size	Sample mean	Sample standard deviation
1	n_1	$\overline{x}_1$	s_1
2	n_2	$\overline{x}_2$	s_2

To do inference about the difference $\mu_1 - \mu_2$ between the means of the two populations, we start from the difference $\overline{x}_1 - \overline{x}_2$ between the means of the two samples.

EXAMPLE 6.7

Does increasing the amount of calcium in our diet reduce blood pressure? Examination of a large sample of people revealed a relationship between calcium intake and blood pressure. The relationship was strongest for black men. Such observational studies do not establish causation. Researchers therefore designed a randomized comparative experiment.

The subjects in part of the experiment were 21 healthy black men. A randomly chosen group of 10 of the men received a calcium supplement for 12 weeks. The control group of 11 men received a placebo pill that looked identical. The experiment was double-blind. The response variable is the decrease in systolic (heart contracted) blood pressure for a subject after 12 weeks, in millimeters of mercury. An increase appears as a negative response.[13]

Take Group 1 to be the calcium group and Group 2 the placebo group. Here are the data for the 10 men in Group 1 (calcium),

$$7 \quad -4 \quad 18 \quad 17 \quad -3 \quad -5 \quad 1 \quad 10 \quad 11 \quad -2$$

and for the 11 men in Group 2 (placebo),

$$-1 \quad 12 \quad -1 \quad -3 \quad 3 \quad -5 \quad 5 \quad 2 \quad -11 \quad -1 \quad -3$$

From the data, calculate the summary statistics:

Group	Treatment	n	$\overline{x}$	s
1	Calcium	10	5.000	8.743
2	Placebo	11	−.273	5.901

The calcium group shows a drop in blood pressure, $\overline{x}_1 = 5.000$, while the placebo group had almost no change, $\overline{x}_2 = -.273$. Is this outcome good evidence that calcium decreases blood pressure in the entire population of healthy black men more than a placebo does?   ◄

   Example 6.7 fits the two-sample setting. We write hypotheses in terms of the mean decreases we would see in the entire population, μ_1 for men taking calcium for 12 weeks and μ_2 for men taking a placebo. The hypotheses are

$$H_0 : \mu_1 = \mu_2$$

$$H_a : \mu_1 > \mu_2$$

We want to test these hypotheses and also estimate the size of calcium's advantage, $\mu_1 - \mu_2$.

Are the assumptions satisfied? Because of the randomization, we are willing to regard the calcium and placebo groups as two independent SRSs. Although the samples are small, we check for serious nonnormality by examining the data. Here is a back-to-back stemplot of the responses. (We have split the stems. Notice that negative responses require -0 and 0 to be separate stems, and that the ordering of leaves out from the stems recognizes that -3 is smaller than -1.)

Calcium		Placebo
	−1	1
5	−0	5
234	−0	33111
1	0	23
7	0	5
10	1	2
87	1	

The placebo responses appear roughly normal. The calcium group has an irregular distribution, which is not unusual when we have only a few observations. There are no outliers, and no departures from normality that prevent use of t procedures.

The natural estimator of the difference $\mu_1 - \mu_2$ is the difference between the sample means:

$$\bar{x}_1 - \bar{x}_2 = 5.000 - (-.273) = 5.273$$

This statistic measures the average advantage of calcium over a placebo. In order to to use it for inference, we must know its sampling distribution.

The sampling distribution of $\bar{x}_1 - \bar{x}_2$

Here are the facts about the sampling distribution of the difference $\bar{x}_1 - \bar{x}_2$ between the sample means of two independent SRSs. These facts can be derived using the mathematics of probability or made plausible by simulation.

- The mean of $\bar{x}_1 - \bar{x}_2$ is $\mu_1 - \mu_2$. That is, the difference of sample means is an unbiased estimator of the difference of population means.
- The variance of the difference is the *sum* of the variances of $\bar{x}_1$ and $\bar{x}_2$, which is

$$\frac{\sigma_1^2}{n_1} + \frac{\sigma_2^2}{n_2}$$

  Note that the *variances* add. The standard deviations do not.
- If the two population distributions are both normal, then the distribution of $\bar{x}_1 - \bar{x}_2$ is also normal.

Because the statistic $\bar{x}_1 - \bar{x}_2$ has a normal distribution, we can standardize it to obtain a standard normal z statistic. Subtract its mean, then *two-sample* divide by its standard deviation to get the **two-sample z statistic**:
z statistic

$$z = \frac{(\bar{x}_1 - \bar{x}_2) - (\mu_1 - \mu_2)}{\sqrt{\dfrac{\sigma_1^2}{n_1} + \dfrac{\sigma_2^2}{n_2}}}$$

Two-sample t procedures

We don't know the population standard deviations σ_1 and σ_2. Following the pattern of the one-sample case, substitute the standard errors $s_i/\sqrt{n_i}$ for the standard deviations $\sigma_i/\sqrt{n_i}$ in the two-sample z statistic. The result *two-sample* is the **two-sample t statistic**:
t statistic

$$t = \frac{(\bar{x}_1 - \bar{x}_2) - (\mu_1 - \mu_2)}{\sqrt{\dfrac{s_1^2}{n_1} + \dfrac{s_2^2}{n_2}}}$$

The statistic t has the same interpretation as any z or t statistic: it says how far $\bar{x}_1 - \bar{x}_2$ is from its mean in standard deviation units. Unfortunately, the two-sample t statistic does *not* have a t distribution. A t distribution replaces a $N(0, 1)$ distribution when we replace just one standard deviation in a z statistic by a standard error. In this case, we replaced two standard deviations by the corresponding standard errors. This does not produce a statistic having a t distribution.

Nonetheless, the two-sample t statistic is used with t critical values in inference for two-sample problems. There are two ways to do this.

Option 1: Use procedures based on the statistic t with critical values from a t distribution with degrees of freedom computed from the data. The degrees of freedom are generally not a whole number. This is a very accurate approximation to the distribution of t.

Option 2: Use procedures based on the statistic t with critical values from the t distribution with degrees of freedom equal to the smaller of $n_1 - 1$ and $n_2 - 1$. These procedures are always conservative for any two normal populations.

Most statistical software systems use the two-sample t statistic with Option 1 for two-sample problems unless the user requests another method. Using this option without software is a bit complicated. We will therefore present the second, simpler, option first. We recommend that you use Option 2 when doing calculations without a computer. If you use a computer package, it should automatically do the calculations for Option 1. Here is a statement of the Option 2 procedures that includes a statement of just how they are "conservative."

THE TWO-SAMPLE t PROCEDURES

Draw an SRS of size n_1 from a normal population with unknown mean μ_1, and draw an independent SRS of size n_2 from another normal population with unknown mean μ_2. The confidence interval for $\mu_1 - \mu_2$ given by

$$(\bar{x}_1 - \bar{x}_2) \pm t^* \sqrt{\frac{s_1^2}{n_1} + \frac{s_2^2}{n_2}}$$

has confidence level *at least C* no matter what the population standard deviations may be. Here t^* is the upper $(1 - C)/2$ critical value for the $t(k)$ distribution with k the smaller of $n_1 - 1$ and $n_2 - 1$.

(continued on next page)

(*continued from previous page*)

To test the hypothesis $H_0 : \mu_1 = \mu_2$, compute the two-sample t statistic

$$t = \frac{\overline{x}_1 - \overline{x}_2}{\sqrt{\dfrac{s_1^2}{n_1} + \dfrac{s_2^2}{n_2}}}$$

and use P-values or critical values for the $t(k)$ distribution. The true P-value or fixed significance level will always be *equal to or less than* the value calculated from $t(k)$ no matter what values the unknown population standard deviations have.

These two-sample t procedures always err on the safe side, reporting *higher* P-values and *lower* confidence than are actually true. The gap between what is reported and the truth is quite small unless the sample sizes are both small and unequal. As the sample sizes increase, probability values based on t with degrees of freedom equal to the smaller of $n_1 - 1$ and $n_2 - 1$ become more accurate.[14] The following examples illustrate the two-sample t procedures.

EXAMPLE 6.8

The medical researchers in Example 6.7 can use the two-sample t procedures to compare calcium with a placebo. The test statistic for the null hypothesis $H_0 : \mu_1 = \mu_2$ is

$$\begin{aligned}
t &= \frac{\overline{x}_1 - \overline{x}_2}{\sqrt{\dfrac{s_1^2}{n_1} + \dfrac{s_2^2}{n_2}}} \\[2mm]
&= \frac{5.000 - (-.273)}{\sqrt{\dfrac{8.743^2}{10} + \dfrac{5.901^2}{11}}} \\[2mm]
&= \frac{5.273}{3.2878} = 1.604
\end{aligned}$$

df = 9

p	.10	.05
t^*	1.383	1.833

There are 9 degrees of freedom, the smaller of $n_1 - 1 = 9$ and $n_2 - 1 = 10$. Because H_a is one-sided on the high side, the P-value is the area to the right of $t = 1.604$ under the $t(9)$ curve. Figure 6.7 illustrates this P-value. Table C shows that it lies between 0.05 and 0.10. The experiment found evidence that calcium reduces blood pressure, but the evidence falls a bit short of the traditional 5% and 1% levels.

For a 90% confidence interval, Table C shows that the $t(9)$ critical value is $t^* = 1.833$. We are 90% confident that the mean advantage of calcium over a placebo, $\mu_1 - \mu_2$, lies in the interval

$$(\bar{x}_1 - \bar{x}_2) \pm t^* \sqrt{\frac{s_1^2}{n_1} + \frac{s_2^2}{n_2}} = [5.000 - (-.273)] \pm 1.833\sqrt{\frac{8.743^2}{10} + \frac{5.901^2}{11}}$$

$$= 5.273 \pm 6.026$$

$$= (-.753, \ 11.299)$$

That the 90% confidence interval covers 0 tells us that we cannot reject $H_0 : \mu_1 = \mu_2$ against the two-sided alternative at the $\alpha = .10$ level of significance. ◀

Sample size strongly influences the P-value of a test. An effect that fails to be significant at a specified level α in a small sample will be significant

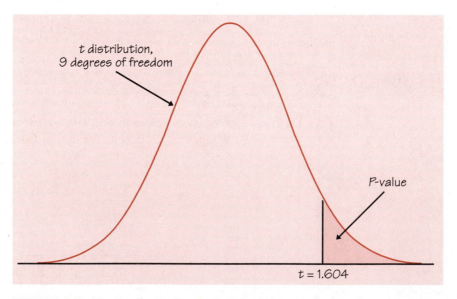

FIGURE 6.7 The P-value in Example 6.8. This example uses the conservative method, which leads to the t distribution with 9 degrees of freedom.

in a larger sample. In the light of the rather small samples in Example 6.8, we suspect that more data might show that calcium has a significant effect. The published account of the study combined these results for blacks with results for whites and adjusted for pretest differences among the subjects. Using this more detailed analysis, the researchers were able to report the P-value $P = .008$.

EXAMPLE 6.9

The Chapin Social Insight Test is a psychological test designed to measure how accurately a person appraises other people. The possible scores on the test range from 0 to 41. During the development of the Chapin test, it was given to several different groups of people. Here are the results for male and female college students majoring in the liberal arts:[15]

Group	Sex	n	$\bar{x}$	s
1	Male	133	25.34	5.05
2	Female	162	24.94	5.44

Do these data support the contention that female and male students differ in average social insight?

Step 1: Hypotheses.   Because we had no specific direction for the male/female difference in mind before looking at the data, we choose the two-sided alternative. The hypotheses are

$$H_0 : \mu_1 = \mu_2$$

$$H_a : \mu_1 \neq \mu_2$$

Step 2: Test statistic.   The two-sample t statistic is

$$t = \frac{\bar{x}_1 - \bar{x}_2}{\sqrt{\frac{s_1^2}{n_1} + \frac{s_2^2}{n_2}}}$$

$$= \frac{25.34 - 24.94}{\sqrt{\frac{5.05^2}{133} + \frac{5.44^2}{162}}}$$

$$= .654$$

Step 3: P-value.   There are 132 degrees of freedom, the smaller of

$$n_1 - 1 = 133 - 1 = 132 \text{ and}$$
$$n_2 - 1 = 162 - 1 = 161$$

Figure 6.8 illustrates the P-value. Find it by comparing 0.654 to critical values for the $t(132)$ distribution and then doubling p because the alternative is two-sided. Degrees of freedom 132 do not appear in Table C, so we use the next smaller table value, degrees of freedom 100. Table C shows that 0.654 does not reach the 0.25 critical value, which is the largest upper tail probability in Table C. The P-value is therefore greater than 0.50. The data give no evidence of a male/female difference in mean social insight score ($t = .654$, df = 132, $P > .5$).   ◄

df = 100

p	.25	.20
t^*	0.677	0.845

The researcher in Example 6.9 did not do an experiment but compared samples from two populations. The large samples imply that the assumption that the populations have normal distributions is of little importance. The sample means will be nearly normal in any case. The major question concerns the population to which the conclusions apply. The student subjects are certainly not an SRS of all liberal arts majors in the country. If they are volunteers from a single college, the sample results may not extend to a wider population.

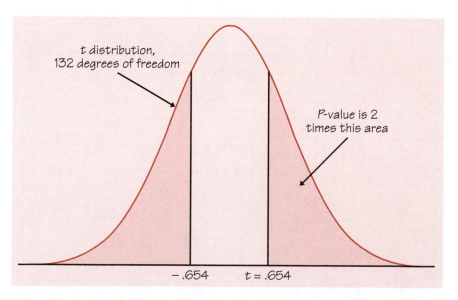

FIGURE 6.8   The P-value in Example 6.9. To find P, find the area above $t = .654$ and double it because the alternative is two-sided.

EXERCISES

6.31   In a study of heart surgery, one issue was the effect of drugs called beta-blockers on the pulse rate of patients during surgery. The available subjects were divided at random into two groups of 30 patients each. One group received a beta-blocker; the other, a placebo. The surgical team recorded the pulse rate of each patient at a critical point during the operation. The treatment group had mean 65.2 beats per minute and standard deviation 7.8. For the control group, the mean was 70.3 and the standard deviation 8.3. The data appear roughly normal.

   (a) Do beta-blockers reduce the pulse rate? State the hypotheses and do a *t* test. Is the result significant at the 5% level? At the 1% level?

   (b) Give a 99% confidence interval for the difference in mean pulse rates.

6.32   In a study of cereal leaf beetle damage on oats, researchers measured the number of beetle larvae per stem in small plots of oats after randomly applying one of two treatments: no pesticide, or malathion at the rate of 0.25 pound per acre. The data appear roughly normal. Here are the summary statistics.[16]

Group	Treatment	n	$\bar{x}$	s
1	Control	13	3.47	1.21
2	Malathion	14	1.36	0.52

   Is there significant evidence at the 1% level that malathion reduces the mean number of larvae per stem? Be sure to state H_0 and H_a.

6.33   A business school study compared a sample of Greek firms that went bankrupt with a sample of healthy Greek businesses. One measure of a firm's financial health is the ratio of current assets to current liabilities, called CA/CL. For the year before bankruptcy, the study found the mean CA/CL to be 1.72565 in the healthy group and 0.78640 in the group that failed. The paper reporting the study says that $t = 7.36$.[17]

   (a) You can draw a conclusion from this t without using a table and even without knowing the sizes of the samples (as long as the samples are not tiny). What is your conclusion? Why don't you need the sample size and a table?

   (b) In fact, the study looked at 33 firms that failed and 68 healthy firms. What degrees of freedom would you use for the t test if you follow the conservative approach recommended for use without software?

Robustness again

The two-sample t procedures are more robust than the one-sample t methods, particularly when the distributions are not symmetric. When the sizes of the two samples are equal and the two populations being compared have distributions with similar shapes, probability values from the t table are quite accurate for a broad range of distributions when the sample sizes are as small as $n_1 = n_2 = 5$.[18] When the two population distributions have different shapes, larger samples are needed.

As a guide to practice, adapt the guidelines given on page 425 for the use of one-sample t procedures to two-sample procedures by replacing "sample size" with the "sum of the sample sizes," $n_1 + n_2$. These guidelines err on the side of safety, especially when the two samples are of equal size. In planning a two-sample study, you should usually choose equal sample sizes. The two-sample t procedures are most robust against nonnormality in this case, and the conservative probability values are most accurate.

EXERCISES

6.34 College financial aid offices expect students to use summer earnings to help pay for college. But how large are these earnings? One college studied this question by asking a sample of students how much they earned. Omitting students who were not employed, there were 1296 responses. Here are the data in summary form.[19]

Group	n	$\bar{x}$	s
Males	675	$1884.52	$1368.37
Females	621	$1360.39	$1037.46

(a) The distribution of earnings is strongly skewed to the right. Nevertheless, use of t procedures is justified. Why?

(b) Give a 90% confidence interval for the difference between the mean summer earnings of male and female students.

(c) Once the sample size was decided, the sample was chosen by taking every 20th name from an alphabetical list of all undergraduates. Is it reasonable to consider the samples as SRSs chosen from the male and female undergraduate populations?

(d) What other information about the study would you request before accepting the results as describing all undergraduates?

6.35   Ordinary corn doesn't have as much of the amino acid lysine as animals need in their feed. Plant scientists have developed varieties of corn that have increased amounts of lysine. In a test of the quality of high-lysine corn as animal feed, an experimental group of 20 one-day-old male chicks ate a ration containing the new corn. A control group of another 20 chicks received a ration that was identical except that it contained normal corn. Here are the weight gains (in grams) after 21 days.[20]

Control				Experimental			
380	321	366	356	361	447	401	375
283	349	402	462	434	403	393	426
356	410	329	399	406	318	467	407
350	384	316	272	427	420	477	392
345	455	360	431	430	339	410	326

(a) Present the data graphically. Are there outliers or strong skewness that might prevent the use of t procedures?

(b) Is there good evidence that chicks fed high-lysine corn gain weight faster? Carry out a test and report your conclusions.

(c) Give a 95% confidence interval for the mean extra weight gain in chicks fed high-lysine corn.

6.36   The Survey of Study Habits and Attitudes (SSHA) is a psychological test that measures the motivation, attitude toward school, and study habits of students. Scores range from 0 to 200. A selective private college gives the SSHA to an SRS of both male and female first-year students. The data for the women are as follows:

154   109   137   115   152   140   154   178   101
103   126   126   137   165   165   129   200   148

Here are the scores of the men:

108   140   114   91   180   115   126   92   169   146
109   132   75   88   113   151   70   115   187   104

(a) Examine each sample graphically, with special attention to outliers and skewness. Is use of a t procedure acceptable for these data?

(b) Most studies have found that the mean SSHA score for men is lower than the mean score in a comparable group of women. Is this true for first-year students at this college? Carry out a test and give your conclusions.

(c) Give a 90% confidence interval for the mean difference between the SSHA scores of male and female first-year students at this college.

More accurate levels in the t procedures*

The two-sample t statistic does not have a t distribution. Moreover, the exact distribution changes as the unknown population standard deviations σ_1 and σ_2 change. However, an excellent approximation is available.

APPROXIMATE DISTRIBUTION OF THE TWO-SAMPLE t STATISTIC

The distribution of the two-sample t statistic is close to the t distribution with degrees of freedom df given by

$$df = \frac{\left(\dfrac{s_1^2}{n_1} + \dfrac{s_2^2}{n_2}\right)^2}{\dfrac{1}{n_1 - 1}\left(\dfrac{s_1^2}{n_1}\right)^2 + \dfrac{1}{n_2 - 1}\left(\dfrac{s_2^2}{n_2}\right)^2}$$

This approximation is quite accurate when both sample sizes n_1 and n_2 are 5 or larger.

The t procedures remain exactly as before except that we use the t distribution with df degrees of freedom to give critical values and P-values.

---

*This section can be omitted unless you are using statistical software and wish to understand what the software does.

EXAMPLE 6.10

In the calcium experiment of Examples 6.7 and 6.8 the data gave

Group	Treatment	n	$\bar{x}$	s
1	Calcium	10	5.000	8.743
2	Placebo	11	−.273	5.901

For improved accuracy, we can use critical points from the t distribution with degrees of freedom df given by

$$\text{df} = \frac{\left(\dfrac{8.743^2}{10} + \dfrac{5.901^2}{11}\right)^2}{\dfrac{1}{9}\left(\dfrac{8.743^2}{10}\right)^2 + \dfrac{1}{10}\left(\dfrac{5.901^2}{11}\right)^2}$$

$$= \frac{116.848}{7.494} = 15.59$$

Notice that the degrees of freedom df is not a whole number.

The conservative 90% confidence interval for $\mu_1 - \mu_2$ in Example 6.8 used the critical value $t^* = 1.833$ based on 9 degrees of freedom. A more exact confidence interval replaces this critical value with the critical value for df = 15.59 degrees of freedom. We cannot find this critical value exactly without using a computer package. For a close approximation, use the next smaller entry (15 degrees of freedom) in Table C. The critical value is $t^* = 1.753$. The 90% confidence interval is now

$$(\bar{x}_1 - \bar{x}_2) \pm t^*\sqrt{\frac{s_1^2}{n_1} + \frac{s_2^2}{n_2}}$$

$$= [5.000 - (-.273)] \pm 1.753\sqrt{\frac{8.743^2}{10} + \frac{5.901^2}{11}}$$

$$= 5.273 \pm 5.764$$

$$= (-.491, \; 11.037)$$

This confidence interval is a bit shorter (margin of error 5.764 rather than 6.026) than the conservative interval in Example 6.8. ◄

As Example 6.10 illustrates, the two-sample t procedures are exactly as before, except that we use a t distribution with more degrees of freedom.

The number df from the box on page 450 is always at least as large as the smaller of $n_1 - 1$ and $n_2 - 1$. On the other hand, df is never larger than the sum $n_1 + n_2 - 2$ of the two individual degrees of freedom. The number of degrees of freedom df is generally not a whole number. There is a t distribution for any positive degrees of freedom, even though Table C contains entries only for whole-number degrees of freedom. Some software packages find df and then use the t distribution with the next smaller whole-number degrees of freedom. Others take care to use $t(\text{df})$ even when df is not a whole number. We do not recommend regular use of this method unless a computer is doing the arithmetic. With a computer, the more accurate procedures are painless, as the following example illustrates.

EXAMPLE 6.11

Poisoning by the pesticide DDT causes convulsions in humans and other mammals. Researchers seek to understand how the convulsions are caused. In a randomized comparative experiment, they compared 6 white rats poisoned with DDT with a control group of 6 unpoisoned rats. Electrical measurements of nerve activity are the main clue to the nature of DDT poisoning. When a nerve is stimulated, its electrical response shows a sharp spike followed by a much smaller second spike. The experiment found that the second spike is larger in rats fed DDT than in normal rats. This finding helped biologists understand how DDT poisoning works.[21]

The researchers measured the height of the second spike as a percent of the first spike when a nerve in the rat's leg was stimulated. For the poisoned rats the results were

12.207   16.869   25.050   22.429   8.456   20.589

The control group data were

11.074   9.686   12.064   9.351   8.182   6.642

Both populations are reasonably normal, as far as can be judged from six observations. The DDT data are much more spread out than the control data. The difference in means is quite large, but in such small samples the sample mean is highly variable. A significance test can help confirm that we are seeing a real effect. Because the researchers did not conjecture in advance that the size of the second spike would increase in rats fed DDT, we use the two-sided alternative:

$$H_0 : \mu_1 = \mu_2$$

$$H_a : \mu_1 \neq \mu_2$$

Here is the output from the SAS statistical software system for these data.[22]

TTEST PROCEDURE

Variable: SPIKE

GROUP	N	Mean	Std Dev	Std Error
DDT	6	17.60000000	6.34014839	2.58835474
CONTROL	6	9.49983333	1.95005932	0.79610839

Variances	T	DF	Prob>|T|
Unequal	2.9912	5.9	0.0247
Equal	2.9912	10.0	0.0135

SAS reports the results of two t procedures: the general two-sample procedure ("Unequal" variances) and a special procedure that assumes that the two population variances are equal. We are interested in the first of these procedures. The two-sample t statistic has the value $t = 2.9912$, the degrees of freedom are df = 5.9, and the P-value from the $t(5.9)$ distribution is 0.0247. There is good evidence that the mean size of the secondary spike is larger in rats fed DDT. ◄

Would the conservative test based on 5 degrees of freedom (both $n_1 - 1$ and $n_2 - 1$ are 5) have given a different result in Example 6.11? The statistic is exactly the same: $t = 2.9912$. The conservative P-value is $2P(T \geq 2.9912)$, where T has the $t(5)$ distribution. Table C shows that 2.9912 lies between the 0.02 and 0.01 upper critical values of the $t(5)$ distribution, so P for the two-sided test lies between 0.02 and 0.04. For practical purposes this is the same result as that given by the software. As this example and Example 6.10 suggest, the difference between the t procedures using the conservative and the approximately correct distributions is rarely of practical importance. That is why we recommend the simpler conservative procedure for inference without a computer.

EXERCISES

6.37   Example 6.11 reports the analysis of data on the effects of DDT poisoning. The software uses the two-sample t test with degrees of freedom given in the box on page 450. Starting from the computer's results for $\bar{x}_i$ and s_i, verify the computer's values for the test statistic $t = 2.99$ and the degrees of freedom df = 5.9.

6.38   What aspects of rowing technique distinguish between novice and skilled competitive rowers? Researchers compared two groups of female competitive rowers: a group of skilled rowers and a group of novices. The researchers measured many mechanical aspects of rowing style as the subjects rowed on a Stanford Rowing Ergometer. One important variable is the angular velocity of the knee, which describes the rate at which the knee joint opens as the legs push the body back on the sliding seat. The data show no outliers or strong skewness. Here is the SAS computer output.[23]

TTEST PROCEDURE

Variable: KNEE

GROUP	N	Mean	Std Dev	Std Error
SKILLED	10	4.18283335	0.47905935	0.15149187
NOVICE	8	3.01000000	0.95894830	0.33903942

Variances	T	DF	Prob>|T|
Unequal	3.1583	9.8	0.0104
Equal	3.3918	16.0	0.0037

(a) The researchers believed that the knee velocity would be higher for skilled rowers. State H_0 and H_a.

(b) What is the value of the two-sample t statistic and its P-value? (Note that SAS provides two-sided P-values. If you need a one-sided P-value, divide the two-sided value by 2.) What do you conclude?

(c) Give a 90% confidence interval for the mean difference between the knee velocities of skilled and novice female rowers.

6.39   The researchers in the previous exercise also wondered whether skilled and novice rowers differ in weight or other physical characteristics. Here is the SAS computer output for weight in kilograms.

```
TTEST PROCEDURE
```

Variable: WEIGHT

GROUP	N	Mean	Std Dev	Std Error
SKILLED	10	70.3700000	6.10034898	1.92909973
NOVICE	8	68.4500000	9.03999930	3.19612240

Variances	T	DF	Prob>|T|
Unequal	0.5143	11.8	0.6165
Equal	0.5376	16.0	0.5982

Is there significant evidence of a difference in the mean weights of skilled and novice rowers? State H_0 and H_a, report the two-sample t statistic and its P-value, and state your conclusion. (Note that SAS provides two-sided P-values. If you need a one-sided P-value, divide the two-sided value by 2.)

The pooled two-sample t procedures[*]

In Example 6.11, the software offered a choice between two t tests. One is labeled for "unequal" variances, the other for "equal" variances. The "unequal" variance procedure is our two-sample t. This test is valid whether or not the population variances are equal. The other choice is a special version of the two-sample t statistic that assumes that the two populations have the same variance. This procedure averages (the statistical term is "pools") the two sample variances to estimate the common population variance. The resulting statistic is called the pooled two-sample t statistic. It is equal to our t statistic if the two sample sizes are the same, but not otherwise. We could choose to use the pooled t for both tests and confidence intervals.

The pooled t statistic has the advantage that it has exactly the t distribution with $n_1 + n_2 - 2$ degrees of freedom *if* the two population variances really are equal. Of course, the population variances are often not equal. Moreover, the assumption of equal variances is hard to check from the data. The pooled t was in common use before software made it easy to use the

---

[*]This is a special topic that is optional.

accurate approximation to the distribution of our two-sample t statistic. Now it is useful only in special situations. We cannot use the pooled t in Example 6.11, for example, because it is clear that the variance is much larger among rats fed DDT.

SUMMARY

The data in a **two-sample problem** are two independent SRSs, each drawn from a separate normally distributed population.

Tests and confidence intervals for the difference between the means μ_1 and μ_2 of the two populations start from the difference $\bar{x}_1 - \bar{x}_2$ of the two sample means. Because of the central limit theorem, the resulting procedures are approximately correct for other population distributions when the sample sizes are large.

Draw independent SRSs of sizes n_1 and n_2 from two normal populations with parameters μ_1, σ_1 and μ_2, σ_2. The **two-sample t statistic** is

$$t = \frac{(\bar{x}_1 - \bar{x}_2) - (\mu_1 - \mu_2)}{\sqrt{\dfrac{s_1^2}{n_1} + \dfrac{s_2^2}{n_2}}}$$

The statistic t does *not* have exactly a t distribution.

For conservative inference procedures to compare μ_1 and μ_2, use the two-sample t statistic with the $t(k)$ distribution. The degrees of freedom k is the smaller of $n_1 - 1$ and $n_2 - 1$. For more accurate probability values, use the $t(\text{df})$ distribution with degrees of freedom df estimated from the data. This is the usual procedure in statistical software.

The **confidence interval** for $\mu_1 - \mu_2$ given by

$$(\bar{x}_1 - \bar{x}_2) \pm t^* \sqrt{\frac{s_1^2}{n_1} + \frac{s_2^2}{n_2}}$$

has confidence level at least C if t^* is the upper $(1 - C)/2$ critical value for $t(k)$ with k the smaller of $n_1 - 1$ and $n_2 - 1$.

Significance tests for $H_0 : \mu_1 = \mu_2$ based on

$$t = \frac{\bar{x}_1 - \bar{x}_2}{\sqrt{\dfrac{s_1^2}{n_1} + \dfrac{s_2^2}{n_2}}}$$

have a true P-value no higher than that calculated from $t(k)$.

The guidelines for practical use of two-sample t procedures are similar to those for one-sample t procedures. Equal sample sizes are recommended.

SECTION 6.2 EXERCISES

In exercises that call for two-sample t procedures, you may use as the degrees of freedom either the smaller of $n_1 - 1$ and $n_2 - 1$ or the more exact value df given in the box on page 450. We recommend the first choice unless you are using a computer. Many of these exercises ask you to think about issues of statistical practice as well as to carry out t procedures.

6.40   The Johns Hopkins Regional Talent Searches give the Scholastic Assessment Tests (intended for high school juniors and seniors) to 13-year-olds. In all, 19,883 males and 19,937 females took the tests between 1980 and 1982. The mean scores of males and females on the verbal test are nearly equal, but there is a clear difference between the sexes on the mathematics test. The reason for this difference is not understood. Here are the data.[24]

Group	$\bar{x}$	s
Males	416	87
Females	386	74

Give a 99% confidence interval for the difference between the mean score for males and the mean score for females in the population that Johns Hopkins searches. Must SAT scores have a normal distribution in order for your confidence interval to be valid? Why?

6.41   A study of iron deficiency in infants compared samples of infants whose mothers chose different ways of feeding them. One group contained breast-fed infants. The children in another group were fed a standard baby formula without any iron supplements. Here are summary results on blood hemoglobin levels at 12 months of age.[25]

Group	n	$\bar{x}$	s
Breast-fed	23	13.3	1.7
Formula	19	12.4	1.8

(a) Is there significant evidence that the mean hemoglobin level is different among breast-fed babies? State H_0 and H_a and carry out a t test. Give the P-value. What is your conclusion?

(b) Give a 95% confidence interval for the mean difference in hemoglobin level between the two populations of infants.

(c) State the assumptions that your procedures in (a) and (b) require in order to be valid.

(d) Is this study an experiment? Why? How does this affect the conclusions we can draw from the study?

6.42   Physical fitness is related to personality characteristics. In one study of this relationship, middle-aged college faculty who had volunteered for a fitness program were divided into low-fitness and high-fitness groups based on a physical examination. The subjects then took the Cattell Sixteen Personality Factor Questionnaire. Here are the data for the "ego strength" personality factor.[26]

Group	Fitness	n	$\bar{x}$	s
1	Low	14	4.64	0.69
2	High	14	6.43	0.43

(a) Is the difference in mean ego strength significant at the 5% level? At the 1% level? Be sure to state H_0 and H_a.

(b) You should be hesitant to generalize these results to the population of all middle-aged men. Explain why.

6.43   A market research firm supplies manufacturers with estimates of the retail sales of their products from samples of retail stores. Marketing managers are prone to look at the estimate and ignore sampling error. An SRS of 75 stores this month shows mean sales of 52 units of a small appliance, with standard deviation 13 units. During the same month last year, an SRS of

53 stores gave mean sales of 49 units, with standard deviation 11 units. An increase from 49 to 52 is a rise of 6%. The marketing manager is happy, because sales are up 6%.

(a) Use the two-sample t procedure to give a 95% confidence interval for the difference between this year and last year in the mean number of units sold at all retail stores.

(b) Explain in language that the manager can understand why he cannot be confident that sales rose by 6%, and that in fact sales may even have dropped.

6.44 A bank compares two proposals to increase the amount that its credit card customers charge on their cards. (The bank earns a percentage of the amount charged, paid by the stores that accept the card.) Proposal A offers to eliminate the annual fee for customers who charge $2400 or more during the year. Proposal B offers a small percent of the total amount charged as a cash rebate at the end of the year. The bank offers each proposal to an SRS of 150 of its credit card customers. At the end of the year, the total amount charged by each customer is recorded. Here are the summary statistics.

Group	n	$\bar{x}$	s
A	150	$1987	$392
B	150	$2056	$413

(a) Do the data show a significant difference between the mean amounts charged by customers offered the two plans? Give the null and alternative hypotheses, and calculate the two-sample t statistic. Obtain the P-value. State your practical conclusions.

(b) The distributions of amounts charged are skewed to the right, but outliers are prevented by the limits that the bank imposes on credit balances. Do you think that skewness threatens the validity of the test that you used in (a)? Explain your answer.

(c) Is the bank's study an experiment? Why? How does this affect the conclusions the bank can draw from the study?

6.45 An educator believes that new reading activities in the classroom will help elementary school pupils improve their reading ability. She arranges for a third-grade class of 21 students to follow these activities for an 8-week period. A control classroom of 23 third graders follows the same curriculum without the activities. At the end of the 8 weeks, all students are given the

Degree of Reading Power (DRP) test, which measures the aspects of reading ability that the treatment is designed to improve. Here are the data.[27]

Treatment					Control				
24	43	58	71	43	42	43	55	26	62
49	61	44	67	49	37	33	41	19	54
53	56	59	52	62	20	85	46	10	17
54	57	33	46	43	60	53	42	37	42
57					55	28	48		

(a) Examine the data with a graph. Are there strong outliers or skewness that could prevent use of the t procedures?

(b) Is there good evidence that the new activities improve the mean DRP score? Carry out a test and report your conclusions.

(c) Although this study is an experiment, its design is not ideal because it had to be done in a school without disrupting classes. What aspect of good experimental design is missing?

6.46   Researchers studying the learning of speech often compare measurements made on the recorded speech of adults and children. One variable of interest is called the voice onset time (VOT). Here are the results for 6-year-old children and adults asked to pronounce the word "bees." The VOT is measured in milliseconds and can be either positive or negative.[28]

Group	n	$\bar{x}$	s
Children	10	-3.67	33.89
Adults	20	-23.17	50.74

(a) The researchers were investigating whether VOT distinguishes adults from children. State H_0 and H_a and carry out a two-sample t test. Give a P-value and report your conclusions.

(b) Give a 95% confidence interval for the difference in mean VOTs when pronouncing the word "bees." Explain why you knew from your result in (a) that this interval would contain 0 (no difference).

6.47   The researchers in the study discussed in Exercise 6.46 looked at VOTs for adults and children pronouncing many different words. Explain why they should not do a separate two-sample t test for each word and conclude that those words with a significant difference (say $P < .05$) distinguish children from adults. (The researchers did not make this mistake.)

The remaining exercises concern the power of the two-sample t test, an optional topic. If you have read Section 5.4 and the discussion of the power of the one-sample t test on pages 428–430, Exercise 6.48 guides you in finding the power of the two-sample t.

6.48   **(Optional)** In Example 6.8 on page 443, a small study of black men suggested that a calcium supplement can reduce blood pressure. Now we are planning a larger clinical trial of this effect. We plan to use 100 subjects in each of the two groups. Are these sample sizes large enough to make it very likely that the study will give strong evidence ($\alpha = .01$) of the effect of calcium if in fact calcium lowers blood pressure by 5 millimeters more than a placebo? To answer this question, we will compute the power of the two-sample t test of

$$H_0 : \mu_1 = \mu_2$$

$$H_a : \mu_1 > \mu_2$$

against the specific alternative $\mu_1 - \mu_2 = 5$. Based on the pilot study reported in Example 6.8, we take 8, the larger of the two observed s-values, as a rough estimate of both the population σ's and future sample s's.

(a) What is the approximate value of the $\alpha = .01$ critical value t^* for the two-sample t statistic when $n_1 = n_2 = 100$?

(b) **Step 1** *Write the rule for rejecting H_0 in terms of $\overline{x}_1 - \overline{x}_2$.* The test rejects H_0 when

$$\frac{\overline{x}_1 - \overline{x}_2}{\sqrt{\dfrac{s_1^2}{n_1} + \dfrac{s_2^2}{n_2}}} \geq t^*$$

Take both s_1 and s_2 to be 8, and n_1 and n_2 to be 100. Find the number c such that the test rejects H_0 when $\overline{x}_1 - \overline{x}_2 \geq c$.

(c) **Step 2** *The power is the probability of rejecting H_0 when the alternative is true.* Suppose that $\mu_1 - \mu_2 = 5$ and that both σ_1 and σ_2 are 8. The power we seek is the probability that $\overline{x}_1 - \overline{x}_2 \geq c$ under these assumptions. Calculate the power.

6.49   **(Optional)** You are planning a larger study of VOTs, based on the pilot study reported in Exercise 6.46. Not all words distinguish children (Group 1) from adults (Group 2) as well as "bees," so you want high power against the alternative that $\mu_1 - \mu_2 = 10$ in a *two-sided t* test. From the pilot study, take 30 as an estimate of σ_1 and s_1, and 50 as an estimate of σ_2 and s_2.

(a) State H_0 and H_a, and write the formula for the test statistic.

(b) Give the $\alpha = .05$ critical value for the test when $n_1 = 100$ and $n_2 = 300$. (Although we recommend equal sample sizes to improve the robustness of t procedures against nonnormality, sample sizes that are proportional to the variances give higher power for the same total number of observations.)

(c) Find the approximate power against the alternative $\mu_1 - \mu_2 = 10$.

6.50   **(Optional)** A bank asks you to compare two ways to increase the use of their credit cards. Plan A would offer customers a cash-back rebate based on their total amount charged. Plan B would reduce the interest rate charged on card balances. The response variable is the total amount a customer charges during the test period. You decide to offer each of Plan A and Plan B to a separate SRS of the bank's credit card customers. In the past, the mean amount charged in a six-month period has been about $1100, with a standard deviation of $400. Will a two-sample t test based on SRSs of 350 customers in each group detect a difference of $100 in the mean amounts charged under the two plans?

(a) State H_0 and H_a, and write the formula for the test statistic.

(b) Give the $\alpha = .05$ critical value for the test when $n_1 = n_2 = 350$.

(c) Calculate the power of the test with $\alpha = .05$, using $400 as a rough estimate of all standard deviations.

6.3   INFERENCE FOR POPULATION SPREAD*

The two most basic descriptive features of a distribution are its center and spread. In a normal population, we measure center and spread by the mean and the standard deviation. We use the t procedures for inference about population means for normal populations, and we know that t procedures are often useful for nonnormal populations as well. It is natural to turn next to inference about the standard deviations of normal populations. Our advice here is short and clear: Don't do it without expert advice.

Avoid inference about standard deviations

There are methods for inference about the standard deviations of normal populations. We will describe the most common such method, the F test

*This section is not required for an understanding of later material, except for Chapter 9.

for comparing the spread of two normal populations. Unlike the t procedures for means, the F test and other procedures for standard deviations are extremely sensitive to nonnormal distributions. This lack of robustness does not improve in large samples. It is difficult in practice to tell whether a significant F-value is evidence of unequal population spreads or simply a sign that the populations are not normal.

The deeper difficulty underlying the very poor robustness of normal population procedures for inference about spread already appeared in our work on describing data. The standard deviation is a natural measure of spread for normal distributions but not for distributions in general. In fact, because skewed distributions have unequally spread tails, no single numerical measure does a good job of describing the spread of a skewed distribution. In summary, the standard deviation is not always a useful parameter, and even when it is (for symmetric distributions), the results of inference are not trustworthy. Consequently, we do not recommend trying to do inference about population standard deviations in basic statistical practice.[29]

It was once common to test equality of standard deviations as a preliminary to performing the pooled two-sample t test for equality of two population means. It is better practice to check the distributions graphically, with special attention to skewness and outliers, and to use the version of the two-sample t featured in Section 6.2. This test does not require equal standard deviations.

The F test for comparing two standard deviations

Because of the limited usefulness of procedures for inference about the standard deviations of normal distributions, we will present only one such procedure. Suppose that we have independent SRSs from two normal populations, a sample of size n_1 from $N(\mu_1, \sigma_1)$ and a sample of size n_2 from $N(\mu_2, \sigma_2)$. The population means and standard deviations are all unknown. The two-sample t test examines whether the means are equal in this setting. To test the hypothesis of equal spread,

$$H_0 : \sigma_1 = \sigma_2$$

$$H_a : \sigma_1 \neq \sigma_2$$

we use the ratio of sample variances. This is the F *statistic*.

THE F STATISTIC AND F DISTRIBUTIONS

When s_1^2 and s_2^2 are sample variances from independent SRSs of sizes n_1 and n_2 drawn from normal populations, the **F statistic**

$$F = \frac{s_1^2}{s_2^2}$$

has the **F distribution** with $n_1 - 1$ and $n_2 - 1$ degrees of freedom when $H_0 : \sigma_1 = \sigma_2$ is true.

The F distributions are a family of distributions with two parameters. The parameters are the degrees of freedom of the sample variances in the numerator and denominator of the F statistic. The numerator degrees of freedom are always mentioned first. Interchanging the degrees of freedom changes the distribution, so the order is important. Our brief notation will be $F(j, k)$ for the F distribution with j degrees of freedom in the numerator and k in the denominator. The F distributions are not symmetric but are right-skewed. The density curve in Figure 6.9 illustrates the shape. Because sample variances cannot be negative, the F statistic takes only positive values, and the F distribution has no probability below 0. The peak of the F density curve is near 1. When the two populations have the same standard

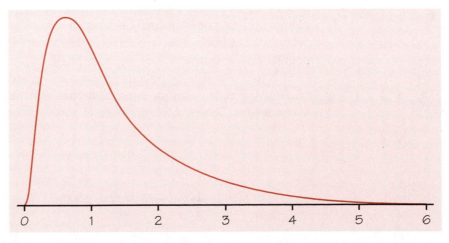

FIGURE 6.9   The density curve for the $F(9, 10)$ distribution. The F distributions are skewed to the right.

deviation, we expect the two sample variances to be close in size, so that F takes a value near 1. Values of F far from 1 in either direction provide evidence against the hypothesis of equal standard deviations.

Tables of F critical points are awkward, because we need a separate table for every pair of degrees of freedom j and k. Table D in the back of the book gives upper p critical points of the F distributions for $p = 0.10$, 0.05, 0.025, 0.01, and 0.001. For example, these critical points for the $F(9, 10)$ distribution shown in Figure 6.9 are

p	.10	.05	.025	.01	.001
F^*	2.35	3.02	3.78	4.94	8.96

The skewness of the F distributions causes additional complications. In the symmetric normal and t distributions, the point with probability 0.05 below it is just the negative of the point with probability 0.05 above it. This is not true for F distributions. We therefore need either tables of both the upper and lower tails or some way to eliminate the need for lower tail critical values. Statistical software that does away with the need for tables is very convenient. If you do not use statistical software, arrange the two-sided F test as follows:

CARRYING OUT THE *F* TEST

Step 1   Take the test statistic to be

$$F = \frac{\text{larger } s^2}{\text{smaller } s^2}$$

This amounts to naming the populations so that Population 1 has the larger of the observed sample variances. The resulting F is always 1 or greater.

Step 2   Compare the value of F with critical values from Table D. Then *double* the significance levels from the table to obtain the significance level for the two-sided F test.

The idea is that we calculate the probability in the upper tail and double it to obtain the probability of all ratios on either side of 1 that are at least as improbable as that observed. Remember that the order of the degrees of freedom is important in using Table D.

EXAMPLE 6.12

Example 6.7 describes a medical experiment to compare the mean effects of calcium and a placebo on the blood pressure of black men. We might also compare the standard deviations to see whether calcium changes the spread of blood pressures among black men. We want to test

$$H_0 : \sigma_1 = \sigma_2$$

$$H_a : \sigma_1 \neq \sigma_2$$

The larger of the two sample standard deviations is $s = 8.743$ from 10 observations. The other is $s = 5.901$ from 11 observations. The F test statistic is therefore

$$F = \frac{\text{larger } s^2}{\text{smaller } s^2} = \frac{8.743^2}{5.901^2} = 2.195$$

Compare the calculated value $F = 2.20$ with critical points for the $F(9, 10)$ distribution. Table D shows that 2.20 is less than the 0.10 critical value of the $F(9, 10)$ distribution, which is $F^* = 2.35$. Doubling 0.10, we know that the P-value for the two-sided test is greater than 0.20. The results are not significant at the 20% level (or any lower level). Statistical software shows that the exact upper tail probability is 0.118, and hence $P = .24$. If the populations were normal, the observed standard deviations would give little reason to suspect unequal population standard deviations. Because one of the populations shows some nonnormality, we cannot be fully confident of this conclusion. ◄

SUMMARY

Inference procedures for comparing the standard deviations of two normal populations are based on the **F statistic**, which is the ratio of sample variances

$$F = \frac{s_1^2}{s_2^2}.$$

If an SRS of size n_1 is drawn from Population 1 and an independent SRS of size n_2 is drawn from Population 2, the F statistic has the **F distribution** $F(n_1 - 1, n_2 - 1)$ if the two population standard deviations σ_1 and σ_2 are in fact equal.

The **F distributions** are skewed to the right and take only values greater than 0. A specific F distribution $F(j, k)$ is fixed by the two **degrees of freedom** j and k.

The two-sided **F test** of $H_0 : \sigma_1 = \sigma_2$ uses the statistic

$$F = \frac{\text{larger } s^2}{\text{smaller } s^2}$$

and doubles the upper tail probability to obtain the P-value.

The F tests and other procedures for inference on the spread of one or more normal distributions are so strongly affected by lack of normality that we do not recommend them for regular use.

SECTION 6.3 EXERCISES

In all exercises calling for use of the F test, assume that both population distributions are very close to normal. The actual data are not always sufficiently normal to justify use of the F test.

6.51 The F statistic $F = s_1^2/s_2^2$ is calculated from samples of size $n_1 = 10$ and $n_2 = 8$. (Remember that n_1 is the numerator sample size.)

(a) What is the upper 5% critical value for this F?

(b) In a test of equality of standard deviations against the two-sided alternative, this statistic has the value $F = 3.45$. Is this value significant at the 10% level? Is it significant at the 5% level?

6.52 The F statistic for equality of standard deviations based on samples of sizes $n_1 = 21$ and $n_2 = 16$ takes the value $F = 2.78$.

(a) Is this significant evidence of unequal population standard deviations at the 5% level? At the 1% level?

(b) Between which two values obtained from Table D does the P-value of the test fall?

6.53 The sample variance for the treatment group in the DDT experiment of Example 6.11 (page 452) is more than 10 times as large as the sample variance for the control group. Calculate the F statistic. Can you reject the hypothesis of equal population standard deviations at the 5% significance level? At the 1% level?

6.54   Exercise 6.38 (page 454) records the results of comparing a measure of rowing style for skilled and novice female competitive rowers. Is there significant evidence of inequality between the standard deviations of the two populations?

(a) State H_0 and H_a.

(b) Calculate the F statistic. Between which two levels does the P-value lie?

6.55   Answer the same questions for the weights of the two groups, recorded in Exercise 6.39 (page 454).

6.56   The data for VOTs of children and adults in Exercise 6.46 (page 460) show quite different sample standard deviations. How statistically significant is the observed inequality?

6.57   Return to the SSHA data in Exercise 6.36 (page 449). We want to know if the spread of SSHA scores is different among women and among men at this college. Use the F test to obtain a conclusion.

CHAPTER REVIEW

This chapter presented t tests and confidence intervals for inference about the mean of a single population and for comparing the means of two populations. The one-sample t procedures do inference about one mean and the two-sample t procedures compare two means. Matched pairs studies use one-sample procedures because you first create a single sample by taking the differences in the responses within each pair.

The t procedures require that the data be random samples and that the distribution of the population or populations be normal. One reason for the wide use of t procedures is that they are not very strongly affected by lack of normality. If you can't regard your data as a random sample, however, the results of inference may be of little value.

Chapter 5 concentrated on the reasoning of confidence intervals and tests. Understanding the reasoning is essential for wise use of the t and other inference methods. The discussion in this chapter paid more attention to practical aspects of using the methods. We saw that there are several versions of the two-sample t, for example. Which one you use depends largely on whether or not you use statistical software. Before you use any inference method, think about the design of the study and examine the data for outliers and other problems.

The chapter exercises are important in this and later chapters. You must now recognize problem settings and decide which of the methods presented in the chapter fits. In this chapter, you must recognize one-sample studies, matched pairs studies, and two-sample studies.

Here are the most important skills you should have after reading this chapter.

A. RECOGNITION

1. Recognize when a problem requires inference about a mean or comparing two means.

2. Recognize from the design of a study whether one-sample, matched pairs, or two-sample procedures are needed.

B. ONE-SAMPLE t PROCEDURES

1. Use the t procedure to obtain a confidence interval at a stated level of confidence for the mean μ of a population.

2. Carry out a t test for the hypothesis that a population mean μ has a specified value against either a one-sided or a two-sided alternative. Use Table C of t critical values to approximate the P-value or carry out a fixed α test.

3. Recognize when the t procedures are appropriate in practice, in particular that they are quite robust against lack of normality but are influenced by outliers.

4. Also recognize when the design of the study, outliers, or a small sample from a skewed distribution make the t procedures risky.

5. Recognize matched pairs data and use the t procedures to obtain confidence intervals and to perform tests of significance for such data.

C. TWO-SAMPLE t PROCEDURES

1. Give a confidence interval for the difference between two means. Use the two-sample t statistic with conservative degrees of freedom if you do not have statistical software. Use software if you have it.

2. Test the hypothesis that two populations have equal means against either a one-sided or a two-sided alternative. Use the two-sample t test with conservative degrees of freedom if you do not have statistical software. Use software if you have it.

3. Recognize when the two-sample t procedures are appropriate in practice.

4. Know that procedures for comparing the standard deviations of two normal populations are available, but that these procedures are risky because they are not at all robust against nonnormal distributions.

CHAPTER 6 REVIEW EXERCISES

6.58   In a study of the effectiveness of a weight-loss program, 47 subjects who were at least 20% overweight took part in the program for 10 weeks. Private weighings determined each subject's weight at the beginning of the program and 6 months after the program's end. The matched pairs t test was used to assess the significance of the average weight loss. The paper reporting the study said, "The subjects lost a significant amount of weight over time, $t(46) = 4.68, p < .01$." It is common to report the results of statistical tests in this abbreviated style.[30]

(a) Why was the matched pairs t test appropriate?

(b) Explain to someone who knows no statistics but is interested in weight-loss programs what the practical conclusion is.

(c) The paper follows the tradition of reporting significance only at fixed levels such as $\alpha = .01$. In fact, the results are more significant than "$p < .01$" suggests. Use Table C to say more about the P-value of the t test.

6.59   Consumers who think a product's advertising is expensive often also think the product must be of high quality. Can other information undermine this effect? To find out, marketing researchers did an experiment. The subjects were 90 women from the clerical and administrative staff of a large organization. All subjects read an ad that described a fictional line of food products called "Five Chefs." The ad also described the major TV commercials that would soon be shown, an unusual expense for this type of product. The 45 women in the control group read nothing else. The 45 in the "undermine group" also read a news story headlined "No Link between Advertising Spending and New Product Quality."

All the subjects then rated the quality of Five Chefs products on a seven-point scale. The study report said, "The mean quality ratings were significantly lower in the undermine treatment ($\overline{X}_A = 4.56$) than in the control treatment ($\overline{X}_C = 5.05; t = 2.64, p < .01$)."[31]

(a) Is the matched pairs t test or the two-sample t test the right test in this setting? Why?

(b) What degrees of freedom would you use for the t statistic you chose in (a)?

(c) The distribution of individual responses is not normal, because there is only a seven-point scale. Why is it nonetheless proper to use a *t* test?

6.60 A major study of alternative welfare programs randomly assigned women on welfare to one of two programs, called "WIN" and "Options." WIN was the existing program. The new Options program gave more incentives to work. An important question was how much more (on the average) women in Options earned than those in WIN. Here is Minitab output for earnings in dollars over a three-year period.[32]

```
TWOSAMPLE T FOR 'OPT' VS 'WIN'

          N      MEAN    STDEV    SE MEAN
OPT    1362      7638      289     7.8309
WIN    1395      6595      247     6.6132

95 PCT CI FOR MU OPT - MU WIN: (1022.90, 1063.10)
```

(a) Give a 99% confidence interval for the amount by which the mean earnings of Options participants exceeded the mean earnings of WIN subjects. (Minitab will give a 99% confidence interval if you instruct it to do so. Here we have only the basic output, which includes the 95% confidence interval.)

(b) The distribution of incomes is strongly skewed to the right but includes no extreme outliers because all the subjects were on welfare. What fact about these data allows us to use *t* procedures despite the strong skewness?

6.61 The one-hole test is used to test the manipulative skill of job applicants. This test requires subjects to grasp a pin, move it to a hole, insert it, and return for another pin. The score on the test is the number of pins the subject inserts in a fixed time interval. A study compared the scores of male college students with those of experienced female industrial workers. Here are the data for the first minute of the test, summarized by the Data Desk statistical software. ("Cases" is the count of subjects.)[33]

Summary statistics for students

Mean 35.1242
Cases 750
StdDev 4.3108

Summary statistics for workers

Mean 37.3234
Cases 412
StdDev 3.8317

(a) We expect that the experienced workers will outperform the students. State the hypotheses for a statistical test of this expectation and perform the test. Give a *P*-value and state your conclusions.

(b) The distribution of scores is slightly skewed to the left. Explain why the procedure you used in (a) is nonetheless acceptable.

(c) One purpose of the study was to develop performance norms for job applicants. Based on the data above, estimate the range that covers the middle 95% of experienced workers. (Be careful! This is not the same as a 95% confidence interval for the mean score of experienced workers.)

6.62 A college golf coach thinks that her players improve their scores in the second round of tournaments because they are less nervous after the first round. Here are the scores of the 12 team members in the two rounds of a tournament.

						Golfer						
	1	2	3	4	5	6	7	8	9	10	11	12
Round 1	89	90	87	95	86	81	102	105	83	88	91	79
Round 2	94	85	89	89	81	76	107	89	87	91	88	80

The coach asks you whether the scores support her opinion. Remember that in golf lower scores are better.

(a) State the hypotheses you would like to test. What test procedure do you plan to use? Explain your choice.

(b) Graph the data in a way that matches your planned analysis. Unfortunately, the graph shows that the *t* procedures cannot be safely used to analyze these data. Why? (There are other procedures that can be used. You should seek expert advice.)

6.63 Nitrites are often added to meat products as preservatives. In a study of the effect of nitrites on bacteria, researchers measured the rate of uptake of an amino acid for 60 cultures of bacteria: 30 growing in a medium to which nitrites had been added and another 30 growing in a standard medium as a control group. Here are the data from this study.

Control			Nitrite		
6,450	8,709	9,361	8,303	8,252	6,594
9,011	9,036	8,195	8,534	10,227	6,642
7,821	9,996	8,202	7,688	6,811	8,766
6,579	10,333	7,859	8,568	7,708	9,893
8,066	7,408	7,885	8,100	6,281	7,689
6,679	8,621	7,688	8,040	9,489	7,360
9,032	7,128	5,593	5,589	9,460	8,874
7,061	8,128	7,150	6,529	6,201	7,605
8,368	8,516	8,100	8,106	4,972	7,259
7,238	8,830	9,145	7,901	8,226	8,552

Examine the data and briefly describe their distribution. Carry out a test of the research hypothesis that nitrites decrease amino acid uptake, and report your results.

6.64 The composition of the earth's atmosphere may have changed over time. To try to discover the nature of the atmosphere long ago, we can examine the gas in bubbles inside ancient amber. Amber is tree resin that has hardened and been trapped in rocks. The gas in bubbles within amber should be a sample of the atmosphere at the time the amber was formed. Measurements on specimens of amber from the late Cretaceous era (75 to 95 million years ago) give these percents of nitrogen:

63.4 65.0 64.4 63.3 54.8 64.5 60.8 49.1 51.0

These values are quite different from the present 78.1% of nitrogen in the atmosphere. Assume (this is not yet agreed on by experts) that these observations are an SRS from the late Cretaceous atmosphere.[34]

(a) Graph the data, and comment on skewness and outliers.

(b) The t procedures will be only approximate in this case. Give a 95% t confidence interval for the mean percent of nitrogen in ancient air.

6.65 A pharmaceutical manufacturer does a chemical analysis to check the potency of products. The standard release potency for cephalothin crystals is 910. An assay of 16 lots gives the following potency data:

897 914 913 906 916 918 905 921
918 906 895 893 908 906 907 901

(a) Check the data for outliers or strong skewness that might threaten the validity of the t procedures.

(b) Give a 95% confidence interval for the mean potency.

(c) Is there significant evidence at the 5% level that the mean potency is not equal to the standard release potency?

6.66   Do various occupational groups differ in their diets? A British study compared the food and drink intake of 98 drivers and 83 conductors of London double-decker buses. The conductors' jobs require more physical activity. The article reporting the study gives the data as "Mean daily consumption (± s. e.)." Some of the study results appear below.[35]

	Drivers	Conductors
Total calories	2821 ± 44	2844 ± 48
Alcohol (grams)	.24 ± .06	.39 ± .11

(a) What does "s. e." stand for? Give $\bar{x}$ and s for each of the four sets of measurements.

(b) Is there significant evidence at the 5% level that conductors and drivers consume different numbers of calories per day?

(c) How significant is the observed difference in mean alcohol consumption?

(d) Give a 90% confidence interval for the mean daily alcohol consumption of London double-decker bus conductors.

(e) Give an 80% confidence interval for the difference in mean daily alcohol consumption between drivers and conductors.

6.67   You look up a census report that gives the populations of all 92 counties in the state of Indiana. Is it proper to apply the one-sample t method to these data to give a 95% confidence interval for the mean population of an Indiana county? Explain your answer.

6.68   The amount of lead in a type of soil, measured by a standard method, averages 86 parts per million (ppm). A new method is tried on 40 specimens of the soil, yielding a mean of 83 ppm lead and a standard deviation of 10 ppm.

(a) Is there significant evidence at the 1% level that the new method frees less lead from the soil?

(b) A critic argues that because of variations in the soil, the effectiveness of the new method is confounded with characteristics of the particular soil specimens used. Briefly describe a better data production design that avoids this criticism.

6.69 High levels of cholesterol in the blood are not healthy in either humans or dogs. Because a diet rich in saturated fats raises the cholesterol level, it is plausible that dogs owned as pets have higher cholesterol levels than dogs owned by a veterinary research clinic. "Normal" levels of cholesterol based on the clinic's dogs would then be misleading. A clinic compared healthy dogs it owned with healthy pets brought to the clinic to be neutered. The summary statistics for blood cholesterol levels (milligrams per deciliter of blood) appear below.[36]

Group	n	$\bar{x}$	s
Pets	26	193	68
Clinic	23	174	44

(a) Is there strong evidence that pets have higher mean cholesterol level than clinic dogs? State the H_0 and H_a and carry out an appropriate test. Give the P-value and state your conclusion.

(b) Give a 95% confidence interval for the difference in mean cholesterol levels between pets and clinic dogs.

(c) Give a 95% confidence interval for the mean cholesterol level in pets.

(d) What assumptions must be satisfied to justify the procedures you used in (a), (b), and (c)? Assuming that the cholesterol measurements have no outliers and are not strongly skewed, what is the chief threat to the validity of the results of this study?

6.70 Exercise 1.35 (page 53) gives 29 measurements of the density of the earth, made in 1798 by Henry Cavendish. Display the data graphically to check for skewness and outliers. Then give an estimate for the density of the earth from Cavendish's data and a margin of error for your estimate.

6.71 Elite distance runners are thinner than the rest of us. Here are data on skinfold thickness, which indirectly measures body fat, for 20 elite runners and 95 ordinary men in the same age group. The data are in millimeters and are given in the form "mean (standard deviation)."[37]

	Runners	Others
Abdomen	7.1 (1.0)	20.6 (9.0)
Thigh	6.1 (1.8)	17.4 (6.6)

Use confidence intervals to describe the difference between runners and typical young men.

6.72   **(Optional)** Do the data in Example 6.9 (page 445) provide evidence of different standard deviations for Chapin test scores in the populations of female and male college liberal arts majors?

(a) State the hypotheses and carry out the test. Software can assess significance exactly, but inspection of the proper table is enough to draw a conclusion.

(b) Do the large sample sizes allow us to ignore the assumption that the population distributions are normal?

The remaining exercises concern a study of air pollution. One component of air pollution is airborne particulate matter such as dust and smoke. To measure particulate pollution, a vacuum motor draws air through a filter for 24 hours. Weigh the filter at the beginning and end of the period. The weight gained is a measure of the concentration of particles in the air. A study of air pollution made measurements every 6 days with identical instruments in the center of a small city and at a rural location 10 miles southwest of the city. Because the prevailing winds blow from the west, we suspect that the rural readings will be generally lower than the city readings, but that the city readings can be predicted from the rural readings. Table 6.2 gives readings taken every 6 days over a 7-month period. The entry NA means that the reading for that date is not available, usually because of equipment failure.[38]

Missing data are common, especially in field studies like this one. We think that equipment failures are not related to pollution levels. If that is true, the missing data do not introduce bias. We can work with the data that are not missing as if they are a random sample of days. We can analyze these data in different ways to answer different questions. For each of the three exercises below, do a careful descriptive analysis with graphs and summary statistics and whatever formal inference is called for. Then present and interpret your findings.

6.73   We want to assess the level of particulate pollution in the city center. Describe the distribution of city pollution levels, and estimate the mean particulate level in the city center. (All estimates should include a statistically justified margin of error.)

6.74   We want to compare the mean level of particulates in the city with the rural level on the same day. We suspect that pollution is higher in the city, and we hope that a statistical test will show that there is significant evidence to confirm this suspicion. Make a graph to check for conditions that might prevent the use of the test you plan to employ. Your graph should reflect

TABLE 6.2	Particulate levels (grams) in two nearby locations				
Day	Rural	City	Day	Rural	City
1	NA	39	19	43	42
2	67	68	20	39	38
3	42	42	21	NA	NA
4	33	34	22	52	57
5	46	48	23	48	50
6	NA	82	24	56	58
7	43	45	25	44	45
8	54	NA	26	51	69
9	NA	NA	27	21	23
10	NA	60	28	74	72
11	NA	57	29	48	49
12	NA	NA	30	84	86
13	38	39	31	51	51
14	88	NA	32	43	42
15	108	123	33	45	46
16	57	59	34	41	NA
17	70	71	35	47	44
18	42	41	36	35	42

the type of procedure that you will use. Then carry out a significance test and report your conclusion. Also estimate the mean amount by which the city particulate level exceeds the rural level on the same day.

6.75 We hope to use the rural particulate level to predict the city level on the same day. Make a graph to examine the relationship. Does the graph suggest that using the least-squares regression line for prediction will give approximately correct results over the range of values appearing in the data? Calculate the least-squares line for predicting city pollution from rural pollution. What percent of the observed variation in city pollution levels does this straight-line relationship account for? On the fourteenth date in the series, the rural reading was 88 and the city reading was not available. What do you estimate the city reading to be for that date? (In Chapter 10, we will learn how to give a margin of error for the predictions we make from the regression line.)

NOTES AND DATA SOURCES

1. This example is based on information in D. L. Shankland et al., "The effect of 5-thio-D-glucose on insect development and its absorption by insects," *Journal of Insect Physiology*, 14 (1968), pp. 63–72.

2. Data from D. L. Shankland, "Involvement of spinal cord and peripheral nerves in DDT-poisoning syndrome in albino rats," *Toxicology and Applied Pharmacology*, 6 (1964), pp. 197–213.

3. The data are from Joan M. Susic, *Dietary Phosphorus Intakes, Urinary and Peritoneal Phosphate Excretion and Clearance in Continuous Ambulatory Peritoneal Dialysis Patients*, M.S. thesis, Purdue University, 1985.

4. Data provided by Joseph Wipf, Department of Foreign Languages and Literatures, Purdue University.

5. Data provided by Timothy Sturm.

6. These recommendations are based on extensive computer work. See, for example, Harry O. Posten, "The robustness of the one-sample *t*-test over the Pearson system," *Journal of Statistical Computation and Simulation*, 9 (1979), pp. 133–149, and E. S. Pearson and N. W. Please, "Relation between the shape of population distribution and the robustness of four simple test statistics," *Biometrika*, 62 (1975), pp. 223–241.

7. Based on I. Cuellar, L. C. Harris, and R. Jasso, "An acculturation scale for Mexican American normal and clinical populations," *Hispanic Journal of Behavioral Sciences*, 2 (1980), pp. 199–217.

8. Data provided by Diana Schellenberg, Purdue University School of Health Sciences.

9. Data from the appendix of D. A. Kurtz (ed.), *Trace Residue Analysis*, American Chemical Society Symposium Series, no. 284, 1985.

10. From S. C. Hand and E. Gnaiger, "Anaerobic dormancy quantified in *Artemia* embryos," *Science*, 239 (1988), pp. 1425–1427.

11. Data provided by Joseph Wipf, Department of Foreign Languages and Literatures, Purdue University.

12. Based on Charles W. L. Hill and Phillip Phan, "CEO tenure as a determinant of CEO pay," *The Academy of Management Journal*, 34 (1991), pp. 707–717.

13. This study is reported in Roseann M. Lyle et al., "Blood pressure and metabolic effects of calcium supplementation in normotensive white and black men," *Journal of the American Medical Association*, 257 (1987), pp. 1772–1776. The data were provided by Dr. Lyle.

14. Detailed information about the conservative t procedures can be found in Paul Leaverton and John J. Birch, "Small sample power curves for the two sample location problem," *Technometrics*, 11 (1969), pp. 299–307; in Henry Scheffé, "Practical solutions of the Behrens-Fisher problem," *Journal of the American Statistical Association*, 65 (1970), pp. 1501–1508; and in D. J. Best and J. C. W. Rayner, "Welch's approximate solution for the Behrens-Fisher problem," *Technometrics*, 29 (1987), pp. 205–210.

15. From H. G. Gough, *The Chapin Social Insight Test*, Consulting Psychologists Press, Palo Alto, Calif., 1968.

16. Based on M. C. Wilson et al., "Impact of cereal leaf beetle larvae on yields of oats," *Journal of Economic Entomology*, 62 (1969), pp. 699–702.

17. From Costas Papoulias and Panayiotis Theodossiou, "Analysis and modeling of recent business failures in Greece," *Managerial and Decision Economics*, 13 (1992), pp. 163–169.

18. See the extensive simulation studies in Harry O. Posten, "The robustness of the two-sample t-test over the Pearson system," *Journal of Statistical Computation and Simulation*, 6 (1978), pp. 295–311, and in Harry O. Posten, H. Yeh, and Donald B. Owen, "Robustness of the two-sample t-test under violations of the homogeneity assumption," *Communications in Statistics*, 11 (1982), pp. 109–126.

19. Data for 1982, provided by Marvin Schlatter, Division of Financial Aid, Purdue University.

20. Based on G. L. Cromwell et al., "A comparison of the nutritive value of *opaque-2*, *floury-2* and normal corn for the chick," *Poultry Science*, 47 (1968), pp. 840–847.

21. This example is loosely based on D. L. Shankland, "Involvement of spinal cord and peripheral nerves in DDT-poisoning syndrome in albino rats," *Toxicology and Applied Pharmacology*, 6 (1964), pp. 197–213.

22. We did not use Minitab or Data Desk in Example 6.11 because these packages shortcut the two-sample t procedure. They calculate the degrees of freedom df using the formula in the box on page 450 but then truncate to the next lower whole-number degrees of freedom to obtain the P-value. The result is slightly less accurate than the P-value from the $t(\text{df})$ distribution.

23. Based on W. N. Nelson and C. J. Widule, "Kinematic analysis and efficiency estimate of intercollegiate female rowers," unpublished manuscript, 1983.

24. From a news article in *Science*, 224 (1983), pp. 1029–1031.

25. From M. F. Picciano and R. H. Deering, "The influence of feeding regimens on iron status during infancy," *The American Journal of Clinical Nutrition*, 33 (1980), pp. 746–753.

26. From A. H. Ismail and R. J. Young, "The effect of chronic exercise on the personality of middle-aged men," *Journal of Human Ergology*, 2 (1973), pp. 47–57.

27. Adapted from Maribeth Cassidy Schmitt, *The Effects of an Elaborated Directed Reading Activity on the Metacomprehension Skills of Third Graders*, Ph.D. dissertation, Purdue University, 1987.

28. From M. A. Zlatin and R. A. Koenigsknecht, "Development of the voicing contrast: a comparison of voice onset time in stop perception and production," *Journal of Speech and Hearing Research*, 19 (1976), pp. 93–111.

29. The problem of comparing spreads is difficult even with advanced methods. Common distribution-free procedures do not offer a satisfactory alternative to the *F* test, because they are sensitive to unequal shapes when comparing two distributions. A good introduction to the available methods is W. J. Conover, M. E. Johnson, and M. M. Johnson, "A comparative study of tests for homogeneity of variances, with applications to outer continental shelf bidding data," *Technometrics*, 23 (1981), pp. 351–361. Modern resampling procedures often work well. See Dennis D. Boos and Colin Brownie, "Bootstrap methods for testing homogeneity of variances," *Technometrics*, 31 (1989), pp. 69–82.

30. Based loosely on D. R. Black et al., "Minimal interventions for weight control: a cost-effective alternative," *Addictive Behaviors*, 9 (1984), pp. 279–285.

31. Based on Amna Kirmani and Peter Wright, "Money talks: perceived advertising expense and expected product quality," *Journal of Consumer Research*, 16 (1989), pp. 344–353.

32. Based on D. Friedlander, *Supplemental Report on the Baltimore Options Program*, Manpower Demonstration Research Corporation, 1987.

33. Based on G. Salvendy, "Selection of industrial operators: the one-hole test," *International Journal of Production Research*, 13 (1973), pp. 303–321.

34. Data from R. A. Berner and G. P. Landis, "Gas bubbles in fossil amber as possible indicators of the major gas composition of ancient air," *Science*, 239 (1988), pp. 1406–1409.

35. From J. W. Marr and J. A. Heady, "Within- and between-person variation in dietary surveys: number of days needed to classify individuals," *Human Nutrition: Applied Nutrition*, 40A (1986), pp. 347–364.

36. From V. D. Bass, W. E. Hoffmann, and J. L. Dorner, "Normal canine lipid profiles and effects of experimentally induced pancreatitis and hepatic necrosis on lipids," *American Journal of Veterinary Research*, 37 (1976), pp. 1355–1357.

37. From M. L. Pollock et al., "Body composition of elite class distance runners," in P. Milvey (ed.), *The Marathon: Physiological, Medical, Epidemiological, and Psychological Studies*, New York Academy of Sciences, 1977, p. 366.

38. Data provided by Matthew Moore.

JANET NORWOOD

The commissioner of labor statistics is one of the nation's most influential statisticians. As head of the Bureau of Labor Statistics, the commissioner supervises the collection and interpretation of data on employment, earnings, and many other economic and social trends.

The data collected by the Bureau of Labor Statistics are often politically sensitive, as when a report released just before an election shows rising unemployment. For this reason, the bureau must remain objective and independent of political influence. To safeguard the bureau's independence, the commissioner is appointed by the president and confirmed by the Senate for a fixed term of four years. The commissioner must have statistical skill, administrative ability, and a facility for working with both Congress and the president.

Janet Norwood served three terms as commissioner, from 1979 to 1991, under three presidents. When she retired, the *New York Times* said (December 31, 1991) that she left with "a near-legendary reputation for nonpartisanship and plaudits that include one senator's designation of her as a 'national treasure.'" Norwood says, "There have been times in the past when commissioners have been in open disagreement with the Secretary of Labor or, in some cases, with the President. We have guarded our professionalism with great care."

Some of the most important statistics produced by the Bureau of Labor Statistics are proportions. The monthly unemployment rate, for example, is the proportion of the labor force that is unemployed this month. Methods for inference about proportions are the topic of this chapter.

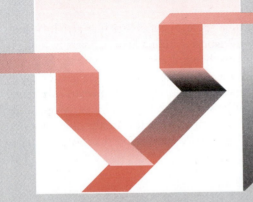

CHAPTER **7**

Inference for
Proportions

INTRODUCTION

We often want to answer questions about the proportion of some outcome in a population or to compare proportions in several populations. Here are some examples that call for inference about population proportions.

EXAMPLE 7.1

How common is behavior that puts people at risk of AIDS? The National AIDS Behavioral Surveys interviewed a random sample of 2673 adult heterosexuals. Of these, 170 had more than one sexual partner in the past year. That's 6.36% of the sample.[1] Based on these data, what can we say about the percent of all adult heterosexuals who have multiple partners? We want to *estimate a single population proportion*. ◄

EXAMPLE 7.2

Do preschool programs for poor children make a difference in later life? A study looked at 62 children who were enrolled in a Michigan preschool in the late 1960s and at a control group of 61 similar children who were not enrolled. At 27 years of age, 61% of the preschool group and 80% of the control group had required the help of a social service agency (mainly welfare) in the previous ten years.[2] Is this significant evidence that preschool for poor children reduces later use of social services? We want to *compare two population proportions*. ◄

EXAMPLE 7.3

What is the relationship between time spent in extracurricular activities and success in a tough course in college? North Carolina State University looked at the 123 students in an introductory chemical engineering course. Students needed a grade of C or better to advance to the next course. The passing rates were 55% for students who spent less than 2 hours per week in extracurricular activities, 75% for those who spent between 2 and 12 hours per week, and 38% for those who spent more than 12 hours per week.[3] Are the differences in passing rates statistically significant? We must *compare more than two population proportions*. ◄

Our study of inference for proportions will follow the same pattern as these examples. Section 7.1 discusses inference for one population proportion, and Section 7.2 presents methods for comparing two proportions. Comparing more than two proportions raises new issues and requires more elaborate methods that also apply to some other inference problems. These methods are the topic of Chapter 8.

7.1   INFERENCE FOR A POPULATION PROPORTION

We are interested in the unknown proportion p of a population that has some outcome. For convenience, call the outcome we are looking for a "success." In Example 7.1, the population is adult heterosexuals, and the parameter p is the proportion who have had more than one sexual partner in the past year. To estimate p, the National AIDS Behavioral Surveys used random dialing of telephone numbers to contact a sample of 2673 people. Of these, 170 said they had multiple sexual partners. The statistic that estimates the parameter p is the **sample proportion**

sample proportion

$$\hat{p} = \frac{\text{count of successes in the sample}}{\text{count of observations in the sample}}$$

$$= \frac{170}{2673} = .0636$$

Read the sample proportion $\hat{p}$ as "p-hat."

EXERCISES

In each of the following settings:

(a) Describe the population and explain in words what the parameter p is.

(b) Give the numerical value of the statistic $\hat{p}$ that estimates p.

7.1   Tonya wants to estimate what proportion of the students in her dormitory like the dorm food. She interviews an SRS of 50 of the 175 students living in the dormitory. She finds that 14 think the dorm food is good.

7.2   Glenn wonders what proportion of the students at his school think that tuition is too high. He interviews an SRS of 50 of the 2400 students at his college. Thirty-eight of those interviewed think tuition is too high.

7.3   A college president says, "99% of the alumni support my firing of Coach Boggs." You contact an SRS of 200 of the college's 15,000 living alumni and find that 76 of them support firing the coach.

Assumptions for inference

As always, inference is based on the sampling distribution of a statistic. We described the sampling distribution of a sample proportion $\hat{p}$ in Section 3 of Chapter 4. The mean is p. That is, the sample proportion $\hat{p}$ is an unbiased estimator of the population proportion p. The standard deviation of $\hat{p}$ is $\sqrt{p(1-p)/n}$. The normal approximation says that for large samples, the distribution of $\hat{p}$ is approximately normal. Figure 7.1 displays this sampling distribution.

Standardize $\hat{p}$ by subtracting its mean and dividing by its standard deviation. The result is a z statistic:

$$z = \frac{\hat{p} - p}{\sqrt{\dfrac{p(1-p)}{n}}}$$

The statistic z has approximately the standard normal distribution $N(0, 1)$. Inference about p uses this z statistic and standard normal critical values.

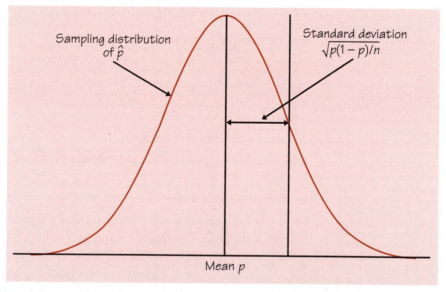

FIGURE 7.1   Select a large SRS from a population that contains proportion p of successes. The sampling distribution of the proportion $\hat{p}$ of successes in the sample is approximately normal. The mean is p and the standard deviation is $\sqrt{p(1-p)/n}$.

However, we need to deal with the fact that we don't know the standard deviation $\sqrt{p(1-p)/n}$ because we don't know p. Here's what we do:

- To test the null hypothesis $H_0 : p = p_0$ that the unknown p has a specific value p_0, just replace p by p_0 in the z statistic.
- In a confidence interval for p, we have no specific value to substitute. In large samples, $\hat{p}$ will be close to p, so replace the standard deviation by the **standard error of $\hat{p}$**

standard error of $\hat{p}$

$$SE = \sqrt{\frac{\hat{p}(1-\hat{p})}{n}}$$

The confidence interval has the form

$$\text{estimate} \pm z^*SE_{\text{estimate}}$$

The normal approximation is accurate if the sample is large and the population proportion p is not too close to 0 or 1. We don't know the value of p. For a test, the null hypothesis states that p has a specified value p_0. For confidence intervals, we use the statistic $\hat{p}$ to estimate the unknown p. So the requirements for using the z procedures for inference about a proportion are stated in terms of p_0 or $\hat{p}$.

ASSUMPTIONS FOR INFERENCE ABOUT A PROPORTION

- The data are an SRS from the population of interest.
- The population is at least 10 times as large as the sample.
- For a test of $H_0 : p = p_0$, the sample size n is so large that both np_0 and $n(1 - p_0)$ are 10 or more. For a confidence interval, n is so large that both the count of successes $n\hat{p}$ and the count of failures $n(1 - \hat{p})$ are 10 or more.

If you have a small sample or a sampling design more complex than an SRS, you can still do inference, but the details are more complicated. Get expert advice.

EXAMPLE 7.4

We want to use the National AIDS Behavioral Surveys data to give a confidence interval for the proportion of adult heterosexuals who have had multiple sexual partners. Does the sample meet the requirements for inference?

- The sampling design was in fact a complex stratified sample, and the survey used inference procedures for that design. The overall effect is close to an SRS, however.
- The number of adult heterosexuals (the population) is much larger than 10 times the sample size, $n = 2673$.
- The counts of "Yes" and "No" responses are much greater than 10:

$$n\hat{p} = (2673)(.0636) = 170$$
$$n(1 - \hat{p}) = (2673)(.9364) = 2503$$

The second and third requirements are easily met. The first requirement, that the sample be an SRS, is only approximately met.   ◀

As usual, the practical problems of a large sample survey pose a greater threat to the AIDS survey's conclusions. Only people in households with telephones could be reached. This is acceptable for surveys of the general population, because about 94% of American households have telephones. However, some groups at high risk for AIDS, like intravenous drug users, often don't live in settled households and are underrepresented in the sample. About 30% of the people reached refused to cooperate. A nonresponse rate of 30% is not unusual in large sample surveys, but it may cause some bias if those who refuse differ systematically from those who cooperate. The survey used statistical methods that adjust for unequal response rates in different groups. Finally, some respondents may not have told the truth when asked about their sexual behavior. The survey team tried hard to make respondents feel comfortable. For example, Hispanic women were interviewed only by Hispanic women, and Spanish speakers were interviewed by Spanish speakers with the same regional accent (Cuban, Mexican, or Puerto Rican). Nonetheless, the survey report says that some bias is probably present:

It is more likely that the present figures are underestimates; some respondents may underreport their numbers of sexual partners and intravenous drug use because of embarrassment and fear of reprisal, or they may forget or not know details of their own or of their partner's HIV risk and their antibody testing history.[4]

Reading the report of a large study like the National AIDS Behavioral Surveys reminds us that statistics in practice involves much more than recipes for inference.

EXERCISES

7.4   In which of the following situations can you safely use the methods of this section to get a confidence interval for the population proportion p? Explain your answers.

(a) Tonya wants to estimate what proportion of the students in her dormitory like the dorm food. She interviews an SRS of 50 of the 175 students living in the dormitory. She finds that 14 think the dorm food is good.

(b) Glenn wonders what proportion of the students at his school think that tuition is too high. He interviews an SRS of 50 of the 2400 students at his college. Thirty-eight of those interviewed think tuition is too high.

(c) In the National AIDS Behavioral Surveys sample of 2673 adult heterosexuals, 0.2% (that's 0.002 as a decimal fraction) had both received a blood transfusion and had a sexual partner from a group at high risk of AIDS. (We want to estimate the proportion p in the population who share these two risk factors.)

7.5   In which of the following situations can you safely use the methods of this section for a significance test? Explain your answers.

(a) You toss a coin 10 times in order to test the hypothesis $H_0 : p = .5$ that the coin is balanced.

(b) A college president says, "99% of the alumni support my firing of Coach Boggs." You contact an SRS of 200 of the college's 15,000 living alumni to test the hypothesis $H_0 : p = .99$.

(c) Do a majority of the 250 students in a statistics course agree that knowing statistics will help them in their future careers? You interview an SRS of 20 students to test $H_0 : p = .5$.

The z procedures

Here are the z procedures for inference about p.

LARGE-SAMPLE INFERENCE FOR A POPULATION PROPORTION

Draw an SRS of size n from a large population with unknown proportion p of successes. An approximate level C confidence interval for p is

$$\hat{p} \pm z^* \sqrt{\frac{\hat{p}(1 - \hat{p})}{n}}$$

where z^* is the upper $(1 - C)/2$ standard normal critical value.

To test the hypothesis $H_0 : p = p_0$, compute the z statistic

$$z = \frac{\hat{p} - p_0}{\sqrt{\dfrac{p_0(1 - p_0)}{n}}}$$

In terms of a variable Z having the standard normal distribution, the approximate P-value for a test of H_0 against

$H_a : p > p_0$ is $P(Z \geq z)$

$H_a : p < p_0$ is $P(Z \leq z)$

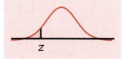

$H_a : p \neq p_0$ is $2P(Z \geq |z|)$

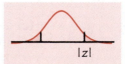

EXAMPLE 7.5

The National AIDS Behavioral Surveys found that 170 of a sample of 2673 adult heterosexuals had multiple partners. That is, $\hat{p} = .0636$. We will act as if the sample were an SRS.

A 99% confidence interval for the proportion p of all adult heterosexuals with multiple partners uses the standard normal critical value $z^* = 2.576$. (Look in the bottom row of Table C for standard normal critical values.) The confidence interval is

$$\hat{p} \pm z^* \sqrt{\frac{\hat{p}(1 - \hat{p})}{n}} = .0636 \pm 2.576 \sqrt{\frac{(.0636)(.9364)}{2673}}$$
$$= .0636 \pm .0122$$
$$= (.0514, .0758)$$

We are 99% confident that the percent of adult heterosexuals who had more than one sexual partner in the past year lies between about 5.1% and 7.6%. ◀

EXAMPLE 7.6

A coin that is balanced should come up heads half the time in the long run. The population for coin tossing contains the results of tossing the coin forever. The parameter p is the probability of a head, which is the proportion of all tosses that give a head. The tosses we actually make are an SRS from this population.

The French naturalist Count Buffon (1707–1788) tossed a coin 4040 times. He got 2048 heads. The sample proportion of heads is

$$\hat{p} = \frac{2048}{4040} = .5069$$

That's a bit more than one-half. Is this evidence that Buffon's coin was not balanced? This is a job for a significance test.

Step 1: Hypotheses. The null hypothesis says that the coin is balanced ($p = .5$). The alternative hypothesis is two-sided, because we did not suspect before seeing the data that the coin favored either heads or tails. We therefore test the hypotheses

$$H_0 : p = .5$$
$$H_a : p \neq .5$$

The null hypothesis gives p the value $p_0 = .5$.

Step 2: Test statistic.   The z test statistic is

$$z = \frac{\hat{p} - p_0}{\sqrt{\frac{p_0(1 - p_0)}{n}}}$$

$$= \frac{.5069 - .5}{\sqrt{\frac{(.5)(.5)}{4040}}} = .88$$

Step 3: P-value.   Because the test is two-sided, the P-value is the area under the standard normal curve more than 0.88 away from 0 in either direction. Figure 7.2 shows this area. From Table A we find that the area below -0.88 is 0.1894. The P-value is twice this area:

$$P = 2(.1894) = .3788$$

Conclusion.   A proportion of heads as far from one-half as Buffon's would happen 38% of the time when a balanced coin is tossed 4040 times. Buffon's result doesn't show that his coin is unbalanced.   ◀

In Example 7.6, we failed to find good evidence against $H_0 : p = .5$. We *cannot* conclude that H_0 is true, that is, that the coin is perfectly

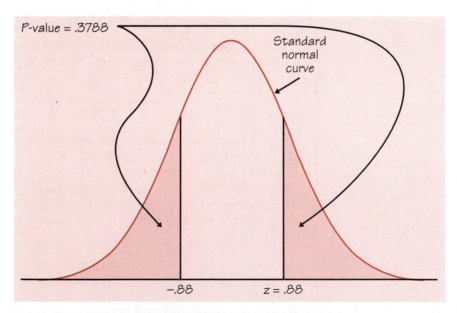

FIGURE 7.2   The P-value for the two-sided test of Example 7.6.

balanced. No doubt p is not exactly 0.5. The test of significance only shows that the results of Buffon's 4040 tosses can't distinguish this coin from one that is perfectly balanced. To see what values of p are consistent with the sample results, use a confidence interval.

EXAMPLE 7.7

The 95% confidence interval for the probability p that Buffon's coin gives a head is

$$\hat{p} \pm z^* \sqrt{\frac{\hat{p}(1 - \hat{p})}{n}} = .5069 \pm 1.960 \sqrt{\frac{(.5069)(.4931)}{4040}}$$

$$= .5069 \pm .0154$$

$$= (.4915, .5223)$$

We are 95% confident that the probability of a head is between 0.4915 and 0.5223.

The confidence interval is more informative than the test in Example 7.6. It tells us that any null hypothesis $H_0 : p = p_0$ for a p_0 between 0.4915 and 0.5223 would not be rejected at the $\alpha = .05$ level of significance. We would not be surprised if the true probability of a head for Buffon's coin were something like 0.51. ◀

EXERCISES

7.6   As part of a quality improvement program, your mail-order company is studying the process of filling customer orders. Company standards say an order is shipped on time if it is sent within 3 working days after it is received. You audit an SRS of 100 of the 5000 orders received in the past month. The audit reveals that 86 of these orders were shipped on time.

(a) Check that you can safely use the methods in this section.

(b) Find a 95% confidence interval for the true proportion of the month's orders that were shipped on time.

7.7   The National AIDS Behavioral Surveys (Example 7.1) also interviewed a sample of adults in the cities where AIDS is most common. The sample included 803 heterosexuals who reported having more than one sexual partner in the past year. We can consider this an SRS of size 803 from the population of all heterosexuals in high-risk cities who have multiple partners. These people risk infection with the AIDS virus. Yet 304 of the respondents said they never use condoms. Is this strong evidence that more than one-third of this population never use condoms?

7.8   In a recent year, 73% of first-year college students responding to a national survey identified "being very well-off financially" as an important personal goal. A state university finds that 132 of an SRS of 200 of its first-year students say that this goal is important.

   (a) Give a 95% confidence interval for the proportion of all first-year students at the university who would identify being well-off as an important personal goal.

   (b) Is there good evidence that the proportion of first-year students at this university who think being very well-off is important differs from the national value, 73%? (Be sure to state hypotheses, give the P-value, and state your conclusion.)

   (c) Check that you could safely use the methods of this section in both (a) and (b).

7.9   Around the year 1900, the English statistician Karl Pearson tossed a coin 24,000 times. He obtained 12,012 heads.

   (a) Test the null hypothesis that Pearson's coin had probability 0.5 of coming up heads versus the two-sided alternative. Give the P-value. Do you reject H_0 at the 1% significance level?

   (b) Find a 99% confidence interval for the probability of heads for Pearson's coin. This is the range of probabilities that cannot be rejected at the 1% significance level.

Choosing the sample size

In planning a study, we may want to choose a sample size that will allow us to estimate the parameter within a given margin of error. We saw earlier (page 341) how to do this for a population mean. The method is similar for estimating a population proportion.

The margin of error in the approximate confidence interval for p is

$$m = z^* \sqrt{\frac{\hat{p}(1 - \hat{p})}{n}}$$

Here z^* is the standard normal critical value for the level of confidence we want. Because the margin of error involves the sample proportion of successes $\hat{p}$, we need to guess this value when choosing n. Call our guess p^*. Here are two ways to get p^*:

1. Use a guess p^* based on a pilot study or on past experience with similar studies. You should do several calculations that cover the range of $\hat{p}$-values you might get.

2. Use $p^* = .5$ as the guess. The margin of error m is largest when $\hat{p} = .5$, so this guess is conservative in the sense that if we get any other $\hat{p}$ when we do our study, we will get a margin of error smaller than planned.

Once you have a guess p^*, the recipe for the margin of error can be solved to give the sample size n needed. Here is the result.

SAMPLE SIZE FOR DESIRED MARGIN OF ERROR

The level C confidence interval for a population proportion p will have a margin of error approximately equal to a specified value m when the sample size is

$$n = \left(\frac{z^*}{m}\right)^2 p^*(1 - p^*)$$

where p^* is a guessed value for the sample proportion. The margin of error will be less than or equal to m if you take the guess p^* to be 0.5.

Which method for finding the guess p^* should you use? The n you get doesn't change much when you change p^* as long as p^* is not too far from 0.5. So use the conservative guess $p^* = .5$ if you expect the true $\hat{p}$ to be roughly between 0.3 and 0.7. If the true $\hat{p}$ is close to 0 or 1, using $p^* = .5$ as your guess will give a sample much larger than you need. So try to use a better guess from a pilot study when you suspect that $\hat{p}$ will be less than 0.3 or greater than 0.7.

EXAMPLE 7.8

Gloria Chavez and Ronald Flynn are the candidates for mayor in a large city. You are planning a sample survey to determine what percent of the voters plan to vote for Chavez. This is a population proportion p. You will contact an SRS of registered voters in the city. You want to estimate p with 95% confidence and a margin of error no greater than 3%, or 0.03. How large a sample do you need?

The winner's share in all but the most lopsided elections is between 30% and 70% of the vote. So use the guess $p^* = .5$. The sample size you need is

$$n = \left(\frac{1.96}{.03}\right)^2 (.5)(1 - .5) = 1067.1$$

You should round the result up to $n = 1068$. (Rounding down would give a margin of error slightly greater than 0.03.) If you want a 2.5% margin of error, we have (after rounding up)

$$n = \left(\frac{1.96}{.025}\right)^2 (.5)(1 - .5) = 1537$$

For a 2% margin of error the sample size you need is

$$n = \left(\frac{1.96}{.02}\right)^2 (.5)(1 - .5) = 2401$$

As usual, smaller margins of error call for larger samples.   ◀

EXERCISES

7.10   A national opinion poll found that 44% of all American adults agree that parents should be given vouchers good for education at any public or private school of their choice. The result was based on a small sample. How large an SRS is required to obtain a margin of error of 0.03 (that is, ±3%) in a 95% confidence interval?

(a) Answer this question using the previous poll's result as the guessed value p^*.

(b) Do the problem again using the conservative guess $p^* = .5$. By how much do the two sample sizes differ?

7.11   PTC is a substance that has a strong bitter taste for some people and is tasteless for others. The ability to taste PTC is inherited. About 75% of Italians can taste PTC, for example. You want to estimate the proportion of Americans with at least one Italian grandparent who can taste PTC. Starting with the 75% estimate for Italians, how large a sample must you test in order to estimate the proportion of PTC tasters within ±0.04 with 95% confidence?

SUMMARY

Tests and confidence intervals for a population proportion p when the data are an SRS of size n are based on the **sample proportion** $\hat{p}$. When n is large, $\hat{p}$ has approximately the normal distribution with mean p and standard deviation $\sqrt{p(1-p)/n}$.

The level C **confidence interval** for p is

$$\hat{p} \pm z^* \sqrt{\frac{\hat{p}(1-\hat{p})}{n}}$$

where z^* is the upper $(1-C)/2$ standard normal critical value.

Tests of $H_0 : p = p_0$ are based on the **z statistic**

$$z = \frac{\hat{p} - p_0}{\sqrt{\dfrac{p_0(1-p_0)}{n}}}$$

with P-values calculated from the standard normal distribution.

These inference procedures are approximately correct when the population is at least 10 times as large as the sample and the sample is large enough to satisfy $n\hat{p} \geq 10$ and $n(1-\hat{p}) \geq 10$ for a confidence interval or $np_0 \geq 10$ and $n(1-p_0) \geq 10$ for a test of $H_0 : p = p_0$.

The **sample size** needed to obtain a confidence interval with approximate margin of error m for a population proportion is

$$n = \left(\frac{z^*}{m}\right)^2 p^*(1-p^*)$$

where p^* is a guessed value for the sample proportion $\hat{p}$, and z^* is the standard normal critical point for the level of confidence you want. If you use $p^* = .5$ in this formula, the margin of error of the interval will be less than or equal to m no matter what the value of $\hat{p}$ is.

SECTION 7.1 EXERCISES

7.12   How likely are patients who file complaints with a health maintenance organization (HMO) to leave the HMO? In one recent year, 639 of the more than 400,000 members of a large New England HMO filed complaints. Fifty-four of the complainers left the HMO voluntarily. (That is, they were not forced to leave by a move or a job change.)[5] Consider this year's complainers as an SRS of all patients who will complain in the future. Give a 90% confidence interval for the proportion of complainers who voluntarily leave the HMO.

7.13   An experiment on the side effects of pain relievers assigned arthritis patients to one of several over-the-counter pain medications. Of the 440 patients who took one brand of pain reliever, 23 suffered some "adverse symptom." Does the experiment provide strong evidence that fewer than 10% of patients who take this medication have adverse symptoms?

7.14   The Gallup Poll asked a sample of 1785 adults, "Did you, yourself, happen to attend church or synagogue in the last 7 days?" Of the respondents, 750 said "Yes." Suppose (it is not, in fact, true) that Gallup's sample was an SRS of all American adults.

(a) Give a 99% confidence interval for the proportion of all adults who attended church or synagogue during the week preceding the poll.

(b) Do the results provide good evidence that less than half of the population attended church or synagogue?

(c) How large a sample would be required to obtain a margin of error of 0.01 in a 99% confidence interval for the proportion who attend church or synagogue? (Use the conservative guess $p^* = .5$, and explain why this method is reasonable in this situation.)

7.15   A study of the survival of small businesses chose an SRS from the telephone directory's Yellow Pages listings of food-and-drink businesses in 12 counties in central Indiana. For various reasons, the study got no response from 45% of the businesses chosen. Interviews were completed with 148 businesses. Three years later, 22 of these businesses had failed.[6]

(a) Give a 95% confidence interval for the percent of all small businesses in this class that fail within three years.

(b) Based on the results of this study, how large a sample would you need to reduce the margin of error to 0.04?

(c) The authors hope that their findings describe the population of all small businesses. What about the study makes this unlikely? What population do you think the study findings describe?

7.16   One-sample procedures for proportions, like those for means, are used to analyze data from matched pairs designs. Here is an example.

Each of 50 subjects tastes two unmarked cups of coffee and says which he or she prefers. One cup in each pair contains instant coffee; the other, fresh-brewed coffee. Thirty-one of the subjects prefer the fresh-brewed coffee. Take p to be the proportion of the population who would prefer fresh-brewed coffee in a blind tasting.

(a) Test the claim that a majority of people prefer the taste of fresh-brewed coffee. State hypotheses and report the z statistic and its P-value. Is your result significant at the 5% level? What is your practical conclusion?

(b) Find a 90% confidence interval for p.

(c) When you do an experiment like this, in what order should you present the two cups of coffee to the subjects?

7.17 An automobile manufacturer would like to know what proportion of its customers are not satisfied with the service provided by their local dealer. The customer relations department will survey a random sample of customers and compute a 99% confidence interval for the proportion who are not satisfied.

(a) From past studies, they believe that this proportion will be about 0.2. Find the sample size needed if the margin of error of the confidence interval is to be about 0.015.

(b) When the sample is actually contacted, 10% of the sample say they are not satisfied. What is the margin of error of the 99% confidence interval?

7.18 You are planning a survey of students at a large university to determine what proportion favor an increase in student fees to support an expansion of the student newspaper. Using records provided by the registrar you can select a random sample of students. You will ask each student in the sample whether he or she is in favor of the proposed increase. Your budget will allow a sample of 100 students.

(a) For a sample of size 100, construct a table of the margins of error for 95% confidence intervals when $\hat{p}$ takes the values 0.1, 0.2, 0.3, 0.4, 0.5, 0.6, 0.7, 0.8, and 0.9.

(b) A former editor of the student newspaper offers to provide funds for a sample of size 500. Repeat the margin of error calculations in (a) for the larger sample size. Then write a short thank-you note to the former editor describing how the larger sample size will improve the results of the survey.

7.2   COMPARING TWO PROPORTIONS

two-sample
problem

In a **two-sample problem**, we want to compare two populations or the responses to two treatments based on two independent samples. When the comparison involves the mean of a quantitative variable, we use the two-sample t methods of Section 6.2. To compare the standard deviations of a variable in two groups, we use (under restrictive conditions) the F statistic described in the optional Section 6.3. Now we turn to methods to compare the proportions of successes in two groups.

We will use notation similar to that used in our study of two-sample t statistics. The groups we want to compare are Population 1 and Population 2. We have a separate SRS from each population or responses from two treatments in a randomized comparative experiment. A subscript shows which group a parameter or statistic describes. Here is our notation:

Population	Population proportion	Sample size	Sample proportion
1	p_1	n_1	$\hat{p}_1$
2	p_2	n_2	$\hat{p}_2$

We compare the populations by doing inference about the difference $p_1 - p_2$ between the population proportions. The statistic that estimates this difference is the difference between the two sample proportions, $\hat{p}_1 - \hat{p}_2$.

EXAMPLE 7.9

To study the long-term effects of preschool programs for poor children, the High/Scope Educational Research Foundation has followed two groups of Michigan children since early childhood. One group of 62 attended preschool as 3- and 4-year-olds. This is a sample from Population 2, poor children who attend preschool. A control group of 61 children from the same area and similar backgrounds represents Population 1, poor children with no preschool. Thus the sample sizes are $n_1 = 61$ and $n_2 = 62$.

One response variable of interest is the need for social services as adults. In the past ten years, 38 of the preschool sample and 49 of the control sample have needed social services (mainly welfare). The sample proportions are

$$\hat{p}_1 = \frac{49}{61} = .803$$

$$\hat{p}_2 = \frac{38}{62} = .613$$

That is, about 80% of the control group uses social services, as opposed to about 61% of the preschool group.

To see if the study provides significant evidence that preschool reduces the later need for social services, we test the hypotheses

$$H_0 : p_1 - p_2 = 0 \quad \text{or} \quad H_0 : p_1 = p_2$$
$$H_a : p_1 - p_2 > 0 \quad \text{or} \quad H_a : p_1 > p_2$$

To estimate how large the reduction is, we give a confidence interval for the difference, $p_1 - p_2$. Both the test and the confidence interval start from the difference of sample proportions:

$$\hat{p}_1 - \hat{p}_2 = .803 - .613 = .190 \qquad \blacktriangleleft$$

The sampling distribution of $\hat{p}_1 - \hat{p}_2$

Here are the facts about the sampling distribution of $\hat{p}_1 - \hat{p}_2$ that we need for inference:

- The mean of $\hat{p}_1 - \hat{p}_2$ is $p_1 - p_2$. That is, the difference of sample proportions is an unbiased estimator of the difference of population proportions.
- The variance of the difference is the *sum* of the variances of $\hat{p}_1$ and $\hat{p}_2$, which is

$$\frac{p_1(1 - p_1)}{n_1} + \frac{p_2(1 - p_2)}{n_2}$$

Note that the *variances* add. The standard deviations do not.

- When the samples are large, the distribution of $\hat{p}_1 - \hat{p}_2$ is approximately normal.

Figure 7.3 displays the distribution of $\hat{p}_1 - \hat{p}_2$. The standard deviation of $\hat{p}_1 - \hat{p}_2$ involves the unknown parameters p_1 and p_2. Just as in the previous section, we must replace these by estimates in order to do inference. And just as in the previous section, we do this a bit differently for confidence intervals and for tests.

Confidence intervals for $p_1 - p_2$

The standard deviation of $\hat{p}_1 - \hat{p}_2$ is the square root of the variance:

$$\sqrt{\frac{p_1(1 - p_1)}{n_1} + \frac{p_2(1 - p_2)}{n_2}}$$

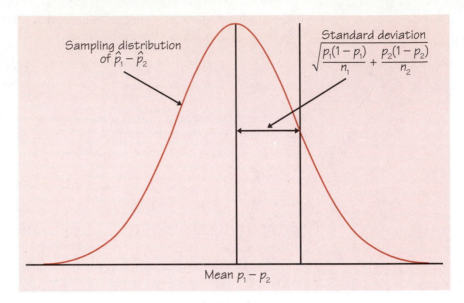

FIGURE 7.3 Select independent SRSs from two populations having proportions of successes p_1 and p_2. The proportions of successes in the two samples are $\hat{p}_1$ and $\hat{p}_2$. When the samples are large, the sampling distribution of the difference $\hat{p}_1 - \hat{p}_2$ is approximately normal.

To obtain a confidence interval, replace the population proportions p_1 and p_2 in this expression by the sample proportions. The result is the **standard error** of the statistic $\hat{p}_1 - \hat{p}_2$:

standard error

$$\text{SE} = \sqrt{\frac{\hat{p}_1(1 - \hat{p}_1)}{n_1} + \frac{\hat{p}_2(1 - \hat{p}_2)}{n_2}}$$

The confidence interval again has the form

$$\text{estimate} \pm z^* \text{SE}_{\text{estimate}}$$

CONFIDENCE INTERVALS FOR COMPARING TWO PROPORTIONS

Draw an SRS of size n_1 from a population having proportion p_1 of successes and draw an independent SRS of size n_2 from another population having proportion p_2 of successes. When n_1 and n_2 are large, an approximate level C confidence interval for $p_1 - p_2$ is

$$(\hat{p}_1 - \hat{p}_2) \pm z^* \text{SE}$$

(continued on next page)

(continued from previous page)

In this formula the standard error SE of $\hat{p}_1 - \hat{p}_2$ is

$$SE = \sqrt{\frac{\hat{p}_1(1 - \hat{p}_1)}{n_1} + \frac{\hat{p}_2(1 - \hat{p}_2)}{n_2}}$$

and z^* is the upper $(1 - C)/2$ standard normal critical value. In practice, use this confidence interval when the populations are at least 10 times as large as the samples and $n_1\hat{p}_1$, $n_1(1 - \hat{p}_1)$, $n_2\hat{p}_2$, and $n_2(1 - \hat{p}_2)$ are all 5 or more.

EXAMPLE 7.10

Example 7.9 describes a study of the effect of preschool on later use of social services. The facts are:

Population	Population description	Sample size	Sample proportion
1	control	$n_1 = 61$	$\hat{p}_1 = .803$
2	preschool	$n_2 = 62$	$\hat{p}_2 = .613$

To check that our approximate confidence interval is safe, look at the counts of successes and failures in the two samples. The smallest of these four quantities is

$$n_1(1 - \hat{p}_1) = (61)(1 - .803) = 12$$

This is larger than 5, so the interval will be accurate.

The difference $p_1 - p_2$ measures the effect of preschool in reducing the proportion of people who later need social services. To compute a 95% confidence interval for $p_1 - p_2$, first find the standard error

$$SE = \sqrt{\frac{\hat{p}_1(1 - \hat{p}_1)}{n_1} + \frac{\hat{p}_2(1 - \hat{p}_2)}{n_2}}$$
$$= \sqrt{\frac{(.803)(.197)}{61} + \frac{(.613)(.387)}{62}}$$
$$= \sqrt{.00642} = .0801$$

The 95% confidence interval is

$$(\hat{p}_1 - \hat{p}_2) \pm z^*\mathrm{SE} = (.803 - .613) \pm (1.960)(.0801)$$
$$= .190 \pm .157$$
$$= (.033, .347)$$

We are 95% confident that the percent needing social services is somewhere between 3.3% and 34.7% lower among people who attended preschool. The confidence interval is wide because the samples are quite small. ◄

EXERCISES

7.19 The 1958 Detroit Area Study was an important investigation of the influence of religion on everyday life. The sample "was basically a simple random sample of the population of the metropolitan area" of Detroit, Michigan. Of the 656 respondents, 267 were white Protestants and 230 were white Catholics.

   The study took place at the height of the cold war. One question asked if the right of free speech included the right to make speeches in favor of communism. Of the 267 white Protestants, 104 said "Yes," while 75 of the 230 white Catholics said "Yes."[7]

   (a) Give a 95% confidence interval for the difference between the proportion of Protestants who agreed that communist speeches are protected and the proportion of Catholics who held this opinion.

   (b) Check that it is safe to use the z confidence interval.

7.20 Exercise 7.12 describes a study of whether patients who file complaints leave a health maintenance organization (HMO). We want to know whether complainers are more likely to leave than patients who do not file complaints. In the year of the study, 639 patients filed complaints, and 54 of these patients left the HMO voluntarily. For comparison, the HMO chose an SRS of 743 patients who had not filed complaints. Twenty-two of these patients left voluntarily.

   (a) How much higher is the proportion of complainers who leave? Give a 90% confidence interval.

   (b) The HMO has more than 400,000 members. Check that you can safely use the methods of this section.

Significance tests for $p_1 - p_2$

An observed difference between two sample proportions can reflect a difference in the populations, or it may just be due to chance variation in random sampling. Significance tests help us decide if the effect we see in the samples is really there in the populations. The null hypothesis says that there is no difference between the two populations:

$$H_0 : p_1 = p_2$$

The alternative hypothesis says what kind of difference we expect.

EXAMPLE 7.11

High levels of cholesterol in the blood are associated with higher risk of heart attacks. Will using a drug to lower blood cholesterol reduce heart attacks? The Helsinki Heart Study looked at this question. Middle-aged men were assigned at random to one of two treatments: 2051 men took the drug gemfibrozil to reduce their cholesterol levels, and a control group of 2030 men took a placebo. During the next five years, 56 men in the gemfibrozil group and 84 men in the placebo group had heart attacks.

The sample proportions who had heart attacks are

$$\hat{p}_1 = \frac{56}{2051} = .0273 \qquad \text{(gemfibrozil group)}$$

$$\hat{p}_2 = \frac{84}{2030} = .0414 \qquad \text{(placebo group)}$$

That is, about 4.1% of the men in the placebo group had heart attacks, against only about 2.7% of the men who took the drug. Is the apparent benefit of gemfibrozil statistically significant? We hope to show that gemfibrozil reduces heart attacks, so we have a one-sided alternative:

$$H_0 : p_1 = p_2$$
$$H_a : p_1 < p_2$$
◀

To do a test, standardize $\hat{p}_1 - \hat{p}_2$ to get a z statistic. If H_0 is true, all the observations in both samples really come from a single population of men of whom a single unknown proportion p will have a heart attack in a five-year period. So instead of estimating p_1 and p_2 separately, we pool the two samples and use the overall sample proportion to estimate the single population parameter p. Call this the ***pooled sample proportion***. It is

pooled sample proportion

$$\hat{p} = \frac{\text{count of successes in both samples combined}}{\text{count of observations in both samples combined}}$$

Use $\hat{p}$ in place of both $\hat{p}_1$ and $\hat{p}_2$ in the expression for the standard error SE of $\hat{p}_1 - \hat{p}_2$ to get a z statistic that has the standard normal distribution when H_0 is true. Here is the test.

SIGNIFICANCE TEST FOR COMPARING TWO PROPORTIONS

To test the hypothesis

$$H_0 : p_1 = p_2$$

first find the pooled proportion $\hat{p}$ of successes in both samples combined. Then compute the z statistic

$$z = \frac{\hat{p}_1 - \hat{p}_2}{\sqrt{\hat{p}(1 - \hat{p})\left(\dfrac{1}{n_1} + \dfrac{1}{n_2}\right)}}$$

In terms of a variable Z having the standard normal distribution, the P-value for a test of H_0 against

$H_a : p_1 > p_2$ is $P(Z \geq z)$

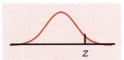

$H_a : p_1 < p_2$ is $P(Z \leq z)$

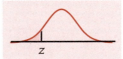

$H_a : p_1 \neq p_2$ is $2P(Z \geq |z|)$

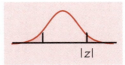

Use these tests in practice when the populations are at least 10 times as large as the samples and $n_1\hat{p}$, $n_1(1 - \hat{p})$, $n_2\hat{p}$, and $n_2(1 - \hat{p})$ are all 5 or more.

EXAMPLE 7.12

The pooled proportion of heart attacks for the two groups in the Helsinki Heart Study is

$$\hat{p} = \frac{\text{count of heart attacks in both samples combined}}{\text{count of subjects in both samples combined}}$$

$$= \frac{56 + 84}{2051 + 2030}$$

$$= \frac{140}{4081} = .0343$$

The z test statistic is

$$z = \frac{\hat{p}_1 - \hat{p}_2}{\sqrt{\hat{p}(1 - \hat{p})\left(\dfrac{1}{n_1} + \dfrac{1}{n_2}\right)}}$$

$$= \frac{.0273 - .0414}{\sqrt{(.0343)(.9657)\left(\dfrac{1}{2051} + \dfrac{1}{2030}\right)}}$$

$$= \frac{-.0141}{.005698} = -2.47$$

The one-sided P-value is the area under the standard normal curve to the left of -2.47. Figure 7.4 shows this area. Table A gives $P = .0068$. Because $P < .01$, the results are statistically significant at the $\alpha = .01$ level. There is strong evidence that gemfibrozil reduced the rate of heart attacks. The large samples in the Helsinki Heart Study helped the study get highly significant results. ◀

EXERCISES

7.21   The 1958 Detroit Area Study was an important investigation of the influence of religion on everyday life. The sample "was basically a simple random sample of the population of the metropolitan area" of Detroit, Michigan. Of the 656 respondents, 267 were white Protestants and 230 were white Catholics.

One question asked whether the government was doing enough in areas such as housing, unemployment, and education; 161 of the Protestants

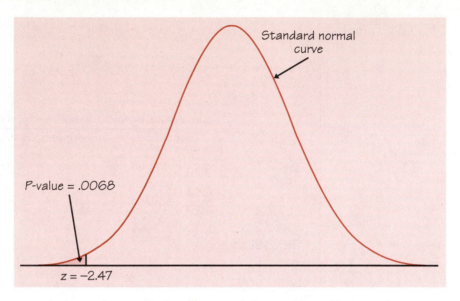

FIGURE 7.4   The *P*-value for the one-sided test of Example 7.12.

and 136 of the Catholics said "No." Is there evidence that white Protestants and white Catholics differed on this issue?

(a) State hypotheses and check that you can safely use the *z* test.

(b) Carry out the test, find the *P*-value, and give your conclusion.

7.22   A study of "adverse symptoms" in users of over-the-counter pain relief medications assigned subjects at random to one of two common pain relievers: acetaminophen and ibuprofen. (Both of these pain relievers are sold under various brand names, sometimes combined with other ingredients.) In all, 650 subjects took acetaminophen, and 44 experienced some adverse symptom. Of the 347 subjects who took ibuprofen, 49 had an adverse symptom. How strong is the evidence that the two pain relievers differ in the proportion of people who experience an adverse symptom?

(a) State hypotheses and check that you can use the *z* test.

(b) Find the *P*-value of the test and give your conclusion.

SUMMARY

We want to **compare the proportions p_1 and p_2 of successes in two populations**. The comparison is based on the difference $\hat{p}_1 - \hat{p}_2$ between the sample proportions of successes. When the sample sizes n_1 and n_2 are large enough, we can use *z* procedures because the sampling distribution of $\hat{p}_1 - \hat{p}_2$ is close to normal.

An approximate level C **confidence interval** for $p_1 - p_2$ is

$$(\hat{p}_1 - \hat{p}_2) \pm z^* \text{SE}$$

where the **standard error** of $\hat{p}_1 - \hat{p}_2$ is

$$\text{SE} = \sqrt{\frac{\hat{p}_1(1 - \hat{p}_1)}{n_1} + \frac{\hat{p}_2(1 - \hat{p}_2)}{n_2}}$$

and z^* is a standard normal critical value.

Significance tests of $H_0 : p_1 = p_2$ use the **pooled sample proportion**

$$\hat{p} = \frac{\text{count of successes in both samples combined}}{\text{count of observations in both samples combined}}$$

and the z **statistic**

$$z = \frac{\hat{p}_1 - \hat{p}_2}{\sqrt{\hat{p}(1 - \hat{p})\left(\dfrac{1}{n_1} + \dfrac{1}{n_2}\right)}}$$

P-values come from the standard normal table.

SECTION 7.2 EXERCISES

7.23 The drug AZT was the first drug that seemed effective in delaying the on-set of AIDS. Evidence for AZT's effectiveness came from a large randomized comparative experiment. The subjects were 1300 volunteers who were in-fected with HIV, the virus that causes AIDS, but did not yet have AIDS. The study assigned 435 of the subjects at random to take 500 milligrams of AZT each day, and another 435 to take a placebo. (The others were as-signed to a third treatment, a higher dose of AZT. We will compare only two groups.) At the end of the study, 38 of the placebo subjects and 17 of the AZT subjects had developed AIDS. We want to test the claim that tak-ing AZT lowers the proportion of infected people who will develop AIDS in a given period of time.

(a) State hypotheses, and check that you can safely use the z procedures.

(b) How significant is the evidence that AZT is effective?

(c) The experiment was double-blind. Explain what this means.

(**Comment**: Medical experiments on treatments for AIDS and other fatal diseases raise hard ethical questions. Some people argue that because AIDS is always fatal, infected people should get any drug that has any hope of helping them. The counterargument is that we will then never find out which drugs really work. The placebo patients in this study were given AZT as soon as the results were in.)

7.24   North Carolina State University looked at the factors that affect the success of students in a required chemical engineering course. Students must get a C or better in the course in order to continue as chemical engineering majors. There were 65 students from urban or suburban backgrounds, and 52 of these students succeeded. Another 55 students were from rural or small-town backgrounds; 30 of these students succeeded in the course.[8]

(a) Is there good evidence that the proportion of students who succeed is different for urban/suburban versus rural/small-town backgrounds? (State hypotheses, give the P-value of a test, and state your conclusion.)

(b) Give a 90% confidence interval for the size of the difference.
(**Comment**: This study did not select two separate samples. A single sample of students was divided after the fact into two groups from different backgrounds. The two-sample z procedures for comparing proportions are valid in such situations. This is an important fact about these methods.)

7.25   The study of small-business failures described in Exercise 7.15 looked at 148 food-and-drink businesses in central Indiana. Of these, 106 were headed by men and 42 were headed by women. During a three-year period, 15 of the men's businesses and 7 of the women's businesses failed. Is there a significant difference between the rate at which businesses headed by men and women fail?
(**Comment**: This study did not select two separate samples. A single sample of businesses was divided after the fact into those headed by men and those headed by women. The two-sample z procedures for comparing proportions are valid in such situations. This is an important fact about these methods.)

7.26   The North Carolina State University study (see Exercise 7.24) also looked at possible differences in the proportions of female and male students who succeeded in the course. They found that 23 of the 34 women and 60 of the 89 men succeeded. Is there evidence of a difference between the proportions of women and men who succeed?

7.27   Different kinds of companies compensate their key employees in different ways. Established companies may pay higher salaries, while new companies may offer stock options that will be valuable if the company succeeds. Do high-tech companies tend to offer stock options more often than other companies? One study looked at a random sample of 200 companies. Of these, 91 were listed in the *Directory of Public High Technology Corporations* and 109 were not listed. Treat these two groups as SRSs of high-tech and non-high-tech companies. Seventy-three of the high-tech companies and 75 of the non-high-tech companies offered incentive stock options to key employees.[9]

(a) Is there evidence that a higher proportion of high-tech companies offer stock options?

(b) Give a 95% confidence interval for the difference in the proportions of the two types of companies that offer stock options.

7.28 Nonresponse to sample surveys may differ with the season of the year. In Italy, for example, many people leave town during the summer. The Italian National Statistical Institute called random samples of telephone numbers between 7 p.m. and 10 p.m. at several seasons of the year. Here are the results for two seasons.[10]

Dates	Number of calls	No answer	Total nonresponse
Jan. 1 to Apr. 13	1558	333	491
July 1 to Aug. 31	2075	861	1174

(a) How much higher is the proportion of "no answers" in July and August compared with the early part of the year? Give a 99% confidence interval.

(b) The difference between the proportions of "no answers" is so large that it is clearly statistically significant. How can you tell from your work in (a) that the difference is significant at the $\alpha = .01$ level?

(c) Use the information given to find the counts of calls which had nonresponse for some reason other than "no answer." Do the rates of nonresponse due to other causes also differ significantly for the two seasons?

7.29 The Physicians' Health Study examined the effects of taking an aspirin every other day. Earlier studies suggested that aspirin might reduce the risk of heart attacks. The subjects were 22,071 healthy male physicians at least 40 years old. The study assigned 11,037 of the subjects at random to take aspirin. The others took a placebo pill. The study was double-blind. Here are the counts for some of the outcomes of interest to the researchers.

	Aspirin group	Placebo group
Fatal heart attacks	10	26
Nonfatal heart attacks	129	213
Strokes	119	98

For which outcomes is the difference between the aspirin and placebo groups significant? (Use two-sided alternatives. Check that you can apply the z test. Write a brief summary of your conclusions.)

7.30   The National Assessment of Educational Progress (NAEP) Young Adult Literacy Assessment Survey interviewed a random sample of 1917 people 21 to 25 years old. The sample contained 840 men, of whom 775 were fully employed. There were 1077 women, and 680 of them were fully employed.[11]

(a) Use a 99% confidence interval to describe the difference between the proportions of young men and young women who are fully employed. Is the difference statistically significant at the 1% significance level?

(b) The mean and standard deviation of scores on the NAEP's test of quantitative skills were $\bar{x}_1 = 272.40$ and $s_1 = 59.2$ for the men in the sample. For the women, the results were $\bar{x}_2 = 274.73$ and $s_2 = 57.5$. Is the difference between the mean scores for men and women significant at the 1% level?

7.31   The Current Population Survey (CPS) is the monthly government sample survey of 60,000 households that provides data on employment in the United States. A study of child-care workers drew a sample from the CPS data tapes. We can consider this sample to be an SRS from the population of child-care workers.[12]

(a) Out of 2455 child-care workers in private households, 7% were black. Of 1191 nonhousehold child-care workers, 14% were black. Give a 99% confidence interval for the difference in the percents of these groups of workers who are black. Is the difference statistically significant at the $\alpha = .01$ level?

(b) The study also examined how many years of school child-care workers had. For household workers, the mean and standard deviation were $\bar{x}_1 = 11.6$ years and $s_1 = 2.2$ years. For nonhousehold workers, $\bar{x}_2 = 12.2$ years and $s_2 = 2.1$ years. Give a 99% confidence interval for the difference in mean years of education for the two groups. Is the difference significant at the $\alpha = .01$ level?

7.32   Never forget that even small effects can be statistically significant if the samples are large. To illustrate this fact, return to the study of 148 small businesses in Exercise 7.25.

(a) Find the proportions of failures for businesses headed by women and businesses headed by men. These sample proportions are quite close to each other. Give the P-value for the z test of the hypothesis that the same proportion of women's and men's businesses fail. (Use the two-sided alternative.) The test is very far from being significant.

(b) Now suppose that the same sample proportions came from a sample 30 times as large. That is, 210 out of 1260 businesses headed by women and 450 out of 3180 businesses headed by men fail. Verify that the proportions of failures are exactly the same as in (a). Repeat the z test for the new data, and show that it is now significant at the $\alpha = .05$ level.

(c) It is wise to use a confidence interval to estimate the size of an effect, rather than just giving a P-value. Give 95% confidence intervals for the difference between the proportions of women's and men's businesses that fail for the settings of both (a) and (b). What is the effect of larger samples on the confidence interval?

CHAPTER REVIEW

Statistical inference always draws conclusions about one or more parameters of a population. When you think about doing inference, ask first what the population is and what parameter you are interested in. The t procedures of Chapter 6 allow us to give confidence intervals and carry out tests about population means. We use the z procedures of this chapter for inference about population proportions.

Inference about population proportions is based on sample proportions. We rely on the fact that a sample proportion has a distribution that is close to normal unless the sample is quite small. All the z procedures in this chapter work well when the samples are large enough. You must check this before using them. Here are the things you should now be able to do.

A. RECOGNITION

1. Recognize when a problem requires inference about a proportion or comparing two proportions.

2. Calculate from sample counts the sample proportion or proportions that estimate the parameters of interest.

B. INFERENCE ABOUT ONE PROPORTION

1. Use the z procedure to give a confidence interval for a population proportion p.

2. Use the z statistic to carry out a test of significance for the hypothesis $H_0 : p = p_0$ about a population proportion p against either a one-sided or a two-sided alternative.

3. Check that you can safely use these z procedures in a particular setting.

C. COMPARING TWO PROPORTIONS

1. Use the two-sample z procedure to give a confidence interval for the difference $p_1 - p_2$ between proportions in two populations based on independent samples from the populations.

2. Use a z statistic to test the hypothesis $H_0 : p_1 = p_2$ that proportions in two distinct populations are equal.

3. Check that you can safely use these z procedures in a particular setting.

CHAPTER 7 REVIEW EXERCISES

7.33   A television news program conducts a call-in poll about a proposed city ban on handgun ownership. Of the 2372 calls, 1921 oppose the ban. The station, following recommended practice, makes a confidence statement: "81% of the Channel 13 Pulse Poll sample opposed the ban. We can be 95% confident that the true proportion of citizens opposing a handgun ban is within 1.6% of the sample result." Is this conclusion justified?

7.34   Some people think that chemists are more likely than other parents to have female children. (Perhaps chemists are exposed to something in their laboratories that affects the sex of their children.) The Washington State Department of Health lists the parents' occupations on birth certificates. Between 1980 and 1990, 555 children were born to fathers who were chemists. Of these births, 273 were girls. During this period, 48.8% of all births in Washington State were girls.[13] Is there evidence that the proportion of girls born to chemists is higher than the state proportion?

7.35   The first law requiring child restraints in motor vehicles went into effect in 1978, in Tennessee. "Before enactment of child restraint laws, most children in the United States were unrestrained when traveling in motor vehicles." The writer of this statement cites a 1974 survey of drivers in Maryland and Virginia that found that "about 93% ($n = 8,933$) of children under the age of 10 were traveling unrestrained."[14] Assuming that the children observed are an SRS from the areas involved, give a 99% confidence interval for the proportion of unrestrained children in 1974.

7.36 Sickle-cell trait is a hereditary condition that is common among blacks and can cause medical problems. Some biologists suggest that sickle-cell trait protects against malaria. That would explain why it is found in people who originally came from Africa, where malaria is common. A study in Africa tested 543 children for the sickle-cell trait and also for malaria. In all, 136 of the children had the sickle-cell trait, and 36 of these had heavy malaria infections. The other 407 children lacked the sickle-cell trait, and 152 of them had heavy malaria infections.[15]

(a) Give a 95% confidence interval for the proportion of all children in the population studied who have the sickle-cell trait.

(b) Is there good evidence that the proportion of heavy malaria infections is lower among children with the sickle-cell trait?

7.37 The National Collegiate Athletic Association (NCAA) requires colleges to report the graduation rates of their athletes. Here are data from a Big Ten university's report.[16]

(a) Forty-five of the 74 athletes admitted in a specific year graduated within six years. Does the proportion of athletes who graduate differ significantly from the all-university proportion, which is 68%?

(b) The graduation rates that year were 21 of 28 female athletes and 24 of 46 male athletes. Is there evidence that a smaller proportion of male athletes than of female athletes graduate within six years?

(c) We are willing to regard athletes admitted in a specific year as an SRS from the large population of athletes the university will admit under its present standards. Explain why you can safely use the z procedures in parts (a) and (b). Then explain why you cannot use these procedures to test whether baseball players (3 out of 5 admitted that year graduated) differ from other athletes.

7.38 A study comparing American and Australian corporations examined a sample of 133 American and 63 Australian corporations. There are the usual practical difficulties involving nonresponse and the question of what population the samples represent. Ignore these issues and treat the samples as SRSs from the United States and Australia. The average percent of revenues from "highly regulated businesses" was 27% for the Australian companies and 41% for the American companies.[17]

(a) The data are given as percents. Explain carefully why comparing the percent of revenues from highly regulated businesses for U.S. and Australian corporations is *not* a comparison of two population proportions.

(b) What test would you use to make the comparison? (Don't try to carry out a test.)

NOTES AND DATA SOURCES

1. Data from Joseph H. Catania et al., "Prevalence of AIDS-related risk factors and condom use in the United States," *Science*, 258 (1992), pp. 1101–1106.

2. The study is reported in William Celis III, "Study suggests Head Start helps beyond school," *New York Times*, April 20, 1993.

3. Data from Richard M. Felder et al., "Who gets it and who doesn't: a study of student performance in an introductory chemical engineering course," *1992 ASEE Annual Conference Proceedings*, American Society for Engineering Education, Washington, D.C., 1992, pp. 1516–1519.

4. The quotation is from page 1104 of the article cited in Note 1.

5. Sara J. Solnick and David Hemenway, "Complaints and disenrollment at a health maintenance organization," *The Journal of Consumer Affairs*, 26 (1992), pp. 90–103.

6. Arne L. Kalleberg and Kevin T. Leicht, "Gender and organizational performance: determinants of small business survival and success," *The Academy of Management Journal*, 34 (1991), pp. 136–161.

7. The Detroit Area Study is described in Gerhard Lenski, *The Religious Factor*, Doubleday, New York, 1961.

8. The data are from the source cited in Note 3.

9. Based on Greg Clinch, "Employee compensation and firms' research and development activity," *Journal of Accounting Research*, 29 (1991), pp. 59–78.

10. Giuliana Coccia, "An overview of non-response in Italian telephone surveys," *Proceedings of the 99th Session of the International Statistical Institute*, 1993, Book 3, pp. 271–272.

11. Francisco L. Rivera-Batiz, "Quantitative literacy and the likelihood of employment among young adults," *Journal of Human Resources*, 27 (1992), pp. 313–328.

12. David M. Blau, "The child care labor market," *Journal of Human Resources*, 27 (1992), pp. 9–39.

13. Eric Ossiander, letter to the editor, *Science*, 257 (1992), p. 1461.

14. William N. Evans and John D. Graham, "An estimate of the lifesaving benefit of child restraint use legislation," *Journal of Health Economics*, 9 (1990), pp. 121–132.

15. A. C. Allison and D. F. Clyde, "Malaria in African children with deficient erythrocyte dehydrogenase," *British Medical Journal*, 1 (1961), pp. 1346–1349.

16. Office of the Registrar, Purdue University, *Summary of the 1992–1993 NCAA Graduation-Rates Disclosure Form, 1985–1986 Academic Year Cohort*, West Lafayette, Ind., 1993.

17. From Noel Capon et al., "A comparative analysis of the strategy and structure of United States and Australian corporations," *Journal of International Business Studies*, 18 (1987), pp. 51–74.

Statistical inference offers more methods than anyone can know well, as a glance at the offerings of any large statistical software package demonstrates. In an introduction, we must be selective. In Parts I and II we laid a foundation for understanding statistics:

- The nature and purpose of data analysis.
- The central ideas of designs for data production.
- The reasoning behind confidence intervals and significance tests.
- Experience applying these ideas in simple settings.

Each of the three chapters of Part III offers a short introduction to a more advanced topic in statistical inference. You may choose to read any or all of them, in any order.

What makes a statistical method "more advanced"? More complex data, for one thing. Our introduction to inference in Part II looked only at methods for inference about a single population parameter and for comparing two parameters. The next step is to compare more than two parameters. Chapter 8 shows how to compare more than two proportions, and Chapter 9 discusses comparing more than two means. In these chapters we have data from three or more samples or experimental treatments, not just one or two samples. Another more complex setting concerns the association between two variables. We describe association by a two-way table if the variables are categorical, or by correlation and regression if they are quantitative. Chapter 8 presents inference for two-way tables. Inference for regression is the topic of Chapter 10. Chapters 8, 9, and 10 together bring our knowledge of inference to the same point that our study of data analysis reached in Chapters 1 and 2.

With greater complexity comes greater reliance on software. In these final three chapters you will more often be interpreting the output of statistical software or, if possible, using software yourself. You can do the calculations needed in Chapter 8 without software or a specialized calculator. In Chapters 9 and 10, the pain is too great and the contribution to learning too small. Fortunately, you can grasp the ideas without step-by-step arithmetic.

Another aspect of "more advanced" methods is new concepts and ideas. This is where we draw the line in deciding what statistical topics we can master in a first course. Chapters 8, 9, and 10 build more elaborate methods on the foundation we have laid without introducing fundamentally new concepts. You can see that statistical practice does need additional big ideas by reading the sections on "the problem of multiple comparisons" in Chapters 8 and 9. But the ideas you already know place you among the world's statistical sophisticates.

PART  III

TOPICS IN INFERENCE

"It was a numbers explosion."

KARL PEARSON

Karl Pearson (1857–1936), a professor at University College in London, had already published nine books before he turned his abundant energy to statistics in 1893. Of course, Pearson didn't really take up statistics, which was not yet a separate field of study. He took up problems of heredity and evolution, which led him into statistics.

Pearson developed a family of curves—we would call them density curves—for describing biological data that don't follow a normal distribution. He then asked how he could test whether one of these curves actually fit a set of data well. In 1900 he invented a method, the chi-square test. Pearson's chi-square test has the honor of being the oldest inference procedure still in use. It is now most often used for problems somewhat different from the one that motivated Pearson, as we will see in this chapter.

After Pearson, statistics was a field of study. Fisher and Neyman in the 1920s and 1930s would provide much of its present form, but here is what the leading historian of statistics says about the origins:

Before 1900 we see many scientists of different fields developing and using techniques we now recognize as belonging to modern statistics. After 1900 we begin to see identifiable statisticians developing such techniques into a unified logic of empirical science that goes far beyond its component parts. There was no sharp moment of birth; but with Pearson and Yule and the growing numbers of students in Pearson's laboratory, the infant discipline may be said to have arrived.[1]

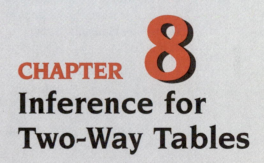

CHAPTER 8
Inference for Two-Way Tables

INTRODUCTION

The two-sample z procedures of Chapter 7 allow us to compare the proportions of successes in two groups, either two populations or two treatment groups in an experiment. What if we want to compare more than two groups? We need a new statistical test. The new test starts by presenting the data in a new way, as a two-way table. Two-way tables have more general uses than comparing the proportions of successes in several groups. As we saw in Section 5 of Chapter 2, they describe relationships between any two categorical variables. The same test that compares several proportions tests whether the row and column variables are related in any two-way table. We will start with the problem of comparing several proportions.

EXAMPLE 8.1

Chronic users of cocaine need the drug to feel pleasure. Perhaps giving them a medication that fights depression will help them stay off cocaine. A three-year study compared an antidepressant called desipramine with lithium (a standard treatment for cocaine addiction) and a placebo. The subjects were 72 chronic users of cocaine who wanted to break their drug habit. Twenty-four of the subjects were randomly assigned to each treatment. Here are the counts and proportions of the subjects who avoided relapse into cocaine use during the study.[2]

Group	Treatment	Subjects	No relapse	Proportion
1	Desipramine	24	14	.583
2	Lithium	24	6	.250
3	Placebo	24	4	.167

The sample proportions of subjects who stayed off cocaine are quite different. The bar chart in Figure 8.1 compares the results visually. Are these data good evidence that the proportions of successes for the three treatments differ in the population of all cocaine users? ◄

The problem of multiple comparisons

Call the population proportions of successes in the three groups $p_1, p_2,$ and p_3. We again use a subscript to remind us which group a parameter or statistic describes. To compare these three population proportions, we might use the two-sample z procedures several times:

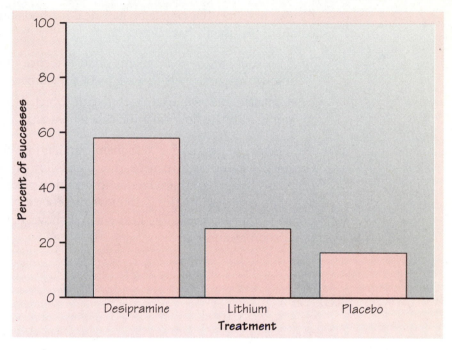

FIGURE 8.1 Bar chart comparing the success rates of three treatments for cocaine addiction, from Example 8.1.

- Test $H_0 : p_1 = p_2$ to see if the success rate of desipramine differs from that of lithium.
- Test $H_0 : p_1 = p_3$ to see if desipramine differs from a placebo.
- Test $H_0 : p_2 = p_3$ to see if lithium differs from a placebo.

The weakness of doing three tests is that we get three *P*-values, one for each test alone. That doesn't tell us how likely it is that *three* sample proportions are spread apart as far as these are. It may be that 0.167 and 0.583 are significantly different if we look at just two groups, but not significantly different if we know that they are the smallest and largest proportions in three groups. As we look at more groups, we expect the gap between the smallest and largest sample proportion to get larger. (Think of comparing the tallest and shortest person in larger and larger groups of people.) We can't safely compare many parameters by doing tests or confidence intervals for two parameters at a time.

The problem of how to do many comparisons at once with some overall measure of confidence in all our conclusions is common in more advanced statistics. We are leaving the elementary parts of statistical inference now that we have met this problem. Statistical methods for dealing with many comparisons usually have two parts:

1. An *overall test* to see if there is good evidence of *any* differences among the parameters that we want to compare.
2. A detailed *follow-up analysis* to decide which of the parameters differ and to estimate how large the differences are.

The overall test is often reasonably straightforward, though still more complex than the tests we met earlier. The follow-up analysis can be quite elaborate. In our basic introduction to statistical practice, we will look only at some overall tests. In this chapter we present a test for comparing several population proportions. The next chapter shows how to compare several population means.

TWO-WAY TABLES

two-way table   The first step in the overall test for comparing several proportions is to arrange the data in a **two-way table** that gives counts for both successes and failures. Here is the two-way table of the cocaine addiction data:

	Relapse	
	No	Yes
Desipramine	14	10
Lithium	6	18
Placebo	4	20

r × c table   We call this a 3 × 2 table because it has 3 rows and 2 columns. A table with *r* rows and *c* columns is an **r × c table**. The table shows the relationship between two categorical variables. The explanatory variable is the treatment (one of three drugs). The response variable is success (no relapse) or failure (relapse). The two-way table gives the counts for all 6 combinations of *cell*   values of these variables. Each of the 6 counts occupies a **cell** of the table.

Expected counts

We want to test the null hypothesis that there are *no differences* among the proportions of successes for addicts given the three treatments:

$$H_0 : p_1 = p_2 = p_3$$

The alternative hypothesis is that there *is* some difference, that not all three proportions are equal:

$$H_a : \text{ not all of } p_1, p_2, \text{ and } p_3 \text{ are equal}$$

The alternative hypothesis is no longer one-sided or two-sided. It is "many-sided," because it allows any relationship other than "all three equal." For example, H_a includes the situation in which $p_2 = p_3$ but p_1 has a different value.

To test H_0, we compare the observed counts in a two-way table with the *expected counts*, the counts we would expect—except for random variation—if H_0 were true. If the observed counts are far from the expected counts, that is evidence against H_0. It is easy to find the expected counts.

EXPECTED COUNTS

The **expected count** in any cell of a two-way table when H_0 is true is

$$\text{expected count} = \frac{\text{row total} \times \text{column total}}{\text{table total}}$$

To understand why this recipe works, think first about just one proportion.

EXAMPLE 8.2

Linda is a basketball player who makes 70% of her free throws. If she shoots 10 free throws in a game, we expect her to make 70% of them, or 7 of the 10. Of course, she won't make exactly 7 every time she shoots 10 free throws in a game. There is chance variation from game to game. But in the long run, 7 of 10 is what we expect. It is, in fact, the *mean* number of shots Linda makes when she shoots 10 times. ◀

In more formal language, if we have n independent tries and the probability of a success on each try is p, we expect np successes. If we draw an SRS of n individuals from a population in which the proportion of successes is p, we expect np successes in the sample. That's the fact behind the formula for expected counts in a two-way table.

Let's apply this fact to the cocaine study. The two-way table with row and column totals is

| | Relapse | | |
	No	Yes	Total
Desipramine	14	10	24
Lithium	6	18	24
Placebo	4	20	24
Total	24	48	72

We will find the expected count for the cell in row 1 (desipramine) and column 2 (relapse). The proportion of relapses among all 72 subjects is

$$\frac{\text{count of relapses}}{\text{table total}} = \frac{\text{column 2 total}}{\text{table total}} = \frac{48}{72} = \frac{2}{3}$$

Think of this as p, the overall proportion of relapses. If H_0 is true, we expect (except for random variation) this same proportion of relapses in all three groups. So the expected count of relapses among the 24 subjects who took desipramine is

$$np = (24)\left(\frac{2}{3}\right) = 16.00$$

This expected count has the form announced in the box:

$$\frac{\text{row 1 total} \times \text{column 2 total}}{\text{table total}} = \frac{(24)(48)}{72}$$

EXAMPLE 8.3

Here are the observed and expected counts side-by-side:

| | Observed | | Expected | |
	No	Yes	No	Yes
Desipramine	14	10	8	16
Lithium	6	18	8	16
Placebo	4	20	8	16

Because 2/3 of all subjects relapsed, we expect 2/3 of the 24 subjects in each group to relapse if there are no differences among the treatments. In fact, desipramine has fewer relapses (10) and more successes (14) than expected. The placebo has fewer successes (4) and more relapses (20). That's another way of saying what the sample proportions in Example 8.1 say more directly: desipramine does much better than the placebo with lithium in between. ◀

EXERCISES

8.1 North Carolina State University studied student performance in a course required by its chemical engineering major. One question of interest is the relationship between time spent in extracurricular activities and whether a student earned a C or better in the course. Here are the data for the 119 students who answered a question about extracurricular activities.[3]

	Extracurricular activities (hours per week)		
	< 2	2 to 12	> 12
C or better	11	68	3
D or F	9	23	5

(a) This is an $r \times c$ table. What are the numbers r and c?

(b) Find the proportion of successful students (C or better) in each of the three extracurricular activity groups. What kind of relationship between extracurricular activities and succeeding in the course do these proportions seem to show?

(c) Make a bar chart to compare the three proportions of successes.

(d) The null hypothesis says that the proportions of successes are the same in all three groups if we look at the population of all students. Find the expected counts if this hypothesis is true, and display them in a two-way table.

(e) Compare the observed counts with the expected counts. Are there large deviations between them? These deviations are another way of describing the relationship you described in (b).

8.2 How are the smoking habits of students related to their parents' smoking? Here are data from a survey of students in eight Arizona high schools.[4]

	Student smokes	Student does not smoke
Both parents smoke	400	1380
One parent smokes	416	1823
Neither parent smokes	188	1168

(a) This is an $r \times c$ table. What are the numbers r and c?

(b) Calculate the proportion of students who smoke in each of the three parent groups. Then describe in words the association between parent smoking and student smoking.

(c) Make a graph to display the association.

(d) Explain in words what the null hypothesis $H_0 : p_1 = p_2 = p_3$ says about student smoking.

(e) Find the expected counts if H_0 is true, and display them in a two-way table similar to the table of observed counts.

(f) Compare the tables of observed and expected counts. Explain how the comparison expresses the same association you see in (b) and (c).

THE CHI-SQUARE TEST

Comparing the sample proportions of successes describes the differences among the three treatments for cocaine addiction. But the statistical test that tells us whether those differences are statistically significant doesn't use the sample proportions. It compares the observed and expected counts. The test statistic that makes the comparison is the *chi-square statistic*.

CHI-SQUARE STATISTIC

The **chi-square statistic** is a measure of how far the observed counts in a two-way table are from the expected counts. The formula for the statistic is

$$X^2 = \sum \frac{(\text{observed count} - \text{expected count})^2}{\text{expected count}}$$

The sum is over all $r \times c$ cells in the table.

The chi-square statistic is a sum of terms, one for each cell in the table. In the cocaine example, 14 of the desipramine group succeeded in avoiding a relapse. The expected count for this cell is 8. So the component of the chi-square statistic from this cell is

$$\frac{(\text{observed count} - \text{expected count})^2}{\text{expected count}} = \frac{(14 - 8)^2}{8}$$

$$= \frac{36}{8} = 4.5$$

Think of the chi-square statistic X^2 as a measure of the distance of the observed counts from the expected counts. Like any distance, it is always zero or positive, and it is zero only when the observed counts are exactly equal to the expected counts. Large values of X^2 are evidence against H_0 because they say that the observed counts are far from what we would expect if H_0 were true. Although the alternative hypothesis H_a is many-sided, the chi-square test is one-sided because any violation of H_0 tends to produce a large value of X^2. Small values of X^2 are not evidence against H_0.

Calculating the expected counts and then the chi-square statistic by hand is a bit time-consuming. As usual, statistical software saves time and always gets the arithmetic right. Some calculators will also find the chi-square statistic from keyed-in data.

EXAMPLE 8.4

Enter the two-way table (the 6 counts) for the cocaine study into the Minitab software package and request the chi-square test. The output appears in Figure 8.2. Most statistical software packages produce chi-square output similar to this.

Minitab repeats the two-way table of observed counts and puts the expected count for each cell below the observed count. It numbers the rows (1, 2, and 3) and the columns (C1 and C2) and also puts in the row and column totals. Then the software calculates the chi-square statistic X^2. For these data, $X^2 = 10.500$. The statistic is a sum of 6 terms, one for each cell in the table. The "ChiSq" display in the output shows the individual terms, as well as their sum. The first term is 4.500, just as we calculated.

The P-value is the probability that X^2 would take a value as large as 10.500 if H_0 were really true. Many software systems give the P-value. Minitab, like many calculators, requires us to ask for the probability of a value of 10.500 or smaller. This probability is 0.9948 (at the bottom of the output), so the P-value

```
Expected counts are printed below observed counts

              C1        C2      Total
     1        14        10        24
            8.00     16.00


     2         6        18        24
            8.00     16.00


     3         4        20        24
            8.00     16.00


Total        24        48        72


ChiSq =   4.500 +  2.250 +
          0.500 +  0.250 +
          2.000 +  1.000 = 10.500

df = 2


     Chisquare 2.
   10.5000      0.9948
```

FIGURE 8.2   Minitab output for that two-way table in the cocaine study. The output gives the observed counts, the expected counts, and the value 10.500 for the chi-square statistic. The last line gives 0.9948 as the probability of a value *less than* 10.500 if the null hypothesis is true. The *P*-value is therefore $1 - .9948$.

is $1 - .9948 = .0052$. The small *P*-value gives us good reason to conclude that there *are* differences among the effects of the three treatments.   ◄

The chi-square test is the overall test for comparing any number of population proportions. If the test allows us to reject the null hypothesis that all the proportions are equal, we then want to do a follow-up analysis that examines the differences in detail. We won't describe how to do a formal follow-up analysis, but you should look at the data to see what specific effects they suggest.

EXAMPLE 8.5

The cocaine study found significant differences among the proportions of successes for three treatments for cocaine addiction. We can see the specific differences in three ways.

Look first at the *sample proportions*:

$$\hat{p}_1 = .583 \qquad \hat{p}_2 = .250 \qquad \hat{p}_3 = .167$$

These suggest that the major difference between the proportions is that desipramine has a much higher success rate than either lithium or a placebo. That is the effect that the study hoped to find.

Next, *compare the observed and expected counts* in Figure 8.2. Treatment 1 (desipramine) has more successes and fewer failures than we would expect if all three treatments had the same success rate in the population. The other two treatments had fewer successes and more failures than expected.

components of chi-square

Finally, Minitab prints under the table the 6 individual "distances" between the observed and expected counts that are added to get X^2. The arrangement of these **components of X^2** is the same as the 3 × 2 arrangement of the table. The largest components show which cells contribute the most to the overall distance X^2. The largest component by far is for the top left cell in the table: desipramine has more successes than would be expected.

All three ways of examining the data point to the same conclusion: desipramine works better than the other treatments. This is an informal conclusion. More advanced methods provide tests and confidence intervals that make this follow-up analysis formal. ◄

EXERCISES

8.3 In Exercise 8.1, you began to analyze data on the relationship between time spent on extracurricular activities and success in a tough course. Figure 8.3 gives Minitab output for the two-way table in Exercise 8.1.

(a) Starting from the table of expected counts, find the 6 components of the chi-square statistic and then the statistic X^2 itself. Check your work against the computer output.

(b) What is the *P*-value for the test? Explain in simple language what it means to reject H_0 in this setting.

(c) Which term contributes the most to X^2? What specific relation between extracurricular activities and academic success does this term point to?

(d) Does the North Carolina State study convince you that spending more or less time on extracurricular activities *causes* changes in academic success? Explain your answer.

```
Expected counts are printed below observed counts

              C1         C2         C3       Total
    1         11         68          3         82
            13.78      62.71       5.51

    2          9         23          5         37
             6.22      28.29       2.49

 Total        20         91          8        119

ChiSq =    0.561 +   0.447 +   1.145 +
           1.244 +   0.991 +   2.538 = 6.926
df = 2
1 cells with expected counts less than 5.0

   Chisquare 2.
   6.9260        0.9687
```

FIGURE 8.3   Minitab output for the study of extracurricular activity and success in a tough course, for Exercise 8.3.

8.4   In Exercise 8.2, you began to analyze data on the relationship between smoking by parents and smoking by high school students. Figure 8.4 gives the Minitab output for the two-way table in Exercise 8.2.

(a) Starting from the table of expected counts, find the 6 components of the chi-square statistic and then the statistic X^2 itself. Check your work against the computer output.

(b) What is the P-value for the test? Explain in simple language what it means to reject H_0 in this setting.

(c) Which two terms contribute the most to X^2? What specific relation between parent smoking and student smoking do these terms point to?

(d) Does the study convince you that parent smoking *causes* student smoking? Explain your answer.

The chi-square distributions

Software usually finds P-values for us. The P-value for a chi-square test comes from comparing the value of the chi-square statistic with critical values for a *chi-square distribution*.

```
Expected counts are printed below observed counts

              C1          C2      Total
    1        400        1380       1780
          332.49     1447.51

    2        416        1823       2239
          418.22     1820.78

    3        188        1168       1356
          253.29     1102.71

Total       1004        4371       5375

ChiSq =  13.709 +   3.149 +
          0.012 +   0.003 +
         16.829 +   3.866 = 37.566
df = 2

    Chisquare 2.
    37.5680      1.0000
```

FIGURE 8.4   Minitab output for the study of parent smoking and student smoking, for Exercise 8.4.

THE CHI-SQUARE DISTRIBUTIONS

The **chi-square distributions** are a family of distributions that take only positive values and are skewed to the right. A specific chi-square distribution is specified by one parameter, called the **degrees of freedom**.

The chi-square test for a two-way table with r rows and c columns uses critical values from the chi-square distribution with $(r-1)(c-1)$ degrees of freedom. The P-value is the area to the right of X^2 under the chi-square density curve.

Figure 8.5 shows the density curves for three members of the chi-square family of distributions. As the degrees of freedom increase, the density curves become less skewed and larger values become more probable. Table E in the back of the book gives critical values for chi-square

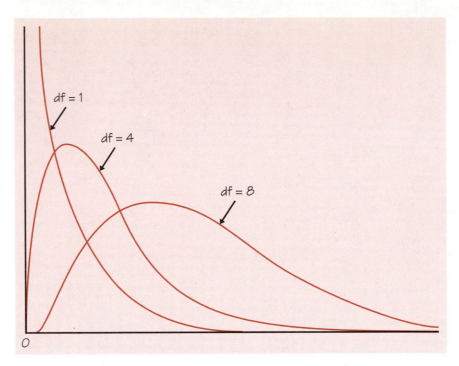

FIGURE 8.5 Density curves for the chi-square distributions with 1, 4, and 8 degrees of freedom. Chi-square distributions take only positive values.

distributions. You can use Table E if software does not give you P-values for a chi-square test.

EXAMPLE 8.6

The two-way table of 3 treatments by 2 outcomes for the cocaine study has 3 rows and 2 columns. That is, $r = 3$ and $c = 2$. The chi-square statistic therefore has degrees of freedom

$$(r - 1)(c - 1) = (3 - 1)(2 - 1) = (2)(1) = 2$$

The computer output in Figure 8.2 gives 2 as the degrees of freedom.

The observed value of the chi-square statistic is $X^2 = 10.500$. Look in the $df = 2$ row of Table E. The value $X^2 = 10.500$ falls between the 0.01 and 0.005 critical values of the chi-square distribution with 2 degrees of freedom. Remember that the chi-square test is always one-sided. So the P-value of $X^2 = 10.500$ is between 0.01 and 0.005. The differences among the three proportions of successes are significant at the $\alpha = .01$ level. ◀

df = 2

p	.01	.005
x^*	9.21	10.60

8.5 The computer output in Figure 8.3 gives the degrees of freedom for the table in Exercise 8.1 as 2.

(a) Verify that this is correct.

(b) The computer gives the value of the chi-square statistic as $X^2 = 6.926$. Between what two entries in Table E does this value lie? What does the table tell you about the P-value?

8.6 The computer output in Figure 8.4 gives the degrees of freedom for the table in Exercise 8.2 as 2.

(a) Verify that this is correct.

(b) The computer gives the value of the chi-square statistic as $X^2 = 37.568$. Where in Table E does this value lie? What does the table tell you about the P-value?

More uses of the chi-square test

Two-way tables can arise in several ways. The cocaine study is an experiment that assigned 24 addicts to each of three groups. Each group is a sample from a separate population corresponding to a separate treatment. The study design fixes the size of each sample in advance, and the data record which of two outcomes occurred for each subject. The null hypothesis of "no difference" among the treatments takes the form of "equal proportions of successes" in the three populations. The next example illustrates a different setting for a two-way table.

EXAMPLE 8.7

A study of the relationship between men's marital status and the level of their jobs used data on all 8235 male managers and professionals employed by a large manufacturing firm. Each man's job has a grade set by the company that reflects the value of that particular job to the company. The authors of the study grouped the many job grades into quarters. Grade 1 contains jobs in the lowest quarter of job grades, and grade 4 contains those in the highest quarter. Here are the data.[5]

		Single	Married	Divorced	Widowed
	1	58	874	15	8
Job	2	222	3927	70	20
grade	3	50	2396	34	10
	4	7	533	7	4

Marital status

Do these data show a statistically significant relationship between marital status and job grade? ◄

In Example 8.7 we do not have four separate samples from the four marital statuses. We have a single group of 8235 men, each classified in two ways, by marital status and job grade. The number of men in each marital status is not fixed in advance but is known only after we have the data. Both marital status and job grade have four levels, so a careful statement of the null hypothesis

H_0: there is no relationship between marital status and job grade

in terms of population parameters is complicated.

In fact, we should probably regard these 8235 men as an entire population rather than as a sample. The data include all the managers and professionals employed by this company. The company no doubt has many special features. Its employees are not necessarily a random sample from any larger population. Nevertheless, we would still like to decide if the relationship between marital status and job grade is statistically significant in the sense that it is too strong to happen just by chance if job grades were handed out at random to men of all marital statuses. "Not likely to happen just by chance if H_0 is true" is the usual meaning of statistical significance. In this example, that meaning makes sense even though we have data on an entire population.

The setting of Example 8.7 is very different from a comparison of several proportions. Nevertheless, we can apply the chi-square test. One of the most useful properties of chi-square is that it tests the hypothesis "the row and column variables are not related to each other" whenever this hypothesis makes sense for a two-way table.

USES OF THE CHI-SQUARE TEST

Use the chi-square test to test the null hypothesis

H_0 : there is no relationship between two categorical variables

when you have a two-way table from any of these situations:

- Independent SRSs from each of several populations, with each individual classified according to one categorical variable. (The other variable says which sample the individual comes from.)

- A single SRS, with each individual classified according to both of two categorical variables.

- An entire population, with each individual classified according to both of two categorical variables.

EXAMPLE 8.8

To analyze the job grade data in Example 8.7, first do the overall chi-square test. The Minitab chi-square output appears in Figure 8.6. The observed chi-square is very large, $X^2 = 67.397$. The probability of a value smaller than this when H_0 is true is 1.0000 (to four decimal places). That is, the P-value is close to 0. We have overwhelming evidence that job grade is related to marital status.

Table E gives a similar result. For a 4×4 table, the degrees of freedom are $(r-1)(c-1) = 9$. Look in the df $= 9$ row of Table E. The largest critical value is 29.67, corresponding to the P-value 0.0005. The observed $X^2 = 67.397$ is beyond that value, so $P < .0005$.

Next, do a follow-up analysis to describe the relationship. As in Section 5 of Chapter 2, we describe a relationship between two categorical variables by comparing percents. Here is a table of the percent of men in each marital status whose jobs have each grade. Each column of this table gives the conditional distribution of job grade among men with a specific marital status. Each column adds to 100% because it accounts for all the men in one marital status.

		Single	Married	Divorced	Widowed
	1	17.2%	11.3%	11.9%	19.1%
Job	2	65.9%	50.8%	55.5%	47.7%
grade	3	14.9%	31.0%	26.9%	23.8%
	4	2.0%	6.9%	5.6%	9.6%
		100%	100%	100%	100%

The bar charts in Figure 8.7 help us compare these four conditional distributions. We see at once that smaller percents of single men have jobs in the higher grades 3 and 4. Not only married men but men who were once married and are now divorced or widowed are more likely to hold higher-grade jobs. Look at the 16 components of the chi-square sum in the computer output for confirmation. The four cells for single men have the four largest components of

```
Expected counts are printed below observed counts

               C1          C2          C3          C4       Total
      1         58         874          15           8         955
             39.08      896.44       14.61        4.87

      2        222        3927          70          20        4239
            173.47     3979.05       64.86       21.62

      3         50        2396          34          10        2490
            101.90     2337.30       38.10       12.70

      4          7         533           7           4         551
             22.55      517.21        8.43        2.81

Total        337        7330         126          42        8235

ChiSq =   9.158 +   0.562 +   0.010 +   2.011 +
         13.575 +   0.681 +   0.407 +   0.121 +
         26.432 +   1.474 +   0.441 +   0.574 +
         10.722 +   0.482 +   0.243 +   0.504 = 67.397
df = 9
2 cells with expected counts less than 5.0

   Chisquare 9.
   67.3970        1.0000
```

FIGURE 8.6 Minitab output for the 4 × 4 table in Example 8.8.

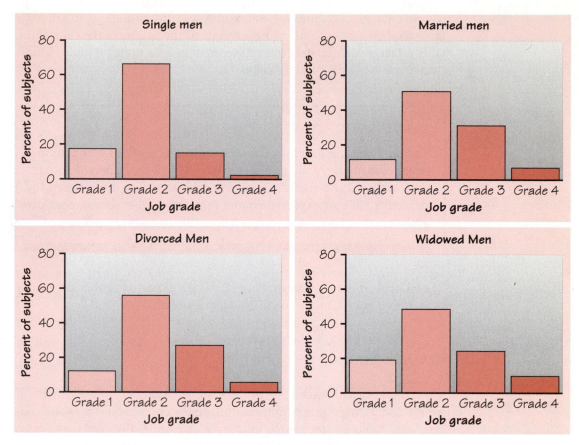

FIGURE 8.7 Bar charts for the data in Example 8.8. Each chart presents the percents of each job grade among men with one marital status.

X^2. Minitab's table of counts shows that the observed counts for single men are higher than expected in grades 1 and 2 and lower than expected in grades 3 and 4.

Of course, this association between marital status and job grade does not show that being single *causes* lower-grade jobs. The explanation might be as simple as the fact that single men tend to be younger and so have not yet advanced to higher grades. ◀

Cell counts required for the chi-square test

The computer output in Figure 8.6 has one more feature. It warns us that the expected counts in two of the 16 cells are less than 5. The chi-square test, like the z procedures for comparing two proportions, is an approximate method that becomes more accurate as the counts in the cells of the table get larger. Fortunately, the approximation is accurate for quite modest counts. Here is a practical guideline.[6]

CELL COUNTS REQUIRED FOR THE CHI-SQUARE TEST

You can safely use the chi-square test with critical values from the chi-square distribution when no more than 20% of the expected counts are less than 5 and all individual expected counts are 1 or greater. In particular, all four expected counts in a 2 × 2 table should be 5 or greater.

Example 8.8 easily passes this test. All the expected counts are greater than 1, and only 2 out of 16 (12.5%) are less than 5.

The chi-square test and the z test

We can use the chi-square test to compare any number of proportions. If we are comparing r proportions and make the columns of the table "success" and "failure," the counts form an $r \times 2$ table. P-values come from the chi-square distribution with $r - 1$ degrees of freedom. If $r = 2$, we are comparing just two proportions. We have two ways to do this: the z test from Section 7.2 and the chi-square test with 1 degree of freedom for a 2 × 2 table. *These two tests always agree.* In fact, the chi-square statistic X^2 is just the square of the z statistic, and the P-value for X^2 is exactly the same as the two-sided P-value for z. We recommend using the z test to compare two proportions, because it gives you the choice of a one-sided test and is related to a confidence interval for $p_1 - p_2$.

EXERCISES

8.7   The North Carolina State study described in Exercise 8.1 also looked at the relationship between student goals and success in getting a C or better in the course. The study report says: "The probability of passing CHE 205 was different for students who would be satisfied with a grade of C or better

(36% of 14 students), B or better (64% of 64), A (90% of 30), creative work beyond an A (82% of 11)."[7]

(a) Use the information given to make a 4 × 2 table of goal by success or failure for the 119 students.

(b) Find the expected counts. The expected counts don't pass our rule of safe practice. Why? Chi-square distribution P-values are nonetheless reasonably accurate.

(c) A statistical calculator gives the chi-square statistic for this table as $X^2 = 14.986$. Give the degrees of freedom. Use Table E to approximate the P-value for this statistic.

(d) Describe briefly the relationship between students' goals and their grades in the course.

8.8   Exercise 7.20 (page 504) compared HMO members who filed complaints with an SRS of members who did not complain. The study actually broke the complainers into two subgroups: those who filed complaints about medical treatment and those who filed nonmedical complaints. Here are the data on the total number in each group and the number who voluntarily left the HMO.

	No complaint	Medical complaint	Nonmedical complaint
Total	743	199	440
Left	22	26	28

(a) Find the percent of each group who left.

(b) Make a two-way table of complaint status by left or not.

(c) Find the expected counts and check that you can safely use the chi-square test.

(d) The chi-square statistic for this table is $X^2 = 31.765$. What null and alternative hypotheses does this statistic test? What are its degrees of freedom? Use Table E to approximate the P-value.

(e) What do you conclude from these data?

8.9   A large study of child care used samples from the data tapes of the Current Population Survey over a period of several years. The result is close to an SRS of child-care workers. The Current Population Survey has three classes of child-care workers: private household, nonhousehold, and preschool teacher. Here are data on the number of blacks among women workers in these three classes.[8]

	Total	Black
Household	2455	172
Nonhousehold	1191	167
Teachers	659	86

(a) What percent of each class of child-care workers is black?

(b) Make a two-way table of class of worker by race (black or other).

(c) Can we safely use the chi-square test? What null and alterative hypotheses does X^2 test?

(d) The chi-square statistic for this table is $X^2 = 53.194$. What are its degrees of freedom. Use Table E to approximate the P-value.

(e) What do you conclude from these data?

8.10 Gastric freezing was once a recommended treatment for ulcers in the upper intestine. Use of gastric freezing stopped after experiments showed it had no effect. One randomized comparative experiment found that 28 of the 82 gastric freezing patients improved, while 30 of the 78 patients in the placebo group improved.[9] We can test the hypothesis of "no difference" between the two groups in two ways: using the two-sample z statistic or using the chi-square statistic.

(a) State the null hypothesis with a two-sided alternative and carry out the z test. What is the P-value from Table A?

(b) Present the data in a 2×2 table. Use the chi-square test to test the hypothesis from (a). Verify that the X^2 statistic is the square of the z statistic. Use Table E to verify that the chi-square P-value agrees with the z result up to the accuracy of the table.

(c) What do you conclude about the effectiveness of gastric freezing as a treatment for ulcers?

SUMMARY

The **chi-square test** for a two-way table tests the null hypothesis that there is no relationship between the row variable and the column variable.

One common use of the chi-square test is to **compare several population proportions**. The null hypothesis states that all of the population proportions are equal. The alternative hypothesis states that they are not all equal but allows any other relationship among the population proportions.

The **expected count** in any cell of a two-way table when H_0 is true is

$$\text{expected count} = \frac{\text{row total} \times \text{column total}}{\text{table total}}$$

The **chi-square statistic** is

$$X^2 = \sum \frac{(\text{observed count} - \text{expected count})^2}{\text{expected count}}$$

The chi-square test compares the value of the statistic X^2 with critical values from the **chi-square distribution** with $(r - 1)(c - 1)$ **degrees of freedom**. Large values of X^2 are evidence against H_0 so the P-value is the area under the chi-square density curve to the right of X^2.

The chi-square distribution is an approximation to the distribution of the statistic X^2. You can safely use this approximation when all expected cell counts are at least 1 and no more than 20% are less than 5.

If the chi-square test finds a statistically significant relationship between the row and column variables in a two-way table, do a follow-up analysis to describe the nature of the relationship. An informal follow-up analysis compares well-chosen percents, compares the observed counts with the expected counts, and looks for the largest **components of chi-square**.

CHAPTER REVIEW

Advanced statistical inference often concerns relationships among several parameters. This chapter begins with the chi-square test for one such relationship: equality of the proportions of successes in any number of populations. The alternative to this hypothesis is "many-sided," because it allows any relationship other than "all equal." The chi-square test is an overall test that tells us whether the data give good reason to reject the hypothesis that all the population proportions are equal. You should always accompany the chi-square test by data analysis to see what kind of inequality is present.

   To do a chi-square test, arrange the counts of successes and failures in a two-way table. Two-way tables can display counts for any two categorical variables, not just successes and failures in r populations. The chi-square test is also more general than just a test for equal population proportions of successes. It tests the null hypothesis that there is "no relationship" between the row variable and the column variable in a two-way table. The chi-square test is actually an approximate test that becomes more accurate as the cell counts in the two-way table increase. Fortunately, chi-square P-values are quite accurate even for small counts.

After studying this chapter, you should be able to do the following.

A. TWO-WAY TABLES

1. Arrange data on successes and failures in several groups into a two-way table of counts of successes and failures in all groups.

2. Use percents to describe the relationship between two categorical variables starting from the counts in a two-way table.

B. INTERPRETING CHI-SQUARE TESTS

1. Locate expected cell counts, the chi-square statistic, and its P-value in output from your software or calculator.

2. Explain what null hypothesis the chi-square statistic tests in a specific two-way table.

3. If the test is significant, use percents, comparison of expected and observed counts, and the components of the chi-square statistic to see what deviations from the null hypothesis are most important.

C. DOING CHI-SQUARE TESTS

1. Calculate the expected count for any cell from the observed counts in a two-way table.

2. Calculate the component of the chi-square statistic for any cell, as well as the overall statistic.

3. Give the degrees of freedom of a chi-square statistic.

4. Use the chi-square critical values in Table E to approximate the P-value of a chi-square test.

CHAPTER 8 REVIEW EXERCISES

If you have access to statistical software or a statistical calculator, use it to speed your analysis of the data in these exercises.

8.11   A study of the career plans of young women and men sent questionnaires to all 722 members of the senior class in the College of Business Administration at the University of Illinois. One question asked which major within the business program the student had chosen. Here are the data from the students who responded.[10]

	Female	Male
Accounting	68	56
Administration	91	40
Economics	5	6
Finance	61	59

(a) Test the null hypothesis that there is no relation between the sex of students and their choice of major. Give a P-value and state your conclusion.

(b) Describe the differences between the distributions of majors for women and men with percents, with a graph, and in words.

(c) Which two cells have the largest components of the chi-square statistic? How do the observed and expected counts differ in these cells? (This should strengthen your conclusions in (b).)

(d) Two of the observed cell counts are small. Do these data satisfy our guidelines for safe use of the chi-square test?

(e) What percent of the students did not respond to the questionnaire? The nonresponse weakens conclusions drawn from these data.

8.12   To study the export activity of manufacturing firms on Taiwan, researchers mailed questionnaires to an SRS of firms in each of five industries that export many of their products. The response rate was only 12.5%, because private companies don't like to fill out long questionnaires from academic researchers. Here are data on the planned sample sizes and the actual number of responses received from each industry.[11]

	Sample size	Responses
Metal products	185	17
Machinery	301	35
Electrical equipment	552	75
Transportation equipment	100	15
Precision instruments	90	12

If the response rates differ greatly, comparisons among the industries may be difficult. Is there good evidence of unequal response rates among the five industries? (Start by creating a two-way table of response or nonresponse by industry.)

8.13   Shopping at secondhand stores is becoming more popular and has even attracted the attention of business schools. A study of customers' attitudes toward secondhand stores interviewed samples of shoppers at two second-hand stores of the same chain in two cities. The breakdown of the respondents by sex is as follows.[12]

	City 1	City 2
Men	38	68
Women	203	150
Total	241	218

Is there a significant difference between the proportions of women customers in the two cities?

(a) State the null hypothesis, find the sample proportions of women in both cities, do a two-sided z test, and give a P-value using Table A.

(b) Calculate the chi-square statistic X^2 and show that it is the square of the z statistic. Show that the P-value from Table E agrees (up to the accuracy of the table) with your result from (a).

(c) Give a 95% confidence interval for the difference between the proportions of women customers in the two cities.

8.14   The study of shoppers in secondhand stores cited in the previous exercise also compared the income distributions of shoppers in the two stores. Here is a two-way table of counts.

Income	City 1	City 2
Under $10,000	70	62
$10,000 to $19,999	52	63
$20,000 to $24,999	69	50
$25,000 to $34,999	22	19
$35,000 or more	28	24

A statistical calculator gives the chi-square statistic for this table as $X^2 = 3.955$. Is there good evidence that customers at the two stores have different income distributions? (Give the degrees of freedom, the P-value, and your conclusion.)

8.15   It seems that the attitude of cancer patients can influence the progress of their disease. We can't experiment with humans, but here is a rat experiment on this theme. Inject 60 rats with tumor cells and then divide them at random into two groups of 30. All the rats receive electric shocks, but rats in Group 1 can end the shock by pressing a lever. (Rats learn this sort of thing quickly.) The rats in Group 2 cannot control the shocks, which presumably makes them feel helpless and unhappy. We suspect that the rats in Group 1 will develop fewer tumors. The results: 11 of the Group 1 rats and 22 of the Group 2 rats developed tumors.[13]

(a) State the null and alternative hypotheses for this investigation. Explain why the z test rather than the chi-square test for a 2×2 table is the proper test.

(b) Carry out the test and report your conclusion.

8.16   Do unregulated providers of child care in their homes follow different health and safety practices in different cities? A study looked at people who regularly provided care for someone else's children in poor areas of three cities. The numbers who required medical releases from parents to allow medical care in an emergency were 42 of 73 providers in Newark, N.J., 29 of 101 in Camden, N.J., and 48 of 107 in South Chicago, Ill.[14]

(a) Use the chi-square test to see if there are significant differences among the proportions of child-care providers who require medical releases in the three cities.

(b) How should the data be produced in order for your test to be valid? (In fact, the samples came in part from asking parents who were subjects in another study who provided their child care. The author of the study wisely did not use a statistical test. He wrote: "Application of conventional statistical procedures appropriate for random samples may produce biased and misleading results.")

8.17   Sample surveys on sensitive issues can give different results depending on how the question is asked. A University of Wisconsin study divided 2400 respondents into 3 groups at random. All were asked if they had ever used cocaine. One group of 800 was interviewed by phone; 21% said they had used cocaine. Another 800 people were asked the question in a one-on-one personal interview; 25% said "Yes." The remaining 800 were allowed to make an anonymous written response; 28% said "Yes."[15] Are there statistically significant differences among these proportions? (State the hypotheses, convert the information given into a two-way table, give the test statistic and its P-value, and state your conclusions.)

8.18   The success of a sample survey can depend on the season of the year. The Italian National Statistical Institute kept records of nonresponse to one of its national telephone surveys. All calls were made between 7 p.m. and 10 p.m. Here is a table of the percents of responses and of three types of nonresponse at different seasons. The percents in each row add to 100% (up to roundoff error).[16]

| Season | Calls made | Successful interviews | Nonresponse | | |
			No answer	Busy signal	Refusal
Jan. 1 to Apr. 13	1558	68.5%	21.4%	5.8%	4.3%
Apr. 21 to June 20	1589	52.4%	35.8%	6.4%	5.4%
July 1 to Aug. 31	2075	43.4%	41.5%	8.6%	6.5%
Sept. 1 to Dec. 15	2638	60.0%	30.0%	5.3%	4.7%

(a) What are the degrees of freedom for the chi-square test of the hypothesis that the distribution of responses varies with the season? (Don't do the test. The sample sizes are so large that the results are sure to be highly significant.)

(b) Consider just the proportion of successful interviews. Describe how this proportion varies with the seasons, and assess the statistical significance of the changes. What do you think explains the changes? (Look at the full table for ideas.)

(c) (Optional) It is incorrect to apply the chi-square test to percents rather than to counts. If you enter the 4 × 4 table of percents above into statistical software and ask for a chi-square test, well-written software should give an error message. (Counts must be whole numbers, so the software should check that.) Try this using your software or calculator, and report the result.

8.19   Continue the analysis of the data in the previous exercise by considering just the proportion of people called who refused to participate. We might think that the refusal rate changes less with the season than, for example, the rate of "no answer." State the hypothesis that the refusal rate does not change with the season. Check that you can safely use the chi-square test. Carry out the test. What do you conclude?

8.20   In 1912 the luxury liner *Titanic*, on its first voyage across the Atlantic, struck an iceberg and sank. Some passengers got off the ship in lifeboats, but many died. Think of the *Titanic* disaster as an experiment in how the people of that time behaved when faced with death in a situation where only some can escape. The passengers are a sample from the population of their peers. Here is information about who lived and who died, by sex and economic status. (The data leave out a few passengers whose economic status is unknown.)[17]

	Men			**Women**	
Status	Died	Survived	Status	Died	Survived
Highest	111	61	Highest	6	126
Middle	150	22	Middle	13	90
Lowest	419	85	Lowest	107	101
Total	680	168	Total	126	317

(a) Compare the percents of men and of women who died. Is there strong evidence that a higher proportion of men die in such situations? Why do you think this happened?

(b) Look only at the women. Describe how the three economic classes differ in the percent of women who died. Are these differences statistically significant?

(c) Now look only at the men and answer the same questions.

8.21   How accurate are pre-election polls of voters' intentions? The 1992 presidential election featured the Republican incumbent, George Bush, the Democrat Bill Clinton, and the third-party candidate Ross Perot. Clinton was elected. Here are results for three polls taken a few days before the election, as well as the actual election result. The poll results don't add to 100% because some voters were undecided.[18]

	Sample size	Bush	Clinton	Perot
ABC News	912	38%	41%	18%
USA Today/CNN	1610	39%	42%	14%
New York Times/CBS	1912	34%	43%	15%
Actual vote		38%	43%	19%

The polls use complex multistage sample designs. For the purposes of this exercise, treat each poll as an SRS of registered voters.

(a) Did the three polls differ significantly in the percent of their samples who favored the winner, Bill Clinton?

(b) For each poll individually, test whether its percent for Ross Perot differs significantly at the $\alpha = .05$ level from the actual election result, which is 19%.

(c) A single test at the $\alpha = .05$ level will wrongly reject H_0 only 5% of the time when H_0 is actually true. In (b) you made three separate tests at the $\alpha = .05$ level. Explain why at least one of three tests will be wrong on more than 5% of the occasions you do three tests.

8.22   Example 2.23 (page 158) presents artificial data that illustrate Simpson's paradox. The data concern the survival rates of surgery patients at two hospitals.

(a) Apply the chi-square test to the data for all patients combined and summarize the results.

(b) Do separate chi-square tests for the patients in good condition and for those in poor condition. Summarize these results.

(c) Are the effects that illustrate Simpson's paradox in this example statistically significant?

8.23   During the 1991 Persian Gulf War, the military's control of news reporting from the combat zone was a public issue. Princeton Survey Research Associates interviewed a random sample of 924 adults in January of 1991. One question asked was

Do you think the military should exert more control over how news organizations report about the war or do you think that most decisions about how to report about the war should be left to news organizations themselves?

(a) Fifty-seven percent of the sample thought the military should have more control. Give a 90% confidence interval for the proportion of all adults who feel this way. (Treat the sample as an SRS.)

(b) Professor Ted Chang of the University of Virginia did an experiment on the effect of the wording of questions in a survey. He asked 174 students the question above; 55% said the military should have more control. Is the student response significantly different from the national response?

(c) Professor Chang asked another 199 students this question:

Do you feel the amount of control the military is exerting over news organizations reporting the war is about right?

Now only 16% said that the military should exert more control. Is this significantly different from the student response to the first wording as given in (b)?

(d) All three samples were also asked the question

To the best of your knowledge are news reports from the Gulf being censored by the American military?

The percents who said "Yes" were 76% in the national sample ($n =$ 924), 67% in the first student sample ($n = 174$), and 65% in the second student sample ($n = 199$). Is there significant evidence that the three populations did not all have the same proportion who thought the news is censored? (The two student samples represent different populations because they were asked different questions about military control of the news, and this might influence their response to other questions.)

8.24   Substances suspected of causing cancer are often tested (in high doses) on mice or rats. How well do results from mice agree with the results of testing the same substances on rats? Here, in a 2×2 table, are the results of testing 249 chemicals on both mice and rats (+ indicates that the substance was found to cause cancer; −, that it was not).[19]

		Rats	
		+	−
Mice	+	111	21
	−	17	100

(a) Explain in words what hypothesis the chi-square test for this table examines.

(b) We are interested in how often results from rats and results from mice for the same chemical agree. Explain why the chi-square test does *not* help assess agreement. Beware the temptation to apply chi-square to every two-way table!

(c) On what percent of the 249 chemicals do mice and rats agree? What percent of chemicals to which mice give a + also get a + from rats? What percent of chemicals that get a − from mice also get a − from rats? These percents help answer our question.

NOTES AND DATA SOURCES

1. Stephen M. Stigler, *The History of Statistics: The Measurement of Uncertainty before 1900*, Belknap Press, Cambridge, Mass., 1986. The quotation is from p. 361.

2. D. M. Barnes, "Breaking the cycle of addiction," *Science*, 241 (1988), pp. 1029–1030.

3. Richard M. Felder et al., "Who gets it and who doesn't: a study of student performance in an introductory chemical engineering course," *1992 ASEE Annual Conference Proceedings*, American Society for Engineering Education, Washington, D.C., 1992, pp. 1516–1519.

4. S. V. Zagona (ed.), *Studies and Issues in Smoking Behavior*, University of Arizona Press, Tucson, 1967, pp 157–180.

5. Sanders Korenman and David Neumark, "Does marriage really make men more productive?" *Journal of Human Resources*, 26 (1991), pp. 282–307.

6. There are many computer studies of the accuracy of chi-square critical values for X^2. For a brief discussion and some references, see Section 3.2.5 of David S. Moore, "Tests of chi-squared type," in Ralph B. D'Agostino and Michael A. Stephens (eds.), *Goodness-of-Fit Techniques*, Marcel Dekker, New York, 1986, pp. 63–95. If the expected cell counts are roughly equal, the chi-square approximation is adequate when the average expected counts are as small as 1 or 2. The guideline given in the text protects against unequal expected counts. For a survey of inference for smaller samples, see Alan Agresti, "A survey of exact inference for contingency tables," *Statistical Science*, 7 (1992), pp. 131–177.

7. The quotation is from p. 1517 of the article cited in Note 3.

8. David M. Blau, "The child care labor market," *Journal of Human Resources*, 27 (1992), pp. 9–39.

9. Lillian Lin Miao, "Gastric freezing: an example of the evaluation of medical therapy by randomized clinical trials," in John P. Bunker, Benjamin A. Barnes, and Frederick Mosteller (eds.), *Costs, Risks, and Benefits of Surgery*, Oxford University Press, New York, 1977, pp. 198–211.

10. Francine D. Blau and Marianne A. Ferber, "Career plans and expectations of young women and men," *Journal of Human Resources*, 26 (1991), pp. 581–607.

11. Erdener Kaynak and Wellington Kang-yen Kuan, "Environment, strategy, structure, and performance in the context of export activity: an empirical study of Taiwanese manufacturing firms," *Journal of Business Research*, 27 (1993), pp. 33–49.

12. William D. Darley, "Store-choice behavior for pre-owned merchandise," *Journal of Business Research*, 27 (1993), pp. 17–31.

13. Adapted from M. A. Visintainer, J. R. Volpicelli, and M. E. P. Seligman, "Tumor rejection in rats after inescapable or escapable shock," *Science*, 216 (1982), pp. 437–439.

14. James R. Walker, "New evidence on the supply of child care," *Journal of Human Resources*, 27 (1991), pp. 40–69.

15. Modified from Felicity Barringer, "Measuring sexuality through polls can be shaky," *New York Times*, April 25, 1993.

16. Giuliana Coccia, "An overview of non-response in Italian telephone surveys," *Proceedings of the 99th Session of the International Statistical Institute*, 1993, Book 3, pp. 271–272.

17. Data provided by Don Bentley, Pomona College.

18. Poll results reported in the *New York Times*, November 1, 1992.

19. Cited by J. K. Haseman in his discussion of D. A. Freedman and H. Zeisel, "From mouse to man: the quantitative assessment of cancer risks," *Statistical Science*, 3 (1988), pp. 3–28. The citation is on p. 34.

W. EDWARDS DEMING

From one point of view, statistics is about understanding variation. Elimination of variation in products and processes is the central theme of statistical quality control. So it is not surprising that a statistician should become the leading guru of quality management. In the final decades of his long life, W. Edwards Deming (1900–1993) was one of the world's most influential consultants to management.

Deming grew up in Wyoming and earned a doctorate in physics at Yale. Working for the U.S. Department of Agriculture in the 1930s, he became acquainted with Neyman's fundamental work on sampling design and with statistical process control, newly invented by Walter Shewhart of AT&T. In 1939 he moved to the Census Bureau as an expert on sampling.

The work that made Deming famous began after he left the government in 1946. He visited Japan to advise on a census but returned to lecture on quality control. He earned a large following in Japan, which named its premier prize for industrial quality after him. As Japan's reputation for manufacturing excellence rose, Deming's fame rose with it. Blunt-spoken and even abrasive, he told corporate leaders that most quality problems are system problems for which management is responsible. He urged breaking down barriers to worker involvement and constant search for causes of variation. Finding the sources of variation is the theme of analysis of variance, the statistical method we introduce in this chapter.

CHAPTER **9**

One-Way Analysis of Variance: Comparing Several Means

INTRODUCTION

We use the two-sample t procedures of Chapter 6 to compare the means of two populations or the mean responses to two treatments in an experiment. Of course, studies don't always compare just two groups. We need a method for comparing any number of means.

EXAMPLE 9.1

Do smaller cars really have better gas mileage? Table 9.1 contains data on the city and highway gas mileage (in miles per gallon, as reported by the Environmental Protection Agency) for a sample of 59 compact, midsize, and large 1994 car models.[1] We want to compare the city gas mileages.

TABLE 9.1 City and highway gas mileage for 1994 car models			
Model	Size	City MPG	Highway MPG
Acura Legend	Compact	19	24
Acura Vigor	Compact	20	26
Audi 100	Compact	18	24
Audi 90	Compact	18	26
BMW 525i	Compact	18	25
Buick Skylark	Compact	22	32
Chevrolet Beretta	Compact	25	31
Chrysler Lebaron	Compact	22	27
Dodge Shadow	Compact	23	30
Ford Tempo	Compact	22	27
Honda Accord	Compact	23	29
Jaguar XJ12	Compact	12	16
Mazda Protege	Compact	23	29
Mazda 323	Compact	26	33
Mercedes-Benz E320	Compact	19	25
Mercedes-Benz E420	Compact	18	24
Mitsubishi Diamante	Compact	18	24
Mitsubishi Galant	Compact	20	26
Mitsubishi Precis	Compact	27	35
Nissan Altima	Compact	21	29
Oldsmobile Achieva	Compact	22	32

TABLE 9.1   (*continued*)

Model	Size	City MPG	Highway MPG
Saturn SL	Compact	26	35
Subaru Legacy	Compact	22	29
Toyota Corolla	Compact	26	29
Volkswagen Golf	Compact	21	28
Volkswagen Jetta	Compact	21	27
BMW 740i	Midsize	16	23
Buick Century	Midsize	25	31
Buick Regal	Midsize	19	29
Cadillac Eldorado	Midsize	16	25
Chevrolet Lumina	Midsize	19	29
Dodge Spirit	Midsize	22	27
Ford Taurus	Midsize	20	29
Ford Thunderbird	Midsize	19	26
Hyundai Sonata	Midsize	21	27
Infiniti Q45	Midsize	17	22
Lexus GS300	Midsize	17	23
Lexus LS400	Midsize	18	23
Lincoln-Mercury Mark VIII	Midsize	18	25
Mazda 626	Midsize	23	31
Mazda 929	Midsize	19	24
Nissan Maxima	Midsize	19	26
Rolls-Royce Silver Spur	Midsize	10	15
Saab 900	Midsize	19	26
Toyota Camry	Midsize	21	28
Volvo 850	Midsize	19	26
Buick LeSabre	Large	19	28
Buick Park Avenue	Large	19	27
Buick Roadmaster	Large	17	25
Cadillac DeVille	Large	16	25
Chevrolet Caprice	Large	18	26
Chrysler Concorde	Large	20	28
Chrysler New Yorker	Large	18	26
Ford LTD	Large	18	25
Lincoln-Mercury Continental	Large	18	26
Mercedes-Benz S320	Large	17	24
Mercedes-Benz S420	Large	15	20
Mercedes-Benz S500	Large	14	19
Saab 9000	Large	17	27

```
        Compact              Midsize              Large

        10 |                 10 | 0   ← Rolls-Royce    10 |
        11 |     ← Jaguar    11 |       Silver Spur     11 |
        12 | 0      XJ12     12 |                        12 |
        13 |                 13 |                        13 |
        14 |                 14 |                        14 | 0
        15 |                 15 |                        15 | 0
        16 |                 16 | 00                      16 | 0
        17 |                 17 | 00                      17 | 000
        18 | 00000           18 | 00                      18 | 0000
        19 | 00              19 | 0000000                 19 | 00
        20 | 00              20 | 0                        20 | 0
        21 | 000             21 | 00                       21 |
        22 | 00000           22 | 0                        22 |
        23 | 000             23 | 0                        23 |
        24 |                 24 |                           24 |
        25 | 0               25 | 0                         25 |
        26 | 000             26 |                           26 |
        27 | 0               27 |                           27 |
```

FIGURE 9.1   Side-by-side stemplots comparing the city gas mileages of compact, midsize, and large cars from Table 9.1.

Figure 9.1 shows side-by-side stemplots of the city mileages for the three types of cars. We used the same stems in all three for easier comparison. It does appear that gas mileage decreases as cars get larger. The plots show a low outlier in the compact group and another in the midsize group. The Jaguar XJ12 and the Rolls-Royce Silver Spur are responsible. These are rather exotic cars, so we will omit these outliers from all further analyses.

Here are the means, standard deviations, and five-number summaries for the three car types, from the Minitab statistical software:

	N	MEAN	MEDIAN	STDEV	MIN	MAX	Q1	Q3
Compact	25	21.600	22.000	2.814	18.000	27.000	19.000	23.000
Midsize	19	19.316	19.000	2.311	16.000	25.000	18.000	21.000
Large	13	17.385	18.000	1.660	14.000	20.000	16.500	18.500

We will use the mean to describe the center of the gas mileage distributions. As we expect, mean gas mileage goes down as we move from compact to midsize and then to large cars. The differences among the means are not large. Are they statistically significant?   ◄

The problem of multiple comparisons

Call the mean city gas mileage for the three populations of cars μ_1 for compact cars, μ_2 for midsize cars, and μ_3 for large cars. The subscript reminds us which group a parameter or statistic describes. To compare these three population means, we might use the two-sample t test several times:

- Test $H_0 : \mu_1 = \mu_2$ to see if the mean miles per gallon for compact cars differs from the mean for midsize cars.
- Test $H_0 : \mu_1 = \mu_3$ to see if compact cars differ from large cars.
- Test $H_0 : \mu_2 = \mu_3$ to see if midsize cars differ from large cars.

The weakness of doing three tests is that we get three P-values, one for each test alone. That doesn't tell us how likely it is that *three* sample means are spread apart as far as these are. It may be that 17.385 and 21.600 are significantly different if we look at just two groups, but not significantly different if we know that they are the smallest and largest means in three groups. As we look at more groups, we expect the gap between the smallest and largest sample mean to get larger. (Think of comparing the tallest and shortest person in larger and larger groups of people.) We can't safely compare many parameters by doing tests or confidence intervals for two parameters at a time.

The problem of how to do many comparisons at once with some overall measure of confidence in all our conclusions is common in more advanced statistics. We are leaving the elementary parts of statistical inference now that we have met this problem. Statistical methods for dealing with many comparisons usually have two parts:

1. An *overall test* to see if there is good evidence of *any* differences among the parameters that we want to compare.
2. A detailed *follow-up analysis* to decide which of the parameters differ and to estimate how large the differences are.

The overall test is often reasonably straightforward, though still more complex than the tests we met in Chapters 6 and 7. The follow-up analysis can be quite elaborate. In our basic introduction to statistical practice, we will look only at some overall tests. Chapter 8 describes an overall test to compare several population proportions. In this chapter we present a test for comparing several population means.

THE ANALYSIS OF VARIANCE F TEST

We want to test the null hypothesis that there are *no differences* among the mean city gas mileages for the three car types:

$$H_0 : \mu_1 = \mu_2 = \mu_3$$

The alternative hypothesis is that there *is* some difference, that not all three population means are equal:

$$H_a: \text{not all of } \mu_1, \mu_2, \text{and } \mu_3 \text{ are equal}$$

The alternative hypothesis is no longer one-sided or two-sided. It is "many-sided," because it allows any relationship other than "all three equal." For example, H_a includes the case in which $\mu_2 = \mu_3$ but μ_1 has a different value. The test of H_0 against H_a is called the ***analysis of variance F test***. Analysis of variance is usually abbreviated as ANOVA.

analysis of variance F test

EXAMPLE 9.2

Enter the city gas mileage data from Table 9.1 (omitting the two outliers) into the Minitab software package and request analysis of variance. The output appears in Figure 9.2. Most statistical software packages produce an ANOVA output similar to this one.

```
ANALYSIS OF VARIANCE
SOURCE       DF          SS          MS         F          P
FACTOR        2       160.96       80.48     13.62      0.000
ERROR        54       319.18        5.91
TOTAL        56       480.14
                                         INDIVIDUAL 95 PCT CI'S FOR MEAN
                                         BASED ON POOLED STDEV
   LEVEL      N        MEAN       STDEV   ----------+---------+---------+------
Compact      25      21.600       2.814                             (----*----)
Midsize      19      19.316       2.311                   (-----*----)
Large        13      17.385       1.660   (------*------)
                                         ----------+---------+---------+------
POOLED STDEV =      2.431                      18.0      20.0      22.0
```

FIGURE 9.2   Minitab output for analysis of variance of the city gas mileage data, for Example 9.2.

For now, look at just two parts of this output. First, check that the sample sizes, means, and standard deviations agree with those in Example 9.1. Then find the *F* test statistic, $F = 13.62$, and its *P*-value. The *P*-value is given as 0.000. This means that *P* is zero to three decimal places, or $P < .001$. There is extremely strong evidence that the three car sizes do not all have the same mean gas mileage.

The *F* test does not say *which* of the three means are significantly different. It appears from our preliminary analysis of the data that midsize and large cars differ by less than do compact and midsize cars. The computer output includes confidence intervals for all three means that suggest the same conclusion. The midsize- and large-car intervals overlap, and the compact-car interval lies a bit above them on the mileage scale. These are 95% confidence intervals for each mean separately. We are not 95% confident that *all three* intervals cover the three means. There are follow-up procedures that provide 95% confidence that we have caught all three means at once, but we won't study them.

Our conclusion: There is strong evidence ($P < .001$) that the means are not all equal, and the most important difference among the means is that compact cars have better gas mileage than midsize and large cars. ◀

Example 9.2 illustrates our approach to comparing means. The ANOVA *F* test (often done by software) assesses the evidence for *some* difference among the population means. In most cases, we expect the *F* test to be significant. We would not undertake a study if we did not expect to find some effect. The formal test is nonetheless important to guard against being misled by chance variation. We will not do the formal follow-up analysis that is often the most useful part of an ANOVA study. Follow-up analysis would allow us to say which means differ and by how much, with (say) 95% confidence that all our conclusions are correct. We rely instead on a preliminary examination of the data to show what differences are present and whether they are large enough to be interesting. The gap of more than 4 miles per gallon between compact cars and large cars *is* large enough to be of practical interest.

EXERCISES

9.1   Table 9.2 gives the results of a study of fish caught in a lake in Finland.[2] We are willing to regard the fish caught by the researchers a random sample of fish in this lake. The weight of commercial fish is of particular interest. Is there evidence that the mean weights of all bream, perch, and roach found in the lake are different? Before doing ANOVA, we must examine the data.

TABLE 9.2 Weights (grams) of three fish species					
Bream		Perch			Roach
13.4	14.8	16.0	15.1	17.5	14.0
13.8	14.1	13.6	15.1	20.9	13.9
15.1	13.7	15.2	15.0	17.6	13.7
13.3	13.3	15.3	14.8	17.6	14.3
15.1	15.1	15.9	14.9	15.9	16.1
14.2	13.8	17.3	15.0	16.2	14.7
15.3	14.8	16.1	15.9	18.1	14.7
13.4	15.0	15.1	13.9	14.5	13.9
13.8	14.1	14.6	15.7	17.8	15.2
13.7	14.9	13.2	14.8	16.8	14.6
14.1	15.5	15.8	17.9	17.0	15.1
13.3	14.3	14.7	14.6	17.6	13.3
12.0	14.3	16.3	15.0	15.6	15.2
13.6	14.9	15.5	15.0	15.4	14.1
13.9	14.7	14.5	15.8	16.1	13.6
15.0		15.0	14.3	16.3	15.4
13.8		15.0	15.4		14.0
13.5		15.0	15.1		15.4
13.3		17.0	17.7		15.6
13.7		16.3	17.7		15.3

(a) Display the distribution of weights for each species of fish with side-by-side stemplots. Are there outliers or strong skewness in any of the distributions?

(b) Find the five-number summary for each distribution. What do the data appear to show about the weights of these species?

9.2   Now we proceed to the ANOVA for the data in Table 9.2. The heaviest perch had six roach in its stomach when caught. This fish may be an outlier and its condition is unusual, so remove this observation from the data. Figure 9.3 gives the Minitab ANOVA output for the data with the outlier removed.

(a) What null hypothesis does the ANOVA F statistic test? State this hypothesis both in words and in symbols.

(b) What is the value of the F statistic? What is its P-value?

(c) Based on your work in this and the previous exercise, what do you conclude about the weights of these species of fish?

```
ANALYSIS OF VARIANCE
SOURCE      DF          SS          MS          F          P
FACTOR       2        60.17       30.08      29.92      0.000
ERROR      107       107.60        1.01
TOTAL      109       167.77
                                        INDIVIDUAL 95 PCT CI'S FOR MEAN
                                        BASED ON POOLED STDEV
   LEVEL       N        MEAN       STDEV   ---+---------+---------+---------+---
bream         35      14.131      0.770   (----*----)
perch         55      15.747      1.186                          (---*---)
roach         20      14.605      0.780           (-------*-----)
                                        ---+---------+---------+---------+---
POOLED STDEV =        1.003              14.00     14.70     15.40     16.10
```

FIGURE 9.3   Minitab output for the data in Table 9.2 of weights of three species of fish, for Exercise 9.2.

9.3   How much corn per acre should a farmer plant to obtain the highest yield? Too few plants will give a low yield. On the other hand, if there are too many plants, they will compete with each other for moisture and nutrients, and yields will fall. To find out, plant at different rates on several plots of ground and measure the harvest. (Be sure to treat all the plots the same except for the planting rate.) Here are data from such an experiment.[3]

Plants per acre	Yield (bushels per acre)			
12,000	150.1	113.0	118.4	142.6
16,000	166.9	120.7	135.2	149.8
20,000	165.3	130.1	139.6	149.9
24,000	134.7	138.4	156.1	
28,000	119.0	150.5		

(a) Make side-by-side stemplots of yield for each number of plants per acre. Find the mean yield for each planting rate. What do the data appear to show about the influence of plants per acre on yield?

(b) ANOVA will assess the statistical significance of the observed differences in yield. What are H_0 and H_a for the ANOVA F test in this situation?

(c) The Minitab ANOVA output for these data appears in Figure 9.4. What does the ANOVA F test say about the significance of the effects you described in (a)?

(d) The observed differences among the mean yields in the sample were quite large. Why are they not statistically significant?

```
ANALYSIS OF VARIANCE ON yield
SOURCE        DF          SS          MS          F          P
Rate          4          600         150        0.50      0.736
ERROR        12         3597         300
TOTAL        16         4197
                                          INDIVIDUAL 95 PCT CI'S FOR MEAN
                                          BASED ON POOLED STDEV
LEVEL         N         MEAN        STDEV    ---+---------+---------+---------+---
  12          4        131.03       18.09       (-----------*-----------)
  16          4        143.15       19.79           (-----------*-----------)
  20          4        146.23       15.07             (----------*-----------)
  24          3        143.07       11.44         (------------*-------------)
  28          2        134.75       22.27     (---------------*---------------)
                                          ---+---------+---------+---------+---
POOLED STDEV =        17.31                 112       128       144       160
```

FIGURE 9.4   Minitab output for yields of corn at five planting rates, for Exercise 9.3.

The idea of analysis of variance

Here is the main idea for comparing means: what matters is not how far apart the sample means are but how far apart they are *relative to the variability of individual observations*. Look at the two sets of boxplots in Figure 9.5. For simplicity, these distributions are all symmetric, so that the mean and median are the same. The centerline in each boxplot is therefore the sample mean. Both figures compare three samples with the same three means. Like the three car types in Example 9.1, the means are different but not very different. Could differences this large easily arise just due to chance, or are they statistically significant?

- The boxplots in Figure 9.5(a) have tall boxes, which show lots of variation among the individuals in each group. With this much variation among individuals, we would not be surprised if another set of samples gave quite different sample means. The observed differences among the sample means could easily happen just by chance.

- The boxplots in Figure 9.5(b) have the same centers as those in Figure 9.5(a), but the boxes are much shorter. That is, there is much less variation among the individuals in each group. It is unlikely that any sample from the first group would have a mean as small as the mean of the second group. Because means as far apart as those

FIGURE 9.5 Boxplots for two sets of three samples each. The sample means are the same in (**a**) and (**b**). Analysis of variance will find a more significant difference among the means in (**b**) because there is less variation among the individuals within those samples.

observed would rarely arise just by chance in repeated sampling, they are good evidence of real differences among the means of the three populations we are sampling from.

This comparison of the two parts of Figure 9.5 is too simple in one way. It ignores the effect of the sample sizes, an effect that boxplots do not show. Small differences among sample means can be significant if the samples are very large. Large differences among sample means may fail to be significant if the samples are very small. All we can be sure of is that for the same sample size, Figure 9.5(b) will give a much smaller *P*-value than Figure 9.5(a). Despite this qualification, the big idea remains: if sample means are far apart relative to the variation among individuals in the same group, that's evidence that something other than chance is at work.

THE ANALYSIS OF VARIANCE IDEA

Analysis of variance compares the variation due to specific sources with the variation among individuals who should be similar. In particular, ANOVA tests whether several populations have the same mean by comparing how far apart the sample means are with how much variation there is within the samples.

one-way ANOVA

It is one of the oddities of statistical language that methods for comparing means are named after the variance. The reason is that the test works by comparing two kinds of variation. Analysis of variance is a general method for studying sources of variation in responses. Comparing several means is the simplest form of ANOVA, called **one-way ANOVA**. One-way ANOVA is the only form of ANOVA that we will study.

The ANOVA F statistic for comparing several means has this form:

$$F = \frac{\text{variation among the sample means}}{\text{variation among individuals in the same sample}}$$

We give more detail later. Because ANOVA is in practice done by software, the idea is more important than the detail. The F statistic can only take values that are zero or positive. It is zero only when all the sample means are identical. Chance variation creates some differences among the sample means even when the population means are equal. In fact, when the null hypothesis is true, we expect F to take values near 1. As the sample means get farther apart, the value of F gets larger. Large values of F are evidence against the null hypothesis H_0 that all population means are the same. Although the alternative hypothesis H_a is many-sided, the ANOVA F test is one-sided because any violation of H_0 tends to produce a large value of F.

F distribution

How large must F be to provide significant evidence against H_0? To answer questions of statistical significance, compare the F statistic with critical values from an **F distribution**. The F distributions are described on page 464 in Chapter 6. A specific F distribution is specified by two parameters: a numerator degrees of freedom and a denominator degrees of freedom. Table D in the back of the book contains critical values for F distributions with various degrees of freedom.

EXAMPLE 9.3

Look again at the computer output for the city gas mileage data in Figure 9.2. The degrees of freedom for the F test appear in the first two rows of the "DF" column. There are 2 degrees of freedom in the numerator and 54 in the denominator.

In Table D, find the numerator degrees of freedom 2 at the top of the table. Then look for the denominator degrees of freedom 54 at the left of the table. There is no entry for 54, so we use the next smaller entry, 50 degrees of freedom. The critical values for 2 and 50 degrees of freedom are df = 2, 50

p	Critical value
0.100	2.41
0.050	3.18
0.025	3.97
0.010	5.06
0.001	7.96

The observed $F = 13.62$ is larger than the upper 0.001 critical value, so $P <$ .001. Figure 9.6 shows the F density curve with 2 and 50 degrees of freedom. The observed $F = 13.62$ lies far to the right on this curve.   ◀

The degrees of freedom of the F statistic depend on the number of means we are comparing and the number of observations in each sample.

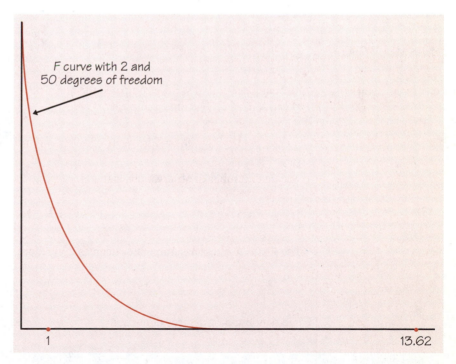

FIGURE 9.6   The density curve of the F distribution with 2 degrees of freedom in the numerator and 50 degrees of freedom in the denominator, with the observed value $F = 13.62$ from Example 9.3 marked. This observed F is highly significant.

That is, the F test does take into account the number of observations. Here are the details.

DEGREES OF FREEDOM FOR THE F TEST

We want to compare the means of I populations. We have an SRS of size n_i from the ith population, so that the total number of observations in all samples combined is

$$N = n_1 + n_2 + \cdots + n_I$$

If the null hypothesis that all population means are equal is true, the ANOVA F statistic has the F distribution with $I - 1$ degrees of freedom in the numerator and $N - I$ degrees of freedom in the denominator.

EXAMPLE 9.4

In Examples 9.1 and 9.2, we compared the mean city gas mileage for three sizes of car, so $I = 3$. The three sample sizes are

$$n_1 = 25 \qquad n_2 = 19 \qquad n_3 = 13$$

The total number of observations is therefore

$$N = 25 + 19 + 13 = 57$$

The ANOVA F test has numerator degrees of freedom

$$I - 1 = 3 - 1 = 2$$

and denominator degrees of freedom

$$N - I = 57 - 3 = 54$$

These degrees of freedom are given in the computer output as the first two entries in the "DF" column in Figure 9.2.   ◄

EXERCISES

9.4   Exercises 9.1 and 9.2 compare the weights of three species of fish.

(a) What are I, the n_i, and N for these data? (Remember that the heaviest perch was removed before doing ANOVA.) Identify these quantities in words, and give their numeric values.

(b) Find the degrees of freedom for the ANOVA F statistic. Check your work against the computer output in Figure 9.3.

(c) For these data, $F = 29.92$. What does Table D tell you about the P-value of this statistic?

9.5   Exercise 9.3 compares the yields for several planting rates for corn.

(a) What are I, the n_i, and N for these data? Identify these quantities in words, and give their numeric values.

(b) Find the degrees of freedom for the ANOVA F statistic. Check your work against the computer output in Figure 9.4.

(c) For these data, $F = .50$. What does Table D tell you about the P-value of this statistic?

9.6   In each of the following situations, we want to compare the mean response in several populations. For each setting, identify the populations and the response variable. Then give I, the n_i, and N. Finally, give the degrees of freedom of the ANOVA F statistic.

(a) Which of four tomato varieties has the highest mean yield? Grow ten plants of each variety and record the yield of each plant in pounds of tomatoes.

(b) A maker of detergents wants to know which of six package designs is most attractive to consumers. Each package is shown to 120 different consumers who rate the attractiveness of the design on a 1 to 10 scale.

(c) An experiment to compare the effectiveness of three weight-loss programs has 32 subjects who want to lose weight. Ten subjects are assigned at random to each of two programs, and the remaining 12 subjects follow the third program. After six months, each subject's change in weight is recorded.

Assumptions for ANOVA

Like all inference procedures, ANOVA is valid only in some circumstances. Here are the requirements for using ANOVA to compare population means.

ANOVA ASSUMPTIONS

- We have I **independent SRSs**, one from each of I populations.
- The ith population has a **normal distribution** with unknown mean μ_i. The means may be different in the different populations. The ANOVA F statistic tests the null hypothesis that all of the populations have the same mean:

$$H_0: \mu_1 = \mu_2 = \cdots = \mu_I$$
$$H_a: \text{not all of the } \mu_i \text{ are equal}$$

- All of the populations have the **same standard deviation** σ, whose value is unknown.

The first two requirements are familiar from our study of the two-sample t procedures for comparing two means. As usual, the design of the data production is the most important foundation for inference. Biased sampling or confounding can make any inference meaningless. If we do not actually draw separate SRSs from each population or carry out a randomized comparative experiment, it is often unclear to what population the conclusions of inference apply. (This is the case in Example 9.1, for example.) ANOVA, like other inference procedures, is often used when random samples are not available. You must judge each use on its merits, a judgment that usually requires some knowledge of the subject of the study in addition to some knowledge of statistics.

Because no real population has exactly a normal distribution, the usefulness of inference procedures that assume normality depends on how sensitive they are to departures from normality. Fortunately, procedures for comparing means are not very sensitive to lack of normality. The ANOVA

robustness F test, like the t procedures, is *robust*. What matters is normality of the sample means, so ANOVA becomes safer as the sample sizes get larger, because of the central limit theorem effect. Remember to check for outliers that change the value of sample means and for extreme skewness. When there are no outliers and the distributions are roughly symmetric, you can safely use ANOVA for sample sizes as small as 4 or 5. (Don't confuse the ANOVA F, which compares several means, with the F statistic of Section 6.3, which compares two standard deviations and is *not* robust against nonnormality.)

The third assumption is annoying: ANOVA assumes that the variability of observations, measured by the standard deviation, is the same in all populations. You may recall from Chapter 6 (page 455) that there is a special version of the two-sample *t* test that assumes equal standard deviations in both populations. The ANOVA *F* for comparing two means is exactly the square of this special *t* statistic. We prefer the *t* test that does not assume equal standard deviations, but for comparing more than two means there is no general alternative to the ANOVA *F*. It is not easy to check the assumption that the populations have equal standard deviations. Statistical tests for equality of standard deviations are very sensitive to lack of normality, so much so that they are of little practical value. You must either seek expert advice or rely on the robustness of ANOVA.

How serious are unequal standard deviations? ANOVA is not too sensitive to violations of the assumption, especially when all samples have the same or similar sizes and no sample is very small. When designing a study, try to take samples of the same size from all the groups you want to compare. The sample standard deviations estimate the population standard deviations, so check before doing ANOVA that the sample standard deviations are similar to each other. We expect some variation among them due to chance. Here is a rule of thumb that is safe in almost all situations.

CHECKING STANDARD DEVIATIONS IN ANOVA

The results of the ANOVA *F* test are approximately correct when the largest sample standard deviation is no more than twice as large as the smallest sample standard deviation.

EXAMPLE 9.5

In the gas mileage study, the sample standard deviations for compact, midsize, and large cars are

$$s_1 = 2.814 \qquad s_2 = 2.311 \qquad s_3 = 1.660$$

These standard deviations easily satisfy our rule of thumb. The three sample sizes are 25, 19, and 13, and we left out the two exotic cars that appeared as outliers in Figure 9.1. We can safely use ANOVA to compare the mean gas mileage for the three sizes of cars.

The report from which Table 9.1 was taken also contained data on 33 subcompact cars. The standard deviation of their city gas mileage is $s_4 = 5.888$ miles per gallon. This is more than twice the smallest standard deviation:

$$\frac{\text{largest } s}{\text{smallest } s} = \frac{5.888}{1.660} = 3.55$$

It would *not* be safe to use ANOVA to compare the mean gas mileage of all four car sizes.

A large standard deviation is often due to skewness in the distribution. Here is a stemplot of the subcompact-car gas mileages:

```
1 | 778888899
2 | 002233334
2 | 556666678899
3 | 3
3 | 6
4 | 4
```

The distribution is skewed to the right. Three models get 33, 36, and 44 miles per gallon in the city. The mean is $\bar{x} = 24.1$ for all 33 small cars, and $\bar{x} = 22.8$ when we omit these three. We are wise not to apply ANOVA to such data. ◀

EXAMPLE 9.6

To detect the presence of harmful insects in farm fields, put up boards covered with a sticky material and examine the insects trapped on the boards. Which colors attract insects best? Experimenters placed six boards of each of four colors at random locations in a field of oats and measured the number of cereal leaf beetles trapped.[4]

Board color	Insects trapped					
Blue	16	11	20	21	14	7
Green	37	32	20	29	37	32
White	21	12	14	17	13	20
Yellow	45	59	48	46	38	47

We would like to use ANOVA to compare the mean numbers of beetles that would be trapped by all boards of each color. Because the samples are small, we plot the data in side-by-side stemplots in Figure 9.7. Computer output for descriptive statistics and ANOVA appears in Figure 9.8. It appears that yellow boards attract by far the most insects ($\bar{x}_4 = 47.167$), with green next ($\bar{x}_2 = 31.167$) and blue and white far behind.

Blue	Green	White	Yellow
0 | 7	0 |	0 |	0 |
1 | 146	1 |	1 | 2347	1 |
2 | 01	2 | 09	2 | 01	2 |
3 |	3 | 2277	3 |	3 | 8
4 |	4 |	4 |	4 | 5678
5 |	5 |	5 |	5 | 9

FIGURE 9.7 Side-by-side stemplots comparing the counts of insects attracted by six boards for each of four board colors, for Example 9.6.

Check that we can safely use ANOVA to test equality of the four means. The largest of the four sample standard deviations is 6.795 and the smallest is 3.764. The ratio

$$\frac{\text{largest } s}{\text{smallest } s} = \frac{6.795}{3.764} = 1.8$$

is less than 2, so these data satisfy our rule of thumb for safe use of ANOVA. The shapes of the four distributions are irregular, as we expect with only 6 observations in each group, but there are no outliers. The ANOVA results will be approximately correct.

There are $I = 4$ groups and $N = 24$ observations overall, so the degrees of freedom for F are

$$\text{numerator: } I - 1 = 4 - 1 = 3$$
$$\text{denominator: } N - I = 24 - 4 = 20$$

This agrees with the computer results. The F statistic is $F = 42.84$, a very large F with P-value $P < .001$. Despite the small samples, the experiment gives very strong evidence of differences among the colors. Yellow boards appear best at attracting leaf beetles.   ◄

EXERCISES

9.7   Do the sample standard deviations for the fish weights in Table 9.2 allow use of ANOVA to compare the mean weights? (Use the computer output in Figure 9.3.)

```
            N     MEAN   MEDIAN   STDEV      MIN      MAX      Q1       Q3
blue        6    14.83   15.00    5.34      7.00    21.00    10.00    20.25
green       6    31.17   32.00    6.31     20.00    37.00    26.75    37.00
white       6    16.17   15.50    3.76     12.00    21.00    12.75    20.25
yellow      6    47.17   46.50    6.79     38.00    59.00    43.25    50.75


ANALYSIS OF VARIANCE
SOURCE      DF         SS          MS          F          P
FACTOR       3      4134.0      1378.0      42.84      0.000
ERROR       20       643.3        32.2
TOTAL       23      4777.3
                                        INDIVIDUAL 95 PCT CI'S FOR MEAN
                                        BASED ON POOLED STDEV
  LEVEL      N        MEAN       STDEV    --+---------+---------+---------+-----
blue         6      14.833      5.345     (---*---)
green        6      31.167      6.306                   (---*---)
white        6      16.167      3.764     (---*---)
yellow       6      47.167      6.795                               (---*---)

                                         --+---------+---------+---------+-----
POOLED STDEV =       5.672                12        24        36        48
```

FIGURE 9.8   Minitab output for comparing the four board colors in Example 9.6.

9.8   Do the standard deviations for the corn yields in Exercise 9.3 allow use of ANOVA to compare the mean yields? (Use the computer output in Figure 9.4.)

9.9   Married men tend to earn more than single men. An investigation of the relationship between marital status and income collected data on all 8235 men employed as managers or professionals by a large manufacturing firm in 1976. Suppose (this is risky) we regard these men as a random sample from the population of all men employed in managerial or professional positions in large companies. Here are descriptive statistics for the salaries of these men.[5]

	Single	Married	Divorced	Widowed
n_i	337	7,730	126	42
$\bar{x}_i$	\$21,384	\$26,873	\$25,594	\$26,936
s_i	\$5,731	\$7,159	\$6,347	\$8,119

(a) Briefly describe the relationship between marital status and salary.

(b) Do the sample standard deviations allow use of the ANOVA F test? (The distributions are skewed to the right. We expect right skewness in income distributions. The investigators actually applied ANOVA to the logarithms of the salaries, which are more symmetric.)

(c) What are the degrees of freedom of the ANOVA F test?

(d) The F test is a formality for these data, because we are sure that the P-value will be very small. Why are we sure?

(e) Single men earn less on the average than men who are or have been married. Do the highly significant differences in mean salary show that getting married raises men's mean income? Explain your answer.

9.10 What factors influence the success of students who plan to study computer science (CS)? Look at all 256 students who entered a major university planning a CS major in a specific year. We are willing to regard these students as a random sample of the students the university CS program will attract in subsequent years. After three semesters of study, some of these students were CS majors, some were majors in another field of science or engineering, and some had left science and engineering or left the university. The table below gives the sample means and standard deviations and the ANOVA F statistics for three variables that describe the students' high school performance. These are three separate ANOVA F tests.[6]

The first variable is a student's rank in the high school class, given as a percentile (so rank 50 is the middle of the class and rank 100 is the top). The next variable is the number of semester courses in mathematics the student took in high school. The third variable is the student's average grade in high school mathematics. The mean and standard deviation appear in a form common in published reports, with the standard deviation in parentheses following the mean.

Group	n	Mean (standard deviation)		
		High school class rank	Semesters of HS math	Average grade in HS math
CS majors	103	88.0 (10.5)	8.74 (1.28)	3.61 (.46)
Sci./Eng. majors	31	89.2 (10.8)	8.65 (1.31)	3.62 (.40)
Other	122	85.8 (10.8)	8.25 (1.17)	3.35 (.55)
F statistic		1.95	4.56	9.38

(a) What null and alternative hypotheses does F test for rank in the high school class? Express the hypotheses both in symbols and in words. The hypotheses are similar for the other two variables.

(b) What are the degrees of freedom for each F?

(c) Check that the standard deviations allow use of all three F tests. The shapes of the distributions also allow use of F. How significant is F for each of these variables?

(d) Write a brief summary of the differences among the three groups of students, taking into account both the significance of the F tests and the values of the means.

SOME DETAILS OF ANOVA*

Now we will give the actual recipe for the ANOVA F statistic. We have SRSs from each of I populations. Subscripts from 1 to I tell us which sample a statistic refers to:

Population	Sample size	Sample mean	Sample std. dev.
1	n_1	$\overline{x}_1$	s_1
2	n_2	$\overline{x}_2$	s_2
$\vdots$	$\vdots$	$\vdots$	$\vdots$
I	n_I	$\overline{x}_I$	s_I

You can find the F statistic from just the sample sizes n_i, the sample means $\overline{x}_i$, and the sample standard deviations s_i. You don't need to go back to the individual observations.

The ANOVA F statistic has the form

$$F = \frac{\text{variation among the sample means}}{\text{variation among individuals}}$$

*This more advanced section is optional if you are using computer software to find the F statistic.

The measures of variation in the numerator and denominator of F are called
mean squares *mean squares*. A mean square is a more general form of a sample variance.
An ordinary sample variance s^2 is an average (or mean) of the squared
deviations of observations from their mean, so it qualifies as a "mean
square."

The numerator of F is a mean square that measures variation among
the I sample means $\bar{x}_1, \bar{x}_2, \ldots, \bar{x}_I$. Call the overall mean response, the mean
of all N observations together, $\bar{x}$. You can find $\bar{x}$ from the I sample means
by

$$\bar{x} = \frac{n_1\bar{x}_1 + n_2\bar{x}_2 + \cdots + n_I\bar{x}_I}{N}$$

The sum of each mean multiplied by the number of observations it represents is the sum of all the individual observations. Dividing this sum by N,
the total number of observations, gives the overall mean $\bar{x}$. The numerator
mean square in F is an average of the I squared deviations of the means of
MSG the samples from $\bar{x}$. It is called the **mean square for groups**, abbreviated
as MSG.

$$\text{MSG} = \frac{n_1(\bar{x}_1 - \bar{x})^2 + n_2(\bar{x}_2 - \bar{x})^2 + \cdots + n_I(\bar{x}_I - \bar{x})^2}{I - 1}$$

Each squared deviation is weighted by n_i, the number of observations it
represents.

The mean square in the denominator of F measures variation among
individual observations in the same sample. For any one sample, the sample variance s_i^2 does this job. For all I samples together, we use an average
of the individual sample variances. It is again a weighted average in which
each s_i^2 is weighted by one fewer than the number of observations it represents, $n_i - 1$. Another way to put this is that each s_i^2 is weighted by its
degrees of freedom $n_i - 1$. The resulting mean square is called the **mean
MSE square for error**, MSE.

$$\text{MSE} = \frac{(n_1 - 1)s_1^2 + (n_2 - 1)s_2^2 + \cdots + (n_I - 1)s_I^2}{N - I}$$

Here is a summary of the ANOVA test.

THE ANOVA F TEST

Draw an independent SRS from each of I populations. The ith population has the $N(\mu_i, \sigma)$ distribution, where σ is the common standard deviation in all the populations. The ith sample has size n_i, sample mean $\bar{x}_i$, and sample standard deviation s_i.

The **ANOVA F statistic** tests the null hypothesis that all I populations have the same mean:

$$H_0 : \mu_1 = \mu_2 = \cdots = \mu_I$$

$$H_a : \text{not all of the } \mu_i \text{ are equal}$$

The statistic is

$$F = \frac{\text{MSG}}{\text{MSE}}$$

The **mean squares** that make up F are

$$\text{MSG} = \frac{n_1(\bar{x}_1 - \bar{x})^2 + n_2(\bar{x}_2 - \bar{x})^2 + \cdots + n_I(\bar{x}_I - \bar{x})^2}{I - 1}$$

and

$$\text{MSE} = \frac{(n_1 - 1)s_1^2 + (n_2 - 1)s_2^2 + \cdots + (n_I - 1)s_I^2}{N - I}$$

When H_0 is true, F has the **F distribution** with $I - 1$ and $N - I$ degrees of freedom.

The denominators in the recipes for MSG and MSE are the two degrees of freedom $I - 1$ and $N - I$ of the F test. The numerators are called *sums of squares*, from their algebraic form. It is usual to present the results *ANOVA table* of ANOVA in an **ANOVA table** like that in the Minitab output. The table has columns for degrees of freedom (DF), sums of squares (SS), and mean squares (MS). Check that each MS entry in Figure 9.8, for example, is the sum of squares SS divided by the degrees of freedom DF in the same row. The F statistic in the "F" column is MSG/MSE. The rows are labeled by sources of variation. In this output, variation among groups is labeled "FACTOR." Other statistical software calls this line "Treatments" or "Groups." Variation among observations in the same group is called "ERROR" by most software. This doesn't mean a mistake has been made. It's a traditional term for chance variation.

Because MSE is an average of the individual sample variances, it is also called the *pooled sample variance*, written as s_p^2. When all I populations have the same population variance σ^2 (ANOVA assumes that they *pooled standard* do), s_p^2 estimates the common variance σ^2. The square root of MSE is the *deviation* **pooled standard deviation** s_p. It estimates the common standard deviation σ of observations in each group. Minitab, like most ANOVA programs, gives the value of s_p as well as MSE. It is the "POOLED STDEV" value in Figure 9.8.

The pooled standard deviation s_p is a better estimator of the common σ than any individual sample standard deviation s_i because it combines (pools) the information in all I samples. We can get a confidence interval for any of the means μ_i from the usual form

$$\text{estimate} \pm t^* \text{SE}_{\text{estimate}}$$

using s_p to estimate σ. The confidence interval for μ_i is

$$\bar{x}_i \pm t^* \frac{s_p}{\sqrt{n_i}}$$

Use the critical value t^* from the t distribution with $N - I$ degrees of freedom, because s_p has $N - I$ degrees of freedom. These are the confidence intervals that appear in the Minitab ANOVA output.

EXAMPLE 9.7

We can do the ANOVA test comparing board colors in Example 9.6 using only the sample sizes, sample means, and sample standard deviations. Minitab gives these in Figure 9.8, but it is not hard to find them with a calculator.

The overall mean of the 24 counts is

$$\bar{x} = \frac{n_1\bar{x}_1 + n_2\bar{x}_2 + \cdots + n_I\bar{x}_I}{N}$$

$$= \frac{(6)(14.833) + (6)(31.167) + (6)(16.167) + (6)(47.167)}{24}$$

$$= \frac{656}{24} = 27.333$$

The mean square for groups is

$$MSG = \frac{n_1(\bar{x}_1 - \bar{x})^2 + n_2(\bar{x}_2 - \bar{x})^2 + \cdots + n_I(\bar{x}_I - \bar{x})^2}{I - 1}$$

$$= \frac{1}{4 - 1}[(6)(14.833 - 27.333)^2 + (6)(31.167 - 27.333)^2$$

$$+ (6)(16.167 - 27.333)^2 + (6)(47.167 - 27.333)^2]$$

$$= \frac{4134.100}{3} = 1378.033$$

The mean square for error is

$$MSE = \frac{(n_1 - 1)s_1^2 + (n_2 - 1)s_2^2 + \cdots + (n_I - 1)s_I^2}{N - I}$$

$$= \frac{(5)(5.345^2) + (5)(6.306^2) + (5)(3.764^2) + (5)(6.795^2)}{24 - 4}$$

$$= \frac{643.372}{20} = 32.169$$

Finally, the ANOVA test statistic is

$$F = \frac{\text{MSG}}{\text{MSE}} = \frac{1378.033}{32.169} = 42.84$$

Our work agrees with the computer output in Figure 9.8. We don't recommend doing these calculations, because tedium and roundoff errors cause frequent mistakes.

The pooled estimate of the standard deviation σ in any group is

$$s_p = \sqrt{\text{MSE}} = \sqrt{32.169} = 5.672$$

A 95% confidence interval for the mean count of insects trapped by yellow boards, using s_p and 20 degrees of freedom, is

$$\bar{x}_4 \pm t^* \frac{s_p}{\sqrt{n_4}} = 47.167 \pm 2.086 \frac{5.672}{\sqrt{6}}$$

$$= 47.167 \pm 4.830$$

$$= 42.34 \text{ to } 52.00$$

This confidence interval appears in the graph in the Minitab ANOVA output in Figure 9.8. ◀

EXERCISES

9.11 Return to the study of fish weights in Table 9.2 and Figure 9.3.

(a) Starting from the sample standard deviations in Figure 9.3, calculate MSE and the pooled standard deviation s_p. Use the computer output to check your work.

(b) Give a 95% confidence interval for the mean weight of perch, using the pooled standard deviation s_p. A graph of this interval appears in the computer output.

9.12 Continue the ANOVA calculations for the fish data. Starting from the sample means in the computer output in Figure 9.3, find the overall mean weight $\bar{x}$. Then find MSG. Finally, combine MSG with MSE from the previous exercise to obtain F.

9.13   Return to the data in Exercise 9.3 on corn yields for different planting rates. Starting from the sample means and standard deviations for the five groups (Figure 9.4), calculate MSE, the overall mean yield $\bar{x}$, and MSG. Use the computer output in Figure 9.4 to check your work.

9.14   Give a 90% confidence interval for the mean yield of corn planted at 20,000 plants per acre under the growing conditions of the experiment described in Exercise 9.3. Use the pooled standard deviation s_p to estimate σ in the standard error.

SUMMARY

One-way analysis of variance (ANOVA) compares the means of several populations. The **ANOVA F test** tests the overall H_0 that all the populations have the same mean. If the F test shows significant differences, examine the data to see where the differences lie and whether they are large enough to be important.

ANOVA assumes that we have an **independent SRS** from each population; that each population has a **normal distribution**; and that all populations have the **same standard deviation**.

In practice, ANOVA is relatively **robust** when the populations are nonnormal, especially when the samples are large. Before doing the F test, check the observations in each sample for outliers or strong skewness. Also verify that the largest sample standard deviation is no more than twice as large as the smallest standard deviation.

When the null hypothesis is true, the **ANOVA F statistic** for comparing I means from a total of N observations in all samples combined has the **F distribution** with $I - 1$ and $N - I$ degrees of freedom.

ANOVA calculations are reported in an **ANOVA table** that gives sums of squares, mean squares, and degrees of freedom for variation among groups and for variation within groups. In practice, we use software to do the calculations.

CHAPTER REVIEW

Advanced statistical inference often concerns relationships among several parameters. This chapter introduces the ANOVA F test for one such relationship: equality of the means of any number of populations. The alternative to this hypothesis is "many-sided," because it allows any relationship other than "all equal." The ANOVA F test is an overall test that tells us whether the data give good reason to reject the hypothesis that all the population means are equal. You should always accompany the test by data analysis to see what kind of inequality is present. Plotting the data in all groups side-by-side is particularly helpful.

ANOVA requires that the data satisfy conditions similar to those for t procedures. The most important assumption concerns the design of the data production. Inference is most trustworthy when we have random samples or responses to the treatments in a randomized comparative experiment. ANOVA assumes that the distribution of responses in each population has a normal distribution. Fortunately, the ANOVA F test shares the robustness of the t procedures, especially when the sample sizes are not very small. Do beware of outliers, which can greatly influence the mean responses and the F statistic.

ANOVA also requires a new assumption: the populations must all have the same standard deviation. Fortunately, ANOVA is not highly sensitive to unequal standard deviations, especially when the samples are similar in size. As a rule of thumb, you can use ANOVA when the largest sample standard deviation is no more than twice as large as the smallest.

After studying this chapter, you should be able to do the following.

A. RECOGNITION

1. Recognize when testing the equality of several means is helpful in understanding data.

2. Recognize that the statistical significance of differences among sample means depends on the sizes of the samples and on how much variation there is within the samples.

3. Recognize when you can safely use ANOVA to compare means. Check the data production, the presence of outliers, and the sample standard deviations for the groups you want to compare.

B. INTERPRETING ANOVA

1. Explain what null hypothesis F tests in a specific setting.

2. Locate the F statistic and its P-value on the output of a computer analysis of variance program.

3. Find the degrees of freedom for the F statistic from the number and sizes of the samples. Use Table D of the F distributions to approximate the P-value when software does not give it.

4. If the test is significant, use graphs and descriptive statistics to see what differences among the means are most important.

CHAPTER 9 REVIEW EXERCISES

9.15   In each of the following situations, we want to compare the mean response in several populations. For each setting, identify the populations and the response variable. Then give I, the n_i, and N. Finally, give the degrees of freedom of the ANOVA F test.

(a) A study of the effects of smoking classifies subjects as nonsmokers, moderate smokers, or heavy smokers. The investigators interview a sample of 200 people in each group. Among the questions is "How many hours do you sleep on a typical night?"

(b) The strength of concrete depends on the mixture of sand, gravel, and cement used to prepare it. A study compares five different mixtures. Workers prepare six batches of each mixture and measure the strength of the concrete made from each batch.

(c) Which of four methods of teaching American Sign Language is most effective? Assign 10 of the 42 students in a class at random to each of three methods. Teach the remaining 12 students by the fourth method. Record the students' scores on a standard test of sign language after a semester's study.

9.16   How do nematodes (microscopic worms) affect plant growth? A botanist prepares 16 identical planting pots and then introduces different numbers of nematodes into the pots. He transplants a tomato seedling into each plot. Here are data on the increase in height of the seedlings (in centimeters) 16 days after planting.[7]

Nematodes	Seedling growth			
0	10.8	9.1	13.5	9.2
1,000	11.1	11.1	8.2	11.3
5,000	5.4	4.6	7.4	5.0
10,000	5.8	5.3	3.2	7.5

(a) Make a table of means and standard deviations for the four treatments. Make side-by-side stemplots to compare the treatments. What do the data appear to show about the effect of nematodes on growth?

(b) State H_0 and H_a for the ANOVA test for these data, and explain in words what ANOVA tests in this setting.

(c) Use computer software to carry out the ANOVA. Report your overall conclusions about the effect of nematodes on plant growth.

9.17   Table 9.1 presents both the city and highway gas mileage for a number of cars of three sizes. We analyzed the city gas mileage data in Examples 9.1 and 9.2. There are significant differences ($P < .001$) among the mean miles per gallon for the three sizes, and the most important difference is that compact cars have better mileage than either of the other types. Now use computer software to analyze the highway gas mileage data and report your conclusions in detail. (Be sure to include graphs of the data and descriptive statistics as well as an ANOVA.)

9.18   **(Optional)** We have two methods to compare the means of two groups: the two-sample t test of Section 6.2 and the ANOVA F test with $I = 2$. We prefer the t test because it allows one-sided alternatives and does not assume that both populations have the same standard deviation. Let us apply both tests to the same data.

There are two types of life insurance companies. "Stock" companies have shareholders, and "mutual" companies are owned by their policy-holders. Take an SRS of each type of company from those listed in a directory of the industry. Then ask the annual cost per $1000 of insurance for a $50,000 policy insuring the life of a 35-year-old man who does not smoke. Here are the data summaries.[8]

	Stock companies	Mutual companies
n_i	13	17
$\bar{x}_i$	$2.31	$2.37
s_i	$0.38	$0.58

(a) Calculate the two-sample t statistic for testing $H_0: \mu_1 = \mu_2$ against the two-sided alternative. Use the conservative method to find the P-value.

(b) Calculate MSG, MSE, and the ANOVA F statistic for the same hypotheses. What is the P-value of F?

(c) How close are the two P-values? (The square root of the F statistic is a t statistic with $N - I = n_1 + n_2 - 2$ degrees of freedom. This is the "pooled two-sample t" mentioned on page 455. So F for $I = 2$ is exactly equivalent to a t statistic, but it is a slightly different t from the one we use.)

9.19   **(Optional)** Carry out the ANOVA calculations (MSG, MSE, and F) required for part (c) of Exercise 9.16. Find the degrees of freedom for F and report its P-value as closely as Table D allows.

NOTES AND DATA SOURCES

1. The data in Table 9.1 are from the U.S. Department of Energy's *1994 Gas Mileage Guide,* October 1993. The table gives data for the basic engine/transmission combination for each model. Models that are essentially identical (such as the Ford Taurus and Mercury Sable) appear only once.

2. The data in Table 9.2 are part of a larger data set in the *Journal of Statistics Education* archive, accessible via Internet. The original source is Pekka Brofeldt, "Bidrag till kaennedom on fiskbestondet i vaara sjoear. Laengelmaevesi," in T. H. Jaervi, *Finlands Fiskeriet,* Band 4, *Meddelanden utgivna av fiskerifoereningen i Finland,* Helsinki, 1917. The data were contributed to the archive (with information in English) by Juha Puranen of the University of Helsinki.

3. The data are from W. L. Colville and D. P. McGill, "Effect of rate and method of planting on several plant characters and yield of irrigated corn," *Agronomy Journal,* 54 (1962), pp. 235–238.

4. Modified from M. C. Wilson and R. E. Shade, "Relative attractiveness of various luminescent colors to the cereal leaf beetle and the meadow spittlebug," *Journal of Economic Entomology,* 60 (1967), pp. 578–580.

5. Sanders Korenman and David Neumark, "Does marriage really make men more productive?" *Journal of Human Resources,* 26 (1991), pp. 282–307.

6. Patricia F. Campbell and George P. McCabe, "Predicting the success of freshmen in a computer science major," *Communications of the ACM,* 27 (1984), pp. 1108–1113.

7. Data provided by Matthew Moore.

8. Mark Kroll, Peter Wright, and Pochera Theerathorn, "Whose interests do hired managers pursue? An examination of select mutual and stock life insurers," *Journal of Business Research,* 26 (1993), pp. 133–148.

SIR FRANCIS GALTON

The least-squares method will happily fit a straight line to any two-variable data. It is an old method, going back to the French mathematician Legendre in about 1805. Legendre invented least squares for use on data from astronomy and surveying. It was Sir Francis Galton (1822–1911), however, who turned "regression" into a general method for understanding relationships. He even invented the word.

Galton was one of the last gentleman scientists, an upper-class Englishman who studied medicine at Cambridge and explored Africa before turning to the study of heredity. He was well connected here also: Charles Darwin, who published *The Origin of Species* in 1859, was his cousin.

Galton was full of ideas but was no mathematician. He didn't even use least squares, preferring to avoid unpleasant computations. But Galton's ideas led eventually to the machinery for inference about regression that we will meet in this chapter. He asked: If people's heights are distributed normally in every generation, and height is inherited, what is the relationship between generations? He discovered a straight-line relationship between the heights of parent and child and found that tall parents tended to have children who were taller than average but less tall than their parents. He called this "regression toward mediocrity." Galton went further: he described inheritance by a straight-line relationship with responses y that have a normal distribution about the line for every fixed input x. This is the model for regression we use in this chapter.

CHAPTER **10**

Inference for Regression

INTRODUCTION

When a scatterplot shows a linear relationship between a quantitative explanatory variable x and a quantitative response variable y, we can use the least-squares line fitted to the data to predict y for a given value of x. Now we want to do tests and confidence intervals in this setting.

EXAMPLE 10.1

The Sanchez household is about to install solar panels to reduce the cost of heating their house. In order to know how much the solar panels help, they record their consumption of natural gas before the panels are installed. Gas consumption is higher in cold weather, so the relationship between outside temperature and gas consumption is important.

Table 10.1 gives data for 16 months.[1] The response variable y is the average amount of natural gas consumed each day during the month, in hundreds of cubic feet. The explanatory variable x is the average number of heating degree-days each day during the month. (One heating degree-day is accumulated for each degree a day's average temperature falls below 65° F. An average temperature of 20° F, for example, corresponds to 45 degree-days.) ◄

We met the Sanchez's natural gas consumption data in Chapter 2. Before attempting inference, examine the data as follows.

TABLE 10.1 Average degree-days and natural gas consumption for the Sanchez household

Month	Degree-days	Gas (100 cu ft)	Month	Degree-days	Gas (100 cu ft)
Nov.	24	6.3	July	0	1.2
Dec.	51	10.9	Aug.	1	1.2
Jan.	43	8.9	Sept.	6	2.1
Feb.	33	7.5	Oct.	12	3.1
Mar.	26	5.3	Nov.	30	6.4
Apr.	13	4.0	Dec.	32	7.2
May	4	1.7	Jan.	52	11.0
June	0	1.2	Feb.	30	6.9

scatterplot 1. Make a *scatterplot.* Plot the explanatory variable x horizontally and the response variable y vertically. Fitting a line only makes sense if the overall pattern of the scatterplot is roughly linear (straight-line).

least-squares line 2. Use a calculator or computer to fit the *least-squares regression line* to the data. This line lies as close as possible to the points (in the sense of least squares) in the vertical (y) direction. We use it to predict y for a given x.

outliers 3. Look for *outliers* and *influential observations.* Outliers are points that lie far from the overall linear pattern. Influential obser-

influential observations vations are points that move the fitted line, usually points that are far out in the x direction and isolated from other points. Inference is not safe if there are influential points, because the results will depend strongly on these few points.

correlation 4. Use a calculator or computer to calculate the *correlation* r and then its square r^2. The squared correlation r^2 describes how well the regression line fits the data. It is the proportion of the observed variation in y that is accounted for by the straight-line relationship of y with x.

EXAMPLE 10.2

Figure 10.1 is a scatterplot of the Sanchez gas consumption data. Degree-days, the explanatory variable, appears on the horizontal scale. There is a strong linear relationship.

Enter the data into a calculator or computer software. The least-squares regression line is

$$\hat{y} = a + bx$$
$$= 1.0892 + .1890x$$

We use the notation $\hat{y}$ to remind ourselves that this line gives *predictions* of y. The predictions usually won't agree exactly with the observed values of y. Drawing the least-squares line on the scatterplot helps us see that there are no strong outliers or influential observations.

The calculator or computer also tells us the correlation is $r = .9953$. So $r^2 = .9906$. The scatterplot shows that almost all the variation in gas consumption is explained by degree-days. In fact, $r^2 = .99$ says that 99% is explained. Prediction of gas consumption from degree-days should be quite accurate. ◄

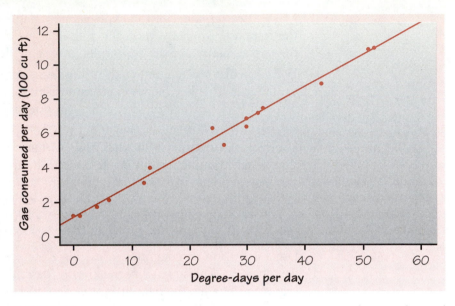

FIGURE 10.1   Scatterplot of natural gas consumption against heating degree-days for the Sanchez household, for Example 10.2. The line is the least-squares regression line for predicting gas consumption from degree-days.

The regression model

The slope *b* and intercept *a* of the least-squares line are *statistics*. They are computed from the sample data and would no doubt be different if we monitored the Sanchez house for another 16 months. To do formal inference, we need to know what unknown *parameters a* and *b* estimate. The

model   parameters appear in a description, often called a **model**, of the process that produces our data. Here are the assumptions that describe the regression model.

ASSUMPTIONS FOR REGRESSION INFERENCE

We have *n* observations on an explanatory variable *x* and a response variable *y*. Our goal is to study or predict the behavior of *y* for given values of *x*.

- For any fixed value of *x*, the response *y* varies according to a normal distribution. Repeated responses *y* are independent of each other.

(continued on next page)

(continued from previous page)

- The mean response μ_y has a straight-line relationship with x:

$$\mu_y = \alpha + \beta x$$

The slope β and intercept α are unknown parameters.
- The standard deviation of y (call it σ) is the same for all values of x. The value of σ is unknown.

true regression line

The heart of this model is that there is an "on the average" straight-line relationship between y and x. The **true regression line** $\mu_y = \alpha + \beta x$ says that the *mean* response μ_y moves along a straight line as the explanatory variable x changes. We can't observe the true regression line. The values of y that we do observe vary about their means according to a normal distribution. If we hold x fixed and take many observations on y, the normal pattern will eventually appear in a stemplot or histogram. In practice, we observe y for many different values of x, and so we see an overall linear pattern with points scattered about it. The standard deviation σ determines whether the points fall close to the true regression line (small σ) or are widely scattered (large σ).

Figure 10.2 shows the regression model in picture form. The line in the figure is the true regression line. The mean of the response y moves along this line as the explanatory variable x takes different values. The normal

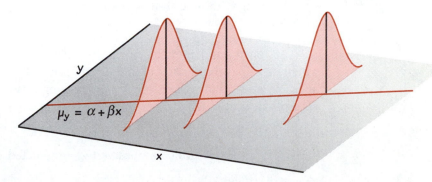

FIGURE 10.2 The regression model. The line is the true regression line, which shows how the mean response μ_y changes as the explanatory variable x changes. For any fixed value of x, the observed response y varies according to a normal distribution having mean μ_y.

curves show how y will vary when x is held fixed at different values. All of the curves have the same σ, so the variability of y is the same for all values of x. You should check the assumptions for inference when you do inference about regression. We will see later how to do that.

INFERENCE ABOUT THE MODEL

The first step in inference is to estimate the unknown parameters α, β, and σ. When the regression model describes our data and we calculate the least-squares line $\hat{y} = a + bx$:

- The slope b of the least-squares line is an unbiased estimator of the true slope β.
- The intercept a of the least-squares line is an unbiased estimator of the true intercept α.

EXAMPLE 10.3

The Sanchez household gas consumption data in Figure 10.1 fit the regression model of scatter about an invisible true regression line quite well. The least-squares line is $\hat{y} = 1.0892 + .1890x$. The slope is particularly important. *A slope is a rate of change.* The true slope β says how much the average gas consumption changes when the number of degree-days x increases by 1. Because $b = .1890$ estimates the unknown β, we estimate that on the average the Sanchez household will use 0.1890 hundred cubic feet more gas per day for each degree the outside temperature drops. ◄

The remaining parameter of the model is the standard deviation σ, which describes the variability of the response y about the true regression line. The least-squares line estimates the true regression line. So the ***residuals*** estimate how much y varies about the true line. Recall that the residuals are the vertical deviations of the data points from the least-squares line:

residuals

$$\text{residual} = \text{observed } y - \text{predicted } y$$
$$= y - \hat{y}$$

There are n residuals, one for each data point. Because σ is the standard deviation of responses about the true regression line, we estimate it by a sample standard deviation of the residuals. We call this sample standard deviation a *standard error* to emphasize that it is estimated from data. The residuals from a least-squares line always have mean zero. That simplifies their standard error.

STANDARD ERROR ABOUT THE LEAST-SQUARES LINE

The **standard error about the line** is

$$s = \sqrt{\frac{1}{n-2} \sum \text{residual}^2}$$

$$= \sqrt{\frac{1}{n-2} \sum (y - \hat{y})^2}$$

Use s to estimate the unknown σ in the regression model.

Because we use the standard error about the line so often in regression inference, we just call it s. Notice that s^2 is an average of the squared deviations of the data points from the line, so it qualifies as a variance. We average the squared deviations by dividing by $n - 2$, the number of data points less 2. It turns out that if we know $n - 2$ of the n residuals, the other *degrees of* two are determined. So $n - 2$ is the ***degrees of freedom*** of s. We first met *freedom* the idea of degrees of freedom in the case of the ordinary sample standard deviation of n observations, which has $n - 1$ degrees of freedom. Now we observe two variables rather than one, and the proper degrees of freedom is $n - 2$ rather than $n - 1$.

Calculating s is unpleasant. You must find the predicted response for each x in your data set, then the residuals, and then s. In practice you will use software that does this arithmetic instantly. Nonetheless, here is an example to make certain you understand what s is.

EXAMPLE 10.4

The first observation in Table 10.1, for the month of November, has $x = 24$ degree-days per day and $y = 6.3$ hundred cubic feet of gas used per day. The predicted gas usage for $x = 24$ is

$$\hat{y} = 1.0892 + .1890x$$
$$= 1.0892 + (.1890)(24) = 5.6252$$

The residual for November's observation is

$$\text{residual} = y - \hat{y}$$
$$= 6.3 - 5.6252 = .6748$$

That is, the November data point on the scatterplot lies 0.6748 above the least-squares line.

Repeat this calculation 15 more times, once for each data point. The 16 residuals are

.6748	.1718	−.3162	.1738	−.7032	.4538	−.1452	.1108
.1108	−.0782	−.1232	−.2572	−.3592	.0628	.0828	.1408

Check the calculations by verifying that the sum of the residuals is zero. It is −0.0002, not quite zero because of roundoff error. Another reason to use software in regression is that roundoff errors in hand calculation can accumulate to make the results inaccurate.

The variance about the line is

$$s^2 = \frac{1}{n-2} \sum \text{residual}^2$$

$$= \frac{1}{16-2}(.6748^2 + .1718^2 + \cdots + .1408^2)$$

$$= \frac{1}{14}(1.6082) = .1149$$

Finally, the standard error about the line is

$$s = \sqrt{.1149} = .3390$$

We rounded each step to four decimal places. As a result, s is off by 0.0001. The correct value is $s = .3389$ to four places. If you do calculations by hand, carry at least one more decimal place than the accuracy you want in the final result.   ◀

We will study several kinds of inference in the regression setting. The standard error s about the line is the key measure of the variability of the responses in regression. It is part of the standard error of all the statistics we will use for inference.

EXERCISES

10.1   *Archaeopteryx* is an extinct beast having feathers like a bird but teeth and a long bony tail like a reptile. Here are the lengths in centimeters of the femur (a leg bone) and the humerus (a bone in the upper arm) for the five fossil specimens that preserve both bones.[2]

$$
\begin{array}{llllll}
\text{Femur} & 38 & 56 & 59 & 64 & 74 \\
\text{Humerus} & 41 & 63 & 70 & 72 & 84
\end{array}
$$

The strong linear relationship between the lengths of the two bones helped persuade scientists that all five specimens belong to the same species.

(a) Examine the data. Make a scatterplot with femur length as the explanatory variable. Use your calculator to obtain the correlation r and the equation of the least-squares regression line. Do you think that femur length will allow good prediction of humerus length?

(b) Explain in words what the slope β of the true regression line says about *Archaeopteryx*. What is the estimate of β from the data? What is your estimate of the intercept α of the true regression line?

(c) Calculate the residuals for the five data points. Check that their sum is 0 (up to roundoff error). Use the residuals to estimate the standard deviation σ in the regression model. You have now estimated all three parameters in the model.

10.2   Good runners take more steps per second as they speed up. Here are the average number of steps per second for a group of top female runners at different speeds. The speeds are in feet per second.[3]

$$
\begin{array}{llllllll}
\text{Speed (ft/s)} & 15.86 & 16.88 & 17.50 & 18.62 & 19.97 & 21.06 & 22.11 \\
\text{Steps per second} & 3.05 & 3.12 & 3.17 & 3.25 & 3.36 & 3.46 & 3.55
\end{array}
$$

(a) You want to predict steps per second from running speed. Make a scatterplot of the data with this goal in mind. Use your calculator to find the correlation r and the equation of the least-squares regression line. Describe the form and strength of the relationship. Do steps per second increase with running speed at a steady rate?

(b) Find the residuals for all 7 data points. Check that their sum is 0 (up to roundoff error).

(c) The model for regression inference has three parameters, which we call α, β, and σ. Estimate these parameters from the data.

Confidence intervals for the regression slope

The slope β of the true regression line is usually the most important parameter in practical regression problems. The slope is the rate of change of the mean response as the explanatory variable increases. We often want to estimate β. The slope b of the least-squares line is an unbiased estimator of β. A confidence interval is more useful because it shows how accurate the estimate b is likely to be. The confidence interval for β has the familiar form

$$\text{estimate} \pm t^*\text{SE}_{\text{estimate}}$$

Because b is our estimate, the confidence interval becomes

$$b \pm t^*\text{SE}_b$$

Here are the details.

CONFIDENCE INTERVAL FOR REGRESSION SLOPE

A level C confidence interval for the slope β of the true regression line is

$$b \pm t^*\text{SE}_b$$

In this recipe, the standard error of the least-squares slope b is

$$\text{SE}_b = \frac{s}{\sqrt{\sum(x - \overline{x})^2}}$$

and t^* is the upper $(1 - C)/2$ critical value from the t distribution with $n - 2$ degrees of freedom.

As advertised, the standard error of b is a multiple of s. Although we give the recipe for this standard error, you should rarely have to calculate it by hand. Regression software gives the standard error SE_b along with b itself.

EXAMPLE 10.5

Figure 10.3 shows the basic output for the gas consumption data from the regression command in the Minitab software package. Most statistical software provides similar output. (Minitab, like other software, produces more than this basic output. When you use software, just ignore the parts you don't need.)

The first line gives the equation of the least-squares regression line. The slope and intercept are rounded off there, so look in the "Coef" column of the table that follows for more accurate values. The intercept $a = 1.0892$ appears in the "Constant" row. The slope $b = .188999$ appears in the "ddays" row because we named the x variable "ddays" when we entered the data.

The next column of output, headed "Stdev," gives standard errors. In particular, $SE_b = .004934$. The standard error about the line, $s = .3389$, appears below the table.

There are 16 data points, so the degrees of freedom are $n - 2 = 14$. A 95% confidence interval for the true slope β uses the critical value $t^* = 2.145$ from the df $= 14$ row of Table C. The interval is

$$b \pm t^*SE_b = .188999 \pm (2.145)(.004934)$$
$$= .1890 \pm .0106$$
$$= .1784 \text{ to } .1996$$

We are 95% confident that each additional degree-day causes the Sanchez household to use between 0.1784 and 0.1996 additional hundred cubic feet of natural gas on the average. ◀

You can find a confidence interval for the intercept α of the true regression line in the same way, using a and SE_a from the "Constant" line of the printout. It is not common to want to estimate α.

```
The regression equation is
gas used = 1.09 + 0.189 ddays

Predictor        Coef        Stdev      t-ratio          P
Constant       1.0892       0.1389         7.84      0.000
ddays        0.188999     0.004934        38.31      0.000

s = 0.3389        R-sq = 99.1%       R-sq(adj) = 99.0%
```

FIGURE 10.3   Minitab regression output for the Sanchez household gas consumption data, for Example 10.5.

Testing the hypothesis of no linear relationship

We can also test hypotheses about the value of the slope β. The most common hypothesis is

$$H_0 : \beta = 0$$

A regression line with slope 0 is horizontal. That is, the mean of y does not change at all when x changes. So this H_0 says that there is *no true linear relationship* between x and y. Put another way, H_0 says that *straight-line dependence on x is of no value for predicting y*. Put yet another way, H_0 says that there is *no correlation* between x and y in the population from which we drew our data. You can use the test for zero slope to test the hypothesis of zero correlation between any two quantitative variables. That's a useful trick.

The test statistic is just the standardized version of the least-squares slope b. It is another t statistic. Here are the details.

SIGNIFICANCE TESTS FOR REGRESSION SLOPE

To test the hypothesis $H_0 : \beta = 0$, compute the t statistic

$$t = \frac{b}{SE_b}$$

In terms of a random variable T having the $t(n - 2)$ distribution, the P-value for a test of H_0 against

$H_a : \beta > 0$ is $P(T \geq t)$

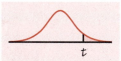

$H_a : \beta < 0$ is $P(T \leq t)$

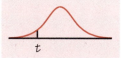

$H_a : \beta \neq 0$ is $2P(T \geq |t|)$

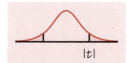

Regression output from statistical software usually gives t and its *two-sided* P-value. For a one-sided test, divide the P-value in the output by 2.

EXAMPLE 10.6

The computer output in Figure 10.3 gives $t = 38.31$ with two-sided P-value 0.000 for the hypothesis $H_0 : \beta = 0$ that there is no linear relationship between degree-days and natural gas used. (The P-value is rounded to three decimal places.) Because the scatterplot shows a very strong relationship, we are not surprised that the P-value is very small. ◀

EXAMPLE 10.7

How well do golfers' scores in the first round of a two-round tournament predict their scores in the second round? Here are data for the 12 members of a college's women's golf team in a recent tournament. (A golf score is the number of strokes required to complete the course, so low scores are better.)

						Golfer						
	1	2	3	4	5	6	7	8	9	10	11	12
Round 1	89	90	87	95	86	81	102	105	83	88	91	79
Round 2	94	85	89	89	81	76	107	89	87	91	88	80

The scatterplot in Figure 10.4 shows a clear linear relationship. There are two unusual points. Golfer 7 had poor scores on both rounds (102 and 107). This point is close to the linear pattern of the other points but may be influential. Golfer 8 is an outlier. She scored 105 on the first round, then improved to 89 on the second round. We will verify later (Example 10.10) that the regression assumptions are quite well satisfied except for these unusual observations.

Figure 10.5 gives part of the output from the regression command in the Data Desk statistical software package. You should be able to interpret this output. The least-squares line for predicting second-round score y from first-round score x is

$$\hat{y} = 26.332 + .688x$$

The t statistic for testing

$$H_0 : \beta = 0$$
$$H_a : \beta \neq 0$$

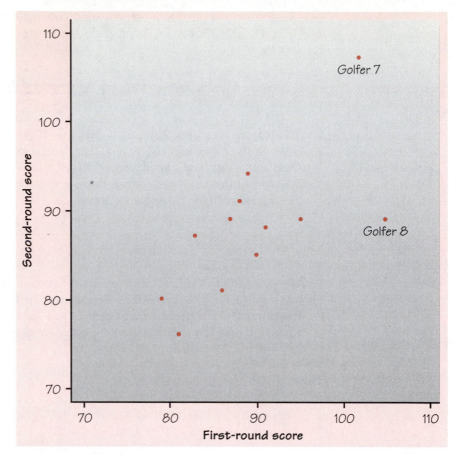

FIGURE 10.4   Scatterplot of the scores of 12 college golfers in two rounds of a tournament, for Example 10.7.

```
Dependent variable is:  round2

R squared = 47.2%
s =  5.974  with  12 - 2 = 10  degrees of freedom

Variable    Coefficient    s.e. of Coeff    t-ratio    prob
Constant    26.3320        20.69            1.27       0.2320
round1      0.687747       0.2300           2.99       0.0136
```

FIGURE 10.5   Data Desk output for the regression of second-round score on first-round score, for Example 10.7.

is $t = 2.99$ with two-sided P-value $P = .0136$. Check that t is the slope $b = .688$ divided by its standard error, $SE_b = .230$.

We think that there is a positive association between the same golfer's scores in two rounds of play. So we want to test the one-sided alternative,

$$H_0 : \beta = 0$$
$$H_a : \beta > 0$$

The value of t is unchanged. The one-sided P is half the two-sided $P : P = .0136/2 = .0068$. Alternatively, we can use Table C. There are $n - 2 = 10$ degrees of freedom. Compare the observed $t = 2.99$ with the entries in the df $= 10$ row of the table. The one-sided P-value lies between 0.005 and 0.01. This agrees with the computer's more exact value. There is (as the plot suggests) strong evidence that second-round score increases with first-round score in a straight-line fashion.   ◀

df = 10

p	.01	.005
t^*	2.764	3.169

EXERCISES

10.3 Figure 10.3 gives computer output for the regression of natural gas consumed by the Sanchez household on degree-days. Use the information in this output to give a 90% confidence interval for the slope β of the true regression line.

10.4 Example 10.7 finds strong evidence that golfers' scores in the second round of tournament play are linearly related to their scores in the first round. Use the computer output in Figure 10.5 to give a 95% confidence interval for the slope β of the true regression line. Then explain in plain language what β means in this setting.

10.5 Exercise 10.1 presents data on the lengths of two bones in five fossil specimens of the extinct beast *Archaeopteryx*. Here is part of the output from the S-PLUS statistical software when we regress the length y of the humerus on the length x of the femur.

	coef	std.err	t.stat	p.value
Intercept	-3.6596	4.4590	-0.8207	0.4719
X	1.1969	0.0751		

(a) What is the equation of the least-squares regression line?

(b) We left out the t statistic for testing $H_0 : \beta = 0$ and its P-value. Use the output to find t.

(c) How many degrees of freedom does t have? Use Table C to approximate the P-value of t against the one-sided alternative $H_a : \beta > 0$.

10.6   Exercise 10.2 presents data on the relationship between the speed of runners (x, in feet per second) and the number of steps y that they take in a second. Here is part of the Data Desk regression output for these data.

```
R squared =   99.8%
s =   0.0091    with 7 - 2 = 5 degrees of freedom

Variable  Coefficient  s.e. of Coeff t-ratio  prob
Constant   1.76608      0.0307        57.6    <0.0001
Speed      0.080284     0.0016        49.7    <0.0001
```

(a) How can you tell from this output, even without the scatterplot, that there is a very strong straight-line relationship between running speed and steps per second?

(b) What parameter in the regression model gives the rate at which steps per second increase as running speed increases? Give a 99% confidence interval for this rate.

10.7   The Leaning Tower of Pisa leans more as time passes. Here are measurements of the lean of the tower for the years 1975 to 1987.[4] The lean is the distance between where a point on the tower would be if the tower were straight and where it actually is. The distances are tenths of a millimeter in excess of 2.9 meters. For example, the 1975 lean, which was 2.9642 meters, appears in the table as 642. We use only the last two digits of the year as our time variable.

Year	75	76	77	78	79	80	81	82	83	84	85	86	87
Lean	642	644	656	667	673	688	696	698	713	717	725	742	757

Here is part of the output from the Data Desk regression procedure with year as the explanatory variable and lean as the response variable.

```
Variable  Coefficient  s.e. of Coeff t-ratio  prob
Constant  -61.1209      25.13        -2.43    0.0333
year        9.31868      0.3099      30.1    <0.0001
```

(a) Plot the data. Briefly describe the shape, strength, and direction of the relationship. The tower is tilting at a steady rate.

(b) The main purpose of the study is to estimate how fast the tower is tilting. What parameter in the regression model gives the rate at which the tilt is increasing, in tenths of a millimeter per year?

(c) We want a 95% confidence interval for this rate. How many degrees of freedom does t have? Find the critical value t^* and the confidence interval.

INFERENCE ABOUT PREDICTION

One of the most common reasons to fit a line to data is to predict the response to a particular value of the explanatory variable. The method is simple: just substitute the value of x into the equation of the line.

EXAMPLE 10.8

In Example 10.2 we found that the least-squares line for predicting gas consumption from degree-days for the Sanchez household is

$$\hat{y} = 1.0892 + .1890x$$

To predict gas consumption at 20 degree-days (that's average temperature 45° F), substitute $x = 20$:

$$\hat{y} = 1.0892 + (.1890)(20)$$
$$= 1.0892 + 3.78 = 4.869$$

We predict that the household will use 4.869 hundred cubic feet of gas per day.

◄

We would like to give a confidence interval that describes how accurate this prediction is. To do that, you must answer these questions: Do you want to predict the *mean* gas consumption for *all* months with 20 degree-days per day? Or do you want to predict the gas consumption in *one specific* month that has 20 degree-days per day? Both of these predictions may be interesting, but they are two different problems. The actual prediction is the same, $\hat{y} = 4.869$ hundred cubic feet. But the margin of error is different for the two kinds of prediction. Individual months with 20 degree-days per day don't all have the same gas consumption. So we need a larger margin of error to pin down one month's result than to estimate the mean response for all months that have 20 degree-days per day.

Write the given value of the explanatory variable x as x^*. In the example, $x^* = 20$. The distinction between predicting a single outcome and predicting the mean of all outcomes when $x = x^*$ determines what margin of error is correct. To emphasize the distinction, we use different terms for the two intervals.

- To estimate the *mean* response, we use a *confidence interval*. It is an ordinary confidence interval for the parameter

$$\mu_y = \alpha + \beta x^*$$

  The regression model says that μ_y is the mean of responses y when x has the value x^*. It is a fixed number whose value we don't know.

prediction interval
- To estimate an *individual* response y, we use a **prediction interval.** A prediction interval estimates a single random response y rather than a parameter like μ_y. The response y is not a fixed number. If we took more observations with $x = x^*$, we would get different responses.

Fortunately, the meaning of a prediction interval is very much like the meaning of a confidence interval. A 95% prediction interval, like a 95% confidence interval, is right 95% of the time in repeated use. "Repeated use" now means that we take an observation on y for each of the n values of x in the original data, and then take one more observation y at the point $x = x^*$. Form the prediction interval from the n observations, then see if it covers the one more y. It will in 95% of all repetitions.

The interpretation of prediction intervals is a minor point. The main point is that it is harder to predict one response than to predict a mean response. Both intervals have the usual form

$$\hat{y} \pm t^* \text{SE}$$

but the prediction interval is wider than the confidence interval. Here are the details.

CONFIDENCE AND PREDICTION INTERVALS FOR REGRESSION RESPONSE

A level C **confidence interval for the mean response** μ_y when x takes the value x^* is

$$\hat{y} \pm t^* \text{SE}_{\hat{\mu}}$$

(continued on next page)

(continued from previous page)

The standard error $SE_{\hat{\mu}}$ is

$$SE_{\hat{\mu}} = s\sqrt{\frac{1}{n} + \frac{(x^* - \bar{x})^2}{\sum(x - \bar{x})^2}}$$

The sum runs over all the observations on the explanatory variable x.

A level C **prediction interval for a single observation** on y when x takes the value x^* is

$$\hat{y} \pm t^* SE_{\hat{y}}$$

The standard error for prediction $SE_{\hat{y}}$ is[5]

$$SE_{\hat{y}} = s\sqrt{1 + \frac{1}{n} + \frac{(x^* - \bar{x})^2}{\sum(x - \bar{x})^2}}$$

In both recipes, t^* is the upper $(1 - C)/2$ critical value of the t distribution with $n - 2$ degrees of freedom.

There are two standard errors: $SE_{\hat{\mu}}$ for estimating the mean response μ_y and $SE_{\hat{y}}$ for predicting an individual response y. The only difference between the two standard errors is the extra 1 under the square root sign in the standard error for prediction. The extra 1 makes the prediction interval wider. Both standard errors are multiples of s. The degrees of freedom are again $n - 2$, the degrees of freedom of s. Calculating these standard errors by hand is a nuisance, which software spares us.

EXAMPLE 10.9

The Sanchez household installs solar panels. The following March, there are 20 degree-days per day. What would their gas usage have been without the solar panels? We want to predict gas consumption in a specific single month when x has the value $x^* = 20$.

Despite the importance of prediction, many statistical software systems neglect it and some omit it entirely. Minitab does prediction but offers only 95% intervals. Here is the output from the prediction option in the Minitab regression command for $x^* = 20$.

```
     Fit Stdev.Fit        95% C.I.            95% P.I.
   4.8692    0.0855   (4.6858, 5.0526)   (4.1193, 5.6191)
```

The "Fit" entry gives the prediction: 4.8692 hundred cubic feet. This agrees with our result in Example 10.8. Minitab gives both 95% intervals. You must choose which one you want. We are predicting a single response, so the prediction interval "95% P.I." is the right choice. We are 95% confident that gas usage without the solar panels for this month would fall between 4.119 and 5.619 hundred cubic feet.

The 95% confidence interval for the mean gas used in all months with 20 degree-days per day, given as "95% C.I.," is narrower.   ◀

By the way, the standard deviation Minitab gives as "Stdev.Fit" is $SE_{\hat{\mu}}$, used for confidence intervals for the mean response. You can use that in the recipe

$$\hat{y} \pm t^* SE_{\hat{\mu}}$$

to estimate the mean response with different confidence levels.

EXERCISES

10.8   Figure 10.3 (page 599) gives computer output for the regression of natural gas consumed by the Sanchez household on degree-days. Use this output to answer the following questions.

(a) In the month of January after solar panels were installed, there were 40 degree-days per day. How much gas do you predict the Sanchez household would have used per day without the solar panels? They actually used 7.5 hundred cubic feet per day. How much gas per day did the solar panels save?

(b) Here is the output from the prediction option in Minitab for 40 degree-days per day. Give a 95% interval for the amount of gas that would have been used this January without the solar panels.

```
Fit   Stdev.Fit        95% C.I.           95% P.I.
8.6492    0.1216     ( 8.3882, 8.9101)   ( 7.8767, 9.4217)
```

(c) Give a 95% interval for the mean gas consumption per day in months with 40 degree-days per day.

10.9   Manatees are large, gentle sea creatures that live along the Florida coast. Many manatees are killed or injured by powerboats. Here are data on powerboat registrations (in thousands) and the number of manatees killed by boats in Florida in the years 1977 to 1990.

Year	Powerboat registrations	Manatees killed	Year	Powerboat registrations	Manatees killed
1977	447	13	1984	559	34
1978	460	21	1985	585	33
1979	481	24	1986	614	33
1980	498	16	1987	645	39
1981	513	24	1988	675	43
1982	512	20	1989	711	50
1983	526	15	1990	719	47

(a) Make a scatterplot showing the relationship between powerboats registered and manatees killed. (Which is the explanatory variable?)

(b) Is the overall pattern roughly linear? Are there clear outliers or strongly influential data points?

(c) Here is part of the output from the Minitab regression command.

```
Predictor        Coef        Stdev      t-ratio        p
Constant       -41.430       7.412       -5.59    0.000
Boats          0.12486     0.01290        9.68    0.000

s = 4.276        R-sq = 88.6%
```

What does $r^2 = .886$ tell you about the relationship between boats and manatees killed?

(d) Explain what the slope β of the true regression line means in this setting. Then give a 90% confidence interval for β.

(e) If Florida decided to freeze powerboat registrations at 700,000, how many manatees do you predict would be killed by boats each year?

(f) Here is the result of asking Minitab to do prediction for $x^* = 700$.

```
Fit   Stdev.Fit      95% C.I.          95% P.I.
45.97      2.06   ( 41.49, 50.46) (  35.63,  56.31)
```

Check that the prediction 45.97 agrees with your result in (e). Then give a 95% interval for the mean number of manatees that would be killed each year if Florida froze boat registrations at 700,000.

10.10 The 95% interval in part (f) of the previous exercise is quite wide. Changing to 90% confidence will give a smaller margin of error. Use the computer output in the previous exercise, along with Table C, to give a 90% interval for the mean number of manatees killed when there are 700,000 power-boats registered.

CHECKING THE REGRESSION ASSUMPTIONS

You can fit a least-squares line to any set of explanatory-response data when both variables are quantitative. If the scatterplot doesn't show a roughly linear pattern, the fitted line may be almost useless. But it is still the line that fits the data best in the least-squares sense. To use regression inference, however, the data must satisfy the regression model assumptions. Before we do inference, we must check these assumptions one by one.

The true relationship is linear.   We can't observe the true regression line, so we will almost never see a perfect straight-line relationship in our data. Look at the scatterplot to check that the overall pattern is roughly linear. A plot of the residuals against x magnifies any unusual pattern. Draw a horizontal line at zero on the residual plot to orient your eye. Because the sum of the residuals is always zero, zero is also the mean of the residuals.

The standard deviation of the response about the true line is the same everywhere.   Look at the scatterplot again. The scatter of the data points about the line should be roughly the same over the entire range of the data. A plot of the residuals against x, with a horizontal line at zero, makes this easier to check. It is quite common to find that as the response y gets larger, so does the scatter of the points about the fitted line. Rather than remaining fixed, the standard deviation σ about the line is changing with x as the mean response changes with x. You cannot safely use our inference recipes when this happens. There is no fixed σ for s to estimate.

The response varies normally about the true regression line.   We can't observe the true regression line. We can observe the least-squares line and the residuals, which show the variation of the response about the fitted line.

The residuals estimate the deviations of the response from the true regression line, so they should follow a normal distribution. Make a histogram or stemplot of the residuals and check for clear skewness or other major departures from normality. Like other t procedures, inference for regression is (with one exception) not very sensitive to minor lack of normality, especially when we have many observations. Do beware of influential observations, which move the regression line and can greatly affect the results of inference.

The exception is the prediction interval for a single response y. This interval relies on normality of individual observations, not just on the approximate normality of statistics like the slope a and intercept b of the least-squares line. The statistics a and b become more normal as we take more observations. This contributes to the robustness of regression inference, but it isn't enough for the prediction interval. We will not study methods that carefully check normality of the residuals, so you should regard prediction intervals as rough approximations.

The assumptions for regression inference are a bit elaborate. Fortunately, it is not hard to check for gross violations. There are ways to deal with violations of any of the regression model assumptions. If your data don't fit the regression model, get expert advice. Checking assumptions uses the residuals. Most regression software will calculate and save the residuals for you.

EXAMPLE 10.10

Example 10.7 shows the regression of the scores of 12 college golfers in the second round of a tournament on their first-round scores. The Data Desk statistical software that did the regression calculations also calculates the 12 residuals. Here they are (Golfers 1 to 12 are in order reading across the rows):

6.45850	−3.22925	2.83399	−2.66798	−4.47826	−6.03953
10.5178	−9.54545	3.58498	4.14625	−.916996	−.664032

The computer has given us more decimal places than we need. The residual plot appears in Figure 10.6. The values of x are on the horizontal axis. The residuals are on the vertical axis, with a horizontal line at zero.

Examine the residual plot to check that the relationship is roughly linear and that the scatter about the line is about the same from end to end. The two unusual data points produce large residuals, but overall there is no clear

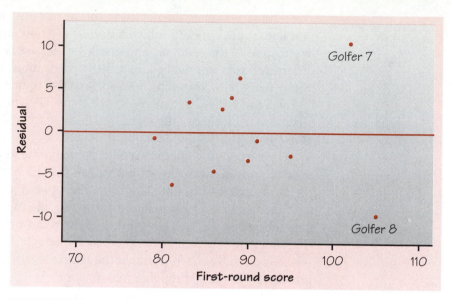

FIGURE 10.6 Plot of the regression residuals for the golf scores data against the explanatory variable, first-round score. The mean of the residuals is always 0.

deviation from the even scatter about the line that should occur (except for chance variation) when the regression assumptions hold.

Now examine the distribution of the residuals for signs of strong nonnormality. Here is a stemplot of the residuals after rounding to the nearest whole number:

$$
\begin{array}{r|l}
-1 & 0 \\
-0 & 643311 \\
0 & 3446 \\
1 & 1 \\
\end{array}
$$

The distribution is quite symmetric. Inference using the assumption that y varies normally will give approximately correct results.

As a reminder that regression is easily influenced by extreme data points, we recalculate the least-squares line leaving out first Golfer 7 and then Golfer 8. The slope is $b = .6877$ with all 12 golfers, $b = .4096$ if we drop Golfer 7, and $b = 1.0696$ if we drop Golfer 8. Study the scatterplot in Figure 10.4 to see why dropping Golfer 7 decreases the slope and dropping Golfer 8 increases it. The data are all correct, so we hesitate to throw out any observations. Because the regression results do depend strongly on two golfers with high first-round scores, we should gather more data before drawing any firm conclusions. ◀

EXAMPLE 10.11

The residual plots in Figure 10.7 illustrate violations of the regression assumptions that require corrective action before using regression. Both plots come from a study of the salaries of major-league baseball players.[6] Salary is the

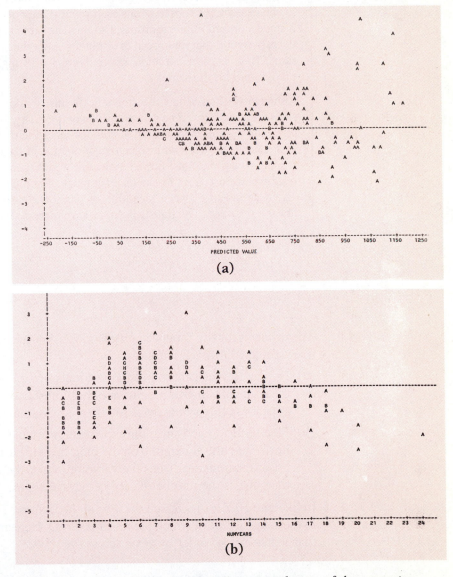

(a)

(b)

FIGURE 10.7 Two residual plots that illustrate violations of the regression assumptions. (**a**) The variation of the residuals is not constant. (**b**) There is a curved relationship between the response variable and the explanatory variable.

response variable. There are several explanatory variables that measure the players' past performance. Regression with more than one explanatory variable is called *multiple regression*. Although interpreting the fitted model is more complex in multiple regression, we check assumptions by examining residuals as usual.

Figure 10.7(a) is a plot of the residuals against the predicted salary $\hat{y}$, produced by the SAS statistical software. When points on the plot overlap, SAS uses letters to show how many observations each point represents. A is one observation, B stands for two observations, and so on. The plot shows a clear violation of the assumption that the spread of responses about the model is everywhere the same. There is more variation among players with high salaries than among players with lower salaries.

Although we don't show a histogram, the distribution of salaries is strongly skewed to the right. Using the *logarithm* of the salary as the response variable gives a more normal distribution and also fixes the unequal-spread problem. It is common to work with some transformation of data in order to satisfy the regression assumptions. But all is not yet well. Figure 10.7(b) plots the new residuals against years in the major leagues. There is a clear curved pattern. The relationship between logarithm of salary and years in the majors is not linear but curved. The statistician must take more corrective action. ◄

EXERCISES

10.11 The residuals for the Sanchez household gas consumption example appear in Example 10.4 (page 596).

(a) Display the distribution of the residuals in a plot. It is hard to assess the shape of a distribution from only 16 observations. Do the residuals appear roughly symmetric? Are there any outliers?

(b) Plot the residuals against the explanatory variable, degree-days. Draw a horizontal line at height 0 on the plot. Is there clear evidence of a nonlinear relationship? Does the variation about the line appear roughly the same as the number of degree-days changes?

10.12 In Exercise 10.7 we regressed the lean of the Leaning Tower of Pisa on year to estimate the rate at which the tower is tilting. Here are the residuals from that regression, in order by years across the rows:

$$
\begin{array}{ccccccc}
4.220 & -3.099 & -.418 & 1.264 & -2.055 & 3.626 & 2.308 \\
-5.011 & .670 & -4.648 & -5.967 & 1.714 & 7.396 &
\end{array}
$$

Use the residuals to check the regression assumptions, and describe your findings. Is the regression in Exercise 10.7 trustworthy?

SUMMARY

Least-squares regression fits a straight line to data in order to predict a response variable y from an explanatory variable x.

The **assumptions for regression inference** say that there is a **true regression line** $\mu_y = \alpha + \beta x$ that describes how the mean response varies as x changes. The observed response y for any x has a normal distribution with mean given by the true regression line and with the same standard deviation σ for any value of x. The parameters of the regression model are the intercept α, the slope β, and the standard deviation σ.

The slope a and intercept b of the least-squares line estimate the slope α and intercept β of the true regression line. To estimate σ, use the **standard error about the line** s.

The standard error s has $n - 2$ **degrees of freedom.** All t procedures in regression inference have $n - 2$ degrees of freedom.

Confidence intervals for the slope of the true regression line have the form $b \pm t^* SE_b$. In practice, use computer software to find the slope b of the least-squares line and its standard error SE_b.

To test the hypothesis that the true slope is zero, use the t statistic $t = b/SE_b$, also given by software. This null hypothesis says that straight-line dependence on x has no value for predicting y. It also says that the population correlation between x and y is zero.

Confidence intervals for the mean response when x has value x^* have the form $\hat{y} \pm t^* SE_{\hat\mu}$. **Prediction intervals** for an individual future response y have a similar form with a larger standard error, $\hat{y} \pm t^* SE_{\hat{y}}$. Computer software often gives these intervals.

CHAPTER REVIEW

When a scatterplot shows a straight-line relationship between an explanatory variable x and a response variable y, we often fit a least-squares regression line to describe the relationship. Use this line to predict y from x. Statistical inference in the regression setting, however, requires more than just an overall linear pattern on a scatterplot.

The regression model says that there is a true straight-line relationship between x and the mean response μ_y. We can't observe this true regression line. The responses y that we do observe vary if we take several observations at the same x. The regression model says that for any fixed x, the responses have a normal distribution. Moreover, the standard deviation σ of this distribution is the same for all values of x.

The standard deviation σ describes how much variation there is in responses y when x is held fixed. Estimating σ is the key to inference about regression. Use the standard error s (roughly, the sample standard deviation of the residuals) to estimate σ. We can then do these types of inference:

- Give confidence intervals for the slope of the true regression line.
- Test the null hypothesis that this slope is zero. This hypothesis says that a straight-line relation between x and y is of no value for predicting y. It is the same as saying that the correlation between x and y in the entire population is zero.
- Give confidence intervals for the mean response for any fixed value of x.
- Give prediction intervals for an individual response y for a fixed value of x.

Here are the skills you should develop from studying this chapter.

A. PRELIMINARIES

1. Make a scatterplot to show the relationship between an explanatory and a response variable.
2. Use a calculator or software to find the correlation and the least-squares regression line.

B. RECOGNITION

1. Recognize the regression setting: a straight-line relationship between an explanatory variable x and a response variable y.
2. Recognize which type of inference you need in a particular regression setting.
3. Inspect the data to recognize situations in which inference isn't safe: a nonlinear relationship, influential observations, strongly skewed residuals in a small sample, or nonconstant variation of the data points about the regression line.

C. DOING INFERENCE USING SOFTWARE OUTPUT

1. Explain in any specific regression setting the meaning of the slope β of the true regression line.

2. Understand computer output for regression. Find in the output the slope and intercept of the least-squares line, their standard errors, and the standard error about the line.

3. Use that information to carry out tests and calculate confidence intervals for β.

4. Explain the distinction between a confidence interval for the mean response and a prediction interval for an individual response.

5. If software gives output for prediction, use that output to give either confidence or prediction intervals.

CHAPTER 10 REVIEW EXERCISES

10.13 Investors ask about the relationship between returns on investments in the United States and investments overseas. Here are data on the total returns on U.S. and overseas common stocks over a 22-year period. (The total return is change in price plus any dividends paid, converted into U.S. dollars. Both returns are averages over many individual stocks.)[7]

Year	Overseas % return	U.S. % return	Year	Overseas % return	U.S. % return
1971	29.6	14.6	1982	−1.9	21.5
1972	36.3	18.9	1983	23.7	22.4
1973	−14.9	−14.8	1984	7.4	6.1
1974	−23.2	−26.4	1985	56.2	31.6
1975	35.4	37.2	1986	69.4	18.6
1976	2.5	23.6	1987	24.6	5.1
1977	18.1	−7.4	1988	28.5	16.8
1978	32.6	6.4	1989	10.6	31.5
1979	4.8	18.2	1990	−23.0	−3.1
1980	22.6	32.3	1991	12.8	30.4
1981	−2.3	−5.0	1992	−12.1	7.6

(a) Make a scatterplot suitable for predicting overseas returns from U.S. returns.

(b) Here is part of the output from the Minitab regression command.

```
Predictor     Coef      Stdev      t-ratio      p
Constant      4.777     5.477      0.87         0.393
US return     0.8130    0.2628     3.09         0.006

s = 20.08     R-sq = 32.4%
```

Find the correlation and r^2. Describe the relationship between U.S. and overseas returns in words, using r and r^2 to make your description more precise. Does the scatterplot show clear outliers or strongly influential observations?

(c) Is there good evidence that there is a linear relationship between U.S. and foreign returns? State hypotheses and give the test statistic and its P-value.

(d) What is the predicted return on overseas stocks in a year when U.S. stocks return 20%?

(e) Here is the output for prediction of overseas returns when U.S. stocks return 20%.

```
Fit     Stdev.Fit         95% C.I.          95% P.I.
21.04        4.66     ( 11.32,   30.76) ( -21.97,   64.04)
```

Check the "Fit" against your result from (d). You think U.S. stocks will return 20% next year. Give a 95% interval for the return on foreign stocks next year if you are right about U.S. stocks. Is the regression prediction useful in practice?

10.14   The computer output in Exercise 10.8 gives 95% intervals for prediction. You want 99% confidence. Use the computer's results for the least-squares line and $SE_{\hat{\mu}}$ to obtain a 99% confidence interval for the mean gas consumption in months that average 40 degree-days per day.

10.15   Exercise 10.13 presents a regression of overseas stock returns on U.S. stock returns based on 22 years' data. The residuals for this regression (in order by years across the rows) are

```
  12.953   16.157    -7.645    -6.514      .379  -21.464  19.339  22.620
 -14.774   -8.437    -3.012   -24.157      .712   -2.337  25.732  49.501
  15.676   10.064   -19.787   -25.257   -16.692  -23.056
```

(a) Plot the residuals against x, the U.S. return. The plot suggests a mild violation of one of the regression assumptions. Which one? The violation is not strong enough to forbid regression inference.

(b) Display the distribution of the residuals in a graph. There is one possible outlier. Circle that point on the residual plot in (a). This point is not an extreme outlier, and redoing the regression without it does not greatly change the results. We are willing to do regression inference for these data.

10.16 The Franklin National Bank failed in 1974. Franklin was one of the 20 largest banks in the nation, and the largest ever to fail. Could regression have detected Franklin's weakened condition in advance? The table below gives the total assets (in billions of dollars) and net income (in millions of dollars) for the 20 largest banks in 1973, the year before Franklin failed.[8] Franklin is bank number 19.

	Bank									
	1	2	3	4	5	6	7	8	9	10
Assets	49.0	42.3	36.3	16.4	14.9	14.2	13.5	13.4	13.2	11.8
Income	218.8	265.6	170.9	85.9	88.1	63.6	96.9	60.9	144.2	53.6

	Bank									
	11	12	13	14	15	16	17	18	19	20
Assets	11.6	9.5	9.4	7.5	7.2	6.7	6.0	4.6	3.8	3.4
Income	42.9	32.4	68.3	48.6	32.2	42.7	28.9	40.7	13.8	22.2

(a) We expect banks with more assets to earn higher income. Make a scatterplot of these data with assets as the explanatory variable. Mark Franklin (Bank 19) with a separate symbol.

(b) The scatterplot suggests that regression calculations will be dominated by the three banks with the highest incomes. Circle these points on your scatterplot. Here is output from a Minitab regression of income on assets for the remaining 17 banks only.

```
Predictor      Coef      Stdev     t-ratio      p
Constant      -0.59      14.48      -0.04      0.968
Assets         5.840      1.363      4.28      0.001

s = 22.59      R-sq = 55.0%
```

(c) A report on this regression analysis says, "There is highly significant evidence that banks' incomes increase as their assets increase ($t = ?$, $df = ?$, $P = ?$)." State the null and alternative hypotheses for this test. Then give the numbers that replace the question marks in the report's summary.

(d) Predict Franklin's income from its assets, and find the residual for Franklin.

(e) Here is output for predicting income for Franklin's assets ($x^* = 3.8$).

```
    Fit   Stdev.Fit        95% C.I.            95% P.I.
  21.60        9.88  (  0.55,  42.66) ( -30.96,  74.17)
```

Give a 95% interval for Franklin's income. Would this prediction have been helpful in showing that Franklin's actual income was lower than we would expect from its assets?

10.17 **(a)** Use the least-squares line from Example 10.7 (page 601) to predict the second-round score for Golfer 1, who shot an 89 on the first round. Her actual second-round score was 94. Find the residual for this data point and check that your work agrees with the first residual given in Example 10.10.

(b) Calculate the sum of the squares of the 12 residuals given in Example 10.10. Divide by $n - 2 = 10$, then take the square root to obtain the standard error s about the line. Check that your work agrees with $s = 5.974$, given by the software in Figure 10.5.

10.18 Call the correlation between first-round and second-round tournament scores in the population of college female competitive golfers ρ. We want to test the null hypothesis of "no correlation":

$$H_0 : \rho = 0$$
$$H_a : \rho \neq 0$$

Regard the 12 golfers in Example 10.7 as an SRS from this population. Then the regression output in Figure 10.5 allows you to find both the sample correlation r that estimates ρ and the P-value for a test of these hypothesis.

(a) What is r? Explain clearly how you found r from the output.

(b) The test of $H_0 : \rho = 0$ is exactly the same as the test of $H_0 : \beta = 0$. What is the P-value of this test against the two-sided alternative?

(c) We expect the scores of the same golfers in two rounds of play to have a positive correlation. What is the P-value for the test of $H_0 : \rho = 0$ against the one-sided alternative $H_a : \rho > 0$?

10.19 **(Optional)** Figure 10.3 (page 599) gives Minitab output for the regression of household natural gas consumption on degree-days. Example 10.5 uses the output to find a 95% confidence interval for the slope β of the true regression line. The intercept α of the true regression line is the average amount of gas the household consumes when there are zero degree-days.

(There are zero degree-days when the average temperature is 65° or above.) Confidence intervals for α have the form

$$a \pm t^* SE_a$$

Use the intercept a of the least-squares line and its standard error, given in the computer output, to find a 95% confidence interval for α. (The degrees of freedom are $n - 2$ once again.)

10.20   **(Optional)** Here is a guide to calculating confidence intervals for the mean response from basic regression output. You can find prediction intervals for an individual response in the same way.

Figure 10.5 gives regression output for predicting second-round score y from first-round score x for college female golfers. We want to predict the mean second-round score for golfers who score 90 in the first round. The box on page 607 shows that we must calculate the standard error

$$SE_{\hat{\mu}} = s \sqrt{\frac{1}{n} + \frac{(x^* - \bar{x})^2}{\sum (x - \bar{x})^2}}$$

(a) Use your calculator to find the mean $\bar{x}$ and the standard deviation s_x of the first-round scores.

(b) What is x^*? Find $(x^* - \bar{x})^2$.

(c) From the recipe for the sample variance, you know that

$$\sum (x - \bar{x})^2 = (n - 1)s_x^2$$

Use this fact to find $\sum (x - \bar{x})^2$.

(d) Now use s from the output with your results in (b) and (c) to find $SE_{\hat{\mu}}$.

(e) Finally, find the predicted second-round score $\hat{y}$ and the 95% confidence interval $\hat{y} \pm t^* SE_{\hat{\mu}}$ for the mean response.

10.21   How well does a car's city gas mileage (as measured by the Environmental Protection Agency) predict its highway gas mileage? Table 9.1 (page 556) records the city and highway mileage for a sample of 59 1994 car models. Statistical software will help you analyze these data.

(a) Make a scatterplot of the data. (Which is the explanatory variable?)

(b) Examine the plot. Is the association positive or negative? Is there an overall linear pattern? How strong is the pattern? Circle the points for the Jaguar XJ12 and Rolls-Royce Silver Spur. These two cars are low outliers in the x direction, but they don't lie outside the overall linear pattern of the data.

(c) Find the equation of the least-squares regression line for predicting highway mileage from city mileage. Draw the line on your plot from (a).

(d) Remove the Jaguar and the Rolls-Royce from the data and find the least-squares regression line for the remaining 57 cars. Plot this line on your scatterplot. Also compare the two values of s, the standard deviation about the line. The two regression lines lie close together and s shows little change. So the two cars removed were not influential. Use all 59 cars for all further analyses.

(e) Find the 59 residuals. Which car has the largest residual in absolute value? Does the regression line predict this car's highway mileage to be higher or lower than the actual value?

(f) Examine the residuals to check the assumptions of the regression model. Are the assumptions reasonably well met?

(g) Explain the meaning of the slope β of the true regression line in this setting. Give a 95% confidence interval for β.

(h) The Volkswagon Passat, a model not in the sample, gets 18 miles per gallon in the city. Predict the Passat's highway mileage. Then give an interval which covers the true highway mileage with 95% confidence.

10.22   Table 10.2 contains data on the size of perch caught in a lake in Finland.[9] Statistical software will help you analyze these data.

(a) We want to know how well we can predict the width of a perch from its length. Make a scatterplot of width against length. There is a strong linear pattern, as expected. The heaviest perch had six newly eaten fish in its stomach and is an outlier in weight. Find this fish on your scatterplot and circle the point. Is this fish an outlier in your plot of width against length?

(b) Find the least-squares regression line to predict width from length.

(c) The length of a typical perch is about $x^* = 25$ centimeters. Predict the mean width of such fish and give a 95% confidence interval.

(d) Examine the residuals. Is there any reason to mistrust inference?

10.23   We can also use the data in Table 10.2 to study the prediction of the weight of a perch from its length.

(a) Make a scatterplot of weight versus length, with length as the explanatory variable. Describe the pattern of the data and any clear outliers.

(b) It is more reasonable to expect the one-third power of the weight to have a straight-line relationship with length than to expect weight

TABLE 10.2   Measurements on 56 perch

Length (centimeters)	Width (centimeters)	Weight (grams)	Length (centimeters)	Width (centimeters)	Weight (grams)
8.4	2.02	16.0	13.7	3.29	13.6
15.0	3.58	15.2	16.2	4.33	15.3
17.4	4.32	15.9	18.0	4.90	17.3
18.7	5.01	16.1	19.0	5.30	15.1
19.6	4.84	14.6	20.0	4.84	13.2
21.0	5.31	15.8	21.0	5.52	14.7
21.0	5.31	16.3	21.3	5.96	15.5
22.0	5.72	14.5	22.0	5.28	15.0
22.0	5.72	15.0	22.0	5.50	15.0
22.0	5.17	17.0	22.5	5.49	15.1
22.5	6.37	15.1	22.7	5.58	15.0
23.0	4.90	14.8	23.5	5.90	14.9
24.0	6.86	14.6	24.0	6.00	15.0
24.6	6.32	15.9	25.0	6.08	13.9
25.6	6.22	15.7	26.5	6.78	14.8
27.3	7.92	17.9	27.5	6.82	15.0
27.5	6.71	15.0	27.5	6.93	15.8
28.0	7.45	14.3	28.7	7.23	15.4
30.0	7.23	15.1	32.8	9.68	17.7
34.5	9.69	17.5	35.0	10.78	20.9
36.5	10.18	17.6	36.0	9.97	17.6
37.0	10.18	15.9	37.0	9.95	16.2
39.0	10.49	18.1	39.0	10.49	14.5
39.0	11.74	17.8	40.0	11.28	16.8
40.0	11.04	17.0	40.0	11.68	17.6
40.0	10.48	15.6	42.0	12.05	15.4
43.0	11.35	16.1	43.0	11.82	16.3
43.5	11.92	17.7	44.0	11.79	16.3

itself to have a straight-line relationship with length. Explain why this is true. (Hint: What happens to weight if length, width, and height all double?)

(c) Use your software to create a new variable that is the one-third power of weight. Make a scatterplot of this new response variable against length. Describe the pattern and any clear outliers.

(d) Is the straight-line pattern in (c) stronger or weaker than that in (a)? Compare the plots and also the values of r^2.

(e) Find the least-squares regression line to predict the new weight variable from length. Predict the mean of the new variable for perch 25 centimeters long, and give a 95% confidence interval.

(f) Examine the residuals from your regressions. Does it appear that any of the regression assumptions are not met?

NOTES AND DATA SOURCES

1. Data provided by Robert Dale, Purdue University.

2. The data are from M. A. Houck et al., "Allometric scaling in the earliest fossil bird, *Archaeopteryx lithographica*," *Science*, 247 (1990), pp. 195–198.

3. Data from R. C. Nelson, C. M. Brooks, and N. L. Pike, "Biomechanical comparison of male and female distance runners," in P. Milvy (ed.), *The Marathon: Physiological, Medical, Epidemiological, and Psychological Studies*, New York Academy of Sciences, 1977, pp. 793–807.

4. Data from G. Geri and B. Palla, "Considerazioni sulle più recenti osservazioni ottiche alla Torre Pendente di Pisa," *Estratto dal Bollettino della Società Italiana di Topografia e Fotogrammetria*, 2 (1988), pp. 121–135. Professor Julia Mortera of the University of Rome provided a translation.

5. Strictly speaking, this quantity is the estimated standard deviation of $\hat{y} - y$, where y is the additional observation taken at $x = x^*$.

6. The data are for 1987 salaries and measures of past performance. They were collected and distributed by the Statistical Graphics Section of the American Statistical Association for an annual data analysis contest. The analysis here was done by Crystal Richard of Purdue University.

7. The U.S. returns are for the Standard & Poor's 500 stock index. The overseas returns are for the Morgan Stanley Europe, Australia, Far East (EAFE) index.

8. Data from D. E. Booth, *Regression Methods and Problem Banks*, COMAP, Inc., 1986.

9. The data in Table 10.2 are part of a larger data set in the *Journal of Statistics Education* archive, accessible via Internet. The original source is Pekka Brofeldt, "Bidrag till kaennedom on fiskbestondet i vaara sjoear. Laengelmaevesi," in T. H. Jaervi, *Finlands Fiskeriet*, Band 4, *Meddelanden utgivna av fiskerifoereningen i Finland*, Helsinki, 1917. The data were contributed to the archive (with information in English) by Juha Puranen of the University of Helsinki.

APPENDIX

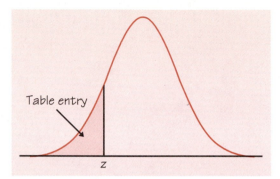

Table entry for z is the area under the standard normal curve to the left of z.

TABLE A	Standard normal probabilities								
z	.00	.01	.02	.03	.04	.05	.06	.07	.08	.09
−3.4	.0003	.0003	.0003	.0003	.0003	.0003	.0003	.0003	.0003	.0002
−3.3	.0005	.0005	.0005	.0004	.0004	.0004	.0004	.0004	.0004	.0003
−3.2	.0007	.0007	.0006	.0006	.0006	.0006	.0006	.0005	.0005	.0005
−3.1	.0010	.0009	.0009	.0009	.0008	.0008	.0008	.0008	.0007	.0007
−3.0	.0013	.0013	.0013	.0012	.0012	.0011	.0011	.0011	.0010	.0010
−2.9	.0019	.0018	.0018	.0017	.0016	.0016	.0015	.0015	.0014	.0014
−2.8	.0026	.0025	.0024	.0023	.0023	.0022	.0021	.0021	.0020	.0019
−2.7	.0035	.0034	.0033	.0032	.0031	.0030	.0029	.0028	.0027	.0026
−2.6	.0047	.0045	.0044	.0043	.0041	.0040	.0039	.0038	.0037	.0036
−2.5	.0062	.0060	.0059	.0057	.0055	.0054	.0052	.0051	.0049	.0048
−2.4	.0082	.0080	.0078	.0075	.0073	.0071	.0069	.0068	.0066	.0064
−2.3	.0107	.0104	.0102	.0099	.0096	.0094	.0091	.0089	.0087	.0084
−2.2	.0139	.0136	.0132	.0129	.0125	.0122	.0119	.0116	.0113	.0110
−2.1	.0179	.0174	.0170	.0166	.0162	.0158	.0154	.0150	.0146	.0143
−2.0	.0228	.0222	.0217	.0212	.0207	.0202	.0197	.0192	.0188	.0183
−1.9	.0287	.0281	.0274	.0268	.0262	.0256	.0250	.0244	.0239	.0233
−1.8	.0359	.0351	.0344	.0336	.0329	.0322	.0314	.0307	.0301	.0294
−1.7	.0446	.0436	.0427	.0418	.0409	.0401	.0392	.0384	.0375	.0367
−1.6	.0548	.0537	.0526	.0516	.0505	.0495	.0485	.0475	.0465	.0455
−1.5	.0668	.0655	.0643	.0630	.0618	.0606	.0594	.0582	.0571	.0559
−1.4	.0808	.0793	.0778	.0764	.0749	.0735	.0721	.0708	.0694	.0681
−1.3	.0968	.0951	.0934	.0918	.0901	.0885	.0869	.0853	.0838	.0823
−1.2	.1151	.1131	.1112	.1093	.1075	.1056	.1038	.1020	.1003	.0985
−1.1	.1357	.1335	.1314	.1292	.1271	.1251	.1230	.1210	.1190	.1170
−1.0	.1587	.1562	.1539	.1515	.1492	.1469	.1446	.1423	.1401	.1379
−0.9	.1841	.1814	.1788	.1762	.1736	.1711	.1685	.1660	.1635	.1611
−0.8	.2119	.2090	.2061	.2033	.2005	.1977	.1949	.1922	.1894	.1867
−0.7	.2420	.2389	.2358	.2327	.2296	.2266	.2236	.2206	.2177	.2148
−0.6	.2743	.2709	.2676	.2643	.2611	.2578	.2546	.2514	.2483	.2451
−0.5	.3085	.3050	.3015	.2981	.2946	.2912	.2877	.2843	.2810	.2776
−0.4	.3446	.3409	.3372	.3336	.3300	.3264	.3228	.3192	.3156	.3121
−0.3	.3821	.3783	.3745	.3707	.3669	.3632	.3594	.3557	.3520	.3483
−0.2	.4207	.4168	.4129	.4090	.4052	.4013	.3974	.3936	.3897	.3859
−0.1	.4602	.4562	.4522	.4483	.4443	.4404	.4364	.4325	.4286	.4247
−0.0	.5000	.4960	.4920	.4880	.4840	.4801	.4761	.4721	.4681	.4641

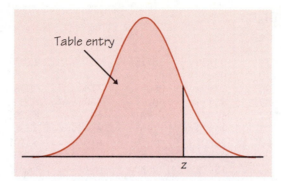

Table entry

Table entry for z is the area under the standard normal curve to the left of z.

TABLE A	Standard normal probabilities (*continued*)								
z	.00	.01	.02	.03	.04	.05	.06	.07	.08	.09
0.0	.5000	.5040	.5080	.5120	.5160	.5199	.5239	.5279	.5319	.5359
0.1	.5398	.5438	.5478	.5517	.5557	.5596	.5636	.5675	.5714	.5753
0.2	.5793	.5832	.5871	.5910	.5948	.5987	.6026	.6064	.6103	.6141
0.3	.6179	.6217	.6255	.6293	.6331	.6368	.6406	.6443	.6480	.6517
0.4	.6554	.6591	.6628	.6664	.6700	.6736	.6772	.6808	.6844	.6879
0.5	.6915	.6950	.6985	.7019	.7054	.7088	.7123	.7157	.7190	.7224
0.6	.7257	.7291	.7324	.7357	.7389	.7422	.7454	.7486	.7517	.7549
0.7	.7580	.7611	.7642	.7673	.7704	.7734	.7764	.7794	.7823	.7852
0.8	.7881	.7910	.7939	.7967	.7995	.8023	.8051	.8078	.8106	.8133
0.9	.8159	.8186	.8212	.8238	.8264	.8289	.8315	.8340	.8365	.8389
1.0	.8413	.8438	.8461	.8485	.8508	.8531	.8554	.8577	.8599	.8621
1.1	.8643	.8665	.8686	.8708	.8729	.8749	.8770	.8790	.8810	.8830
1.2	.8849	.8869	.8888	.8907	.8925	.8944	.8962	.8980	.8997	.9015
1.3	.9032	.9049	.9066	.9082	.9099	.9115	.9131	.9147	.9162	.9177
1.4	.9192	.9207	.9222	.9236	.9251	.9265	.9279	.9292	.9306	.9319
1.5	.9332	.9345	.9357	.9370	.9382	.9394	.9406	.9418	.9429	.9441
1.6	.9452	.9463	.9474	.9484	.9495	.9505	.9515	.9525	.9535	.9545
1.7	.9554	.9564	.9573	.9582	.9591	.9599	.9608	.9616	.9625	.9633
1.8	.9641	.9649	.9656	.9664	.9671	.9678	.9686	.9693	.9699	.9706
1.9	.9713	.9719	.9726	.9732	.9738	.9744	.9750	.9756	.9761	.9767
2.0	.9772	.9778	.9783	.9788	.9793	.9798	.9803	.9808	.9812	.9817
2.1	.9821	.9826	.9830	.9834	.9838	.9842	.9846	.9850	.9854	.9857
2.2	.9861	.9864	.9868	.9871	.9875	.9878	.9881	.9884	.9887	.9890
2.3	.9893	.9896	.9898	.9901	.9904	.9906	.9909	.9911	.9913	.9916
2.4	.9918	.9920	.9922	.9925	.9927	.9929	.9931	.9932	.9934	.9936
2.5	.9938	.9940	.9941	.9943	.9945	.9946	.9948	.9949	.9951	.9952
2.6	.9953	.9955	.9956	.9957	.9959	.9960	.9961	.9962	.9963	.9964
2.7	.9965	.9966	.9967	.9968	.9969	.9970	.9971	.9972	.9973	.9974
2.8	.9974	.9975	.9976	.9977	.9977	.9978	.9979	.9979	.9980	.9981
2.9	.9981	.9982	.9982	.9983	.9984	.9984	.9985	.9985	.9986	.9986
3.0	.9987	.9987	.9987	.9988	.9988	.9989	.9989	.9989	.9990	.9990
3.1	.9990	.9991	.9991	.9991	.9992	.9992	.9992	.9992	.9993	.9993
3.2	.9993	.9993	.9994	.9994	.9994	.9994	.9994	.9995	.9995	.9995
3.3	.9995	.9995	.9995	.9996	.9996	.9996	.9996	.9996	.9996	.9997
3.4	.9997	.9997	.9997	.9997	.9997	.9997	.9997	.9997	.9997	.9998

TABLE B   Random digits

Line							
101	19223	95034	05756	28713	96409	12531	42544	82853
102	73676	47150	99400	01927	27754	42648	82425	36290
103	45467	71709	77558	00095	32863	29485	82226	90056
104	52711	38889	93074	60227	40011	85848	48767	52573
105	95592	94007	69971	91481	60779	53791	17297	59335
106	68417	35013	15529	72765	85089	57067	50211	47487
107	82739	57890	20807	47511	81676	55300	94383	14893
108	60940	72024	17868	24943	61790	90656	87964	18883
109	36009	19365	15412	39638	85453	46816	83485	41979
110	38448	48789	18338	24697	39364	42006	76688	08708
111	81486	69487	60513	09297	00412	71238	27649	39950
112	59636	88804	04634	71197	19352	73089	84898	45785
113	62568	70206	40325	03699	71080	22553	11486	11776
114	45149	32992	75730	66280	03819	56202	02938	70915
115	61041	77684	94322	24709	73698	14526	31893	32592
116	14459	26056	31424	80371	65103	62253	50490	61181
117	38167	98532	62183	70632	23417	26185	41448	75532
118	73190	32533	04470	29669	84407	90785	65956	86382
119	95857	07118	87664	92099	58806	66979	98624	84826
120	35476	55972	39421	65850	04266	35435	43742	11937
121	71487	09984	29077	14863	61683	47052	62224	51025
122	13873	81598	95052	90908	73592	75186	87136	95761
123	54580	81507	27102	56027	55892	33063	41842	81868
124	71035	09001	43367	49497	72719	96758	27611	91596
125	96746	12149	37823	71868	18442	35119	62103	39244
126	96927	19931	36809	74192	77567	88741	48409	41903
127	43909	99477	25330	64359	40085	16925	85117	36071
128	15689	14227	06565	14374	13352	49367	81982	87209
129	36759	58984	68288	22913	18638	54303	00795	08727
130	69051	64817	87174	09517	84534	06489	87201	97245
131	05007	16632	81194	14873	04197	85576	45195	96565
132	68732	55259	84292	08796	43165	93739	31685	97150
133	45740	41807	65561	33302	07051	93623	18132	09547
134	27816	78416	18329	21337	35213	37741	04312	68508
135	66925	55658	39100	78458	11206	19876	87151	31260
136	08421	44753	77377	28744	75592	08563	79140	92454
137	53645	66812	61421	47836	12609	15373	98481	14592
138	66831	68908	40772	21558	47781	33586	79177	06928
139	55588	99404	70708	41098	43563	56934	48394	51719
140	12975	13258	13048	45144	72321	81940	00360	02428
141	96767	35964	23822	96012	94591	65194	50842	53372
142	72829	50232	97892	63408	77919	44575	24870	04178
143	88565	42628	17797	49376	61762	16953	88604	12724
144	62964	88145	83083	69453	46109	59505	69680	00900
145	19687	12633	57857	95806	09931	02150	43163	58636
146	37609	59057	66967	83401	60705	02384	90597	93600
147	54973	86278	88737	74351	47500	84552	19909	67181
148	00694	05977	19664	65441	20903	62371	22725	53340
149	71546	05233	53946	68743	72460	27601	45403	88692
150	07511	88915	41267	16853	84569	79367	32337	03316

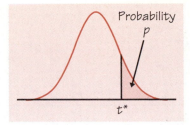

Table entry for p and C is the critical value t^* with probability p
lying to its right and probability C lying between $-t^*$ and t^*.

TABLE C t distribution critical values

df	\.25	.20	.15	.10	.05	.025	.02	.01	.005	.0025	.001	.0005
1	1.000	1.376	1.963	3.078	6.314	12.71	15.89	31.82	63.66	127.3	318.3	636.6
2	0.816	1.061	1.386	1.886	2.920	4.303	4.849	6.965	9.925	14.09	22.33	31.60
3	0.765	0.978	1.250	1.638	2.353	3.182	3.482	4.541	5.841	7.453	10.21	12.92
4	0.741	0.941	1.190	1.533	2.132	2.776	2.999	3.747	4.604	5.598	7.173	8.610
5	0.727	0.920	1.156	1.476	2.015	2.571	2.757	3.365	4.032	4.773	5.893	6.869
6	0.718	0.906	1.134	1.440	1.943	2.447	2.612	3.143	3.707	4.317	5.208	5.959
7	0.711	0.896	1.119	1.415	1.895	2.365	2.517	2.998	3.499	4.029	4.785	5.408
8	0.706	0.889	1.108	1.397	1.860	2.306	2.449	2.896	3.355	3.833	4.501	5.041
9	0.703	0.883	1.100	1.383	1.833	2.262	2.398	2.821	3.250	3.690	4.297	4.781
10	0.700	0.879	1.093	1.372	1.812	2.228	2.359	2.764	3.169	3.581	4.144	4.587
11	0.697	0.876	1.088	1.363	1.796	2.201	2.328	2.718	3.106	3.497	4.025	4.437
12	0.695	0.873	1.083	1.356	1.782	2.179	2.303	2.681	3.055	3.428	3.930	4.318
13	0.694	0.870	1.079	1.350	1.771	2.160	2.282	2.650	3.012	3.372	3.852	4.221
14	0.692	0.868	1.076	1.345	1.761	2.145	2.264	2.624	2.977	3.326	3.787	4.140
15	0.691	0.866	1.074	1.341	1.753	2.131	2.249	2.602	2.947	3.286	3.733	4.073
16	0.690	0.865	1.071	1.337	1.746	2.120	2.235	2.583	2.921	3.252	3.686	4.015
17	0.689	0.863	1.069	1.333	1.740	2.110	2.224	2.567	2.898	3.222	3.646	3.965
18	0.688	0.862	1.067	1.330	1.734	2.101	2.214	2.552	2.878	3.197	3.611	3.922
19	0.688	0.861	1.066	1.328	1.729	2.093	2.205	2.539	2.861	3.174	3.579	3.883
20	0.687	0.860	1.064	1.325	1.725	2.086	2.197	2.528	2.845	3.153	3.552	3.850
21	0.686	0.859	1.063	1.323	1.721	2.080	2.189	2.518	2.831	3.135	3.527	3.819
22	0.686	0.858	1.061	1.321	1.717	2.074	2.183	2.508	2.819	3.119	3.505	3.792
23	0.685	0.858	1.060	1.319	1.714	2.069	2.177	2.500	2.807	3.104	3.485	3.768
24	0.685	0.857	1.059	1.318	1.711	2.064	2.172	2.492	2.797	3.091	3.467	3.745
25	0.684	0.856	1.058	1.316	1.708	2.060	2.167	2.485	2.787	3.078	3.450	3.725
26	0.684	0.856	1.058	1.315	1.706	2.056	2.162	2.479	2.779	3.067	3.435	3.707
27	0.684	0.855	1.057	1.314	1.703	2.052	2.158	2.473	2.771	3.057	3.421	3.690
28	0.683	0.855	1.056	1.313	1.701	2.048	2.154	2.467	2.763	3.047	3.408	3.674
29	0.683	0.854	1.055	1.311	1.699	2.045	2.150	2.462	2.756	3.038	3.396	3.659
30	0.683	0.854	1.055	1.310	1.697	2.042	2.147	2.457	2.750	3.030	3.385	3.646
40	0.681	0.851	1.050	1.303	1.684	2.021	2.123	2.423	2.704	2.971	3.307	3.551
50	0.679	0.849	1.047	1.299	1.676	2.009	2.109	2.403	2.678	2.937	3.261	3.496
60	0.679	0.848	1.045	1.296	1.671	2.000	2.099	2.390	2.660	2.915	3.232	3.460
80	0.678	0.846	1.043	1.292	1.664	1.990	2.088	2.374	2.639	2.887	3.195	3.416
100	0.677	0.845	1.042	1.290	1.660	1.984	2.081	2.364	2.626	2.871	3.174	3.390
1000	0.675	0.842	1.037	1.282	1.646	1.962	2.056	2.330	2.581	2.813	3.098	3.300
z^*	0.674	0.841	1.036	1.282	1.645	1.960	2.054	2.326	2.576	2.807	3.091	3.291
	50%	60%	70%	80%	90%	95%	96%	98%	99%	99.5%	99.8%	99.9%

Confidence level C

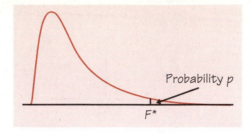

Table entry for p is the critical value F^* with probability p lying to its right.

TABLE D F distribution critical values

	p				Degrees of freedom in the numerator				
		1	2	3	4	5	6	7	8
1	.100	39.86	49.50	53.59	55.83	57.24	58.20	58.91	59.44
	.050	161.45	199.50	215.71	224.58	230.16	233.99	236.77	238.88
	.025	647.79	799.50	864.16	899.58	921.85	937.11	948.22	956.66
	.010	4052.2	4999.5	5403.4	5624.6	5763.6	5859	5928.4	5981.1
	.001	405284	500000	540379	562500	576405	585937	592873	598144
2	.100	8.53	9.00	9.16	9.24	9.29	9.33	9.35	9.37
	.050	18.51	19.00	19.16	19.25	19.30	19.33	19.35	19.37
	.025	38.51	39.00	39.17	39.25	39.30	39.33	39.36	39.37
	.010	98.50	99.00	99.17	99.25	99.30	99.33	99.36	99.37
	.001	998.50	999.00	999.17	999.25	999.30	999.33	999.36	999.37
3	.100	5.54	5.46	5.39	5.34	5.31	5.28	5.27	5.25
	.050	10.13	9.55	9.28	9.12	9.01	8.94	8.89	8.85
	.025	17.44	16.04	15.44	15.10	14.88	14.73	14.62	14.54
	.010	34.12	30.82	29.46	28.71	28.24	27.91	27.67	27.49
	.001	167.03	148.50	141.11	137.10	134.58	132.85	131.58	130.62
4	.100	4.54	4.32	4.19	4.11	4.05	4.01	3.98	3.95
	.050	7.71	6.94	6.59	6.39	6.26	6.16	6.09	6.04
	.025	12.22	10.65	9.98	9.60	9.36	9.20	9.07	8.98
	.010	21.20	18.00	16.69	15.98	15.52	15.21	14.98	14.80
	.001	74.14	61.25	56.18	53.44	51.71	50.53	49.66	49.00
5	.100	4.06	3.78	3.62	3.52	3.45	3.40	3.37	3.34
	.050	6.61	5.79	5.41	5.19	5.05	4.95	4.88	4.82
	.025	10.01	8.43	7.76	7.39	7.15	6.98	6.85	6.76
	.010	16.26	13.27	12.06	11.39	10.97	10.67	10.46	10.29
	.001	47.18	37.12	33.20	31.09	29.75	28.83	28.16	27.65
6	.100	3.78	3.46	3.29	3.18	3.11	3.05	3.01	2.98
	.050	5.99	5.14	4.76	4.53	4.39	4.28	4.21	4.15
	.025	8.81	7.26	6.60	6.23	5.99	5.82	5.70	5.60
	.010	13.75	10.92	9.78	9.15	8.75	8.47	8.26	8.10
	.001	35.51	27.00	23.70	21.92	20.80	20.03	19.46	19.03
7	.100	3.59	3.26	3.07	2.96	2.88	2.83	2.78	2.75
	.050	5.59	4.74	4.35	4.12	3.97	3.87	3.79	3.73
	.025	8.07	6.54	5.89	5.52	5.29	5.12	4.99	4.90
	.010	12.25	9.55	8.45	7.85	7.46	7.19	6.99	6.84
	.001	29.25	21.69	18.77	17.20	16.21	15.52	15.02	14.63
8	.100	3.46	3.11	2.92	2.81	2.73	2.67	2.62	2.59
	.050	5.32	4.46	4.07	3.84	3.69	3.58	3.50	3.44
	.025	7.57	6.06	5.42	5.05	4.82	4.65	4.53	4.43
	.010	11.26	8.65	7.59	7.01	6.63	6.37	6.18	6.03
	.001	25.41	18.49	15.83	14.39	13.48	12.86	12.40	12.05

Degrees of freedom in the denominator

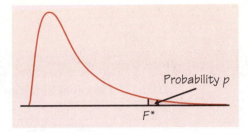

Table entry for p is the critical value F^* with probability p lying to its right.

TABLE D F distribution critical values (*continued*)

		Degrees of freedom in the numerator							
	p	9	10	15	20	30	60	120	1000
1	.100	59.86	60.19	61.22	61.74	62.26	62.79	63.06	63.30
	.050	240.54	241.88	245.95	248.01	250.10	252.20	253.25	254.19
	.025	963.28	968.63	984.87	993.10	1001.4	1009.8	1014	1017.7
	.010	6022.5	6055.8	6157.3	6208.7	6260.6	6313	6339.4	6362.7
	.001	602284	605621	615764	620908	626099	631337	633972	636301
2	.100	9.38	9.39	9.42	9.44	9.46	9.47	9.48	9.49
	.050	19.38	19.40	19.43	19.45	19.46	19.48	19.49	19.49
	.025	39.39	39.40	39.43	39.45	39.46	39.48	39.49	39.50
	.010	99.39	99.40	99.43	99.45	99.47	99.48	99.49	99.50
	.001	999.39	999.40	999.43	999.45	999.47	999.48	999.49	999.50
3	.100	5.24	5.23	5.20	5.18	5.17	5.15	5.14	5.13
	.050	8.81	8.79	8.70	8.66	8.62	8.57	8.55	8.53
	.025	14.47	14.42	14.25	14.17	14.08	13.99	13.95	13.91
	.010	27.35	27.23	26.87	26.69	26.50	26.32	26.22	26.14
	.001	129.86	129.25	127.37	126.42	125.45	124.47	123.97	123.53
4	.100	3.94	3.92	3.87	3.84	3.82	3.79	3.78	3.76
	.050	6.00	5.96	5.86	5.80	5.75	5.69	5.66	5.63
	.025	8.90	8.84	8.66	8.56	8.46	8.36	8.31	8.26
	.010	14.66	14.55	14.20	14.02	13.84	13.65	13.56	13.47
	.001	48.47	48.05	46.76	46.10	45.43	44.75	44.40	44.09
5	.100	3.32	3.30	3.24	3.21	3.17	3.14	3.12	3.11
	.050	4.77	4.74	4.62	4.56	4.50	4.43	4.40	4.37
	.025	6.68	6.62	6.43	6.33	6.23	6.12	6.07	6.02
	.010	10.16	10.05	9.72	9.55	9.38	9.20	9.11	9.03
	.001	27.24	26.92	25.91	25.39	24.87	24.33	24.06	23.82
6	.100	2.96	2.94	2.87	2.84	2.80	2.76	2.74	2.72
	.050	4.10	4.06	3.94	3.87	3.81	3.74	3.70	3.67
	.025	5.52	5.46	5.27	5.17	5.07	4.96	4.90	4.86
	.010	7.98	7.87	7.56	7.40	7.23	7.06	6.97	6.89
	.001	18.69	18.41	17.56	17.12	16.67	16.21	15.98	15.77
7	.100	2.72	2.70	2.63	2.59	2.56	2.51	2.49	2.47
	.050	3.68	3.64	3.51	3.44	3.38	3.30	3.27	3.23
	.025	4.82	4.76	4.57	4.47	4.36	4.25	4.20	4.15
	.010	6.72	6.62	6.31	6.16	5.99	5.82	5.74	5.66
	.001	14.33	14.08	13.32	12.93	12.53	12.12	11.91	11.72
8	.100	2.56	2.54	2.46	2.42	2.38	2.34	2.32	2.30
	.050	3.39	3.35	3.22	3.15	3.08	3.01	2.97	2.93
	.025	4.36	4.30	4.10	4.00	3.89	3.78	3.73	3.68
	.010	5.91	5.81	5.52	5.36	5.20	5.03	4.95	4.87
	.001	11.77	11.54	10.84	10.48	10.11	9.73	9.53	9.36

Degrees of freedom in the denominator

TABLE D F distribution critical values (continued)

	p	1	2	3	4	5	6	7	8
9	.100	3.36	3.01	2.81	2.69	2.61	2.55	2.51	2.47
	.050	5.12	4.26	3.86	3.63	3.48	3.37	3.29	3.23
	.025	7.21	5.71	5.08	4.72	4.48	4.32	4.20	4.10
	.010	10.56	8.02	6.99	6.42	6.06	5.80	5.61	5.47
	.001	22.86	16.39	13.90	12.56	11.71	11.13	10.70	10.37
10	.100	3.29	2.92	2.73	2.61	2.52	2.46	2.41	2.38
	.050	4.96	4.10	3.71	3.48	3.33	3.22	3.14	3.07
	.025	6.94	5.46	4.83	4.47	4.24	4.07	3.95	3.85
	.010	10.04	7.56	6.55	5.99	5.64	5.39	5.20	5.06
	.001	21.04	14.91	12.55	11.28	10.48	9.93	9.52	9.20
12	.100	3.18	2.81	2.61	2.48	2.39	2.33	2.28	2.24
	.050	4.75	3.89	3.49	3.26	3.11	3.00	2.91	2.85
	.025	6.55	5.10	4.47	4.12	3.89	3.73	3.61	3.51
	.010	9.33	6.93	5.95	5.41	5.06	4.82	4.64	4.50
	.001	18.64	12.97	10.80	9.63	8.89	8.38	8.00	7.71
15	.100	3.07	2.70	2.49	2.36	2.27	2.21	2.16	2.12
	.050	4.54	3.68	3.29	3.06	2.90	2.79	2.71	2.64
	.025	6.20	4.77	4.15	3.80	3.58	3.41	3.29	3.20
	.010	8.68	6.36	5.42	4.89	4.56	4.32	4.14	4.00
	.001	16.59	11.34	9.34	8.25	7.57	7.09	6.74	6.47
20	.100	2.97	2.59	2.38	2.25	2.16	2.09	2.04	2.00
	.050	4.35	3.49	3.10	2.87	2.71	2.60	2.51	2.45
	.025	5.87	4.46	3.86	3.51	3.29	3.13	3.01	2.91
	.010	8.10	5.85	4.94	4.43	4.10	3.87	3.70	3.56
	.001	14.82	9.95	8.10	7.10	6.46	6.02	5.69	5.44
25	.100	2.92	2.53	2.32	2.18	2.09	2.02	1.97	1.93
	.050	4.24	3.39	2.99	2.76	2.60	2.49	2.40	2.34
	.025	5.69	4.29	3.69	3.35	3.13	2.97	2.85	2.75
	.010	7.77	5.57	4.68	4.18	3.85	3.63	3.46	3.32
	.001	13.88	9.22	7.45	6.49	5.89	5.46	5.15	4.91
50	.100	2.81	2.41	2.20	2.06	1.97	1.90	1.84	1.80
	.050	4.03	3.18	2.79	2.56	2.40	2.29	2.20	2.13
	.025	5.34	3.97	3.39	3.05	2.83	2.67	2.55	2.46
	.010	7.17	5.06	4.20	3.72	3.41	3.19	3.02	2.89
	.001	12.22	7.96	6.34	5.46	4.90	4.51	4.22	4.00
100	.100	2.76	2.36	2.14	2.00	1.91	1.83	1.78	1.73
	.050	3.94	3.09	2.70	2.46	2.31	2.19	2.10	2.03
	.025	5.18	3.83	3.25	2.92	2.70	2.54	2.42	2.32
	.010	6.90	4.82	3.98	3.51	3.21	2.99	2.82	2.69
	.001	11.50	7.41	5.86	5.02	4.48	4.11	3.83	3.61
200	.100	2.73	2.33	2.11	1.97	1.88	1.80	1.75	1.70
	.050	3.89	3.04	2.65	2.42	2.26	2.14	2.06	1.98
	.025	5.10	3.76	3.18	2.85	2.63	2.47	2.35	2.26
	.010	6.76	4.71	3.88	3.41	3.11	2.89	2.73	2.60
	.001	11.15	7.15	5.63	4.81	4.29	3.92	3.65	3.43
1000	.100	2.71	2.31	2.09	1.95	1.85	1.78	1.72	1.68
	.050	3.85	3.00	2.61	2.38	2.22	2.11	2.02	1.95
	.025	5.04	3.70	3.13	2.80	2.58	2.42	2.30	2.20
	.010	6.66	4.63	3.80	3.34	3.04	2.82	2.66	2.53
	.001	10.89	6.96	5.46	4.65	4.14	3.78	3.51	3.30

Degrees of freedom in the numerator

Degrees of freedom in the denominator

TABLE D   *F* distribution critical values (*continued*)

			Degrees of freedom in the numerator							
		p	9	10	15	20	30	60	120	1000
Degrees of freedom in the denominator	9	.100	2.44	2.42	2.34	2.30	2.25	2.21	2.18	2.16
		.050	3.18	3.14	3.01	2.94	2.86	2.79	2.75	2.71
		.025	4.03	3.96	3.77	3.67	3.56	3.45	3.39	3.34
		.010	5.35	5.26	4.96	4.81	4.65	4.48	4.40	4.32
		.001	10.11	9.89	9.24	8.90	8.55	8.19	8.00	7.84
	10	.100	2.35	2.32	2.24	2.20	2.16	2.11	2.08	2.06
		.050	3.02	2.98	2.85	2.77	2.70	2.62	2.58	2.54
		.025	3.78	3.72	3.52	3.42	3.31	3.20	3.14	3.09
		.010	4.94	4.85	4.56	4.41	4.25	4.08	4.00	3.92
		.001	8.96	8.75	8.13	7.80	7.47	7.12	6.94	6.78
	12	.100	2.21	2.19	2.10	2.06	2.01	1.96	1.93	1.91
		.050	2.80	2.75	2.62	2.54	2.47	2.38	2.34	2.30
		.025	3.44	3.37	3.18	3.07	2.96	2.85	2.79	2.73
		.010	4.39	4.30	4.01	3.86	3.70	3.54	3.45	3.37
		.001	7.48	7.29	6.71	6.40	6.09	5.76	5.59	5.44
	15	.100	2.09	2.06	1.97	1.92	1.87	1.82	1.79	1.76
		.050	2.59	2.54	2.40	2.33	2.25	2.16	2.11	2.07
		.025	3.12	3.06	2.86	2.76	2.64	2.52	2.46	2.40
		.010	3.89	3.80	3.52	3.37	3.21	3.05	2.96	2.88
		.001	6.26	6.08	5.54	5.25	4.95	4.64	4.47	4.33
	20	.100	1.96	1.94	1.84	1.79	1.74	1.68	1.64	1.61
		.050	2.39	2.35	2.20	2.12	2.04	1.95	1.90	1.85
		.025	2.84	2.77	2.57	2.46	2.35	2.22	2.16	2.09
		.010	3.46	3.37	3.09	2.94	2.78	2.61	2.52	2.43
		.001	5.24	5.08	4.56	4.29	4.00	3.70	3.54	3.40
	25	.100	1.89	1.87	1.77	1.72	1.66	1.59	1.56	1.52
		.050	2.28	2.2	2.09	2.01	1.92	1.82	1.77	1.72
		.025	2.68	2.61	2.41	2.30	2.18	2.05	1.98	1.91
		.010	3.22	3.13	2.85	2.70	2.54	2.36	2.27	2.18
		.001	4.71	4.56	4.06	3.79	3.52	3.22	3.06	2.91
	50	.100	1.76	1.73	1.63	1.57	1.50	1.42	1.38	1.33
		.050	2.07	2.03	1.87	1.78	1.69	1.58	1.51	1.45
		.025	2.38	2.32	2.11	1.99	1.87	1.72	1.64	1.56
		.010	2.78	2.70	2.42	2.27	2.10	1.91	1.80	1.70
		.001	3.82	3.67	3.20	2.95	2.68	2.38	2.21	2.05
	100	.100	1.69	1.66	1.56	1.49	1.42	1.34	1.28	1.22
		.050	1.97	1.93	1.77	1.68	1.57	1.45	1.38	1.30
		.025	2.24	2.18	1.97	1.85	1.71	1.56	1.46	1.36
		.010	2.59	2.50	2.22	2.07	1.89	1.69	1.57	1.45
		.001	3.44	3.30	2.84	2.59	2.32	2.01	1.83	1.64
	200	.100	1.66	1.63	1.52	1.46	1.38	1.29	1.23	1.16
		.050	1.93	1.88	1.72	1.62	1.52	1.39	1.30	1.21
		.025	2.18	2.11	1.90	1.78	1.64	1.47	1.37	1.25
		.010	2.50	2.41	2.13	1.97	1.79	1.58	1.45	1.30
		.001	3.26	3.12	2.67	2.42	2.15	1.83	1.64	1.43
	1000	.100	1.64	1.61	1.49	1.43	1.35	1.25	1.18	1.08
		.050	1.89	1.84	1.68	1.58	1.47	1.33	1.24	1.11
		.025	2.13	2.06	1.85	1.72	1.58	1.41	1.29	1.13
		.010	2.43	2.34	2.06	1.90	1.72	1.50	1.35	1.16
		.001	13.13	2.99	2.54	2.30	2.02	1.69	1.49	1.22

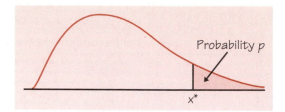

Table entry for p is the critical value x^* with probability p lying to its right.

TABLE E	Chi-square distribution critical values

df	.25	.20	.15	.10	.05	.025	.02	.01	.005	.0025	.001	.0005
1	1.32	1.64	2.07	2.71	3.84	5.02	5.41	6.63	7.88	9.14	10.83	12.12
2	2.77	3.22	3.79	4.61	5.99	7.38	7.82	9.21	10.60	11.98	13.82	15.20
3	4.11	4.64	5.32	6.25	7.81	9.35	9.84	11.34	12.84	14.32	16.27	17.73
4	5.39	5.99	6.74	7.78	9.49	11.14	11.67	13.28	14.86	16.42	18.47	20.00
5	6.63	7.29	8.12	9.24	11.07	12.83	13.39	15.09	16.75	18.39	20.51	22.11
6	7.84	8.56	9.45	10.64	12.59	14.45	15.03	16.81	18.55	20.25	22.46	24.10
7	9.04	9.80	10.75	12.02	14.07	16.01	16.62	18.48	20.28	22.04	24.32	26.02
8	10.22	11.03	12.03	13.36	15.51	17.53	18.17	20.09	21.95	23.77	26.12	27.87
9	11.39	12.24	13.29	14.68	16.92	19.02	19.68	21.67	23.59	25.46	27.88	29.67
10	12.55	13.44	14.53	15.99	18.31	20.48	21.16	23.21	25.19	27.11	29.59	31.42
11	13.70	14.63	15.77	17.28	19.68	21.92	22.62	24.72	26.76	28.73	31.26	33.14
12	14.85	15.81	16.99	18.55	21.03	23.34	24.05	26.22	28.30	30.32	32.91	34.82
13	15.98	16.98	18.20	19.81	22.36	24.74	25.47	27.69	29.82	31.88	34.53	36.48
14	17.12	18.15	19.41	21.06	23.68	26.12	26.87	29.14	31.32	33.43	36.12	38.11
15	18.25	19.31	20.60	22.31	25.00	27.49	28.26	30.58	32.80	34.95	37.70	39.72
16	19.37	20.47	21.79	23.54	26.30	28.85	29.63	32.00	34.27	36.46	39.25	41.31
17	20.49	21.61	22.98	24.77	27.59	30.19	31.00	33.41	35.72	37.95	40.79	42.88
18	21.60	22.76	24.16	25.99	28.87	31.53	32.35	34.81	37.16	39.42	42.31	44.43
19	22.72	23.90	25.33	27.20	30.14	32.85	33.69	36.19	38.58	40.88	43.82	45.97
20	23.83	25.04	26.50	28.41	31.41	34.17	35.02	37.57	40.00	42.34	45.31	47.50
21	24.93	26.17	27.66	29.62	32.67	35.48	36.34	38.93	41.40	43.78	46.80	49.01
22	26.04	27.30	28.82	30.81	33.92	36.78	37.66	40.29	42.80	45.20	48.27	50.51
23	27.14	28.43	29.98	32.01	35.17	38.08	38.97	41.64	44.18	46.62	49.73	52.00
24	28.24	29.55	31.13	33.20	36.42	39.36	40.27	42.98	45.56	48.03	51.18	53.48
25	29.34	30.68	32.28	34.38	37.65	40.65	41.57	44.31	46.93	49.44	52.62	54.95
26	30.43	31.79	33.43	35.56	38.89	41.92	42.86	45.64	48.29	50.83	54.05	56.41
27	31.53	32.91	34.57	36.74	40.11	43.19	44.14	46.96	49.64	52.22	55.48	57.86
28	32.62	34.03	35.71	37.92	41.34	44.46	45.42	48.28	50.99	53.59	56.89	59.30
29	33.71	35.14	36.85	39.09	42.56	45.72	46.69	49.59	52.34	54.97	58.30	60.73
30	34.80	36.25	37.99	40.26	43.77	46.98	47.96	50.89	53.67	56.33	59.70	62.16
40	45.62	47.27	49.24	51.81	55.76	59.34	60.44	63.69	66.77	69.70	73.40	76.09
50	56.33	58.16	60.35	63.17	67.50	71.42	72.61	76.15	79.49	82.66	86.66	89.56
60	66.98	68.97	71.34	74.40	79.08	83.30	84.58	88.38	91.95	95.34	99.61	102.7
80	88.13	90.41	93.11	96.58	101.9	106.6	108.1	112.3	116.3	120.1	124.8	128.3
100	109.1	111.7	114.7	118.5	124.3	129.6	131.1	135.8	140.2	144.3	149.4	153.2

Solutions to Selected Exercises

CHAPTER 1

1.1 "Region" is categorical; all others are quantitative.

1.3 **(a)** Roughly symmetric, though it might be viewed as SLIGHTLY skewed to the right.

   (b) About 15%.

   (c) Smallest: between -70% and -60%; largest: between 100% and 110%.

   (d) 23%.

1.5 Lightning: centered at noon (or "somewhere from 11:30 to 12:30"). Spread: 7 to 17 (or "6:30am to 5:30pm"). Shakespeare: centered at 4, spread from 1 to 12.

1.7 Outlier: 200. Center: between 137 and 140. Spread (ignoring outlier): 101 to 178.

1.9 **(b)** Prices increased throughout the entire period.

   (c) Prices rose most quickly from 1979–80, and most slowly from about 1970–73, with a similar rate of increase in 1985–86.

1.11 Slightly skewed to the right, centered at 4.

1.13 The distribution is skewed to the left, and centered at 46. 60 is *not* an outlier.

1.15 **(a)** Splitting stems is a good idea in this situation. The plot shows the distribution to be fairly symmetric, with a high outlier of 4.7 (4.69).

   (b) The time plot shows no clear trend.

1.19 Splitting stems is a good idea here. The distribution seems to be skewed to the right.

1.21 (a) $\bar{x} = 141.058$.

(b) Without the outlier, $\bar{x}^* = 137.588$. The outlier makes the mean larger.

1.23 $M = 138.5$. The median is smaller than the mean (141.058), because with the outlier included, the distribution is skewed to the right.

1.25 The median is the smaller number ($490,000)—the distribution is skewed to the right, which increases the mean but not the median.

1.27 (a) There is no particular skewness, so M should be about the same as $\bar{x}$.

(b) Five-number summary: 42, 51, 55, 58, 69; $\bar{x} = 54.833$

(c) Between Q_1 and Q_3: 51 to 58.

1.29 (a) $\bar{x} = 5.4$.

(b) $\sum (x_i - \bar{x})^2 = 2.06$; $s^2 = 0.412$; $s = 0.6419$.

1.31 The stemplot reveals two peaks—one around 470 and one around 520—with a valley in between. The mean and median fall between these two peaks.

1.33 There seems to be little difference between beef and meat hot dogs, but poultry hot dogs are generally lower in calories than the other two.

1.35 A stemplot (or histogram) shows the distribution to be fairly symmetrical, with a low outlier of 4.88—$\bar{x}$ and s should be reasonable in this setting. $\bar{x} = 5.4479$ and $s = 0.22095$; $\bar{x}$ is our best estimate of the earth's density.

1.37 The distribution is clearly skewed to the right, with the top two or three salaries as outliers. Use the five-number summary: 109, 158, 635, 2300, 6200.

1.39 (a) Mean.

(b) Median.

1.41 (a) 1, 1, 1, 1. (Or 2, 2, 2, and 2, etc.)

(b) 0, 0, 10, 10.

(c) For (a), any four identical numbers will have $s = 0$. The answer for (b) is unique.

1.43 (a) 20%.

(b) 60%.

(c) 50%.

1.45 Peak of curve should be at 69, with marks at 66.5 and 71.5.

1.47 (a) 50%.

(b) 2.5%.

(c) 110 ± 50, or 60 to 160.

1.49 **(a)** 0.9978.

(b) $1 - 0.9978 = 0.0022$.

(c) $1 - 0.0485 = 0.9515$.

(d) $0.9978 - 0.0485 = 0.9493$.

1.51 **(a)** About 11.51%.

(b) About 88.5%.

(c) About 72.2 inches.

1.53 $\bar{x} = 5.4479$ and $s = 0.22095$. About 75.8% (22 out of 29) fall inside $\bar{x} \pm s$, 96.6% (28/29) fall inside $\bar{x} \pm 2s$, and all fall inside $\bar{x} \pm 3s$.

1.55 **(a)** About 16%.

(b) About 68%.

1.57 Cobb: $z = 4.15$; Williams: $z = 4.26$; Brett: $z = 4.07$. Williams z-score is highest.

1.59 **(a)** About 0.84.

(b) About 0.385.

1.61 **(a)** About 5.2%.

(b) About 55%.

(c) Approximately 279 days or longer.

1.63 **(a)** At about ± 1.28.

(b) 64.5 ± 3.2, or 61.3 and 67.7.

1.65 **(a)** Stemplot is symmetric with no *obvious* outliers (although 10.17 and 9.75 seem to be unusually high, and 6.75 is extraordinarily low).

(b) Outliers show up more clearly in time plot.

(c) $\bar{x} = 8.3628$ and $s = 0.4645$.

(d) Within $\bar{x} \pm s$: 25 (64.1%); within $\bar{x} \pm 2s$: 37 (94.9%); 100% fall within $\bar{x} \pm 3s$.

1.67 DiMaggio: 12, 20.5, 30, 32, 46. Mantle: 13, 21 (or 20.5), 28.5, 37 (or 37.75), 54. The first three numbers in both summaries are similar, but Mantle's Q_3 and maximum are higher—he apparently had higher "big seasons" than DiMaggio.

1.69 **(a)** Distribution is clearly skewed to the right (as expected). Main peak occurs from 50 to 150—the guinea pigs which lived over 500 days are apparent outliers.

(b) The skewness makes the mean larger than the median.

(c) 43, 82.25, 102.5, 153.75, 598. The skewness shows up in the difference between Q_3 and the maximum; it is much larger than the other differences between successive numbers.

1.71 $IQR = 11.40$; this guideline gives the interval -20.05 to 25.55, which would make the nine "Low" and "High" numbers outliers.

1.73 No apparent trend supports either Julie's or John's position; however, one might observe a trend toward having a greater age *spread* in recent years.

1.77 Below 1.7: about 0.052; between 1.7 and 2.1: about 0.078.

1.79 At least 3.42; less than 2.58.

CHAPTER 2

2.1 Age 6 height is explanatory, and age 16 height is the response. Both are quantitative.

2.3 "Treatment"—old or new—is the (categorical) explanatory variable. Survival time is the (quantitative) response variable.

2.5 **(a)** Positive association.
  (b) Linear.
  (c) The relationship is fairly strong, allowing for reasonably accurate prediction. About 50 manatee deaths per year would arise from 700,000 boat registrations.

2.7 **(a)** Body mass is the explanatory variable.
  (b) Positive association, linear, moderately strong.
  (c) The description of the male subjects' plot is much the same, though the scatter appears to be greater. The males typically have larger values for both variables.

2.9 **(a)** Lowest: about 107 calories (with about 145 mg of sodium); highest: about 195 calories, with about 510 mg of sodium.
  (b) There is a positive association; high-calorie hot dogs tend to be high in salt, and low-calorie hot dogs tend to have low sodium.
  (c) The lower left point is an outlier. Ignoring this point, the remaining points seem to fall roughly on a line. The relationship is moderately strong.

2.11 Either variable could go on the vertical axis. The plot shows a strong positive linear relationship, with no outliers; there appears to be only one species represented.

2.13 **(a)** There should be a positive association because money spent for teacher salaries is part of the education budget; more money spent per pupil would typically translate to more money spent overall.
  (c) The plot shows a positive, approximately linear relationship.
  (d) California, spending $4826 per pupil, with median teacher salary $39,600.

(e) The mountain states are clustered down in the lower left: they spend lower-than-average amounts per student, and have low median teacher salaries.

2.15 (a) The means are (in the order given) 47.167, 15.667, 31.5, and 14.833.

(b) Yellow seems to be the most attractive, and green is second. White and blue are poor attractors.

(c) Positive or negative association make no sense here because color is a categorical variable (what is an "above-average" color?).

2.17 (a) With x as femur length and y as humerus length: $\bar{x} = 58.2$, $s_x = 13.20$; $\bar{y} = 66.0$, $s_y = 15.89$; $r = 0.994$.

2.19 Positive but not near 1 (there is positive association but a fair amount of scatter).

2.21 (a) The plot shows a strong positive linear relationship, with little scatter, so we expect that r is close to 1.

(b) r would not change—it is computed from unitless standardized values.

2.23 (a) Both correlations should be positive; the men's may be slightly smaller since those points are more spread out.

(b) Women: $r_w = 0.87645$; Men: $r_m = 0.59207$.

(c) Women: $\bar{x}_w = 43.03$; Men: $\bar{x}_m = 53.10$. The difference in means has no effect on the correlation.

(d) There would be no change, since standardized measurements are dimensionless.

2.25 (a) These points are close to the x-axis, located on the far left and right.

(b) The transformed points are in the middle horizontally; three at the top and three at the bottom.

(c) $r = 0.25310$ for both sets of data.

2.27 $r = -0.17162$—close to zero because the relationship is a curve rather than a line.

2.29 (a) $a = 1.0892$ and $b = 0.1890$, as given.

(b) $\bar{x} = 22.31$, $s_x = 17.74$; $\bar{y} = 5.306$, $s_y = 3.368$; $r = 0.99526$. Except for roundoff error, we again find $b = 0.1890$ and $a = 1.0892$.

2.31 (b) The slope is close to 1—meaning that the strength after 28 days is *approximately* (strength after one week) plus 1389 psi. In other words, we expect the extra three weeks to add about 1400 psi of strength to the concrete.

(c) 4557 psi.

2.33 **(a)** Put speed on the horizontal axis.

(b) There is a very strong positive linear relationship; $r = 0.9990$.

(c) $\hat{y} = 1.76608 + 0.080284x$.

(d) $r^2 = 0.998$, so nearly all (99.8%) the variation in steps per second is explained by the linear relationship.

(e) The regression line would be different; the line in (c) minimizes the sum of the squared *vertical* distances on the graph, while this new regression would minimize the squared *horizontal* distances. r^2 would remain the same, however.

2.35 **(b)** The line is clearly *not* a good predictor of the actual data—it is too high in the middle and too low on each end.

(c) The sum is -0.01.

2.37 **(a)** Without Child 19, $\hat{y}^* = 109.305 - 1.1933x$. Child 19 is not very influential, since removing this data point does not change the line substantially.

(b) With all children, $r^2 = 0.410$; without Child 19, $r^2 = 0.572$. With Child 19's high Gesell score removed, there is less scatter around the regression line—more of the variation is explained by the regression.

2.39 **(a)** y (weight) $= 100 + 40x$ grams.

(c) When $x = 104$, $y = 4260$ grams, or about 9.4 pounds. The regression line is only reliable for "young" rats; rats do not grow at a constant rate throughout their entire life.

2.41 **(b)** $\hat{y} = 71.950 + 0.38333x$.

(c) When $x = 40$ months, $\hat{y} = 87.2832$; when $x = 60$, $\hat{y} = 94.9498$.

(d) Sarah's rate: 0.38 cm/month; normal rate: 0.5 cm/month $(0.5 = 6/(60 - 48))$.

2.43 When $x = 480$, $\hat{y} = 255.95$ cm $= 100.77$ in $= 8.4$ feet!

2.45 **(a)** About 69.4% of the variation is explained ($r^2 = 0.694$).

(b) The sum is zero.

(c) The residuals change from negative to positive in 1991. In that year, the regression line changes from overestimating to underestimating.

2.47 **(a)** $\hat{y} = 1166.93 - 0.58679x$.

(b) The farm population decreased about 590 thousand (0.59 million) people per year. The regression line explains 97.7% of the variation.

(c) $-782, 100$—but a population must be greater than or equal to 0.

2.49 The correlation would be smaller because there is much more variation among the individual data points.

2.51 Age is the lurking variable: we would expect both quantities— shoe size and reading level—to increase as a child ages.

2.53 (a) $r^2 = 0.925$—more than 90% of the variation in one SAT score can be explained through a linear relationship with the other score.

(b) The correlation would be much smaller, since individual students have much more variation between their scores. By averaging—or, as in this case, taking the median of—the scores of large groups of students, we muffle the effects of these individual variations.

2.55 The explanatory variable is whether or not a student has taken at least two years of foreign language, and the score on the test is the response. The lurking variable is the students' English skills *before* taking (or not taking) the foreign language: students who have a good command of English early in their high school career are more likely to choose (or be advised to choose) to take a foreign language.

2.57 Apparently drivers are typically larger and heavier men than conductors—and are therefore more predisposed to health problems such as heart disease.

2.59 27.0%, 24.4%, 16.2%, 13.5%, and 19.0% (total is 100.1% due to rounding).

2.61 13.9%, 12.3%, 18.8%, 28.1%, and 42.2%. The percentage of people who did not finish high school increases fairly steadily with age.

2.63 Among 35 to 44 year-olds: 12.3% never finished high school, 37.5% finished high school, 22.7% had some college, and 27.5% completed college. This is more like the 25–34 age group than the 65 and over group.

2.65 Two possible answers: Row 1–30, 20; Row 2–30, 20; and Row 1–10, 40; Row 2–50, 0.

2.67 (a) Of students with two smoking parents, 22.5% smoke; with one smoking parent, 18.6% smoke; among students with no smoking parents, only 13.9% smoke.

(c) It appears that children of smokers are more likely to smoke—even more so when both parents are smokers.

2.69 (a) White defendants: 19 got the death penalty; 141 did not. For black defendants, the numbers are 17 vs. 149.

(b) Overall, 11.9% of white defendants and 10.2% of black defendants get the death penalty. However, for white victims, the percentages are 12.6% and 17.5% (respectively); when the victim is black, they are 0% and 5.8%.

(c) In cases involving white victims, 14% of defendants got the death penalty; when the victim was black, only 5.4% of defendants were sentenced to death. White defendants killed whites 94.3% of the time—but are less likely to get the death penalty than blacks who killed whites.

2.71 Apparently women are more likely to be in fields which pay less overall (to both men and women).

2.73 **(a)** 704,000.

(b) 2,065,000.

(c) Roundoff error.

2.75 **(a)** 34.1%.

(b) 36.9%.

2.77 **(a)** 7.2%.

(b) 10.1% of restrained children were injured vs. 15.4% of unrestrained children.

2.79 **(a)** 58.3% of desipramine users did not have a relapse, while 25.0% of lithium users and 16.7% of those who received placebos succeeded in breaking their addictions.

(b) Since the addicts were assigned randomly to their treatments, causation *seems* to be indicated.

2.81 **(b)** There is no clear relationship.

(c) Generally, those with short incubation periods are more likely to die.

(d) Person 6—the 17-year-old with a short incubation (20 hours) who survived—merits extra attention. He or she is also the youngest in the group by far. Among the other survivors, one (person 17) had an incubation period of 28 hours, and the rest had incubation periods of 43 hours or more.

2.83 Both plots show a strong negative association—the logarithmic transformation is perhaps *slightly* more linear.

2.85 **(a)** The straight-line relationship is moderately strong; $r^2 = 88.6\%$.

(b) When $x = 700$ (thousand), $\hat{y} = 46$.

(d) -2.26 lies more than 2, but less than 3, standard deviations from the mean (0). This means that it is among the 5% most extreme values.

2.87 In the aspirin group, 1.26% had heart attacks (0.09% were fatal, 1.17% were not), and 1.08% had strokes. Among those who took placebos, 2.17% had heart attacks (0.24% fatal, 1.93% non-fatal), and 0.89% suffered from strokes. Based on these numbers, it appears that the aspirin group had a slight advantage in heart attacks, but was possibly worse in incidence of stroke. This was an experiment, so a cause-and-effect relationship seems to be indicated—*but* this might not apply outside of the population (healthy male doctors over 40) that was sampled.

2.89 **(a)** The plot shows a positive linear association between the two variables—meaning, for example, that cars with above-average city mileage also do well on the highway. There is some evidence that the points for large cars appear higher on the graph, indicating that for a given city mileage, large cars tend to have a slightly higher highway mileage.

(b) The Rolls Royce Silver Spur has unusually low mileage (city and highway).

(c) The relationship is fairly strong for both car types: for mid-size cars, $r = 0.80475$; for large cars, $r = 0.84261$.

(d) For midsize cars: $\hat{y} = 7.28248 + 0.98263x$. For large cars: $\hat{y} = 2.21774 + 1.32258x$. Over the part of the graph where most of the points are found, the regression line for large cars lies above the line for midsize cars, which agrees with the impression mentioned in (a).

(e) $\hat{y} = 26.9$ mile per gallon for a midsize car; $\hat{y} = 28.7$ for a large car.

CHAPTER 3

3.1 The (desired) population is employed adult women, the sample is the 48 club members who returned the survey.

3.3 Only persons with an strong opinion on the subject—strong enough that they are willing to spend the time, and at least 50 cents—will respond to this advertisement.

3.5 Starting with 01 and numbering down the columns, one chooses 04–Bonds, 10–Fleming, 17–Liao, 19–Naber, 12–Goel, and 13–Gomez.

3.7 Label from 001 to 440; select 400, 077, 172, 417, 350, 131, 211, 273, 208, and 074.

3.9 Label midsize accounts 001–500, and small accounts from 0001–4400. Select 417, 494, 322, 247, and 097 for the midsize group, and then 3698, 1452, 2605, 2480, and 3716 for the small group.

3.11 The higher no-answer was probably the second period—more families are likely to be gone for vacations, etc. Nonresponse of this type might underrepresent those who are more affluent (and are able to travel).

3.13 The increased sample size gives more accurate information about the population.

3.15 (a) An individual is a small business; the specific population is "eating and drinking establishments" in the large city.
(b) An individual is an adult; the Congressman's constituents are the *desired* population. The letter-writers are a voluntary sample and do not represent that population well.
(c) Individual: auto insurance claim; the population is all claims filed in a given month.

3.17 The call-in poll is faulty in part because it is a voluntary sample. Furthermore, even a small charge like 50 cents can discourage some people from calling in—especially poor people. Reagan's Republican policies appealed to upper-class voters, who would be less concerned about a 50-cent charge than lower-class voters who might favor Carter.

3.19 Number the bottles across the rows from 01 to 25, then select 12–B0986, 04–A1101, and 11–A2220. (If numbering is done down columns instead, the sample will be A1117, B1102, and A1098.)

3.21 (a) False—if it were true, then after looking at 39 digits, we would know whether or not the 40th digit was a 0, contrary to property 2.
(b) True—there are 100 pairs of digits 00 through 99, and all are equally likely.
(c) False—0000 is just as likely as any other string of four digits.

3.23 It is *not* an SRS, because some samples of size 250 have no chance of being selected (e.g., a sample containing 250 women).

3.25 A smaller sample gives less information about the population. "Men" constituted only about one-third of our sample, so we know less about that group than we know about all adults.

3.27 (a) The liners are the experimental units.
(b) Factor: heat applied to the liners; levels: 250°F, 275°F, 300°F, and 325°F.
(c) The force required to open the package is the response variable.

3.29 **(a)** The experimental units are the batches of the product; the yield of each batch is the response variable.

(b) There are two factors: temperature (with 2 levels) and stirring rates (with 3 levels), for a total of 6 treatments. Assign 50°C and 60 rpm to treatment group 1, etc.

(c) Since two experimental units will be used for each treatment, we need 12.

3.31 The diagram should show random allocation of 5 liner pairs to each of 4 groups, then applying the appropriate level of heat to each group, then measuring the force required.

3.33 Number the liners from 01 to 20. Group 1: 16, 04, 19, 07, and 10; Group 2: 13, 15, 05, 09, and 08; Group 3: 18, 03, 01, 06, and 11. The others are in Group 4.

3.35 The possible differences between the two years would confound the effects of the treatments. For example, if this summer is warmer, the customers may run their air conditioners more often.

3.37 There almost certainly was *some* difference between the sexes and between blacks and whites; the difference between men and women was so large that it is unlikely to be due to chance. For black and white students, however, the difference was small enough that it could be attributed to random variation.

3.39 **(a)** If only the new drug is administered, and the subjects are then interviewed, their responses will not be useful, because there will be nothing to compare them to: How much "pain relief" does one expect to experience?

(b) Randomly assign 20 patients to each of three groups: Group 1, the placebo group; Group 2, the aspirin group; and Group 3, which will receive the new medication. After treating the patients, ask them how much pain relief they feel, and then compare the average pain relief experienced by each group.

(c) The subjects should certainly not know what drug they are getting—a patient told that she is receiving a placebo, for example, will probably not expect any pain relief.

(d) Yes—presumably, the researchers would like to conclude that the new medication is better than aspirin. If it is not double-blind, the interviewers may subtly influence the subjects into giving responses that support that conclusion.

3.41 **(a)** Split the subjects by gender. Randomly allocate the 24 women into 6 groups of 4 each, and allocate the men into 6 groups of 2. Each group receives the appropriate treatment (1–6), and then the subjects' attitudes about the product, etc., are measured.

(b) Number the women from 01 to 24, and the men from 01 to 12. First we look for 20 women's numbers, and find Group 1: 12, 13, 04, 18; Group 2: 19, 24, 23, 16; Group 3: 02, 08, 17, 21; Group 4: 10, 05, 09, 06; Group 5: 01, 20, 22, and 07. The remaining women are group 6. For the men, we find Group 1: 05, 09; Group 2: 07, 02; Group 3: 01, 08; Group 4: 11, 06; Group 5: 12, 04; the rest go into Group 6.

3.43 The randomization will vary with the starting line in Table B. *Completely randomized design*: Randomly assign 10 students to "Group 1" (which has the trend-highlighting software) and the other 10 to "Group 2" (which does not). Compare the performance of Group 1 with that of Group 2. *Matched pairs design*: Each student does the activity twice, once with the software and once without. Randomly decide (for each student) whether they have the software the first or second time. Compare performance with the software and without it. *Alternate matched pairs design*: Again, all students do the activity twice. Randomly assign 10 students to Group 1 and 10 to Group 2. Group 1 uses the software the first time; Group 2 uses the software the second time.

3.45 **(a)** For each subject, administer both tests in a randomly chosen order.

(b) Take 22 digits from Table B. If the first digit is even, subject 1 takes the BI first; if it is odd, he or she takes the ARSMA first.

3.47 **(a)** It is an experiment (albeit a poorly designed one), because a treatment (herbal tea) is imposed on the subjects.

(b) No, it is a study—the scores on the English test are merely observed for the various subjects.

3.49 **(a)** Subjects: physicians; factor: medication (with two levels—aspirin and placebo); response: observing health, specifically whether the subjects have heart attacks or not.

(b) The diagram should show random allocation into two groups of 11,000 each; one group receives aspirin, the other, the placebo. Then observe heart attacks.

3.51 **(a)** Randomly assign 20 men to each of two groups. Record each subject's blood pressure, then apply the treatments: a calcium supplement for Group 1, and a placebo for Group 2. After sufficient time has passed, measure blood pressure again and observe any change.

(b) Number from 01 to 40 down the columns. Group 1 is 18–Howard, 20–Imrani, 26–Maldonado, 35–Tompkins, 39–Willis, 16–Guillen, 04–Bikalis, 21–James, 19–Hruska, 37–Tullock, 29–O'Brian, 07–Cranston, 34–Solomon, 22–Kaplan, 10–Durr, 25–Liang, 13–Fratianna, 38–Underwood, 15–Green, and 05–Chen.

3.53 Responding to a placebo does not imply that the complaint was not "real"—38% of the placebo group in the gastric freezing experiment improved, and those patients really had ulcers. The placebo effect is a *psychological* response, but it may make an actual *physical* improvement in the patient's health.

3.55 (a) Explanatory: treatment method; response: survival times.
(b) No treatment is actively imposed; the women (or their doctors) chose which treatment to use.
(c) Doctors may make the decision of which treatment to recommend based in part on how advanced the case is—for example, some might be more likely to recommend the older treatment for advanced cases, in which case the chance of recovery is lower.

3.57 (a) Label the students from 0001 to 3478.
(b) Taking four digits at a time gives 2940, 0769, 1481, 2975, and 1315.

3.59 (a) One possible population: all full-time undergraduate students in the fall term on a list provided by the Registrar.
(b) A stratified sample with 125 students from each year is one possibility.
(c) Mailed questionnaires might have high non-response rates. Telephone interviews exclude those without phones, and may mean repeated calling for those that are not home. Face-to-face interviews might be more costly than your funding will allow.

3.61 (a) Diagram should show random allocation into 6 groups (2 batches each). Each group is given the appropriate treatment (temperature/stirring rate combination), then the yield is observed.
(b) The first 10 numbers (between 01 and 12) are 06, 09, 03, 05, 04, 07, 02, 08, 10, and 11. So the 6th and 9th batches will receive treatment 1; batches 3 and 5 will be processed with treatment 2, etc.

3.63 Each subject should taste both kinds of cheeseburger, in a randomly selected order, and then be asked about their preference. Both burgers should have the same "fixings" (ketchup, mustard, etc.). Since some subjects might be able to identify the cheeseburgers by appearance, one might need to take additional steps (such as blindfolding, or serving only the center part of the burger) in order to make this a true "blind" experiment.

3.65 It means that the correlation is large enough (presumably, though not necessarily, in the positive direction) that it is unlikely to have occurred by chance.

CHAPTER 4

4.1 7.2% is a statistic.

4.3 48% is a statistic; 52% is a parameter.

4.7 **(b)** The histogram actually does *not* appear to have a normal shape. The sampling distribution is quite normal in appearance, but even a sample of size 100 does not *necessarily* show it.
    (c) The mean of $\hat{p}$ is 0.0981. The bias seems to be small.
    (d) The mean of the sampling distribution should be $p = 0.10$.
    (e) The mean would still be 0.10, but the spread would be smaller.

4.9 **(a)** Since the smallest number of total tax returns (i.e., the smallest population) is still more than 100 times the sample size, the variability will be the (approximately) the same for all states.
    (b) Yes, it will change—states with larger populations will have larger samples, which will decrease the variability for those states.

4.11 There are *21* 0s among the first 200, so $\hat{p} = 0.105$.

4.13 **(a)** 0.1.
    (b) 0.5.

4.15 **(a)** Use digits 0 and 1 (or any other 2 of the 10 digits) to represent the presence of egg masses. Reading the first 10 digits from line 116, for example, gives YNNNN NNYNN—2 square yards with egg masses, 8 without—so $\hat{p} = 0.2$.
    (c) The mean would be $p = 0.2$.
    (d) 0.4.

4.17 Assuming that the poll's sample size was less than 780,000—10% of the population of New Jersey—the variability would be practically the same for either population.

4.19 **(a)** Most answers should fall between 0.3 and 0.7.

(b) The *exact* probability is $\frac{2}{3}$; most answers should fall between 0.47 and 0.87.

4.21 **(a)** 0.65.

(b) 0.38.

(c) 0.62.

4.23 $18/38 = 9/19$, or about 0.4737.

4.25 **(a)** 75.2%.

(b) All probabilities are between 0 and 1, and they add up to 1.

(c) 0.983.

(d) 0.976.

(e) "$X \geq 9$" or "$X > 8$." The probability is 0.931.

4.27 5/26, or about 0.1923.

4.29 **(a)** 10,000.

(b) $1/10000 = 0.0001$.

(c) There are 10 (0000, 1111, 2222, ..., 9999); the probability that you get one is $10/10000 = 1/1000 = 0.001$.

4.21 $\mu = 4.5$—the balancing point for the histogram (halfway between 4 and 5).

4.33 $\sigma = 2.8723$; make marks at about 1.63 and 7.37. The numbers within one standard deviation of μ are 2, 3, 4, 5, 6, and 7; the probability is 0.60.

4.35 **(a)** The area of a triangle is $\frac{1}{2}bh = \frac{1}{2}(2)(1) = 1$.

(b) $P = 0.5$.

(c) $P = 0.125$.

4.37 $\mu = 2.25$—lower than a B. $\sigma = 1.1779$.

4.39 $\mu = 11.251$, $\sigma = 1.563$.

4.41 **(a)** $P(\hat{p} \geq 0.16) = P(Z \geq 1.087) = 0.1385$.

(b) $P(0.14 \leq \hat{p} \leq 0.16) = P(1.087 \leq Z \leq 1.087) = 0.7230$.

4.43 **(a)** $\mu = p = 0.4$, $\sigma = 0.0116$.

(b) The population (US adults) is considerably larger than 10 times the sample size.

(c) $np = 714$, $n(1 - p) = 1071$—both are much bigger than 10.

(d) $P(0.37 < \hat{p} < 0.43) = P(-2.586 < Z < 2.586) = 0.9904$. Over 99% of all samples should give $\hat{p}$ within 3% of the true population proportion.

4.45 For $n = 300$: $\sigma = 0.02828$, $P = 0.7108$. For $n = 1200$: $\sigma = 0.01414$, $P = 0.9660$. For $n = 4800$: $\sigma = 0.00707$, $P \doteq 1$. Larger sample sizes give more accurate results (the sample proportions are more likely to be close to the true proportion).

4.47 **(a)** 0.86 (86%).

    (b) We use the normal approximation (Rule of Thumb 2 is *just* satisfied—$n(1 - p) = 10$). The standard deviation is 0.03, and $P(\hat{p} \le 0.86) \doteq P(Z \le -1.33) = 0.0918$.

    (c) Even when the claim is correct, there will be some variation in sample proportions. In particular, in about 10% of samples we can expect to observe 86 or fewer orders shipped on time.

4.49 **(a)** $np = 1190$.

    (b) $P(< 1200 \text{ students accepted}) = P(\hat{p} < \frac{1200}{1700}) \doteq P(Z < 0.5293) = 0.7019$.

    (c) $P(\frac{1500}{1700} < \hat{p} < \frac{1250}{1700}) \doteq P(-2.117 < Z < 3.176) = 0.9847$.

4.51 **(a)** $np = 66$ and $n(1 - p) = 234$; $P(\hat{p} > 0.20) = P(Z > -0.8362) = 0.7985$.

    (b) $P(\hat{p} > 0.30) = P(Z > 3.345) = 0.0004$.

4.53 **(a)** $np = 80$.

    (b) Still assuming that $p = 0.04$, $P(\hat{p} \ge \frac{75}{2000}) = P(Z \ge -0.57) = 0.7157$.

4.55 It may be binomial if we assume that there are no twins or other multiple births among the next 20 (these would not be independent), and that for all births, the probability that the baby is female is the same (requirement 4).

4.57 No—since she receives instruction after incorrect answers, her probability of success is likely to increase.

4.59 **(a)** 0, 1, ..., 5.

    (b) $P(X = 0) = \binom{5}{0}(.25)^0(.75)^5 = 0.23730$; $P(X = 1) = 0.39551$; $P(X = 2) = 0.26367$; $P(X = 3) = 0.08789$; $P(X = 4) = 0.01465$; $P(X = 5) = 0.00098$.

4.61 $P(X = 10) = \binom{20}{10}(0.8)^{10}(0.2)^{10} = 0.00203$.

4.63 $\mu = \frac{5}{4} = 1.25$, $\sigma = \sqrt{\frac{15}{16}} = 0.96825$.

4.65 **(a)** $\mu = 16$ (if $p = 0.8$).

    (b) $\sigma = \sqrt{3.2} = 1.78885$.

    (c) $p = 0.9$: $\sigma = 1.34164$. $p = 0.99$: $\sigma = 0.44497$. As p approaches 1, σ approaches 0.

4.67 **(a)** $n = 20$ and $p = 0.25$.

    (b) $\mu = 5$.

    (c) $\binom{20}{5}(0.25)^5(0.75)^{15} = 0.20233$.

4.69 **(a)** That all are assessed as truthful: $P = 0.06872$; the probability that at least one is reported to be a liar is $1 - 0.06872 = 0.93128$.

    (b) $\mu = 2.4$, $\sigma = \sqrt{1.92} = 1.38564$.

4.71 **(a)** P(card is black) $= P$(card is red) $= \frac{26}{52} = 0.5$.

(b) Now P(card is black) $= \frac{25}{51} = 0.49020$, and P(card is red) $= \frac{26}{51} = 0.50980$.

(c) This time P(card is black) $= \frac{26}{51} = 0.50980$, and P(card is red) $= \frac{25}{51} = 0.49020$.

4.73 The mean for 4.17(b) is the population mean from 4.17(a), namely -3.5%. The standard deviation is $\sigma/\sqrt{n} = 26\%/\sqrt{5} = 11.628\%$.

4.75 Mean: 40.125, standard deviation: 0.001; normality is not needed.

4.77 **(a)** $P(X \geq 21) = P(Z \geq \frac{21-18.6}{5.9}) = 0.3421$.

(b) $P(\bar{x} \geq 21) = P(Z \geq \frac{21-18.6}{5.9/\sqrt{50}}) = 0.0020$.

4.79 **(a)** $N(123, 0.04619)$.

(b) $P(Z > 21.65)$—essentially 0.

4.81 $\bar{x}$ (the mean return) has approximately a $N(9\%, 4.174\%)$ distribution; $P(\bar{x} > 15\%) = P(Z > 1.437) = 0.9247$; $P(\bar{x} < 5\%) = P(Z < -0.9583) = 0.1690$.

4.83 **(a)** $N(55000, 4500/\sqrt{8}) = N(55000, 1591)$.

(b) $P(Z < -2.011) = 0.0222$.

4.85 **(a)** $P(Z < -1.5) = 0.0668$.

(b) $P(Z < -3) = 0.0013$.

4.87 $L = \mu - 1.645\sigma/\sqrt{n} = 12.513$.

4.89 Center: 0.8750; control limits: 0.8750 ± 0.0016, i.e., 0.8734 and 0.8766.

4.91 The mean of sample number 7 fell below the lower control limit (2.2037)—that would have been the time to correct the process.

4.93 **(a)** Center: 10; control limits: 7.922 and 12.078.

(b) There are no runs that should concern us here—we would only be concerned with a run of samples with mean *less than* 10. Lot 13 signals that the process is out of control. The two samples that follow the bad one are fine, so it may be that whatever caused the low average for the 13th sample was an isolated incident. The operator should investigate to see if there is some explanation, and try to remedy the situation if necessary.

4.95 **(a)** About 43%: P(width < 2.8 or width > 3.2) $= P(Z < -0.1913$ or $Z > 2.4472) = 0.4241 + 0.0072 = 0.4313$.

(b) Center: 3; control limits: 2.797 to 3.203.

4.97 $c = 3.090$. (Table A suggests 3 possible answers—3.08, 3.09, or 3.10.)

4.99 Center: p; control limits: $p \pm 3\sqrt{\frac{p(1-p)}{n}}$.

4.101 $P(\hat{p} > 0.50) = P(Z > \frac{0.05 - 0.45}{0.02225}) = P(Z > 2.247) = 0.01231$.

4.103 (a) $P(Z > \frac{105 - 100}{15}) = P(Z > \frac{1}{3}) = 0.36944$.

(b) Mean: 100; standard deviation: 1.93649.

(c) $P(Z > \frac{105 - 100}{1.93649}) = P(Z > 2.5820) = 0.00491$.

(d) The answer to (a) could be quite different; (b) would be the same (it does not depend on normality at all). The answer we gave for (c) would be still be fairly reliable because of the central limit theorem.

4.105 (a) No—a count assumes only whole-number values.

(b) $N(1.5, 0.02835)$.

(c) $P(\overline{x} > \frac{1075}{700}) = P(Z > 1.2599) = 0.10386$.

4.107 $P(5067$ or more heads$) = P(\hat{p} \geq 0.5067) = P(Z \geq 1.34) = 0.0901$. If Kerrich's coin was "fair," we would see 5067 or more heads in about 9% of all repetitions of the experiment of flipping the coin 10,000 times, or about once every 11 attempts. This is *some* evidence against the coin being fair, but it is not by any means overwhelming.

CHAPTER 5

5.1 (a) 44% to 50%.

(b) We do not have information about the whole population; we only know about a small sample. We expect our sample to give us a good estimate of the population value, but it will not be exactly correct.

(c) The procedure used gives an estimate within 3 percentage points of the true value in 95% of all samples.

5.3 (a) Mean: 280; standard deviation: 1.89737.

(c) 2 standard deviations—3.8 points.

(e) 95%.

5.5 11.78 ± 0.77, or 11.01 to 12.55 years.

5.7 (a) The distribution is slightly skewed to the right.

(b) 224.002 ± 0.029, or 223.973 to 224.031.

5.9 (a) 0.8354 to 0.8454.

(b) 0.8275 to 0.8533.

(c) Increasing confidence makes the interval longer.

5.11 (a) 271.4 to 278.6.

(b) 90%: 272.0 to 278.0; 99%: 270.3 to 279.7.

(c) 90%: 3.0; 95%: 3.6; 99%: 4.7. Margin of error goes up with increasing confidence.

5.13 **(a)** 10.00209 to 10.00251.

(b) 22 (21.64).

5.15 35 (34.57).

5.17 **(a)** The interval was based on a method that gives correct results 95% of the time.

(b) With a margin of error of 2%, the true value of p could be as low as 49%, so the confidence interval contains some values of p which give the election to Ford.

(c) The proportion in question is not random—discussing probabilities about this proportion has little meaning. The "probability" the politician asked about is either 1 or 0: either a majority favors Carter, or they don't.

5.19 1.888 to 2.372.

5.21 **(a)** The intended population is "the American public"; the population which was actually sampled was "citizens of Indianapolis (with listed phone numbers)."

(b) Food stores: 15.22 to 22.12; Mass merchandisers: 27.77 to 36.99; Pharmacies: 43.68 to 53.52.

(c) The confidence intervals do not overlap at all; in particular, the *lower* confidence limit of the rating for pharmacies is higher than the *upper* confidence limit for the other stores. This indicates that the pharmacies are *really* higher.

5.23 505 residents.

5.25 $657.14 to $670.86. (The sample size is not used; $\sigma_{estimate}$ is given.)

5.27 **(a)** $N(31\%, 1.518\%)$.

(b) The lower percentage lies out in the low tail of the curve, while 30.2% is fairly close to the middle. If H_0 is true, observing a value like 30.2% would not be surprising, but 27.6% is unlikely, and therefore provides evidence against H_0.

5.29 $H_0 : \mu = \$42,500$; $H_a : \mu > \$42,500$.

5.31 $H_0 : \mu = 2.6$; $H_a : \mu \neq 2.6$.

5.33 The P-values are 0.2991 and 0.0125, respectively.

5.35 **(a)** A $N(0, 5.3932)$ density.

(b) 0.1004.

(c) Not significant at $\alpha = 0.05$. The study gives *some* evidence of increased compensation, but it is not very strong.

5.37 Comparing men's and women's earnings for our sample, we observe a difference so large that it would only occur in 3.8% of all samples if men and women actually earned the same amount. Based on this, we conclude that men earn more. While there is almost certainly *some* difference between earnings of black and white students in our sample, it is relatively small—if blacks and whites actually earn the same amount, we would still observe a difference as big as what we saw almost half (47.6%) of the time.

5.39 (a) $H_0 : \mu = 300$ vs. $H_a : \mu < 300$.
      (b) $z = -0.7893$.
      (c) $P = 0.2150$—do not reject H_0.

5.41 (a) Yes, because $z > 1.645$.
      (b) Yes, because $z > 2.326$.

5.43 (a) $N(0, 0.11339)$.
      (b) $\bar{x} = 0.27$ lies out in the tail of the curve, while 0.09 is fairly close to the middle. Assuming H_0 is true, observing a value like 0.09 would not be surprising, but 0.27 is un-likely, and therefore provides evidence against H_0.
      (c) $P = 0.4274$.
      (d) $P = 0.0173$.

5.45 $H_0 : \mu = 18$ vs. $H_a : \mu < 18$.

5.47 (a) No, because $|z| < 1.960$.
      (b) No, because $|z| < 1.645$.

5.49 P is between 0.005 and 0.01 (in fact, $P = 0.0078$).

5.51 $P = 0.1292$. Although this sample showed *some* difference in market share between pioneers with and without patents or trade secrets, the difference was small enough that it could have arisen merely by chance: it would occur in about 13% of all samples even if there is *no* difference between the two types of pioneer companies.

5.53 When a test is significant at the 1% level, it means that if the null hypothesis is true, outcomes similar to those seen are expected to occur less than once in 100 repetitions of the experiment or sampling. "Significant at the 5% level" means we have observed something which occurs in less than 5 out of 100 repetitions (when H_0 is true). Something that occurs "less than once in 100 repetitions" also occurs "less than 5 times in 100 repetitions," so significance at the 1% level implies significance at the 5% level (or any higher level).

5.55 (a) $z = 1.64$; not significant at 5% level ($P = 0.0505$).
      (b) $z = 1.65$; significant at 5% level ($P = 0.0495$).

5.57 $n = 100$: 452.24 to 503.76. $n = 1000$: 469.85 to 486.15. $n = 10000$: 475.42 to 480.58.

5.59 (a) No—in a sample of size 500, we expect to see about 5 people who have a "P-value" of 0.01 or less. These four *might* have ESP, or they may simply be among the "lucky" ones we expect to see.

(b) Repeat the procedure on these four to see if they again perform well.

5.61 We might conclude that customers prefer design A, but perhaps not "strongly." Because the sample size is so large, this statistically significant difference may not be of any practical importance.

5.63 (a) H_0 : the patient is ill; H_a : the patient is healthy. A Type I error means a false negative—clearing a patient who should be referred to a doctor. A Type II error is a false positive—sending a healthy patient to the doctor.

5.65 (a) 0.50.

(b) 0.1841.

(c) 0.0013.

5.67 (a) Reject H_0 if $\bar{x} \le 297.985$, so the power against $\mu = 299$ is 0.2037.

(b) 0.9926.

(c) It would be greater—290 is easier to distinguish from $\mu_0 = 300$.

5.69 (a) Reject if $\bar{x} \ge 0.87011$ or $\bar{x} \le 0.84989$.

(b) Power: 0.8935.

(c) $1 - 0.8935 = 0.1065$.

5.71 P(Type I error) $= 0.05$. P(Type II error) $= 1 - 0.9926 = 0.0074$.

5.73 A test having low power may do a good job of not incorrectly rejecting the null hypothesis, but it is likely to accept H_0 even when some alternative is correct, simply because it is difficult to distinguish between H_0 and "nearby" alternatives.

5.75 (a) 141.6 to 148.4.

(b) $H_0 : \mu = 140$ vs. $H_a : \mu > 140$; $z = 2.421$; P-value is about 0.0077. Reject H_0.

(c) We must assume that the 15 cuttings in our sample are an SRS. Since our sample is not too large, the population should be normally distributed, or at least not extremely nonnormal.

5.77 (a) $H_0 : \mu = 32$ vs. $H_a : \mu > 32$.

(b) $z = 1.8639$; P-value is 0.0312. This is strong evidence against H_0—observations this extreme would only occur in about 3 out of 100 samples if H_0 were true.

5.79 (a) Margin of error decreases.

(b) The P-value decreases (the evidence against H_0 becomes stronger).

(c) Power increases (the test becomes better at distinguishing between H_0 and H_a).

5.81 No—"$P = 0.03$" *does* mean that H_0 is unlikely, but only in the sense that the evidence (from the sample) would not occur very often if H_0 were true. P is a probability associated with the sample, not the null hypothesis; H_0 is either true or it isn't.

5.83 (a) The difference observed in the study would occur in less than 1% of all samples if the two populations actually have the same proportion.

(b) The interval is constructed using a method that is correct (i.e., contains the actual proportion) 95% of the time.

(c) No—treatments were not randomly assigned, but instead were chosen by the mothers. Mothers who choose to attend a job training program may be more inclined to get themselves out of welfare.

CHAPTER 6

6.1 $37/\sqrt{4} = 18.5$.

6.3 (a) 2.145.

(b) 0.688.

6.5 (a) 14.

(b) 1.82 is between 1.761 ($p = 0.05$) and 2.145 ($p = 0.025$).

(c) The P-value is between 0.025 and 0.05 (in fact, $P = 0.0451$).

(d) $t = 1.82$ is significant at $\alpha = 0.05$ but not at $\alpha = 0.01$.

6.7 (a) $\bar{x} = 1.75$ and $s = 0.1291$, so $SE(\bar{x}) = 0.06455$.

(b) 1.598 to 1.902.

6.9 $H_0 : \mu = 1.3$ vs. $H_a : \mu > 1.3$; $t = 6.9714$; P is between 0.005 and 0.0025 (in fact, $P = 0.003$). This is very strong evidence against H_0; reject it in favor of H_a.

6.11 (a) Randomly assign 12 (or 13) into a group which will use the right-hand knob first; the rest should use the left-hand knob first. Alternatively, for each student, randomly select which knob he or she should use first.

(b) Let μ_R be the mean right-hand thread time for all right-handed people (or students), and μ_L be the mean left-hand thread time. Then $\mu = \mu_R - \mu_L$ is the difference between right-handed times and left-handed times; the hypotheses are $H_0 : \mu = 0$ (no difference) and $H_a : \mu < 0$ (i.e., $\mu_R < \mu_L$).

(c) $\bar{x} = -13.32$, $SE(\bar{x}) = 4.5872$, $t = -2.9037$, and $P = 0.0039$. Reject H_0 in favor of H_a.

6.13 (a) 1.54 to 1.80.

(b) We are told the distribution is symmetric; because the scores range from 1 to 5, the possibility for skewness is limited. In this situation, the assumption that the 17 Mexicans are an SRS from the population is the most crucial.

6.15 (a) The distribution is slightly skewed, but there are no apparent outliers.

(b) $H_0 : \mu = 224$ vs. $H_a : \mu \neq 224$. $t = 0.12536$ and $P = 0.9019$, so we have very little evidence against H_0.

6.17 (a) Approximately 2.403 (from Table C), or 2.405 (using software).

(b) Using $t^* = 2.403$: Reject H_0 if $t > 2.403$, which means $\bar{x} > 36.70$.

(c) The power against $\mu = 100$ is 0.99998—basically 1. A sample of size 50 should be quite adequate.

6.19 (a) 9.

(b) $P = 0.0255$; it lies between 0.05 and 0.025.

6.21 (a) 111.22 to 118.58.

(b) We assume that the 27 members of the placebo group can be viewed as an SRS, and that the distribution of seated systolic BP in this population is normal, or at least not too nonnormal. Since the sample size is somewhat large, the procedure should be valid as long as the data show no outliers and no strong skewness.

6.23 (a) Standard error of the mean.

(b) $s = 0.01\sqrt{3} = 0.01732$.

(c) $0.84 \pm 0.0292 = 0.8108$ to 0.8692.

6.25 (a) For each subject, randomly choose which test to administer first. Or, randomly assign 11 subjects to the "ARSMA first" group, and the rest to the "BI first" group.

(b) $t = 4.27$; $P < 0.001$, so we reject $H_0 : \mu = 0$ in favor of $H_a : \mu \neq 0$.

(c) 0.1292 to 0.3746.

6.27 We know the data for *all* presidents; we know about the whole population, not just a sample. (We might want to try to make statements about future presidents, but doing so from this data would be highly questionable; they can hardly be considered an SRS from the population.)

6.29 **(a)** (3)—two samples.

(b) (2)—matched pairs.

6.31 **(a)** $H_0 : \mu_1 = \mu_2$ vs. $H_a : \mu_1 < \mu_2$, where μ_1 is the beta-blocker population mean pulse rate and μ_2 is the placebo mean pulse rate. $t = -2.4525$; use a $t(29)$ distribution, which gives $P = 0.01022$. This makes the result significant at 5% but not at 1%.

(b) -10.8311 to 0.6311 for $\mu_1 - \mu_2$.

6.33 **(a)** If k (the degrees of freedom) is reasonably large, the $t(k)$ distribution looks enough like the $N(0, 1)$ distribution that for $t = 7.36$, we can conclude that the P-value is tiny (based on the 68–95–99.7 rule), so the result is significant.

(b) Use a $t(32)$ distribution.

6.35 **(a)** Both stemplots show no outliers; the experimental data is perhaps slightly left-skewed, but not enough to keep us from using the t procedures.

(b) $H_0 : \mu_1 = \mu_2$ vs. $H_a : \mu_1 < \mu_2$ ($\mu_1 =$ control weight gain, ...). $P = 0.0116$ (from a $t(19)$ distribution), so we reject H_0.

(c) 5.58 to 67.72 grams.

6.39 $H_0 : \mu_{\text{skilled}} = \mu_{\text{novice}}$ vs. $H_a : \mu_s \neq \mu_n$. Use $t = 0.5143$; its P-value is 0.6165. There is no significant difference in weight between skilled and novice rowers.

6.41 **(a)** $H_0 : \mu_{\text{breast-fed}} = \mu_{\text{formula}}$ vs. $H_a : \mu_{\text{bf}} \neq \mu_{\text{f}}$. Conservative: 18 d.f.; more exact method: 37.6 d.f. $t = 1.654$; $P = 0.1155$ (conservative) or $P = 0.1065$ (more exact); in either case, the results are not significant at the 10% level.

(b) $t(18)$: $t^* = 2.101$; -0.2434 to 2.0434. $t(37.6)$: $t^* = 2.0251$; -0.2021 to 2.0021.

(c) We assume that both groups are SRSs of mothers of both types, and that both distributions are normal (or at least don't depart too much from normality) and have no outliers.

(d) This is not an experiment—the mothers chose the feeding method. This may confound the conclusions, since other factors may have affected this choice.

6.43 **(a)** Use either a $t(52)$ or a $t(121.9)$ distribution. These give, respectively, the intervals -1.279 to 7.279, and -1.216 to 7.216.

(b) The samples taken by the market research firm do not give complete information about *all* stores. When we allow for the possible variation that might reasonably occur at the stores not included in the samples, we find that the actual sales might have dropped by 1.3 units, or could have risen by as much as 7.3 units.

6.45 **(a)** Stemplots show little skewness, but the control group has one moderate outlier (85). Nonetheless, the t procedures should be fairly reliable since the total sample size is 44.

(b) $H_0 : \mu_t = \mu_c$ vs $H_a : \mu_t < \mu_c$; $t = 2.311$. Using $t(20)$ and $t(37.9)$ distributions, P equals 0.0158 and 0.0132, respectively; reject H_0.

(c) Randomization was not really possible, because existing classes were used—the researcher could not shuffle the students around.

6.47 If they did (for example) 20 tests at the 5% level of significance, they might see 1 or 2 significant differences even when all null hypotheses are true.

6.49 **(a)** $H_0 : \mu_1 = \mu_2$ vs. $H_a : \mu_1 \neq \mu_2$; $t = (\overline{x}_1 - \overline{x}_2)/\sqrt{\frac{s_1^2}{n_1} + \frac{s_2^2}{n_2}}$.

(b) $t^* \doteq 1.984$, from a $t(100)$ distribution.

(c) Reject H_0 when $|\overline{x}_1 - \overline{x}_2| \geq 8.260$; power against $\mu_1 - \mu_2 = 10$: 0.662.

6.51 **(a)** $F^* = 3.68$.

(b) Not significant at either 10% or 5% (in fact, $P = 0.1166$).

6.53 $F = 10.57$; P-value (for the two-sided alternative) is between 0.02 and 0.05 ($P = 0.0217$)—so this is significant at 5% but not at 1%.

6.55 **(a)** $H_0 : \sigma_s = \sigma_n$ vs. $H_a : \sigma_s \neq \sigma_n$.

(b) $F = 2.196$; P-value is greater than 0.20 ($P = 0.2697$).

6.57 $F = 1.5443$; the P-value is 0.3725, so the difference is not statistically significant.

6.59 **(a)** Two-sample t test—the two groups of women are (presumably) independent.

(b) Use a $t(44)$ distribution.

(c) The sample sizes are large enough that nonnormality has little effect on the reliability of the procedure.

6.61 **(a)** $H_0 : \mu_1 = \mu_2$ vs. $H_a : \mu_1 < \mu_2$; $t = -8.947$; $P \doteq 0$ (however one chooses the degrees of freedom). Reject H_0 and conclude that the workers have higher output.

(b) The t procedures are robust against skewness when the sample sizes are large.

(c) Insertions for experienced workers have (approximately) a $N(\overline{x}_2, s_2)$ distribution; the 68–95–99.7 rule tells us that 95% of all workers can insert between $\overline{x}_2 - 2s_2 = 29.66$

and $\bar{x}_2 + 2s_2 = 44.99$ pins in the allotted time. (Or use $\bar{x}_2 \pm 1.96s_2$.)

6.63 Both stemplots are reasonably symmetrical, though the nitrite group may be slightly left-skewed. There are no *extreme* outliers. $H_0 : \mu_c = \mu_n$ vs. $H_a : \mu_c > \mu_n$; $t = 0.8909$ and P equals 0.1902 (using a $t(29)$ distribution) or 0.1884 (with 56.8 d.f.). In either case, the difference is not significant.

6.65 **(a)** The stemplot is reasonably symmetrical given the small sample size. There are no outliers.
(b) 903.23 to 912.27.
(c) No, because 910 falls inside the 95% confidence interval.

6.67 No—you have information about all Indiana counties (not just a sample).

6.69 **(a)** $H_0 : \mu_1 = \mu_2$ vs. $H_a : \mu_1 > \mu_2$; $t = 1.1738$, so $P = 0.1265$ (using $t(22)$) or 0.123453 (using $t(43.3)$). Not enough evidence to reject H_0.
(b) -14.57 to 52.57 (using $t(22)$), or -13.64 to 51.64 (using $t(43.3)$).
(c) 165.53 to 220.47.
(d) We are assuming that we have two SRSs from each population, and that underlying distributions are normal. It is unlikely that we have random samples from either population, especially among pets.

6.71 Using 95% confidence—for abdomen skinfold: -15.5 to -11.5 (using $t(19)$), or -15.4 to -11.6 (using $t(103.6)$). For thigh skinfold: -12.95 to -9.65 (using $t(19)$), or -12.86 to -9.74 (using $t(106.4)$).

6.73 We have 30 available observations; the distribution is right-skewed, with a high outlier of 123. A 95% confidence interval for the mean city particulate level is 54.07 ± 7.25, or 46.82 to 61.32. If we discard the outlier, we get 51.69 ± 5.57, or 46.12 to 57.26.

Confidence intervals should be used with caution here—the outlier makes t procedures suspect; without the outlier, we may underestimate the actual mean.

6.75 The plot shows a strong positive linear relationship; only one observation—(51,69)—deviates slightly from the pattern. The regression line $\hat{y} = -2.580 + 1.0935x$ explains 95.1% of the variation in the data. When $x = 88$, we predict $\hat{y} = 93.65$.

The point (108,123) is potentially influential (although it does not seem to deviate from the pattern of the other points).

Computing the regression line without this point gives $\hat{y} = 1.963 + 0.9942x$ and $r^2 = 92.1\%$, and $\hat{y} = 89.45$ when $x = 88$.

CHAPTER 7

7.1 (a) Population: the 175 dorm residents; p is the proportion who like the food.

(b) $\hat{p} = 0.28$.

7.3 (a) The population is the 15,000 alumni, and p is the proportion who support the president's decision.

(b) $\hat{p} = 0.38$.

7.5 (a) No—np_0 and $n(1 - p_0)$ are less than 10 (they both equal 5).

(b) No—the expected number of failures is less than 10 $(n(1 - p_0) = 2)$.

(c) Yes—we have an SRS, the population is more than 10 times as large as the sample, and $np_0 = n(1 - p_0) = 10$.

7.7 $\hat{p} = 0.3786$, $z = 2.72$, and $P = 0.0033$—reject $H_0 : p = \frac{1}{3}$ in favor of $H_a : p > \frac{1}{3}$.

7.9 (a) $\hat{p} = 0.5005$, $z = 0.1549$, and $P = 0.8769$—do not reject H_0.

(b) 0.4922 to 0.5088.

7.11 450.2—round up to 451.

7.13 $z = -3.337$ and $P < 0.0005$—very strong evidence against H_0 in favor of $H_a : p < 0.1$.

7.15 (a) 0.0913 to 0.2060.

(b) 304.

(c) The sample comes from a limited area in Indiana, focuses on only one kind of business, and leaves out any businesses not in the Yellow Pages (there might be a few of these; perhaps they are more likely to fail). It is more realistic to believe that this describes businesses that match the above profile; it *might* generalize to food-and-drink establishments elsewhere, but probably not to (e.g.) hardware stores and other types of business.

7.17 (a) 4719.

(b) 0.01125.

7.19 (a) -0.0208 to 0.1476.

(b) The population-to-sample ratio is certainly large enough, and the smallest count in any category is 75—much larger than 5.

7.21 (a) $H_0 : p_1 = p_2$ vs. $H_a : p_1 \neq p_2$. The population-to-sample ratio is large enough, and the smallest count is 94 (Catholics answering "Yes").

(b) $\hat{p}_1 = 0.6030$, $\hat{p}_2 = 0.5913$, $\hat{p} = 0.5976$; $z = 0.2650$, and $P = 0.7910$—H_0 is quite plausible given this sample.

7.23 (a) $H_0 : p_1 = p_2$ vs. $H_a : p_1 > p_2$; the populations are much larger than the samples, and 17 (the smallest count) is greater than 5.

(b) $\hat{p} = 0.0632$, $z = 2.926$, and $P = 0.0017$—the difference is statistically significant.

(c) Neither the subjects nor the researchers who had contact with them knew which subjects were getting which drug— if anyone had known, they might confound the outcome by letting their expectations or biases affect the results.

7.25 $H_0 : p_1 = p_2$ vs. $H_a : p_1 \neq p_2$; $P = 0.6981$—insufficient evidence to reject H_0.

7.27 (a) $H_0 : p_1 = p_2$ vs. $H_a : p_1 > p_2$; $P = 0.0335$—reject H_0 (at the 5% level).

(b) -0.0053 to 0.2336.

7.29 The population-to-sample ratio is large enough, and the smallest count is 10. Fatal heart attacks: $z = -2.67$, $P = 0.0076$. Non-fatal heart attacks: $z = -4.58, P < 0.000005$. Strokes: $z = 1.43, P = 0.1525$. The proportions for both kinds of heart attacks were significantly different; the stroke proportions were not.

7.31 (a) -9.91% to -4.09%. 0 is not in this interval; reject $H_0 : p_1 = p_2$ at the 1% level.

(b) -0.7944 to -0.4056. 0 is not in this interval; reject $H_0 : \mu_1 = \mu_2$ at the 1% level.

7.33 No—the data is not based on an SRS, and thus the z procedures are not reliable in this case. In particular, a voluntary response sample is typically biased.

7.35 92.3% to 93.7%.

7.37 (a) No—$P = 0.1849$.

(b) Yes—$P = 0.0255$.

(c) The population is (presumably) very large, so the ratio of population-to-sample is big enough. Also, all the counts— 45 and 29 in (a); 21, 7, 24, and 22 in (b)—are bigger than 5. The counts for baseball players are too small.

CHAPTER 8

8.1 **(a)** 2×3.

(b) 55.0%, 74.7%, and 37.5%. Some (but not too much) time spent in extracurricular activities seems to be beneficial.

(d) Expected counts—C or better: 13.78, 62.71, 5.51; D or F: 6.22, 28.29, 2.49.

(e) The first and last columns have lower numbers than we expect in the "passing" row (and higher numbers in the "failing" row), while the middle column has this reversed—more passed than we would expect if all proportions were equal.

8.3 **(b)** $P = 0.0313$. Rejecting H_0 means that we conclude that there is a relationship between hours spent in extracurricular activities and performance in the course.

(c) The highest contribution comes from ">12 hours of extracurricular activities, D or F in the course". Too much time spent on these activities seems to hurt academic performance.

(d) No—this study demonstrates association, not causation. Certain types of students may tend to spend a moderate amount of time in extracurricular activities and also work hard on their classes—one does not necessarily cause the other.

8.5 **(a)** $(r - 1)(c - 1) = (2 - 1)(3 - 1) = 2$.

(b) $X^2 = 6.926$ lies between 5.99 and 7.38; therefore, P is between 0.05 and 0.025.

8.7 **(a)** First row: 5, 9; Second: 41, 23; Third: 27, 3; Fourth: 9, 2.

(b) First: 9.65, 4.35; Second: 44.10, 19.90; Third: 20.67, 9.33; Fourth: 7.58, 3.42. 25% (2 out of 8) of the expected counts are less than 5, which goes against our guidelines.

(c) 3 d.f.; P-value is between 0.0025 and 0.001 (in fact, $P = 0.0018$).

(d) Students with high goals show a higher proportion of passing grades than those who merely wanted to pass.

8.9 **(a)** 7.01%, 14.02%, and 13.05%.

(b) 172, 2283; 167, 1024; 86, 573.

(c) 242.36, 2212.64; 117.58, 1073.42; 65.06, 593.94. Expected counts are all much bigger than 5, so the chi-square test is safe. H_0 : there is no relationship between worker class and race vs. H_a : there is some relationship.

(d) 2 d.f.; $P < 0.0005$ (basically 0).

(e) Black female child-care workers are more likely to work in non-household or preschool positions.

8.11 (a) $X^2 = 10.827$ (3 d.f.); $P = 0.0127$; significant at the 5% level.

(b) Females choosing each field: 30.22%, 40.44%, 2.22%, and 27.11%. Males: 34.78%, 24.84%, 3.73%, and 36.65%. The biggest difference between women and men is in Administration: a higher percentage of women chose this major. Meanwhile, a greater proportion of men chose other fields, especially Finance.

(c) The largest chi-square components are the two from the "Administration" row. Many more women than we expect (91 actual, 76.36 expected) chose this major, while only 40 men chose this (54.64 expected).

(d) The "Economics" row had expected counts of 6.41 and 4.59, respectively. Only the second number is less than 5, which is only one eighth (12.5%) of the counts in the table—the chi-square procedure is acceptable.

(e) 386 responded, so 46.54% did not respond.

8.13 (a) $H_0 : p_1 = p_2$, where p_1 and p_2 are the proportions of women customers in each city. $\hat{p}_1 = 0.8423$, $\hat{p}_2 = 0.6881$, $z = 3.9159$, and $P = 0.00009$.

(b) $X^2 = 15.334$, which equals z^2. With 1 d.f., Table E tells us that $P < 0.0005$; a statistical calculator gives $P = 0.00009$.

(c) 0.0774 to 0.2311.

8.15 (a) $H_0 : p_1 = p_2$ vs. $H_a : p_1 < p_2$. The z test must be used because the chi-square procedure will not work for a one-sided alternative.

(b) $z = -2.8545$ and $P = 0.0022$. Reject H_0; there is strong evidence in favor of H_a.

8.17 H_0 : all proportions are equal vs. H_a : some proportions are different. Phone interviews: 168 yes, 632 no; one-on-one: 200 and 600; Anonymous: 224 and 576. $X^2 = 10.619$ with 2 d.f., and $P = 0.0049$—good evidence against H_0, so we conclude that contact method makes a difference in response.

8.19 H_0 : all refusal proportions are equal. Actual counts: 67, 1491; 86, 1503; 135, 1940; 124, 2514. Expected counts: 81.67, 1476.33; 83.29, 1505.71; 108.77, 1966.23; 138.28, 2499.72. The smallest expected count is 81.67, so the chi-square test is safe. $X^2 = 11.106$ (3 d.f.) and $P = 0.0112$—pretty strong evidence against H_0. The two largest components of X^2 are the July–Aug refusals (which were higher than expected) and the Jan–April refusals (which were low). Perhaps more people had "other things to do" in the summer, and had less pressing business during the winter.

8.21 (a) No; $X^2 = 1.051$ with 2 d.f., which gives $P = 0.5913$.

(b) ABC News: $z = -0.7698$; $P = 0.4414$ (not significant). USA Today/CNN: $z = -5.1140$; $P \doteq 0$ (significant). New York Times/CBS: $z = -4.4585$; $P \doteq 0$ (significant).

(c) An individual test will be wrong for only 5% of all samples. Imagine doing three tests in a row: Assuming the first test comes out correct (which it does 95% of the time), there is still a 5% chance that the next test will come out wrong, etc. Altogether, all three will be correct only 85.7%($= 0.95^3$) of the time.

8.23 (a) 0.5432 to 0.5968.

(b) No: $z = 0.4884$, $P = 0.6253$. Or, using the exact counts ($527/924$ national, $96/174$ student), $z = 0.4548$, $P = 0.6492$—again, not at all significant.

(c) Yes: $z = 7.9215$, P is essentially 0.

(d) Actual counts—National group: 702 yes, 222 no; First student group: 117, 57; Second student group: 129, 70. Expected counts—National group: 675.37, 248.63; First student group: 127.18, 46.82; Second student group: 145.45, 53.55. $X^2 = 13.847$ with 2 d.f.; $P < 0.001$. Both student groups were less likely to believe that the military was censoring the news.

CHAPTER 9

9.1 (a) The distribution for perch is slightly higher than the other two, and has an extreme high outlier (20.9); the bream distribution has a mild low outlier (12.0). Otherwise there is no strong skewness.

(b) Bream: 12.0, 13.6, 14.1, 14.9, 15.5; Perch: 13.2, 15.0, 15.55, 16.675, 20.9; Roach: 13.3, 13.925, 14.65, 15.275, 16.1. The most important difference seems to be that perch are larger than the other two fish. It also appears that (typically) bream *may* be slightly smaller than roach.

9.3 (a) Mean yields: 131.03, 143.15, 146.23, 143.07, 134.8. The mean yields first increase with plant density, then decrease; the greatest yield occurs at or around 20,000 plants per acre. This is also reflected (to a lesser extent) in the stemplots.

(b) $H_0 : \mu_1 = \mu_2 = \mu_3 = \mu_4 = \mu_5$ (all plant densities give the same mean yield per acre) vs. H_a : not all means are the same.

(c) $F = 0.50$ and $P = 0.736$. The differences are not significant.

(d) The sample sizes were small, which means there is a lot of potential variation in the outcome.

9.5 (a) I, the number of populations, is 5; the sample sizes are $n_1 = 4$, $n_2 = 4$, $n_3 = 4$, $n_4 = 3$, and $n_5 = 2$; the total sample size is $N = 17$.

(b) numerator ("factor"): $I - 1 = 4$, denominator ("error"): $N - I = 12$.

(c) Since $F < 2.48$, the smallest critical value for an $F(4, 12)$ distribution in Table D, we conclude that $P > 0.100$.

9.7 Yes:

$$\frac{\text{largest } s}{\text{smallest } s} = \frac{1.186}{0.770} = 1.54$$

9.9 (a) The biggest difference is that single men earn considerably less than men who have been or are married. Widowed and married men earn the most; divorced men earn about $1300 less (on the average); single men are $4000 below that.

(b) Yes: $\frac{8119}{5731} = 1.42$.

(c) 3 and 8231.

(d) The sample sizes are so large that even small differences would be found to be significant; we have some fairly large differences.

(e) No—single men are likely to be younger than men in the other categories. This means that typically they have less experience, and have been with their companies less time than the others, and so have not received as many raises, etc.

9.11 (a) MSE $= \frac{107.67}{107} = 1.0063$; $s_p = \sqrt{\text{MSE}} = 1.003$.

(b) Use $t^* = 1.984$ from a $t(100)$ distribution (since $t(107)$ is not available): $15.747 \pm t^* s_p \sqrt{55} = 15.479$ to 16.015. Using software, we find that for a $t(107)$ distribution, $t^* = 1.982$; this rounds to the same interval.

9.13 MSE $= \frac{3595.7}{12} = 299.64$—this agrees with Minitab's output (except for roundoff error). $\bar{x} = \frac{2380.35}{17} = 140.02$. MSG $= \frac{600.18}{4} = 150.04$.

9.15 **(a)** Populations: nonsmokers, moderate smokers, heavy smokers; response variable: hours of sleep per night. $I = 3$, $n_1 = n_2 = n_3 = 200$, $N = 600$; 2 and 597 d.f.

(b) Populations: different concrete mixtures; response variable: strength. $I = 5$, $n_1 = \cdots = n_5 = 6$, and $N = 30$; 4 and 25 d.f.

(c) Populations: teaching methods; response variable: test scores. $I = 4$, $n_1 = n_2 = n_3 = 10$, $n_4 = 12$, and $N = 42$; 3 and 38 d.f.

9.17 Side-by-side stemplots show that the Jaguar XJ12 and the Rolls-Royce Silver Spur are again low outliers, so we omit them. The Mercedes-Benz S420 and S500 are also somewhat low among large cars. However, they are not as extreme as the other two, so one might decide to keep them in.

The mean and standard deviation for each size are (respectively) 28.240 and 3.345, 26.316 and 2.689, and 25.077 and 2.753 (or 26.091 and 1.300 if the two Mercedes are omitted).

Analysis of variance is significant at the 1% level (or the 5% level, when the Mercedes are omitted). The confidence intervals show some overlap, but suggest a conclusion similar to that for city mileage—the most important difference is that compact cars have better average mileage than the other two types.

9.19 $\bar{x} = \frac{128.5}{16} = 8.031$; MSG $= 33.55$; MSE $= 2.78$; $F = 12.08$. Table D places the P-value at less than 0.001; software gives $P = 0.0006$.

CHAPTER 10

10.1 **(a)** $r = 0.99415$ and $\hat{y} = -3.660 + 1.19690x$. The scatterplot shows a strong linear relationship, which is confirmed by r.

(b) β represents how much increase we can expect in humerus length when femur length increases by 1 cm. b (the estimate of β) is 1.1969; $a = -3.660$.

(c) The residuals are -0.82262, -0.36682, 3.04248, -0.94202, and -0.91102; the sum is 0. $s = \sqrt{3.92843} = 1.9820$.

10.3 Using a $t(14)$ distribution: $0.188999 \pm (1.761)(0.004934) = 0.1803$ to 0.1977.

10.5 **(a)** $\hat{y} = -3.6596 + 1.1969$.

(b) $t = \frac{1.1969}{0.0751} = 15.94$.

(c) 3 d.f.; $P < 0.001$.

10.7 (a) The plot shows a strong positive linear relationship.

(b) β (the slope) is this rate; the estimate is listed as the coefficient of 'year': 9.31868.

(c) 11 d.f.; $t^* = 2.201$; $9.31868 \pm (2.201)(0.3099) = 8.6366$ to 10.0008.

10.9 (a) Powerboat registrations is explanatory.

(b) The plot shows a moderately strong positive linear relationship; there are no clear outliers or strongly influential points.

(c) $r^2 = 88.6\%$ indicates that much, but not all, of the variation in manatee deaths is explained by powerboat registrations.

(d) β is the number of additional manatee deaths we can expect when there are 1000 additional powerboat registrations. Using a $t(12)$ distribution: $0.12486 \pm (1.782)(0.01290) = 0.1019$ to 0.1478.

(e) $\hat{y} = 45.972$ (about 46 manatee deaths per year).

(f) Use the confidence interval: 41.49 to 50.46.

10.11 (a) The stemplot does not show any *major* asymmetry, and has no particular outliers.

(b) The plot does not suggest a nonlinear relationship. There is *some* indication that there may be less variation at the high and low ends of the plot, but nothing too strong—there are too few observations to make any judgments about that.

10.13 (b) $r^2 = 0.324$; $r = \sqrt{0.324} = 0.569$—use the *positive* square root. The regression of overseas returns on US returns explains about $1/3$ (32.4%) of the variation in overseas returns; the relationship is positive. There is one outlier (in 1986) and one potentially influential point (in 1974).

(c) $H_0 : \beta = 0$ vs. $H_a : \beta \neq 0$; $t = 3.09$, and $P = 0.006$ (so we reject H_0).

(d) $\hat{y} = 21.037$ percent.

(e) Use the prediction interval: -21.97 to 64.04 percent. In practice, this is of little value—the interval includes everything from a 20% loss to a 60% gain.

10.15 (a) It appears that the variation about the line is greater for larger values of x—on the left side of the plot, the residuals are less spread out.

(b) Round residuals to whole numbers first.

10.17 (a) $\hat{y} = 26.3320 + (0.687747)(89) = 87.5415$, residual $= 6.4585$.

(b) $\sum \text{residual}^2 = 356.885$; $s = \sqrt{35.6885} = 5.974$.

10.19 $t^* = 2.145$ from a $t(14)$ distribution: $1.0892 \pm (2.145)(0.1389)$ $= 0.7913$ to 1.3871.

10.21 (a) City mileage is the explanatory variable.

(b) There is a fairly strong positive linear association.

(c) $\hat{y} = 6.178 + 1.03764x$.

(d) $\hat{y}^* = 7.745 + 0.96251x$. Original s was 1.631; without Jaguar and Rolls-Royce, $s = 1.592$.

(e) The Toyota Corolla has the largest absolute residual at -4.157—the predicted value is higher than the actual value.

(f) The stemplot shows only the mild low outlier for the Corolla; it has no other disturbing departures from normality. Plotting residuals vs. city mileage shows a slight problem in that all the residuals for low x values are negative.

(g) β is the average increase in highway mileage for each increase of 1 unit in city mileage. Confidence interval (50 d.f.): $1.03764 \pm (2.009)(0.06476) = 0.9075$ to 1.1677.

(h) $\hat{y} = 24.856$; we want a prediction interval, which software gives as 21.555 to 28.157.

10.23 (a) The plot shows only a *very* slight pattern (weight increases with length). There is a definite outlier: fish #40, the "heavy fish" mentioned in 10.22.

(b) We would expect weight to increase more-or-less linearly with volume—if we double volume, we double weight; if we triple the volume, we triple the weight, etc. When all dimensions (length, width, and height) are doubled, the *volume* of an object increases by a factor of $8 = 2^3$. Similarly, if we triple all dimensions, volume (and, approximately, weight) increases by a factor of $27 = 3^3$. It then makes sense that the cube root (i.e., the one-third power) of the weight increases at an approximately linear rate with length.

(c) The second plot does not really look any more linear; there is still a lot a scatter that obscures any obvious patterns. There are no apparent outliers (except for the heavy fish).

(d) There is very little change in appearance of the plots, or in r^2: using the original weight variable, $r^2 = 0.2346$; with weight$^{1/3}$, $r^2 = 0.2389$. In fact, if we omit the outlier from both computations, we actually find that r^2 *decreases*: 0.2515 before, 0.2502 after.

(e) With all the fish: $\hat{y} = 2.40306 + 0.0038135x$; $\hat{y} = 2.49839$ when $x = 25$; confidence interval: $2.49839 \pm t^* SE_{\hat{\mu}} = 2.49839 \pm (2.005)(0.00870) = 2.48094$ to 2.51585.

   Without fish #40: $\hat{y} = 2.40899 + 0.0034570x$; $\hat{y} = 2.49541$ when $x = 25$; confidence interval: $2.49541 \pm (2.006)(0.00772) = 2.47994$ to 2.51089.

(f) A stemplot and a plot of residuals vs. length show no gross violations of the assumptions, except for the high outlier for fish #40; as we saw in part (e), however, omitting that fish makes little difference in our line.

Index of Symbols

Symbol	Page	What the Symbol Represents
a	122	intercept of the least-squares regression line
α	361	fixed significance level for a test
α	593	intercept of the true regression line
b	122	slope of the least-squares regression line
β	593	slope of the true regression line
C	332	confidence level
df	414	degrees of freedom
F	464	F statistic
F	566	ANOVA F statistic
$F(j, k)$	464	F distribution with j and k degrees of freedom
F^*	465	critical value from an F distribution
H_a	356	alternative hypothesis
H_0	356	null hypothesis
I	568	number of means compared in ANOVA
m	341	margin of error
M	38	median of a set of data
MSE	578	mean square for error in ANOVA
MSG	577	mean square for groups in ANOVA
μ	59	mean of a density curve
μ	259	mean of a random variable
μ	294	mean of a population
μ_y	593	mean response in regression
n	36	number of observations in a set of data
N	568	total number of observations in ANOVA
$N(\mu, \sigma)$	63	normal distribution with mean μ and standard deviation σ

$N(0, 1)$	65	standard normal distribution
p	268	population proportion
P	359	P-value of a test
$\hat{p}$	268	sample proportion
Q_1, Q_3	42	quartiles of a set of data
r	111	correlation
s	47	standard deviation of a set of data
s	595	standard error about the least-squares line
s_p	579	pooled standard deviation in ANOVA
SE	413	standard error of a statistic
SE_b	598	standard error of regression slope
$SE_{\hat{\mu}}$	606	standard error of regression mean response
$SE_{\hat{y}}$	606	standard error for prediction
σ	59	standard deviation of a density curve
σ	260	standard deviation of a random variable
σ	294	standard deviation of a population
σ	593	standard deviation about the true regression line
Σ	36	summation (add them all up)
t	409	one-sample t statistic
t	443	two-sample t statistic
t	600	t statistic for regression slope
t^*	412	critical value from a t distribution
$t(k)$	410	t distribution with k degrees of freedom
$\bar{x}$	36	mean of a set of data
X^2	528	chi-square statistic
$\hat{y}$	122	predicted value of y
z	64	standardized observation
z	365	one-sample z test statistic for a mean
z	490	one-sample z test statistic for a proportion
z	506	two-sample z test statistic for proportions
z^*	334	standard normal critical value
Z	270	statistic having a standard normal distribution

Index of Procedures
(In Order of Appearance in the Text)

Index